Animal Tissue Techniques

A Series of Books in Biology

EDITORS Donald Kennedy
Roderic B. Park

Third Edition

Animal Tissue
Techniques

Gretchen L. Humason

OAK RIDGE ASSOCIATED UNIVERSITIES

W. H. Freeman and Company
San Francisco

Printed in the United States of America

International Standard Book Number: 0-7167-0692-X

Library of Congress Catalog Card Number: 77-172241

1 2 3 4 5 6 7 8 9

Preface

This book of basic and standard histological procedures (and some specialized techniques) was designed to meet the diverse needs of premedical students, medical technicians, zoology majors, and research assistants. Most histological reactions follow a logical and specific sequence, and I have attempted to include simplified discussions of the basic methods that are applicable both to normal and to pathological conditions in zoology and medicine.

It is not intended that this text should be a complete reference book on histology; the experienced worker knows of numerous such tomes, as well as journals that specialize in histology and related topics. However, special methods of wide usage and exceptional merit are included, particularly those that are not overly complicated or unpredictable. It is hoped that the technician, once familiar with the material covered here, will watch the literature for modifications and improvements of standard techniques; in this way, with this book as a foundation, his work can be kept up to date and, perhaps, simplified.

Methods for fixation are fairly well established, with only occasional variations. The section on fixation presented herein is as modern as I can make it, and it includes a brief description of the chemicals employed. Old staining techniques continue to be perfected and new ones developed; I have tried to include the best of these and, for the sake of the student, to adapt them to the standard three-hour laboratory period and to the kinds of equipment most widely available. Some special methods that are more time-consuming have been included for special projects and research. They have been simplified, wherever possible, to serve as introductory techniques for the student who plans to proceed to more complicated techniques later.

Some instructors may not agree with the way in which I have organized the text, but to me it is a logical one. Thus, fixation is treated first, because it is usually the first process in tissue preparation; this is followed by embedding in some kind of medium, sectioning on a microtome, mounting sections on slides, and, finally, staining them with the help of a microscope. A logical arrangement of staining methods is hard to come by, so I have followed my own inclinations: some sections are organized by related tissues, others by related methods. The latter was considered desirable for such processes as silver impregnation, metachromasia, and the use of Schiff reagent. The final chapters include such specialized techniques as histochemistry, chromosome preparation, autoradiography, and invertebrate mounts. Wherever possible, I have referred to my own experience with these methods to help students succeed with their first efforts, and I have included modifications that might appeal to other adventurous technicians.

This book is in four parts. Part I covers those basic procedures and general considerations with which every tissue technician should be familiar. Part II provides detailed information about specific staining methods for most tissues. An instructor might choose a few favorite methods from this section to round out a course, while the professional technician will find here most of the specific methods required on the job. Part III deals with special procedures, those that are special in the sense that they are not common in most laboratories, although they may be very important in some. Although the discussion of some of these procedures is brief, references have been cited extensively for the benefit of those who might wish to refer to more thorough discussions. Part IV is devoted largely to laboratory aids and the preparation of solutions—useful information in any laboratory.

In the third edition, *Animal Tissue Techniques* has been extensively revised and updated. Many of the changes have been to improve its usefulness for graduate and undergraduate teaching. The typography has been altered and the design improved with an eye to making the book more readable and, hence, more useful to students and technicians alike.

Too, the list of references has been carefully emended to cover recent important publications in the field.

To have included everything necessary to satisfy everyone and still to have kept the price of the book within the means of the average student would have been impossible. Some topics, necessarily, have been treated only in passing. The electron microscope, for example, is much too specialized for students in beginning technique classes, and an entire book could be devoted to instructing students in its operation alone. The topic of photomicrography is equally complex. Methods for preparing plastic whole mounts have not been included; excellent leaflets on the subject are published by the companies that supply the materials necessary for their preparation. Good color photographs are helpful, but they are also, unfortunately, expensive—even a few of them can add appreciably to the cost of a book. In my teaching, I have used a demonstration set of slides to help my students recognize proper staining. The set started with a few of my own slides, and it was gradually enlarged by additions from the students in my classes. The students were happy to contribute examples of their best work, and the collection eventually increased to several hundred excellent slides. Other instructors might consider building a study collection of slides in the same way.

I have derived invaluable personal satisfaction from my association with students. I am grateful to them for helping me to develop my tolerance and patience—two qualities that are essential in my profession. I am grateful to them, too, for what they have helped me to learn, for there is no surer way to master a subject than to teach it to others. One former student in particular should receive credit for her encouragement and for prodding me toward writing this book—Marlies Natzler of the University of California at Los Angeles.

Grateful acknowledgements are also due to Marvin Linke, Jeanne Simmons, and Leta Burleson, the three artists who contributed to the three editions of this book; to Julie Langham, for help with photography; to Nellie M. Bilstad, for valuable suggestions; to the Cytogenetics Division of Oak Ridge Associated Universities, for information about late developments in chromosome preparation; to the Zoology Department of the University of California at Los Angeles for the lessons I learned there as a student, a departmental technician, and a lecturer; and to Dr. C. C. Lushbaugh, for his continued encouragement.

October 1971   *Gretchen L. Humason*

Contents

PART III HISTOCHEMISTRY AND MISCELLANEOUS SPECIAL PROCEDURES

PART **IV** SOLUTION PREPARATION AND
GENERAL LABORATORY AIDS

PART I

BASIC PROCEDURES

CHAPTER 1

Fixation

As soon as a tissue ceases to be alive, its cells start to change. Multiplying bacteria begin to destroy them, and the process of autolysis (self-digestion) by contained enzymes begins to dissolve them. The activity of these enzymes is reversed from that in live cells; instead of synthesizing amino acids into proteins, they begin to split proteins into amino acids. These amino acids diffuse out of the cells; as a result cell proteins are no longer coagulable by chemical reagents. These cell changes are called postmortem conditions and must be prevented if tissue is to be examined in the laboratory.

The prevention of postmortem conditions is the primary objective of tissue preparation, but it is also necessary to treat tissue to differentiate the solid phase of the protoplasm from the aqueous phase, to change cell parts into materials that will remain insoluble during subsequent treatment, and to protect cells from distortion and shrinkage when subjected to such fluids as alcohol and hot paraffin. Other important objectives of tissue preparation are to improve the staining potential of tissue parts and to alter their refractive indices for better visibility.

The procedure used to meet these requirements is called fixation, and the fluids used are called fixatives or fixing solutions. Any fixative should:

1. Penetrate rapidly to prevent postmortem changes.
2. Coagulate cell contents into insoluble substances.
3. Protect tissue against shrinkage and distortion during dehydration, embedding, and sectioning.
4. Allow cell parts to be made selectively and clearly visible.

Some fixatives may have a mordanting effect on the tissue—that is, they combine insolubly with it—and enhance the attachment of dyes and proteins to each other.

Tissues should be placed in fixatives as soon as possible after death. If delay is unavoidable, they should be put in a refrigerator, thus reducing autolysis and putrefaction to a minimum until the fixative can be applied.

Because a single chemical seldom has all the qualities of a good fixative, a fixing solution is rarely composed of only one chemical—familiar exceptions are formalin and glutaraldehyde. Most reliable fixatives contain one or more coagulant chemicals and one or more noncoagulant chemicals. Coagulants change the spongework of proteins into meshes through which paraffin can easily pass, thus forming a tissue of the proper consistency for sectioning. They also strengthen the protein linkages against breaking down during later procedures. Used alone, however, coagulants may form too coarse a network for the best cytological detail or may induce the formation of artificial structures (artifacts). Noncoagulants produce fewer artifacts, but if used alone they give the tissue a poor consistency for embedding.

Thus the most efficient fixing fluids are combinations of protein coagulants and protein noncoagulants. It should be emphasized, however, that no fixing solution is ideal: all chemical agents cause some chemical change in the protein structure of the cells. The choice of fixative will depend on the type of investigation to be undertaken, and an effort should be made to discover what adverse effect the fixing agent may have on the cellular components under study.

Because they contain ingredients that act upon each other, many mixtures are most efficient when made up fresh. The individual ingredients can usually be made up into stock solutions which can then be mixed together immediately before use. Among the frequently used chemicals are formaldehyde, glutaraldehyde, ethyl alcohol, acetic acid, picric acid, potassium dichromate, mercuric chloride, chromic acid, and osmium tetroxide (osmic acid). Since every chemical has its own set of advantages and disadvantages, each component should, whenever possible, compensate for a defect in some other component. For example, in the widely used fixative, Bouin solution: (1) Formaldehyde fixes the cytoplasm, but in such a manner that it retards paraffin penetration. It fixes chromatin poorly and makes cytoplasm basophilic. (2) Picric acid coagulates cytoplasm so that it admits paraffin, leaves the tissue soft, fixes chromatin, and makes the cytoplasm acidophilic. Its disadvantages are that it shrinks the tissue and that it makes chromatin acidophilic. (3) Acetic acid compensates for the defects of both formaldehyde and picric acid.

Another example: Birge and Tibbitts (1961), by adding 0.7% sodium chloride to two fixing solutions (formalin and Bouin), reduced the amount of shrinkage caused in nuclei and cytoplasm.

When the future use of a tissue is in doubt or if it is to be stored for an indefinite time, formalin is usually the conservative choice for a fixative; it permits secondary fixation (postfixation) and will not harden excessively. If the primary objective of tissue preparation is the compilation of a simple anatomical study of cell components, then routine fixatives can be used: formalin, Gomori, Susa, Zenker, Helly, or Bouin. Special fixatives for cell inclusions are Carnoy, Flemming, Champy, Helly, Schaudinn, Regaud, and others. For histochemistry the researcher is limited to aldehydes, acetone, or ethyl alcohol (page 416).

Most fixing solutions are named after the person originating them, e.g., Zenker and Bouin. If the same man originated more than one combination of chemicals, additional means of designating them have been used: Flemming weak and strong solutions, Allen B3, B15 series, etc. A few fixatives have been named arbitrarily: the name Susa was coined by Heidenhain from the first two letters of the words *su*blimate and *sä*ure.

CHEMICALS COMMONLY USED IN FIXATIVES

Acetic Acid

Acetic acid (CH_3COOH) is one of the oldest fixatives on record: in the eighteenth century vinegar (4–10% acetic acid content) was used to preserve hydras. In modern techniques it is rarely used alone but is an important component of many fixing solutions because of its efficient fixing action on the nucleus and its rapid penetration. It fixes the nucleoproteins, but not the proteins of the cytoplasm. It does not harden the tissue; actually it prevents some of the hardening that, without it, might be induced by subsequent alcohol treatment. In some techniques, however, acetic acid must be avoided because it dissolves out certain cell inclusions, such as Golgi and mitochondria, and some metals, such as calcium. Many lipids are miscible with acetic acid or are soluble in it. It neither fixes nor destroys carbohydrates. Pure concentrated acetic acid is called glacial acetic acid because it is solid at temperatures below 17°C.

Acids in general cause swelling in tissues—in collagen in particular—by breaking down some of the cross linkages between protein molecules and by releasing lyophilic radicles that associate with water molecules. An acid's swelling action can be a desirable property because it counteracts some of the shrinkage caused by most fixing chemicals. To curtail swelling after fixation with acetic or trichloracetic acid solutions, tissues should be transferred to an alcoholic washing solution rather than to water.

Acetone

Acetone (CH_3COCH_3) is used only for tissue enzymes, such as phosphatases and lipases. It is used cold and penetrates slowly. Only small pieces of tissue are fixed in this chemical.

Chromium Trioxide (chromic acid)

Crystalline chromium trioxide (CrO_3) is called chromic acid when it is added to water, usually in a 0.5% amount. Chromic acid is a valuable fixative but is rarely used alone. It penetrates slowly, hardens moderately, causes some shrinkage, forms vacuoles in the cytoplasm, and often leaves the nuclei in abnormal shapes. It is a fine coagulant of nucleoproteins and increases the stainability of the nuclei. It oxidizes polysaccharides and converts them into aldehydes—an action forming the basis of the Bauer histochemical test for glycogen and other polysaccharides. To fix water-soluble polysaccharides, it is better to use acetic acid and then posttreat them with chromic acid.

Chromic acid can also be used to partially oxidize fats to make them insoluble in lipid solvents. The oxidizing action may go too far, however, and potassium dichromate, which acts in a similar fashion, is safer for this purpose and is therefore more commonly used.

Excess chromic acid must be washed out, because it may later be reduced (undesirably, for our purposes) to green chromic oxide (Cr_2O_2). Because formalin and alcohol are reducing agents, they must not be mixed with chromic acid until immediately before use.

Alcohols

Alcohol cannot be used as a fixative for lipids because it makes them soluble. It does not fix carbohydrates, but neither does it extract mucins, glycogen, iron, and calcium. Alcohol is seldom used alone, although it is occasionally used for fixing enzymes.

Ethyl alcohol (ethanol, C_2H_5OH) hardens tissue but causes serious shrinkage. It is a strong cytoplasmic coagulant but does not fix chromatin. When alcohol is used, nucleic acid is transformed into a soluble precipitate and is lost in subsequent solutions and during staining.

Methyl alcohol (methanol, CH_3OH) is also used as a fixative, principally for hematologic tissues.

Aldehydes

Formaldehyde (HCHO) is a gas sold as a solution (approximately 40%), in water in which form it is known as formalin. The use of the term formol is incorrect, since a terminal *-ol* designates an alcohol or phenol. Unless the author of a technique specifies the dilution of his formalin in terms of actual formaldehyde content, dilutions must be made from the commercial product; e.g., a 10% solution would be 10 volumes of concentrated formalin (40% formaldehyde-saturated water) to 90 volumes of water.

Formaldehyde, if left standing for long periods, may either polymerize to form paraformaldehyde or oxidize into formic acid. A white precipitate in a stock solution indicates polymerization; the solution has been weakened. Cares (1945) suggests the following correction: shake the solution to suspend the sediment; pour into Mason jars and seal them tightly; autoclave at 15 lbs. for 30 minutes; and cool. This should produce a clear solution. More dilute solutions (such as 10%) tend to oxidize more readily than do stock solutions. Marble chips placed in the jars will keep the solution neutralized, or the solution may be buffered (page 15).

Formalin, when reacting with proteins, appears to form links between adjacent protein chains. The final effect therefore depends on the number of chemical reactions taking place between the formalin and the reactive groups of the protein molecules. Although some of these protein linkages will be broken down by washing in water, excess formalin must be removed; enough formalin-bound proteins will remain to react with future reagents when they are applied. Histologists frequently do not realize that the quantity of irreversibly bound protein drops as the pH of the solution rises above pH 10. For this reason, formalin reacts most efficiently as a buffered solution around the neutral point of pH 7.5–8.0.

Formalin has a moderate speed of penetration, but its action is slow. Although formalin preserves the cells adequately, it may not protect them completely unless it is given a long time to harden the tissue: shrinkage may take place if dehydration, clearing, and infiltration are started before the hardening action is complete.

Formalin is a good fixative for lipids; it does not dissolve lipoids or fats. It does not fix soluble carbohydrates, and it does dissolve some glycogen and urea.

So-called formalin pigment may appear in tissues rich in blood. This pigment is formed when hematein of the hemoglobin escapes from red blood corpuscles before or at death and reacts with the formalin. Its formation may be prevented by a short period of fixation in formalin followed by a prolonged soaking in 5% mercuric chloride. Once formed, the pigment can be removed in a solution of 1% potassium hydroxide in 80% alcohol or in picric acid dissolved in alcohol (Baker 1958). A 30-minute treatment in either of these solutions is recommended (page 284).

Glutaraldehyde and paraformaldehyde have special uses in histochemistry and electron microscopy (see pages 417–419). See Ericsson and Biberfeld (1967) for studies on aldehyde fixation.

Mercuric Chloride (corrosive sublimate)

Mercuric chloride ($HgCl_2$) is usually applied as a saturated aqueous solution (approximately 7%) and is acidic in action, owing to the release of H+ and Cl− ions in water. It is a powerful protein precipitant and forms intermolecular mercury links between SH, carboxyl, and amino groups. It penetrates reasonably well, but not as rapidly as acetic acid. It shrinks tissue less than the other protein coagulants, hardens it moderately, and distorts the cells less than other fixatives. It is excellent for mucin.

One disadvantage of mercuric chloride is that it deposits in the tissue a precipitate of uncertain chemical composition. This precipitate is in part crystalline (needle-shaped)—perhaps mercurous chloride—and in part metallic mercury (amorphous, irregular lumps) (Figure 20,a-b). It can be removed by iodine, the following reaction probably taking place, $2HgCl + I_2 = HgCl_2 + HgI_2$, and any metallic mercury being converted into mercuric iodide. Mercuric iodide is soluble in alcohol, but the brown color of iodine may remain in the tissue. This can be removed by prolonged soaking in 70% alcohol, or more quickly by treatment with sodium thiosulfate, 5% aqueous. A further disadvantage is that mercuric chloride crystals inhibit freezing, making it difficult to prepare good frozen sections.

Most stains react brilliantly on tissue fixed in mercuric chloride. Chromatin stains strongly with basic stains and lakes; cytoplasmic structures react equally well with acidic or basic stains.

Osmium Tetroxide (osmic acid)

In solution, usually 1% aqueous, osmium tetroxide (OsO_4) takes up a molecule of water and becomes H_2OsO_5 erroneously but commonly called osmic acid, as it is in this book. (Osmic acid would be H_2OsO_4.) The substance is chemically neutral, is not an acid, and cannot be isolated. Baker (1958) suggested that it might be named hydrogen per-persomate.

In solution, ionization of osmic acid is so minute that the pH is almost exactly that of the distilled water used in making the solution. The solution penetrates poorly and leaves tissue soft, the tissue later becoming so friable in paraffin that it sections badly. Osmic acid preserves the cytoplasm and nuclei, but, although it increases the stainability of chromatin in basic dyes, it reduces the stainability of the cytoplasm. Fats and other lipids reduce osmium tetroxide or combine with it to form a dark compound. Thus, fat sites become black and insoluble in absolute alcohol, cedarwood oil, and paraffin, but they remain soluble in xylene, toluene, and benzene, and if left more than 5 minutes in xylene or toluene become colorless. Osmic acid is not a fixative for carbohydrates.

When fixation is complete, excess osmic acid must be washed out of the tissue or it will reduce to an insoluble precipitate during treatment in alcohol. Since osmic acid is also easily reduced by light and heat, it must be stored in a cool, dark place.

Osmic acid is a valuable chemical for electron microscopy, and many tissues are postfixed in this solution. Good preservation can be achieved with the controlled pH of a buffer solution which can be a veronal, symcollidine, or phosphate buffer (considered a physiological solution) (Glauert 1965). See page 453 for the use of osmium black.

Picric Acid

Picric acid [$C_6H_2(NO_2)_3OH$] is usually used in a saturated solution, 0.9–1.2% aqueous. It is an excellent protein coagulant, forming protein picrates that have a strong affinity for acid dyes. However, it penetrates slowly, causes extreme shrinkage, and offers no protection against subsequent shrinkage. The shrinkage has been found to be close to 50% of the original tissue volume by the time the tissue has undergone paraffin infiltration. In spite of its acidity, it does not cause swelling. It does not dissolve lipids, nor does it fix carbohydrates, but it is recommended as a fixative for glycogen. It is a desirable constituent of many fixatives because it does not harden tissue, but it cannot be used alone because of the shrinkage it causes. Acetic acid is frequently used with it to counteract this undesirable characteristic.

Potassium Dichromate

Potassium dichromate ($K_2Cr_2O_7$), often incorrectly called potassium bichromate, is a noncoagulant of proteins; it makes them more basic in action but dissolves nucleoproteins. Chromosomes are therefore poorly shown, if at all. Potassium dichromate leaves tissues soft and in poor condition for paraffin sectioning. One valuable use, however, is in preparations for fixation of mitochondria; it makes the lipid components insoluble in lipid solvents. After fixation, tissues may be soaked in 3% aqueous potassium dichromate solution to insure that lipids will be well preserved.

Potassium dichromate can be mixed with mercuric chloride, picric acid, and osmic acid, but it reacts with formalin and must not be mixed with it until immediately before use. If acetic acid is added, mitochondria will be lost, but chromosomes will be more clearly shown than they are when the chromate is used alone. It is usually necessary to wash out the dichromate with water in order to prevent the formation of an insoluble precipitate of Cr_2O_3; this will form if the tissue is placed directly in alcohol.

Trichloracetic Acid

Trichloracetic acid (CCl_3COOH) is never used alone and is similar to acetic acid in action. It swells tissues and has a slight decalcifying property. As was noted under acetic acid, washing should be done in alcoholic solutions instead of in water.

"Indifferent Salts"

Baker (1958) applied this term to a group of chemicals (sodium sulfate, sodium chloride, and others), whose action is not clearly understood. Zenker and Helly solutions often have sodium sulfate added. Sometimes sodium chloride is added to formalin or mercuric chloride fixatives, particularly if marine forms are to be fixed or if it is necessary to hold shrinkage to a minimum.

For detailed discussions of fixation, see Baker (1945, 1958), Birge and Tibbitts (1961), Pearse (1968), and Wolman (1955). See page 416 for fixatives used in histochemistry.

MACERATION

Some tissues are extremely dense and cannot be manipulated while fresh. When working with these it may be desirable to separate the individual fibers of a muscle or nerve, and this is simplified by maceration. Some macerating fluids are not fixatives; these must usually be followed by some form of fixation. Other maceration fluids rely on the differential effect of weak fixatives on interstitial substances to separate the main fibers of a tissue. For example, if a fixative contains a connective tissue solvent, the solution will produce dissociation along with fixation. If a small piece of Bouin-fixed muscle is shaken in water or in 70% alcohol, the block of tissue will separate into individual fibers. Sublimate–acetic acid does not perform quite as well, but the acetic acid dissolves some connective tissue.

The following are the most commonly used macerating fluids (Hale 1958; McClung 1939 or 1950).

1. 30% alcohol: 24 hours or longer (up to 4 days).
2. Formalin, 1 part in 10% salt (NaCl) solution, 100 parts: 24 hours or longer.
3. 1% sodium chloride: 24 hours or longer.
4. Chromic acid, 0.2% aqueous: 24 hours.
5. Nitric acid, 20% aqueous: 24 hours. Particularly recommended for smooth muscle of bladder.
6. Boric acid, saturated solution in saline (sea water for marine forms), plus 2 drops Lugol iodine solution (page 532) each 25 ml: 2–3 days.
7. Potassium hydroxide, 33% aqueous. Good for isolation of smooth and striated muscle. After 1–1½ hours, tease tissue apart with needles.
8. Maceration by enzymes. Good for connective tissue, reticulum (Galigher 1934). Place frozen sections in:

pancreatin siccum	5.0 g
sodium bicarbonate	10.0 g
distilled water	100.0 ml

Wash thoroughly and stain.

9. 0.01% osmic acid dissociates and fixes muscle fibers in a few days. 1% acetic acid added to it causes quicker dissociation and almost as good fixation. Use only small pieces of tissue.

FIXING THE TISSUE

First consideration in the choice of fixative should be the purpose to be served in preparing the tissue for future use. Is a routine all-purpose fixative adequate, or must some special part of a cell be preserved? An aqueous fixing fluid will dissolve glycogen, and an alcoholic one will remove lipids. Thought should be given to the rate of penetration of the fluid and the density of the tissue to be fixed; obviously an extremely dense tissue will not fix well in a fixative that penetrates slowly and poorly. If fixatives of poor penetration are used, the pieces of tissue must be as small as possible. In any case, pieces should never be any larger than is absolutely necessary—the smaller the bulk, the better the fixation.

The hardening effect of fixatives should be considered. A fixative that produces excessive hardening may lead to difficulties with liver and muscle, and a fixative with the desired attributes but with less hardening effect should be used. If there is any doubt concerning future needs for a tissue, fix it in formalin; this can be followed by postfixation treatments.

Use a large volume of fixing fluid—at least ten times the volume of the tissue if possible. Remove the tissues from the animal and place them in the fixative as rapidly as is feasible, thereby keeping postmortem changes at a minimum. In most cases, do not attempt to fix an entire organ; it will be too large to allow rapid and complete penetration of the fixative, particularly if it is covered by a tough membrane. An ideal piece would be 1 to 2 cm in diameter; place a larger piece in the fixative for 15 to 30 minutes, then trim it to smaller size and return it to the fixative. This sometimes is necessary when tissue is very soft or easily crushed. Trim the piece with a new razor blade or sharp scalpel. This will cause less damage than the squeezing action of scissors. Do not crush or tear the tissue while removing it; such material is worthless. Never allow tissue to become dry before placing it in the solution; keep it moist with normal saline. Wash off accumulated blood with normal saline, as blood retards the penetration of a fixative. After placing the tissue in the solution, shake the container gently several times to make certain that the fluid has reached all surfaces and that pieces are not sticking to the bottom or sides. A chunk of glass wool placed in the container will aid in keeping the tissue free of the bottom.

Thin pieces of tissue that show muscular contraction or that may turn inside out (tissues of the gastrointestinal tract are particularly likely to do this) may be placed on thick filter paper, outside wall against the paper, and dropped in the fixative. Tiny, easily lost specimens—biopsies, bone marrow, and so on—may be wrapped in lens paper or coffee filter paper.

The length of time required for complete fixation depends on the rate of penetration and action of the fixative. Most coagulant fixatives produce complete fixation as fast as they can penetrate the tissue. But some fixatives, such as formalin, exhibit progressive improvement in fixing action after the tissue is completely penetrated. Prolonged action in this case improves the condition of the tissue and is rarely harmful. Occasionally some type of postfixation treatment is advisable (page 27).

WASHING THE TISSUE

After fixation is accomplished, the excess of many fixatives must be washed out of the tissue to prevent interference with subsequent processes. Often washing is done with running water; sometimes the tissue may be carried directly to alcohol, 50% or higher. Some technicians maintain, for instance, that if tissue prepared with Bouin fixative is washed in water, it loses some of its soluble picrates and should therefore be transferred from the fixative directly to 50% alcohol. When freezing is planned, formalin-fixed tissue may be washed briefly in water, but if necessary it can be taken from the fixing solution directly to the freezing microtome. Alcohol inhibits freezing, and if it is used it must be thoroughly washed out before any attempt to freeze the tissue.

When tissue has been fixed with a mercuric chloride solution, additional treatment is necessary. After washing in 50% alcohol, transfer the tissue to 70% alcohol containing enough iodine-alcohol (saturated solution of iodine in 70% alcohol) to give the solution a deep brown color. Leave the tissue in this solution from 5 to 8 hours, but no longer. If during this time the solution loses color, add some more iodine-alcohol. The iodine removes some, but probably not all, of the excess mercuric chloride precipitate in the tissue. Later, when staining the sections, iodine-alcohol or Lugol solution (page 532) must be included in the staining series to eliminate the remaining crystals that would otherwise persist as dark clumps and needles in the finished slides.

After washing, the tissue may be stored for several weeks or months in

70% alcohol, but it is always safer to dehydrate and embed it as soon as possible. Storage in alcohol for long periods of time (a year or longer) tends to reduce the stainability of tissues, as do long immersion in chromate and decalcifying solutions, or the presence of traces of acid or iodine.

Do not use corked bottles for fixing or for alcoholic solutions; extractives from corks can be injurious to tissues. Do not use metal caps on bottles of fixatives containing mercuric chloride.

FIXÁTIVES AND THEIR USES

Routine Fixatives and Fixing Procedures for General Microanatomy

Bouin Fixative

Fix for at least 24 hours—several weeks cause no damage.

picric acid, saturated aqueous	75.0 ml
formalin, concentrated	25.0 ml
glacial acetic acid	5.0 ml

Because of acetic acid content, do not use for cytoplasmic inclusions. Large vacuoles often form in tissues. Wash in 50% alcohol. The yellow color must disappear before sections are stained. It is usually removed in the alcohol series, but, if not, treat slides in 70% alcohol plus a few drops of saturated lithium carbonate until the yellow disappears.

Bouin-Duboscq (Alchoholic Bouin) Fixative (Pantin 1946)

Fix for 1–3 days.

80% ethyl alcohol	150.0 ml
formalin, concentrated	60.0 ml
glacial acetic acid	15.0 ml
picric acid crystals	1.0 gm

Prepare just before using. This is better than Bouin for tissues difficult to penetrate. Go directly to 95% alcohol.

Buffered Formalin

May remain in fixative indefinitely; action is progressive.

10% formalin	1000.0 ml
sodium acid phosphate ($NaH_2PO_4 \cdot H_2O$)	4.0 g
anhydrous disodium phosphate (Na_2HPO_4)	6.5 g

Wash in water.

Formalin (Baker 1944)

May remain in fixative indefinitely; action is progressive.

calcium chloride	1.0 g
cadmium chloride	1.0 g
formalin, concentrated	10.0 ml
distilled water	100.0 ml

Wash in water.

Formalin (Baker 1958)

May remain in fixative indefinitely; action is progressive.

formalin, concentrated	10.0 ml
calcium chloride (anhydrous), 10% aqueous solution (10 gm/100 water)	10.0 ml
distilled water	80.0 ml

Wash in water.

Glutaraldehyde, see page 417.

Gomori 1-2-3 Fixative[1]

May remain in fixative 2–3 weeks.

formalin, concentrated	1 part
mercuric chloride, saturated aqueous	2 parts
distilled water	3 parts

Excellent quick penetration, good for general cell morphology. Wash tissue 6–8 hours. Posttreat for mercuric chloride.

---

[1] Personal communication C. C. Lushbaugh.

Helly Fixative (Zenker formol)

Fix for 6–24 hours. If tissue seems to harden excessively, follow a maximum of 18 hours in fixative with 12–24 hours in 3% potassium dichromate.

potassium dichromate	2.5 g
mercuric chloride	4.0–5.0 g
sodium sulfate (may be omitted; see Indifferent	
Salts, page 10)	1.0 g
distilled water	100.0 ml
formalin, concentrated (add just before using)	5.0 ml

Formalin reduces dichromate and should not be left in stock solutions. The above stock, without formalin, can be used for Zenker fixative, page 17. Excellent for bone marrow, blood-forming organs, and intercalated discs. Wash tissue in running water overnight; posttreat for mercuric chloride.

Maximow modification (Romeis 1948)

Add 10% of formalin instead of 5%, and, for some techniques, 10% of osmic acid (fix in a dark container).

Hollande Bouin Fixative (Romeis 1948)

Fix for 8 hours to 3 days.

copper acetate	2.5 g
picric acid crystals	4.0 g
formalin, concentrated	10.0 ml
distilled water	100.0 ml
glacial acetic acid	1.5 ml

Dissolve copper acetate in water without heat; add picric acid slowly while stirring. When dissolved, filter and add formalin and acetic acid. Keeps indefinitely. Wash for several hours in 2 or 3 changes of distilled water. Hartz (1947) recommends this for fixation of calcified areas as in lymph nodes or fat necroses. It is a good general fixative.

Orth Fixative (Gatenby and Beams 1950 and Galigher 1934)

Fix for 12 hours at room temperature or 3 hours at 37°C.

formalin, concentrated	10.0 ml
potassium dichromate	2.5 g
sodium sulfate	1.0 g
distilled water	100.0 ml

Mix fresh. Wash in running water overnight. A good routine fixative, also for glycogen and fat.

Lillie variation (1965):

formalin, concentrated	10.0 ml
2.5% potassium dichromate	100.0 ml

Paraformaldehyde, see page 418.

Stieve Fixative (Romeis 1948)

Fix for 24 hours.

mercuric chloride, saturated aqueous	76.0 ml
formalin, concentrated	20.0 ml
glacial acetic acid	4.0 ml

Similar in effect to Susa, but simpler to prepare. Penetrates rapidly, good for large pieces. Time not critical. Go directly to 70% alcohol. Posttreat for mercuric chloride.

Susa Fixative (Romeis 1948)

Fix for 24 hours.

mercuric chloride saturated in 0.6% NaCl	50.0 ml
trichloracetic acid	2.0 g
glacial acetic acid	4.0 ml
formalin, concentrated	20.0 ml
distilled water	30.0 ml

Good substitute for Zenker if potassium dichromate is not required; Susa hardens tissue less. Rapid penetration. Go directly to 70% alcohol. Post-treat for mercuric chloride.

Zenker Fixative

Fix for 4–24 hours. If tissue seems to harden excessively, follow a maximum of 18 hours in Zenker with 12–24 hours in 3% potassium dichromate.

potassium dichromate	2.5 g
mercuric chloride	4.0–5.0 g
sodium sulfate (may be omitted, see Indifferent Salts, page 10)	1.0 g
distilled water	100.0 ml
glacial acetic acid	5.0 ml

Excellent general fixative, fairly rapid penetration. Wash in running water overnight. Posttreat for mercuric chloride. Zenker may be substituted for Helly if acetic acid is not added to the stock solution.

Fixatives for Cell Inclusions and Special Techniques

Altmann Fixative (Gatenby and Beams 1950)

Fix for 12 hours.

5% potassium dichromate	10.0 ml
2% osmic acid	10.0 ml

Good for fat and mitochondria. Wash in running water overnight.

Altmann Fixative (Gatenby and Beams 1950)

Fix for 2 hours; do not fix longer or material will swell.

chromium potassium sulfate (chrome alum)	3.0 g
formalin, concentrated	30.0 ml
glacial acetic acid	2.0 ml
distilled water	238.0 ml

Good for yolk-rich material and insect larvae. Does not harden and gives good cytological detail. Wash in 70% alcohol or water.

Carnoy Fixative (Gatenby and Beams 1950)

Fix for 3–6 hours.

Formula A

acetic acid	20.0 ml
absolute ethyl alcohol	60.0 ml

Formula B (the true Carnoy)

glacial acetic acid	10.0 ml
absolute ethyl alcohol	60.0 ml
chloroform	30.0 ml

Chloroform is said to make action more rapid. Important fixative for glycogen and Nissl substance, but dissolves most other cytoplasmic elements. Wash 2–3 hours in absolute alcohol to remove chloroform, particularly if embedding in nitrocellulose.

Methacarn (Puchtler et al. 1970)

absolute methyl alcohol	60.0 ml
chloroform	30.0 ml
glacial acetic acid	10.0 ml

For embedding, transfer through two changes of methyl alcohol (2–3 hours and 4 hours); two changes methyl benzoate (3 hours and 1 hour); methyl benzoate-xylene or toluene (1 hour); xylene or toluene; infiltrate and embed. Do not wash tissues in water or saline before fixation and do not contaminate with water during processing; shrinkage and alterations of tissue structure will result. Do not overexpose to alcohol; it causes shrinkage and hardening. Store over a weekend in methacarn or methyl benzoate. Methacarn is recommended over Carnoy, since it causes less hardening.

Carnoy-Lebrun Fixative (Lillie 1965)

Fix for 3–6 hours.

absolute ethyl alcohol	15.0 ml
glacial acetic acid	15.0 ml
chloroform	15.0 ml
mercuric chloride crystals	4.0 g

This is a very fine fixative, but it does not keep; mix fresh. Penetrates rapidly. Wash well in alcohol to remove chloroform.

Champy Fixative (Gatenby and Beams 1950)

Fix for 12 hours.

potassium dichromate, 3% aqueous (3 g/100 ml water)	7.0 ml
chromic acid, 1% aqueous (1 g/100 ml water)	7.0 ml
osmic acid, 2% aqueous (2 g/100 ml water)	4.0 ml

Prepare immediately before use. Good for cytological detail, mitochondria, lipids, etc. Penetrates poorly; use only for small pieces. Wash in running water overnight.

Dafano Fixative (Romeis 1948)

Fix for 12–24 hours.

cobalt nitrate	1.0 g
formalin, concentrated	15.0 ml
distilled water	100.0 ml

Wash in water. Good for Golgi apparatus.

Flemming Fixatives

Fix for 12–24 hours.

Strong solution

chromic acid, 1% aqueous (1 g/100 ml water)	15.0 ml
osmic acid, 2% aqueous (2 g/100 ml water)	4.0 ml
glacial acetic acid	1.0 ml

Mix just before using. Acts slowly: use small pieces with no blood on outside. Wash in running water for 24 hours. High acetic content makes it poor for cytoplasm fixation.

Modification for cytoplasmic study

chromic acid, 1% aqueous (1 g/100 ml water)	15.0 ml
osmic acid, 2% aqueous (2 g/100 ml water)	4.0 ml
glacial acetic acid	2–3 drops

Weak solution

chromic acid, 1% aqueous (1 g/100 ml water)	25.0 ml
osmic acid, 2% aqueous (2 g/100 ml water)	5.0 ml
acetic acid, 1% aqueous (1 ml/99 ml water)	10.0 ml
distilled water	60.0 ml

Lewitsky saline modification (Baker 1958)

Omit acetic acid and replace the water with 0.75% sodium chloride.

Flemming solutions are good for mitotic figures, but are not suitable for general work because they penetrate poorly, harden excessively, blacken material, and interfere with the action of many stains. The weak solution is a fine fixative for minute, delicate objects. For dense tissues use the strong solution. Iron hematoxylin is excellent following these fixatives.

Formol-Alcohol (Tellyesniczky) (Lillie 1965)

Fix for 1–24 hours.

70% ethyl alcohol	100.0 ml
formalin, concentrated	5.0 ml
glacial acetic acid	5.0 ml

Good for insects and crustaceans. Widely used by botanists. Transfer to 85% alcohol.

Gendre Fluid (Lillie 1965)

Fix 1–4 hours at 25°C.

95% ethyl alcohol saturated with picric acid	80.0 ml
formalin, concentrated	15.0 ml
glacial acetic acid	5.0 ml

Good for glycogen. Wash in several changes of 80% alcohol.

Gilson Fixative (Gatenby and Beams 1950)

Fix for 24 hours; may be left several days.

nitric acid, concentrated	15.0 ml
glacial acetic acid	4.0 ml
mercuric chloride crystals	20.0 g
60% ethyl alcohol	100.0 ml
distilled water	880.0 ml

Good for invertebrate material. Does not give a good histological picture; shrinks cytoplasm badly. Wash in 50% alcohol. Posttreat for mercuric chloride. Good for beginners, easy to work with.

Johnson Fixative (Gatenby and Beams 1950)

Fix for 12 hours.

potassium dichromate, 2.5% aqueous (2.5 g/ 100 ml water)	70.0 ml
osmic acid, 2% aqueous (2 g/100 ml water)	10.0 ml
platinum chloride, 1% aqueous (1 g/100 ml water)	15.0 ml
glacial acetic acid	5.0 ml

Shrinks spongy protoplasm less than Flemming. Wash in water.

Hermann modification (Gray 1954)

Fix 12–16 hours.

platinum chloride, 1% aqueous (1 g/100 ml water)	15.0 ml
glacial acetic acid	1.0 ml
osmic acid, 2% aqueous (2 g/100 ml water)	2.0–4.0 ml

Better for protoplasm than chromic mixtures. Good staining of nuclei, but not of plasma. Some shrinkage of chromatin. Without acetic acid it is good for mitochondria.

Kolmer Fixative

Fix for 24 hours.

potassium dichromate, 5% aqueous (5 g/100 ml water)	20.0 ml
10% formalin	20.0 ml
glacial acetic acid	5.0 ml
trichloracetic acid, 50% aqueous (50 g/100 ml water)	5.0 ml
uranyl acetate, 10% aqueous (10 g/100 ml water)	5.0 ml

Good for entire eye (Walls 1938) or nerve tissue, due to presence of uranium salts. Wash in running water.

Lavdowsky Fixative (Gray 1954)

Fix for 12–24 hours.

distilled water	80.0 ml
95% ethyl alcohol	10.0 ml
chromic acid, 2% aqueous (2 g/100 ml water)	10.0 ml
glacial acetic acid	0.5 ml

Good for glycogen (Swigart et al. 1960). Transfer to 80% alcohol.

Navashin Fixative (Randolph 1935)

Fix for 24 hours.

Solution A

chromic acid	1.0 g
glacial acetic acid	10.0 ml
distilled water	90.0 ml

Solution B

formalin, concentrated	40.0 ml
distilled water	60.0 ml

Mix equal parts of A and B just before using. At end of 6 hours change to a new solution for another 18 hours. Useful for preserving cellular detail in plant materials; as good as Flemming on root tips and less erratic. Transfer to 75% alcohol.

Perenyi Fixative (Galigher 1934)

Fix for 12–24 hours.

chromic acid, 1% aqueous (1 g/100 ml water)	15.0 ml
nitric acid, 10% aqueous (10 ml/90 ml water)	40.0 ml
95% ethyl alcohol	30.0 ml
distilled water	15.0 ml

Good for eyes. When fixing the entire eye always make a small hole near the ciliary body so the fluids of both chambers can exchange with the fixing fluid. For the best fixing results, inject a little of the fixative into the chambers. Decalcifies small deposits of calcium; good fixative for calcified arteries and glands. Trichromes stain poorly; hematoxylin is satisfactory. Wash in 50% or 70% alcohol.

Regaud Fixative (Kopsch) (Romeis 1948)

Fix for 4–24 hours.

potassium dichromate, 3% aqueous (3 g/100 ml water)	40.0 ml
formalin, concentrated	10.0 ml

Mix immediately before use. Recommended for mitochondria and cytoplasmic granules. Tends to harden. Follow fixation by chromating several days in 3% potassium dichromate, which should be renewed every 24 hours. Wash in running water overnight.

Rossman Fixative (Lillie 1965)

Fix for 12–24 hours.

absolute alcohol, saturated with picric acid	90.0 ml
formalin, concentrated	10.0 ml

Good for glycogen. Wash in 95% alcohol.

Sanfelice Fixative (Baker 1958)

Fix for 4–6 hours.

chromic acid, 1% aqueous (1 g/100 ml water) ..	80.0 ml
formalin, concentrated	40.0 ml
glacial acetic acid	5.0 ml

Mix immediately before use. Good for chromosomes and mitotic spindles. Fix small pieces. Produces less final shrinkage than others of this type. Wash in running water 6–12 hours.

Schaudinn Fixative (Kessel 1925)

Fix smears for 10–20 minutes at 40°C.

mercuric chloride, saturated aqueous	66.0 ml
95% ethyl alcohol	33.0 ml
glacial acetic acid	5.0–10.0 ml

Recommended for protozoan fixation, and for smears on slides or in bulk. Not for tissue; produces excessive shrinkage. Transfer directly to 50% or 70% alcohol. Posttreat for mercuric chloride.

Sinha Fixative

Specimens may remain in fixative 4–6 days.

picric acid, saturated in 90% alcohol	75.0 ml
formalin, concentrated	25.0 ml
nitric acid, concentrated	8.0 ml

Sinha (1953) adds that 5% mercuric chloride may be included; he probably means 5 g per 100.0 ml of above solution. Recommended for insects; softens hard parts with no damage to internal structure. Transfer directly to 95% alcohol.

Smith Fixative (Galigher 1934)

Fix for 24 hours.

potassium dichromate	5.0 g
formalin, concentrated	10.0 ml
distilled water	87.5 ml
glacial acetic acid	2.5 ml

Mix immediately before use. Good for yolk-rich material (Laufer 1949). Wash in running water overnight.

The *p*H of some fixatives changes after mixing and again after tissue is added. This may be a factor worth considering when stainability or silver impregnation is important (Freeman et al. 1955).

Fixation by Perfusion

Perfusion (forceful flooding of tissue) is advantageous only for tissue that requires rapid fixation but is not readily accessible for rapid removal. A prime example is the central nervous system. Many organs are not adequately fixed by this method because the perfusion fluid may be carried away from the cells rather than to them.

Special equipment necessary for perfusion includes a glass cannula that fits the specific aorta to be used, and rubber tubing to connect the cannula to the perfusion bottle.

When the animal is dead or under deep anesthesia, cut the large vessels in the neck and drain out as much blood as possible. Expose the pericardium by cutting the costal cartilages and elevating the sternum. Cut the pericardium and reflect it back to expose the large arteries. Free part of the aorta from the surrounding tissue and place a moistened ligature behind it. Make a small slit directed posteriorly in the wall of the aorta and insert the moistened cannula. Bring the two ends of the ligature together and tie the cannula firmly in place. Cut open the right atrium to permit escape of blood and other fluids.

Precede fixation with an injection of a small amount of saline (50–100 ml) to wash out residual blood before it attaches to the vessel walls. Fill just the rubber tubing leading from the perfusion bottle to the cannula. (Separate the saline from the fixative with a clamp near the attachment of the rubber tubing to the bottle.) If a formalin-dichromate fixative is being used, substitute 2.5% potassium dichromate for normal saline. Fill the perfusion bottle with fixative (500–1000 ml depending on size of animal) warmed to body temperature.

When ready to start perfusion, hold the bottle at table level and open the clamp on the rubber tubing. Gradually raise the bottle to a height of 4 to 5 feet, at which point enough pressure will be exerted to force out most of the blood. After 5 minutes, open the abdomen and examine the organ to be perfused. If the surface vessels are still filled with blood and the organ has not begun to take on the color of the perfusing solution, it is possible that the perfusion has failed. However, stubborn cases may require 10 to 30 minutes to perfuse. Whenever the blood color is absent, perfusion is complete, but it should be continued long enough to keep shrinkage to a minimum.

Observe these suggestions: (1) the cannula used should be as large as possible to permit as rapid a flow as possible. This aids in washing out blood ahead of the fixative. (2) If the head alone is to be fixed, clamp off the thoracic duct, and if the brain and spinal cord are to be fixed, clamp off intestinal vessels. The fixative is then directed toward brain and spinal cord. (3) When the perfusion bottle is being filled, allow some of the fluid to flow through the rubber tubing and cannula to release air bubbles. This can also be done with the saline. Air bubbles will block the perfusion. (4) Do not allow the injection pressure to exceed the blood pressure; artifacts will result.

If liver is to be perfused—particularly if by glutaraldehyde, which penetrates slowly—perfusion should be done through the portal vein followed by perfusion through the hepatic vein (Fahimi 1967).

If only a small piece of an organ is to be fixed, a modified and easier perfusion is usually adequate. Inject the organ with a hypodermic syringe of fixative, and immediately after injection, cut out a small piece of tissue close to the injection site and immerse in the same type of fixative.

Lillie (1954b) lists two disadvantages of perfusion: (1) the blood content of the vessels is lost, and (2) perfusion is not possible if postmortem clotting is present. But he does favor perfusion as the outstanding method for brain fixation, saying that immersion of the whole brain without perfusion "can only be condemned." He suggests the following as the preferred method of fixation for topographic study, if whole brain perfusion is not possible.

1. Cut a single transverse section anterior to oculomotor roots and interior margin of anterior colliculi, separating the cerebrum from midbrain and hindbrain.
2. Make a series of transverse sections through the brain stem and cerebellum (5–10 mm intervals) leaving part of meninges uncut to keep slices in position.
3. Separate two cerebral hemispheres by a sagittal section. On the sagittal surface identify points through which sections can be cut to agree with standard frontal sections. Make cuts perpendicular to sagittal surface. Cut rest of brain at 10 mm intervals.
4. Fix in a large quantity of solution.

Less shrinkage occurs in the brain after perfusion than after immersion without perfusion. Lodin et al. (1967) recommend formalin-saline (1:1) as the best fixative, ethanol (not tertiary butyl alcohol) for dehydration, and embedding in celloidin or epoxy (or frozen cut sections). Paraffin embedding results in more shrinkage than the above methods; if it is used, try vacuum to reduce the shrinkage. Mounting on slides with 80% ethanol, rather than with water, will recover a good portion of the shrinkage in the sections.

For additional references on perfusion see Bensley and Bensley (1938), Emmel and Cowdry (1964), and Lillie (1965). See also Koenig et al. (1945) and Eayrs (1950).

POSTFIXATION TREATMENTS

Chromatization

Chromatization improves preservation and staining, particularly of mitochondria and the myelin of nervous tissue.

Place the fixed tissue in a 2.5–3% aqueous potassium dichromate solution (2.5–3 g/100 ml water): overnight for small gross specimens (1–2 cm); 2–3 days for larger ones; 1–2 hours for sections on slides before staining (may be left overnight).

Wash thoroughly in running water: overnight for large gross tissues; 15–30 minutes for slides.

Deformalization

The removal of bound formalin is frequently necessary in silver impregnation methods, such as Ramón y Cajal and the del Río-Hortega methods, and the Feulgen technique.

Lhotka and Ferreira Method (1950)

1. Wash tissue blocks in distilled water: 15 minutes.
2. Transfer through 2 changes of chloral hydrate, 20% aqueous (20 g/100 ml water): 24 hours each.
3. Wash in distilled water: 15 minutes.
4. Proceed to any method of staining. With silver stains, time of impregnation needs to be lengthened: Ramón y Cajal to 2 weeks; del Río-Hortega to 24 hours.

Krajian and Gradwohl Method, on Slides (1952)

1. Place in ammonia water (40 drops ammonia in 100.0 ml water): 1 hour.

2. Wash in running water: 1 hour.
3. Fix in special fixative if necessary: 1 hour.
4. Wash in running water: 1 hour.
5. Proceed to stain.

Removal of Mercuric Chloride Crystals

Iodine Method, see page 13.

Gonzalez Method (1959*a*)

Gonzalez suggests the following if it is necessary to avoid the mordant action of iodine.

1. After fixation, wash briefly or immerse directly in Cellosolve (ethyl-ene–glycol–monoethyl ether): 24–48 hours, 3 changes.
2. Follow with toluene: 1–2 hours.
3. Infiltrate and embed.

DECALCIFICATION

Calcium deposits may be so heavily concentrated in the tissue that they interfere with sectioning and result in torn sections and nicks on the knife edge. If deposits are sparse, overnight soaking of blocked tissue in water will soften the deposits sufficiently for sectioning (page 61). Heavy deposits may be removed by any of several methods, but do not leave tissues in any of the fluids longer than necessary.

If any doubt arises about the completion of decalcification, check for calcium by the following method.

To 5 ml of the solution containing the tissue, add 1 ml of 5% sodium or ammonium oxalate. Allow to stand 5 minutes. If a precipitate forms, decalcification is not complete. A clear solution indicates it is complete. Sticking needles in the tissue to check hardness is a sloppy technique that can damage the cells.

An excellent decalcifying fluid, RKO,[2] can be purchased by the gallon. It is rapid in action and staining is good following its use.

---

[2] DuPage Kinetic Laboratories, Downers Grove, Ill.

Acid Reagents

Following the use of an acid for decalcification, prevent swelling in the tissue and impaired staining reactions by transferring into 70% alcohol: 3–4 hours or overnight. If acidity will interfere with staining, treat with 2% aqueous lithium carbonate solution, or 5% sodium sulfate: 6–12 hours, then into 70% alcohol. Chromate treatment (page 27) also improves staining.

Because calcium ions are soluble at pH 4.5, buffer solutions may be used. They are slower than other methods, but cause no perceptible tissue damage.

Formic acid A

formic acid	5.0–25.0 ml
formalin, concentrated	5.0 ml
distilled water to make	100.0 ml

With 5 ml formic acid content, 2–5 days. If increased to 25 ml less time is required, but there is some loss of cellular detail.

Formic acid B

formic acid, 50% aqueous (50 ml/50 ml water)	50.0 ml
sodium citrate, 15% aqueous (15 g/100 ml water)	50.0 ml

Nitric acid A

formalin, concentrated	10.0 ml
nitric acid (68–70% sp. gr. 1.41)	5.0 ml
distilled water	85.0 ml

Nitric acid B

nitric acid (68–70%, sp. gr. 1.41)	50.0 ml
distilled water	950.0 ml

Citrate-citric acid buffer, pH 4.5 (Culling 1957)

citric acid monohydrate, 7% aqueous (7 g/100 ml water)	5.0 ml
ammonium citrate, anhydrous, 7.45% aqueous (7.45 g/100 ml water)	95.0 ml

zinc sulfate, 1% aqueous (1 g/100 ml water) ...   0.2 ml
chloroform ....   few drops

Kristensen fluid (1948), pH 2.2

8 N formic acid (see page 531) ....   50.0 ml
1 N sodium formate ....   50.0 ml

Treat for 24 hours. Wash in running water for 24 hours. Highly recommended.

Pros and Cons in Use of Acid Reagents

Schajowicz and Cabrini (1955) observe that strong acids, such as nitric acid, alter the histochemical behavior of bone and cartilage, and must be used with care. Formic acid and citrate do not have this disadvantage if used for only a few days. Morris and Benton (1956) found 1–2 M hydrochloric acid the most rapid decalcifier (approximately 3 hours). They found that this produced adequate staining reactions if slides were mordanted in 5% aqueous ammonium chloride for 30 minutes before staining.

Electrolytic Method of Decalcification

Electrolytic decalcification is based on the theory that there is an attraction of Ca ion to a negative electrode. The material to be decalcified is suspended by a platinum wire, becoming the anode, and a second platinum wire forms the cathode; electrolysis then frees the calcium ions from the material.[3] Apparently decalcification is caused by the increased temperature during electrolysis, since Lucas (1952) found no evidence that the passage of the electric current itself accelerated decalcification. Lillie et al. (1951) demonstrated that an increase in temperature increases the speed of decalcification; Culling (1957) agrees with Lillie et al. but notes that heat causes severe swelling and does not recommend it for decalcifying. Lillie et al. (1951) note that the best results for staining are obtained by decalcification at room temperature or cooler.

Case (1953) considers the electrolytic method inferior to the use of either nitric or formic acid. I do not recommend electrolysis and prefer the RKO.

---

[3] Bone decalcifier (portable): Chicago Apparatus Co., Catalog no. 28–712; or Lipshaw Co.

Chelating Agents for Decalcification

Chelating agents offer the advantage of maintaining good fixation and sharp staining potential during decalcification. They are organic compounds that have the power of binding certain metals, such as calcium and iron. Commercial preparations of the disodium salt of ethylene diamine tetracetic acid (EDTA)—Versene (Dow Chemical Co.) or Sequestrene (Geigey Chemical Co.)—are the most commonly used agents. The method does have two disadvantages: the tissue tends to harden, and decalcification is slow.

Hilleman and Lee (1953) recommend 200 ml of a 5.5% solution of either Versene or Sequestrene in 10% formalin, for pieces 40 × 10 × 10 mm. Decalcification may require up to 3 weeks. Renew the solution at the end of each week. Transfer directly to 70% alcohol.

Vacek and Plackova (1959) report that a 0.5 M solution of EDTA at pH 8.2–8.5 yields better results in silver methods than does decalcification with acids.

Schajowicz and Cabrini (1955) consider Versene the better of the two solutions. The solution should be adjusted to pH 7.0 with NaOH and HCl. Hematoxylin and eosin stain as usual, but glycogen is lost, and alkaline phosphatase has to be reactivated after the use of chelating agents.

Decalcification Combined with Fixation

Lillie Fluid (1965)

This is used for 1–2 days.

picric acid, 1–2% aqueous (1–2 g/100 ml water)	85.0 ml
formalin, concentrated	10.0 ml
formic acid, 90–95% aqueous (90–95 ml/10–5 ml water)	5.0 ml

To extract some of the yellow: 2–3 days in 70–80% alcohol.

Lillie Alternate Fluid (1965)

Add 5% of 90% formic acid to Zenker fixative.

McNamara et al. Fluid (1940)

Solution A

mercuric chloride	10.0 g
distilled water	300.0 ml

Dissolve the mercuric chloride in the distilled
water by heating. Cool. Add:

trichloracetic acid	30.0 g
distilled water	100.0 ml

Solution B

nitric acid, concentrated	5.0 ml
95% ethyl alcohol	50.0 ml
formalin, concentrated	40.0 ml

Mix solutions A and B. Change daily until bone is soft. If more than 7
days are required, nuclear staining will be impaired. Wash in running
water: 24 hours. Dehydrate and embed.

Perenyi Fluid

This is also a fixative, page 23. Good for small deposits. Little hardening
effect. Excellent cytological detail. Good for calcified arteries, and glands
such as thyroid: 12–24 hours. Wash in 50–70% alcohol.

Schmidt Fluid (1956)

Schmidt uses the following for 24–48 hours, pH 7–9:

4% formalin (4 ml/96 ml water) plus 1 g	
sodium acetate	100.0 ml
disodium versenate (EDTA)	10.0 g

No washing necessary. Transfer tissue directly into 70% alcohol, dehydrate
and embed as usual.

Decalcification Combined with Dehydration

Jenkin Fluid (Culling 1957)

absolute ethyl alcohol	73.0 ml
distilled water	10.0 ml

chloroform .... 10.0 ml
acetic acid .... 3.0 ml
hydrochloric acid, concentrated .... 4.0 ml

The swelling action of the acid is counteracted by the inclusion of alcohol. Large amounts of solution should be used, 40–50 times bulk of tissue. After decalcification, transfer to absolute alcohol for several changes, then clear and embed.

OTHER METHODS OF TISSUE PREPARATION

The major portion of this book is devoted to sectioning methods for preparing tissue for staining, because of the complexity and quantity of such methods. The student, however, should recognize that there are other means of examining tissues.

Exceedingly thin membranes can be examined directly by mounting them in glycerol or other aqueous media. Considerable detail can be observed with reduced light or under the phase microscope. Sometimes a bit of stain can be added to sharpen or differentiate certain elements. More permanent preparations can be secured by fixation.

"Touch" preparations (impression films) are made by pressing the cut surface of fresh tissue against a dry slide. Cells adhere to the surface and can be examined unstained, or the slide can be immediately immersed in a fixative and then stained.

Occasionally, sections of relatively tough tissue are cut freehand for examination.

Smears are one of the commonest devices for simple slide preparation, such as blood and bone marrow (page 245), Papanicolaou (page 454), fecal (page 506), and chromosomes (squash preparations, a modification of smears, page 465).

"Cell blocks"—concentrated clusters of individual cells or grouped cells—are described in detail on page 44.

CHAPTER 2

Dehydration:
Preparation for Embedding

Tissues fixed in aqueous solutions maintain a high water content, which can hinder later processing. Except in special cases (freezing methods, water-soluble waxes, and special cell contents), the tissue must be dehydrated (water removed) before some steps in processing can be successful.

During fixing and washing, tissues lack the ideal consistency for sectioning, or cutting thin slices of a few microns in thickness. They may be soft, or they may contain hollow spaces (lumen) and be easily deformed by sectioning. If the cells are pierced by a knife, their fluid content may be released, causing them to collapse. To preclude these problems, it is necessary to replace the fluids in the tissue by a medium that hardens to a firm, easily sectioned material. The cells are filled intracellularly (impregnated or infiltrated) and enclosed extracellularly (embedded) with the medium and are thereby protected during handling. Universally used media for this purpose are paraffin and nitrocellulose and variations of these. Other media, less frequently used, are gelatin, water-soluble and ester waxes, and plastics.

Various conditions determine the choice of medium. Paraffin is suitable for most histological and embryological purposes when thin sections (1–15μ) are required. Nitrocellulose and plastics can also be used, but serial sections are made more easily with paraffin-impregnated tissue, and paraffin preparation requires much less time. Impregnating with nitrocellulose has distinct advantages when it is desirable to avoid heat and when a tissue becomes hard too readily or is too large for the paraffin technique. With nitrocellulose, shrinkage is kept at a minimum, whereas with paraffin it can amount to as much as 20%. Gelatin can be used for extremely friable tissue in the freezing technique, and water-soluble waxes are used when alcohols, hydrocarbons, and the like must be avoided. Ester waxes are recommended for hard and smooth tissues and when it is necessary to avoid hardening agents, such as hydrocarbons.

Before embedding the tissue in either paraffin, nitrocellulose, or ester waxes, all water must be removed from it. This dehydration is usually achieved by immersing the tissue in a series of solutions of ethyl alcohol (ethanol) in water, with gradually increasing percentages of alcohol. Changing through solutions of 30%, 50%, 70%, 80%, 95%, and absolute alcohol is said to reduce some of the shrinkage occurring in the tissue. If time does not permit such a series, the 30% and 80% steps, and even the 50% one, may be eliminated without great damage to the tissue. The time required for each step depends on the size of the object—30 minutes to 2 hours, maybe 3 hours in extreme cases. A second change of absolute alcohol should be included to insure complete removal of water; a total of 2–3 hours in both changes should be ample, even for large pieces. Too long an exposure to either 95% or absolute alcohol tends to harden the material, making it difficult to section.

There are other agents which are just as successful dehydrants as ethyl alcohol. The ideal dehydrating fluid would be one that would mix in any proportion with water, ethyl alcohol, xylene, or paraffin. Two such solutions are dioxane (page 39) and tertiary butyl alcohol (butanol). Absolute butyl alcohol is miscible with paraffin, and after infiltrating tissues with a warm (50°C) butyl alcohol–paraffin mixture, infiltration with pure paraffin can follow. If isobutyl alcohol is used, there may be some difficulty with impregnations, probably because of the limited miscibility of this alcohol with paraffin and water. The butyl alcohols have the added disadvantage of a disagreeable odor. With both tertiary and isobutyl alcohol there is more shrinkage than with other alcohols. Isopropyl (99%) alcohol (isopropanol) is an excellent substitute for ethyl alcohol and is sufficiently waterfree for use as absolute alcohol. Actually isopropyl alcohol produces less shrinkage and hardening than ethyl alcohol, and is free from Internal Revenue restrictions. There are two disadvantages to the use of isopropyl alcohol for dehydrating purposes: (1) it cannot be used on tissues that are to be embedded in nitrocellulose, since nitrocellulose is practically insoluble in it; (2) most dyes are not soluble in it so, with a few exceptions, it cannot be used during staining procedures. Solox,[1] ethanol denatured with methanol, is an inexpensive substitute for pure ethanol or isopropanol.

For the preparation of dilutions of ethyl alcohol, it is customary to use 95% alcohol and dilute it with distilled water in the following manner: if a 70% solution is required, measure 70 parts of 95% alcohol and add 25 parts of distilled water to make 95 parts of 70% dilution. In other words, into a 100-ml graduated cylinder pour 95% alcohol to the 70-ml mark, and then add distilled water up to the 95-ml mark.

Absolute alcohol is not exactly 100%, but may contain as much as 1 or 2% water. If the water content is no higher than this, the absolute alcohol is considered 100% for practical purposes in microtechnique. If it is necessary to make certain that the water content is no more than 2%, add a few ml of the alcohol to a few ml of toluene or xylene. If turbidity persists, there is more than 2% water present. But if the mixture remains clear, the alcohol is satisfactory as an absolute grade.

Dilutions of isopropyl alcohol (99%) can be handled as a 100% solution; that is, for 70% use 30 ml of water to 70 ml of alcohol, etc.

If distilled water is not provided in the laboratory building, a Barnstead Still [2] can provide a sufficient amount of water for the average microtechnique laboratory. The stills are available in sizes that produce $\frac{1}{2}$ to 10 gallons of water per hour. A special model will produce 30 gallons per hour. Deionized water[3] is a satisfactory substitute for distilled water in some solutions.

See chapter 3, page 39, for solutions that both dehydrate and "clear" tissues.

Special Treatment for Small, Colorless Tissues

A small, colorless tissue often seems to disappear into the opaqueness of the paraffin. An easy answer to this problem is to add some eosin to the last change of 95% alcohol; the tissue can then be seen more readily and oriented with greater facility. (However, eosin cannot be added if isopropyl alcohol is being used, since eosin is not soluble in isopropyl.) The eosin will not interfere with future staining; it is lost in the hydration series following deparaffinization.

---

[1] Industrial Chemicals, Inc., Richmond, Virginia.
[2] Barnstead Still and Sterilizer Company, Boston, Mass.
[3] Barnstead Demineralizer, cartridges containing ion-exchange resin.

CHAPTER 3

Clearing, Infiltrating, and Embedding: Paraffin Method

CLEARING

In most techniques that require dehydration and infiltration or impregnation, an intermediary step is necessary between the two. Because the alcohol used for dehydration will not dissolve or mix with molten paraffin (exception, tertiary butyl alcohol), the tissue must be immersed in some fluid miscible with both alcohol and paraffin before infiltration can take place.

"Clearing" may seem a strange name for this intermediary step, but it describes a special property of the reagents that are used. They remove or *clear* opacity from dehydrated tissues, making them transparent. Blocks of tissue appear to deepen in color; they seem almost crystalline, never milky.

Three hydrocarbons—benzene, toluene, and xylene—are reagents commonly used for clearing. However, if the tissue contains much cartilage or is fibrous or muscular, it is wise to avoid these because of their tendency to harden such tissue. Xylene, formerly one of the most widely used reagents, has the highest tendency of the three to harden such tissue and therefore is not recommended.

An additional problem with these three reagents is their rapid rate of evaporation; unless they are handled quickly, they evaporate out of the tissue and leave air pockets that will not infiltrate with paraffin. D. Tully (M.I.T. Education Research Center) suggests correctly that the rate of evaporation of these reagents is more directly correlated to their vapor pressures than to their boiling points. (Boiling points are: xylene, 142°C; benzene, 80°C; and toluene, 111°C.) Xylene has a vapor pressure of only 5 mm Hg, but, as noted above, hardens tissue excessively. Benzene, with a vapor pressure of 80 mm Hg, has the highest rate of evaporation of the three, although it hardens least. Toluene, then, of the three reagents, is probably the safest; it has a vapor pressure of 25 mm Hg, thus eliminating some of the hazards of evaporation, and it does not harden as excessively as xylene.

Chloroform is used in many laboratories but has outstanding disadvantages. It desiccates some tissues, connective tissue in particular, and has a boiling point of 61°C and a vapor pressure of 160 mm Hg, making it highly volatile and difficult to use. Aniline can be used with good results but is difficult to remove during infiltration. A mixture of equal parts of methyl salicylate (oil of wintergreen) and aniline followed by pure methyl salicylate, and then methyl salicylate-paraffin offers quicker and surer results. Methyl salicylate can also be used alone. Other oils, such as bergamot, clove, creosote, terpinol, can be used. Amyl acetate and Cellosolve (ethylene–glycol–monoethyl ether) do not harden excessively; however, Cellosolve is highly volatile.

Oils, including those above, can be used for clearing if tissue hardening is a serious problem. Cedarwood oil is well known and is relatively safe for the beginner, but it has disadvantages. Overnight immersion is usually required to insure complete replacement of the alcohol in the tissue. Also, as with all oils, all traces must be removed before infiltration, which is sometimes difficult since the oil may have a boiling point in the 200's and be slow moving out of the tissue. Oil will move out of the tissue more rapidly with the addition of an equal part of toluene. Furthermore, cedarwood oil and the absolute alcohol which must be used with it are both expensive.

During the use of xylene, toluene, or benzene for clearing, if the solution becomes turbid, it means that that water is present and the tissue is not completely dehydrated and may shrink. The only remedy is to return the tissue to absolute alcohol to eliminate the water and then to place it in a fresh clearing solution. Embedded tissue containing water can shrink as much as 50% and cause difficulties in sectioning and in mounting sections.

DEHYDRATION AND CLEARING COMBINATIONS

Dioxane Method

Any procedure that shortens the preparation time of tissues to be embedded has merit and finds favor among technicians. Graupner and Weissberger in 1931 proposed the use of dioxane (diethyl dioxide) which dehydrates and clears tissue in a minimum of steps. Dioxane is miscible with water, alcohol, hydrocarbons, and paraffin. It seems to eliminate some shrinkage and hardening, and is a relatively inexpensive method, because fewer solutions are required. (Dioxane, however, is far more expensive than alcohol.)

Some technicians warn of disadvantages to the use of dioxane. Conn (1939) cautions that dioxane is cumulatively toxic and should not be used by anyone with liver or kidney trouble. Navaquez (1935) reports that dioxane toxicity in animals is relatively low, that there is no cumulative effect, and that a tolerance for it develops. Cases noted in humans resulted from a period of heavy exposure to the vapor, and poisoning was acute rather than chronic. Fairly et al. (1936) say there is some evidence that the toxic effects are due to oxidation productions—oxalic acid and diglycolic acid—which are considered nontoxic to humans. Perhaps it is well to be cautious with dioxane: use it only in a well-ventilated room; avoid inhalation and unnecessary contact with hands; and keep dioxane containers tightly closed at all times.

Another disadvantage of dioxane is that it frequently contains water. Stowell (1941) condemned it, saying that some brands contained about 10% water and other impurities; he reported shrinkage as high as 35% with dioxane. If water is present, tissue shrinkage will appear during infiltration and may be as high as 50%. Dioxane solution is cloudy with 1% water content. Many technicians place anhydrous calcium chloride ($CaCl_2$) in the bottom of the dioxane container during dehydration to aid in removing all the water. However, Mossman (1937) observes that dioxane reacts with calcium chloride and swells as though water were present; he suggests the use of calcium oxide (CaO), but feels that the use of either is unnecessary.

If a fixative containing potassium dichromate is used, tissues must be washed thoroughly before using dioxane or the dichromate will crystallize.

Miller Schedule for Dioxane (1937)

1. Move tissue directly from fixative into 3 parts dioxane, 1 part distilled water: $\frac{1}{2}$–1 hour.
2. Dioxane: 1 hour.
3. Fresh dioxane: 2 hours.
4. Paraffin, 1 part, to 1 part dioxane (in paraffin oven): 2–4 hours.
5. Paraffin: overnight.
6. Embed.

Tetrahydrofuran (THF) for Dehydration

Haust (1958, 1959) recommends tetrahydrofuran (THF) for dehydration for several reasons. It mixes with water in all proportions; it also mixes with paraffin depending on the temperature, becoming increasingly miscible as the temperature rises. It is miscible with nearly all solvents and can be used as a solvent for mounting media. Most dyes are not soluble in THF, but iodine, mercuric chloride, acetic acid, and picric acid are soluble in it. THF has a low boiling point of 65°C, so that it evaporates rapidly and must be kept in a tightly closed container at all times. It has a tendency to form peroxides, but storing it in amber bottles lessens this problem. All mixtures of THF should be stored at 4°C (Salthouse 1958).

Haust Method

Following fixation, proceed to step 1.

1. 1 part THF to 1 part water: 2 hours.
2. THF, 3 changes: 2 hours each.
3. 1 part THF to 1 part paraffin, 53°–54°C: 2 hours.
4. Paraffin: 2 hours.
5. Embed.

The highest quality THF can be obtained from Eastman Kodak Co. and Fisher Scientific Co. *The Laboratory*, clinical edition (Fisher Scientific Co., vol. 28, no. C4, 1960), suggests that, when it is necessary to re-purify THF, it is simpler and safer to pass it through a column of activated alumina than to use metallic sodium.

INFILTRATING WITH PARAFFIN

Paraffin is considered to be either soft or hard. The melting point of soft paraffin lies in the 50–52°C or 53–55°C ranges; that of the hard, in the 56–58°C or 60–68°C ranges. The choice of melting point depends upon the thickness at which the tissue is to be sectioned or upon the type of tissue— hard paraffin for hard tissues and soft paraffin for soft ones. If relatively thick sections are to be cut, use a soft paraffin; otherwise the sections will not adhere to each other in a ribbon. If thinner sections are desired (5–7μ), use a paraffin in the 56–58° grade. For extremely thin sections of less than 5μ, the best results can sometimes be obtained with a hard paraffin of 60–68° melting point. The sections will retain their shape and size without excessive compression and will ribbon better than if the paraffin is softer. Temperature can also influence the choice of paraffin; hot weather will force the use of a harder paraffin than can be used in cool weather. If it is impractical to stock more than one kind of paraffin, that with a 56–58°C melting point is usually the best choice.

Tissuemat[1] is recommended for its uniformity. Paraplast[2] is an excellent embedding medium, a rigidly controlled mixture of paraffin and several plastic polymers of regulated molecular weights: it is smooth, requires no ice for embedding, and cuts well with less compression than paraffin. I have discovered that Paraplast does not absorb enough water to soften hard tissues before sectioning (page 61). This obstacle can be overcome by infiltrating with Tissuemat (56–58°) and embedding in Paraplast.

Normally, tissues are transferred directly from clearer to paraffin. Keep the oven temperature just high enough to maintain the paraffin in a melted state, no higher. This lessens the danger of overheating tissue, which can harden and shrink. The paraffin must be fully molten to infiltrate the tissue effectively; partly melted, "mushy" paraffin penetrates poorly, if at all. Melted paraffin that has been kept in a warm oven for several days or weeks is better for infiltrating and embedding than freshly melted paraffin. After 30 minutes to 1 hour in the first bath, the tissue is removed to a fresh dish of paraffin for a similar length of time. Two changes of paraffin are sufficient for most requirements, but for some tissues that are difficult to infiltrate—such as horny skin, bone, or brain—a third change may be necessary, and the total time of infiltration may need to be extended to as much as 6 hours, or even overnight. Fortunately, such cases are rare.

---

[1] Fisher Scientific Co.   [2] Curtin Scientific Co.

However, when infiltrating friable tissues, the following preliminary step must be added. After the tissue is well cleared, begin to saturate the solution with fine shavings of paraffin until some of the paraffin remains undissolved. Leave the tissue in the saturated clearer for 4–6 hours, or overnight for large pieces. Then with a warm spatula transfer the tissue to melted paraffin.

The use of a vacuum oven for infiltrating will remove air from some tissues (lung) and eliminate holes in the final paraffin block (Weiner 1957; Luna and Ballou 1959).

All suggestions concerning the use of paraffin also apply to Tissuemat and Paraplast.

EMBEDDING (BLOCKING) WITH PARAFFIN

As soon as the tissue is thoroughly infiltrated with paraffin, it is ready to be embedded; the paraffin is allowed to solidify around and within the tissue. The tissue is placed in a small container or paper box already filled with melted paraffin, and the whole is cooled rapidly in water. Before transferring the tissue, warm the instruments that manipulate it. This will prevent congealing of paraffin on metal surfaces. Handle the tissue and paraffin as rapidly as possible to prevent the paraffin from solidifying before the tissue is oriented in it. Orientation is important. If the tissue is placed in a known position and carefully marked with a slip of paper in the hardening paraffin (Figure 7, *a*, page 56), it is easy to determine the proper surface for sectioning. If a paper box is used, the orienting mark may be made directly on one of its flaps.

Each technician eventually adopts his or her own pet embedding mold or container. A few suggestions: petri dishes, syracuse watch glasses, shallow stender dishes, test tubes to concentrate solid contents of tissue or body fluids, and a neat little dish for tiny pieces—a miniature syracuse watch glass (watch glasses, U.S. Bureau of Plant Industry Model, 20 mm inside diameter, 8 mm deep).[3] Lightly coat the insides of glass dishes with glycerol; then the solidified paraffin block loosens readily. Cast lead L's (Diamond embedding boxes)[4] when placed on a small flat metal rectangle can be adjusted to almost any size for embedding, and, being metal, they cool the paraffin more quickly than glass. Also they break loose immediately from the paraffin. The Pop-out Embedding Mold[5] opens to release the paraffin block.

---

[3] A. H. Thomas Co. no. 9850.   [4] Lipshaw Co. nos. 334; 358; 359.
[5] Lipshaw Co. no. 335 A–D.

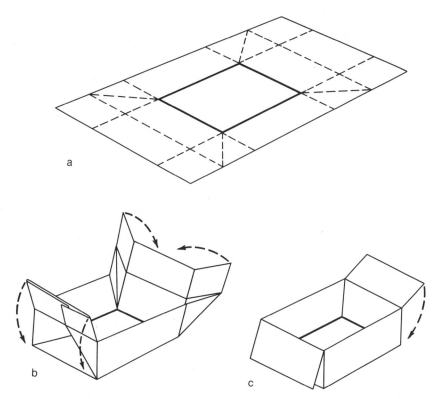

a

b

c

FIGURE 1. *Folding bond paper or aluminum foil to make a box for paraffin embedding.*

Paper or foil boxes may be fashioned as shown in Figure 1. Perhaps the one advantage of these is that they and the data recorded on them can be left on blocks during storage. The Peel-A-Way Disposable Embedding Molds[6] are made of lightweight plastic, available in five sizes, and are easily broken at the corners so the sides can be peeled down and the mold pulled away. Perforated tabs to fit can be purchased. Various devices such as refrigerator trays with dividers can be used for embedding a number of large pieces of tissue. Sections of aluminum[4] are designed to interlock and form 20 or 30 compartments. Embedding rings and molds are parts of the Tissue-Tek[7] system. Tims[8] of Lab-Line is still another embedding

---

[6] Ames Co., Elkhart, Ind.
[7] Ames Co., Elkhart, Ind., Scientific Products, and Fisher Scientific Co.
[8] Curtin Scientific Co.

container. The literature and catalogs are full of ideas; the above are just a few of them.

When small amounts of paraffin are ready to be solidified, they can be cooled immediately in water, preferably at a temperature of 10–15°C, or even at cool room temperature. (Paraplast does not require this treatment.) Make certain that a solidified scum has formed over each potential block before dipping it below the surface of the water. Water colder than 10°C causes the block to contract too strongly, and it may crack. Furthermore, normal crystals may form in the center, but abnormal crystals may form in the periphery of the block. The perfect block is one in which the paraffin crystals are contiguous, and the paraffin appears clear and homogeneous. Paraffin may contain 7–15% air dissolved in it and will appear clear when that air is distributed evenly through its mass, but pockets of air produce milky spots, a condition called "crystallization." Slow hardening of the block in the air or too rapid cooling may cause this effect, particularly in large blocks. Quick hardening of outer surfaces will trap the air. When these problems arise, cool the blocks from the bottom, and with a hot instrument keep the upper surface melted for a short time, thus releasing some of the air. Then blow across the top until a scum forms, and ease the block into water. This treatment also prevents excessive shrinkage in the center of the block and the enclosed tissue (Dempster 1944c).

If the paraffin does crystallize, difficulty may be encountered during sectioning; the only remedy is to return the block to melted paraffin, allow it to remelt, and repeat the embedding process. Experienced technicians soon learn how fast to handle the paraffin to reduce crystallization to a minimum.

Thoroughly hardened paraffin blocks can be stored indefinitely without injury to the tissue, but they must be kept in a cool place where they cannot soften.

Embedding Cellular Contents of Body Fluid "Cell Blocks"

"Cell blocks" are clusters of individual cells which have been concentrated and embedded for sectioning. The process for embedding them is as follows:

1. Collect the material in centrifuge tubes and add fixative: 1 hour, or overnight. Agitate occasionally.
2. Concentrate by centrifuging (preferably in small tubes); decant and

add water or alcohol depending on requirement of fixative. Loosen material at bottom of tube and stir with small glass rod: 5–10 minutes, or longer.

3. Centrifuge and decant. Add next solution and stir thoroughly. Dehydrate in this manner, and clear: approximately 10–15 minutes for each step, depending on size of particles.
4. Add melted paraffin and place tube upright in a glass or beaker in oven. Stir slightly with warm instrument to work paraffin to bottom of tube: 30 minutes. Stir a bit once during this period.
5. Cool tube.
6. Break tube. Place paraffin block in tiny paper box in a small dish of paraffin in the oven. Leave only long enough for block to begin to soften. Cool quickly.

Clarke (1947) suggests alternate steps 5 and 6.

5. Warm tube carefully in water, slightly warmer than the paraffin. As soon as paraffin against the glass melts, tip tube and allow paraffin block to slide out.
6. Mount block in a small hole dug in a square of paraffin and blend two together with a warm needle or spatula.

Farnsworth (1956) transfers minute pieces with a pipette to a piece of lens paper lying on a blotter. The latter absorbs the fluid, then the lens paper bearing the tissue is laid flat on the surface of melted paraffin. She embeds in depression slides wiped with glycerine.

Arnold (1952) concentrates small organisms in agar and carries through with the usual processes (see also page 504).

The Filtrator Coffee Paper method for bone marrow (page 264) can be adapted to use with many body fluid cells.

For other methods of handling clumps of cells, see Del Vecchio et al. (1959*b*), DeWitt et al. (1957), McCormick (1959*b*), Seal (1956), Taft and Arizaga-Cruz (1960).

TIMING SCHEDULE FOR PARAFFIN METHOD

The timing that follows is for tissue blocks ±10 mm in size—smaller pieces will require less time; larger pieces, more time.

Schedule Using Ethyl Alcohol

1. Fix overnight or 24 hours (Bouin, Gomori, Susa, Stieve, formalin, Zenker.—If tissue fixed in Zenker tends to harden, do not leave it overnight in the fixative; transfer to 3–5% aqueous potassium dichromate overnight, then proceed as usual on schedule).
2. Wash in water, running if possible: 6–8 hours or overnight. (Exceptions: Bouin, Gomori, Susa, and Stieve fixed tissue can be transferred directly to 50% or 70% alcohol without washing.)
3. Transfer to 50% ethyl alcohol: 1 hour.
4. Transfer to 70% ethyl alcohol: 1 hour. (See special iodine treatment for mercuric chloride fixatives, page 13.)
5. Transfer to 95% ethyl alcohol: 1 to $1\frac{1}{2}$ hours.
6. Transfer to absolute ethyl alcohol no. 1: $\frac{1}{2}$ to 1 hour.
7. Transfer to absolute ethyl alcohol no. 2: $\frac{1}{2}$ to 1 hour.
8. Transfer to toluene no. 1: $\frac{1}{2}$ to 1 hour.
9. Transfer to toluene no. 2: $\frac{1}{2}$ to 1 hour.
10. Transfer to melted paraffin no. 1: $\frac{1}{2}$ to 1 hour.
11. Transfer to melted paraffin no. 2: $\frac{1}{2}$ to 1 hour.
12. Embed.

Alternate Schedule Using Isopropyl Alcohol

1–4. Same as for ethyl alcohol.
5. Transfer to isopropyl alcohol no. 1: $\frac{1}{2}$ to 1 hour.
6. Transfer to isopropyl alcohol no. 2: $\frac{1}{2}$ to 1 hour.
7. Transfer to toluene no. 1: $\frac{1}{2}$ to 1 hour.
8. Transfer to toluene no. 2: $\frac{1}{2}$ to 1 hour.
9. Transfer to melted paraffin no. 1: $\frac{1}{2}$ to 1 hour.
10. Transfer to melted paraffin no. 2: $\frac{1}{2}$ to 1 hour.
11. Embed.

As noted in chapter 2 (page 35), if the tissue is well hardened by the fixative, it is not necessary to dehydrate the tissue through a number of graduated steps, such as 50%, 60%, 70%, 80%, 95%, and absolute alcohol. Even the elimination of the 50% and 60% steps is possible, when time presents a problem; also the use of isopropyl alcohol considerably shortens the schedule. (See Lampros 1962, for a rapid method—(3–6 hours)—that uses dioxane.)

In passing tissues from one fluid to another, use the decantation method to avoid excessive manipulation with forceps and reduce injury to the tissue. After pouring off a solution, drain the tissue bottle briefly against a paper towel or cleansing tissue to pull off as much as possible of the discarded solution. This reduces contamination of the new solution. Since 95% alcohol, absolute alcohol, clearers, and melted paraffin all contribute to hardening tissue, avoid leaving it in any of these fluids for longer than the maximum time recommended above (preferably only for the minimum period), and never overnight.

AUTOMATIC TISSUE PROCESSORS

Most large laboratories now accomplish the foregoing processes with a machine, such as the Autotechnicon.[9] The changes are controlled with a timing device, and the tissues are shifted automatically through a series of beakers or other types of container. The timing device can be preset to handle changes during the night so that the tissues will be ready for embedding in the morning. Small metal or plastic receptacles with snap-on lids hold the tissues and labels, and are deposited in a basket that clips into the bottom of the lid of the instrument. When the time arrives for removal of the tissues to a new solution, the lid rises and rotates to lower the basket into the next container. The two final beakers are thermostatically controlled paraffin baths. A technician can set the timing device for any interval desired—15 minutes, 30 minutes, 1 or more hours, etc.—over a period of 24 hours. The newest models have a clock that can control the instrument over a weekend or for several days.

There are several tissue processors on the market in addition to the Autotechnicon. There is a Lipshaw model, and the Tissuematon of the Fisher Scientific Company. Shandon Scientific Company has a single-station model as well as a multiple-station type.

---

[9] Technicon Co.

CHAPTER 4

Microtomes and Microtome Knives

MICROTOMES

The first instrument for cutting sections was made by Cummings in 1770. It was a hand model that held the specimen in a cylinder and raised it for sectioning with a screw. In 1835, Pritchard adapted the instrument to a table model by fastening it to a table with a clamp and cutting across the section with a two-handled knife. These instruments were called cutting machines until Chevalier introduced the name *microtome* in 1839. Sliding cutting machines were developed in 1798, rotary microtomes in 1883 and 1886. The Spencer Lens Company manufactured the first clinical microtome in 1901. The large Spencer rotary microtome with increased precision became available in 1910. (Richards 1949.)

The following types of microtomes are used in American laboratories:

1. The rotary microtome for paraffin sections and cryostat sectioning—the most widely used. Special rustproof rotary microtomes are manufactured for use with cryostats.
2. The sliding microtome for nitrocellulose (celloidin) and plastic sections: not always the most practical method; slow and expensive, but

often unexcelled for hard and large objects such as eyes, bone, and cartilage; also for cases in which shrinkage must be kept to a minimum.

3. The clinical (freezing) microtome for unembedded tissues: quick, cheap; required for some processes in which certain cell components must be retained, such as fat and enzymes; used for immediate diagnosis.

4. The ultrathin-sectioning microtome for sections thinner than 1μ for electron microscopy: only for special techniques and not commonly found in student laboratories.

5. The base sledge microtome for exceedingly large tissues e.g., brains, and hard blocks of tissues: only for special techniques and not commonly found in student laboratories.

6. The Faust Scientific Microtome is a small inexpensive instrument that clamps to the table top and cuts sections of about 25μ with a razor blade.

7. The Smith and Farquhar Tissue Sectioner[1] is one of two instruments for cutting unfrozen fresh or fixed tissue with minimum disruption of fine structure and minimum loss of enzymatic activity. It has a cutting range of $5-230\mu$, it cuts 50–200 strokes per minute, and injector razor blades are used in it. The Mark II Vibratome[2] uses a vibrating blade sectioning in a liquid bath to lubricate the blade and prevent heat buildup. Smith (1970) compares the two microtomes, and it is advisable to consult his article before purchasing one of these models.

Microtomes should always be kept well oiled to prevent parts from wearing unnecessarily or sticking. Either of these defects can cause imperfect sectioning (sections of varying thicknesses). Obtain advice from an expert about the parts to be kept oiled, and consult the booklet accompanying the instrument. The best oil for this purpose is Bear Brand Household Oil, formerly Pike Oil.[3]

MICROTOME KNIVES

There are three familiar types of microtome knives:

1. The plane-edge for frozen sections and paraffin ribbons.
2. The biconcave used sometimes for paraffin work.
3. The plane-concave for celloidin and plastic, sometimes for paraffin.

---

[1] Sorvall, Norwalk, Conn.   [2] Oxford Labs., San Mateo, Calif.
[3] Behr-Manning, Troy, N.Y., a division of Norton Co.

Because knives seem to demand hours of attention, they often become the technician's nightmare, and keeping them in optimum condition is difficult.

Theoretically, a perfect cutting edge is the juncture of two plane smooth surfaces meeting at as small an angle as is feasible—ideally 14° as suggested by Dr. Lorimer Rutty (Krajian and Gradwohl 1952). These cutting surfaces are called the cutting facets. The cutting edge of a very sharp knife, when examined by reflected light under ×100 magnification, appears as a fine discontinuous line. It may vary slightly in width, but it should show only a slight reflection, a narrow, straight, bright line. At a magnification of about 500 times, the edge will have a finely serrated appearance. The fineness or coarseness of these serrations depends on the degree of success in sharpening the knife. The facets are determined by the "back," which is slipped on the knife during sharpening to raise it just enough to form these facets (Figure 2). Every knife must have its own back. Never interchange backs. The back must fit tightly enough so that it does not move about on the knife. When it shows considerable wear, buy a new one, because as the back wears the cutting facets are widened. Do not ram the knife into the back; hold the knife in one hand and the back in the other and work it on gradually with a rocking motion.

The importance of taking good care of a knife cannot be overemphasized. Clean it after use; materials such as blood and water, if left on the knife edge, corrode it. Clean the knife with xylene on soft cleansing tissue and wipe it dry. Do not strike the edge with hard objects, such as a section lifter or dissecting needle; the edge can be damaged or dented.

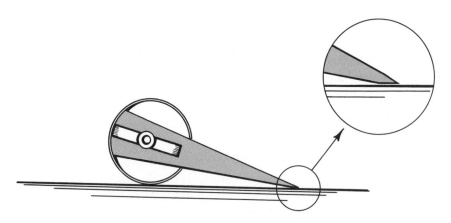

FIGURE 2. *Microtome knife bevel as controlled by knife back.*

Sharpening Knives

The glass plate has had the longest and most successful use as a sharpening instrument. In 1857, von Mohl used a thick glass plate; a rotating glass plate was developed by von Apathy in 1913, and a vertical one by Long in 1927. Leather strops came into use in 1922, followed a couple of years later by Carborundum hones, and then by finer grained hones and strops of canvas or leather. Many argue against a strop, saying that a knife can be honed sharp enough for good sectioning. If, however, there are fine serrations, gentle stropping frequently can remove them and improve the cutting ability of the knife.

If a glass plate is used, it should be at least $\frac{3}{16}$ of an inch thick, approximately 14 inches long, and an inch or two wider than a microtome knife. Levigated alumina (approximately 1%) added to a neutral soap solution is excellent for sharpening; the polishing should be done with diamantine—a small amount. A final stropping will remove fine serrations (Uber 1936).

The Fanz knife sharpener[4] uses a revolving glass disc over which the knife is swept back and forth.

Some pointers on the use and care of hones and strops may be of help to beginners. Never apply oil to hones, only water and neutral soap solution or Lava soap. Never allow the hone to dry while it is in use; always keep it wet with water or water plus soap. Two types of hones can be used, a coarse one for fast grinding of the knife edge when the nicks are deep, and a polishing hone to remove serrations and small amounts of metal left by the preliminary honing. A yellow Belgian hone is one of the best.

There are two patterns of microtome knife honing. Each has the same number of strokes. As in Figure 3, follow course 1, moving from bottom left to top right, roll knife over on its back, and return over same path in the opposite direction along course 2, moving from top right to bottom left; repeat both strokes three or four times. Move knife toward right into the position shown in figure 4 and follow course 3, moving from bottom right to top left, and return from top left to bottom right along course 4 —same number of times as in previous step. Continue this pattern of strokes on hone until deep nicks and scratches are eliminated.

When all nicks and deep serrations have been removed (check under dissecting scope), the knife is ready to be stropped, a final polishing action. Do not use a sagging hammock-like strop, unless it can be held tightly flat so that it does not round the knife edge. The best type is

---

[4] Arthur H. Thomas Co.

FIGURE 3. *First set of strokes for microtome knife honing.*

mounted on a felt pad on a hardwood block. This allows a bit of cushion, but on a solid surface. With a soft cloth keep the strop surface clean and free of dust. Rubbing the leather with the hand improves the texture. Do not use mineral oil on it. If the leather becomes dry, work in neat's-foot oil, working over small areas at a time, not the entire strop. Buff with soft towelling at once; do not allow the oil to sit on the leather, lest it soak in or leave a gummy substance.

The Lipshaw Co. offers several block-strop combinations; the revolving one is especially handy.

First use a honing strop, usually made of pigskin with a fine abrasive embedded in it. Then follow with the finishing strop, a fine-grain horsehide. The two patterns for stropping are similar to those for honing, but the cutting edge moves away from the strop (Figures 5 and 6). Repeat

FIGURE 4. *Second set of strokes for microtome knife honing.*

these strokes on the honing strop until fine abrasions caused by the hone
have been smoothed away, and then polish with a few strokes on the
finishing strop. Only a dozen strokes on the latter should be necessary if
the preliminary honing and stropping have been properly done. While
honing and stropping, use both hands, left hand on the back and adjacent
knife surface, and the right hand on the handle. Press *lightly*, guiding with
both hands and with uniform pressure from both. Always use the same
pressure on the forward strokes as on the return ones. Never hurry honing
or stropping; use at least one full second for each stroke.

Most laboratories have replaced the "old fashioned" hone and strop for
sharpening knives with mechanical knife sharpeners. All scientific supply
houses list one or more varieties of sharpeners, and some excellent

FIGURE 5. *First set of strokes for microtome knife stropping.*

sharpeners are now available.[5] Because most technicians develop a preference for one type, I offer no recommendation.

If the cost of microtome knives and their maintenance is prohibitive for a class of beginning students, the razor blade holders[6] are useful. A student can use a razor blade once and throw it away if the edge has become dull, or he can sharpen it on an old knife strop.

The life of the sharp edge of a good knife can be prolonged if the technician has another knife for the preliminary trimming of paraffin blocks.

---

[5] American Optical Corp. Scientific Products, Hacker Instruments.

[6] Spencer Razor Blade Holder, American Optical Co.; Stainless Steel Razor Blade Holder, International Equipment Co.; diSPo blades and holder, Scientific Products; Disposable histo-knife, Lab-Line Biomedical Products Inc.

FIGURE 6. *Second set of strokes for microtome knife stropping.*

After undesirable parts of the tissue block have been trimmed away, the good knife can be substituted for the old one, and the required sections collected.

Marengo (1967) describes an adjustment of the AO sharpener to produce edges of greater keenness, with a facet bevel between 25° and 30°, instead of the usual 30 to 31°.

CHAPTER 5

Paraffin Sectioning and Mounting

SECTIONING

The embedded blocks (Figure 7,*a*) are trimmed into squares or rectangles, depending on the shape of the tissue, with two edges parallel (Figure 7,*b*). The two short or side edges need not be parallel (sometimes with

FIGURE 7. *Untrimmed* (a) *and trimmed* (b) *paraffin tissue blocks.*

FIGURE 8. *Paraffin tissue blocks:* (a) *mounting on an object disc;*
(b) *mounted on an object disc or wooden block and ready
for sectioning.*

advantage as will be indicated later). Wooden blocks or metal object discs, such as those sold by supply houses,[1] are covered with a layer of paraffin. (*Suggestion:* Use a one-inch brush which can be kept in a beaker of melted paraffin in the paraffin oven.) Then a heated instrument (spatula or slide) is held between the paraffin on the wooden block or metal disc and the

---

[1] Lipshaw Co. no. 800.

undersurface of the tissue block (Figure 8, *a*). When both surfaces are melted, the instrument is withdrawn and the tissue block pressed firmly into the paraffin. Support object discs in holes in a length of wood or plastic, 1 inch thick. Make holes slightly larger than the diameter of the object disc's stem, $1\frac{1}{4}$ inches apart and approximately 1 inch deep. If wooden blocks are used, they will support themselves. After the paraffin has cooled, the tissue block is ready to be sectioned (Figure 8,*b*).

Some molds are designed to form paraffin blocks that can be clamped unmounted in the microtome.[2] Although these unmounted blocks can be used for large pieces of tissue, they will not do for small trimmed blocks of small pieces of tissue, which are more easily handled if they are mounted on an object disc or a wooden block. Trimmed blocks also create a closer alignment of the sections in a ribbon; this is particularly desirable in serial sectioning. Paraffin or Paraplast blocks may be kept in ice water until the technician is ready to begin sectioning. At least 10 to 15 minutes in ice water is necessary for thin sections. Palmer and McDonald (1963) describe a cooling device for the microtome which keeps the knife and block cold throughout the sectioning.

Set a rotary microtome for section thickness (5 or 6μ). Raise the tissue carrier and place in it the object disc with its mounted tissue block. Tighten the clamp of the tissue carrier onto the stem of the object disc. If using a wooden block, allow it to extend about $\frac{1}{4}$ inch beyond the metal clamps; this will prevent the paraffin block from breaking loose from the wooden block when the clamps are tightened.

Insert the microtome knife and tighten its clamps. The knife must be held in the clamps at a proper angle for optimal sectioning, producing a minimal amount of compression, and allowing the sections to adhere to each other. Many suggestions can be made concerning angle determination, but in most instances the technician finds the answer after a process of trial and error. The cutting facets are small as determined by the back. When placed in the microtome, the knife must be tilted just enough so the cutting facet next to the block clears the surface of the block. If the tilt is not sufficient (Figure 9,*a*), the surface of the block is pressed down with the wedging effect of the facet, and no section results. At the next stroke of the knife, this compression increases, and finally a thick section is cut, a composite of the compressed sections. Too great a tilt of the knife makes it scrape through like a chisel, rather than cut through the tissue (Figure 9,*b*).

Turn the feed screw handle (left side of instrument), moving the feed

---

[2] The Johns Hopkins Microtome Object Clamp, American Optical Corp., can be used for unmounted blocks.

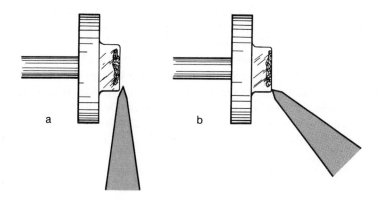

FIGURE 9. *Paraffin knife tilt:* (a) *insufficient;* (b) *excessive.*

mechanism backward or forward until the face of the tissue block barely touches the cutting edge of the knife. With an old knife start trimming into the block until the desired area is reached. Change to a good knife and readjust tissue carrier if necessary. Never touch the edge of the knife with anything hard. Fragments of paraffin can be flicked off with a camel's hair brush or removed with a fingertip. Scratches appearing in sections often can be remedied by rubbing the knife edge with a fingertip —*upward,* unless you like cut fingers. Also, this motion on the back of the knife will remove bits of paraffin, which can cause scratches. Cleansing tissue dipped in xylene is also helpful. (*Warning:* Discard the first section after cleaning the knife; it probably is a thick one.) When all sectioning is completed, use xylene to clean the knife; leave no corrosive material on its edge.

Section with an easy rhythm; never rush, or the result is likely to be sections of varying thicknesses. As the sections move down on the knife they form a "ribbon," each section adhering to the one that precedes it as well as the one that follows it. During the cutting of the sections, a bit of heat is generated, enough to soften the paraffin and cause the individual sections to stick to each other. As the ribbon forms, hold it away from the knife with a camel's hair brush and ease it forward so that the sections do not remain on the knife. This is advisable to prevent the sections from bunching and piling on each other. They can stack high enough to topple over the edge and get caught between the tissue carrier and the next stroke of the knife. The parallel edges of the paraffin block must be cut clean and parallel. If the edges have not been trimmed but remain as the original sides formed by the mold, a ribbon will not form.

In dry weather, static is frequently a problem during sectioning. Woolen and nylon clothing tends to create more static than cotton clothing, particularly when in contact with plastic-covered chairs. The sections stick to the knife or to parts of the microtome. They break apart and stay bunched up instead of lying flat. The friction of the knife as it crosses the paraffin block forms static electricity on its surface. An inexpensive little instrument, the Reco Neutra-Stat,[3] may be used to relieve this situation. An alpha-radiating static eliminator strip in a slot on the head of this instrument irradiates the air with harmless alpha particles which ionize the air and discharge static from the block surface.

The ribbons formed during sectioning can be mounted directly on slides, floated on a water bath, or laid in order in a box. It is particularly convenient to lay the ribbons in order in the box when the slides cannot be made immediately or when numerous sections—serial sections, for instance —have to be cut. Shallow hosiery boxes make handy containers, and if painted black inside with India ink, provide an excellent dark background for the sections. If the shorter edges of the block have not been trimmed parallel, a serrated edge along the ribbon indicates the exact position of each individual section, minimizing the danger of cutting through one. The sections can be stored in boxes until all required slides have been mounted. Valuable sectioning time is conserved by this means.

For thin sections of $2-5\mu$, the Jung plane-concave knife,[4] well sharpened, is an excellent knife to use. Compression is reduced to a minimum. It is also superior for nitrocellulose sectioning. See Collins (1969) concerning an etched microtome knife for improved sectioning.

Suggestion: Use a one-inch paint brush to free the microtome of paraffin litter after sectioning.

Difficulties Encountered in Sectioning: Suggested Remedies[5]

1. Ribbons are crooked.
    a. Wedge-shaped sections caused by poor trimming; sides of paraffin block are not parallel or not parallel to edge of knife.
    b. Part of knife edge may be dull; try another part of it.
    c. Uneven hardness of the paraffin; one side may be softer than the other, or contain areas of crystallization; re-embed.

---

[3] No. 61–579, Model M, E. Machlett & Son, New York, N.Y., or Lipshaw Co.
[4] American Optical Corp.
[5] Modified from Richards (1949).

2. Sections fail to form ribbons (usually due to hardness of paraffin).
   a. Use softer paraffin (lower melting point).
   b. Blow on knife to warm it or dip it in warm water.
   c. Cut thinner sections.
   d. Place table lamp near knife and block to warm them both.
   e. Resharpen knife.
   f. Lessen tilt of knife and clean edge.
   g. Dip block in softer paraffin and retrim so a layer of this paraffin surrounds original block.
3. Sections are wrinkled or compressed.
   a. Resharpen knife; a dull knife compresses badly.
   b. Paraffin too soft; re-embed in harder paraffin.
   c. Cool block and knife.
   d. Increase tilt of knife.
   e. Clean edge of knife with finger or xylene; remove any paraffin collected there.
   f. Tissue is not completely infiltrated,* or too much crystallization is present.
   g. Soak tissue block in water,† ½–1 hour or overnight if necessary (Figure 11, page 67).
4. Ribbons are split or scratched longitudinally (Figure 21, page 139).
   a. Nick in knife; move to another part of edge or resharpen knife.
   b. Knife dirty or gritty along edge.
   c. Dirt or hard particles in tissue or in paraffin; crystals from fixing solution not adequately removed; filter stock paraffin, or decalcify or desilicify tissue.
   d. Decrease tilt of knife.
   e. Tissue too hard; soak in water.†
5. Tissue crumbles or falls out of paraffin.

---

* Poor infiltration is usually caused by traces of water or alcohol. Correct by first removing paraffin that is present. Soak in toluene for 2 or 3 hours (or more); change twice, then place in absolute alcohol for 1 or more hours. This should remove all traces of water. Clear again in toluene (check against milky appearance); re-infiltrate and embed.

† When soaking in water is recommended, the cut face of the tissue is exposed to tap water for ½–1 hour, or overnight for stubborn cases and when time is of no concern. I have found this treatment completely satisfactory, but some technicians advocate the addition of glycerine (1 part to 9 parts water) or 60% ethyl alcohol instead of water (Lendrum 1944). The fluid works in through the cut tissue surface and softens tough parts. (*Exception:* Do not soak nervous system tissue at any time and lymph nodes and fatty tissue only briefly. Paraplast will not absorb water.) Diegenbach (1970) adds a fabric softener (1:9) to the water.

     a. Poor infiltration;* re-infiltrate and re-embed.

     b. Not completely dehydrated.

     c. Not completely dealcoholized.

     d. Too long in paraffin bath or too hot while there; soak in water.†

     e. Clearing fluid made tissue too brittle; soak in water.†

6. Sections cling to block instead of knife.

     a. Knife dull or dirty.

     b. Increase tilt of knife.

     c. Paraffin too soft or room too warm; try harder paraffin or cool block.

     d. Infiltrating paraffin too hot, or too long exposure to solutions that harden; soak in water.†

7. Tissue makes scratching noise while sectioning.

     a. Tissue too hard; paraffin too hot or too long exposure to solutions that harden; soak in water.†

     b. Crystals in tissue; fixing reagents not adequately removed by washing; calcium or silicon deposits present. (See pages 28 and 486.)

8. Knife rings as it passes over tissue.

     a. Knife tilted too much or too little.

     b. Tissue too hard; soak in water.†

     c. Knife blade too thin; try a heavier one.

9. Sections curl, fly about, or stick to things, owing to static electricity from friction during cutting, especially in weather of low humidity.

     a. Increase humidity in room by boiling water in open pan.

     b. Ground microtome to water pipe.

     c. Postpone sectioning until weather is more humid; early morning sectioning often is best.

     d. See page 60 for suggestions concerning clothing, furniture, and static eliminator.

10. Sections are skipped or vary in thickness.

     a. Microtome in need of adjustment or new parts.

     b. Tighten all parts, including knife holder and object holder clamp.

     c. Large or tough blocks of tissue may spring the knife; soak them in water.†

     d. Knife tilt too great or too little.

---

    * See first footnote on page 61.   † See second footnote on page 61.

MOUNTING

Since the embedding medium, that is, paraffin or Paraplast, usually has to be removed from the sections to allow uniform staining throughout the cells, the tissue sections must be affixed to some solid surface or they will disintegrate during removal of the embedding medium. The sections are, therefore, mounted on glass slides which support the tissue and also permit the mounting of many sections in sequence.

Slides and Cover Glasses

Most microscopic materials (sections, whole mounts, smears, touch preparations) are mounted on glass slides and are covered with microscopic cover glasses. A noncorrosive quality of each should be purchased to insure a nonreacting surface. The slides may be obtained in the regular 3×1 inch size or in 3×2 inch for larger tissue sections. The brands labeled "Cleaned, ready for use," save many technician hours that otherwise would be devoted to cleaning slides. In addition, they permit a uniform surface for blood smears, loose cells, and touch preparations.

Cover glasses are manufactured in circles, squares, and rectangles, ranging in thickness from no. 0 to no. 3. Thickness no. 0 is used rarely, usually only when an oil immersion objective rides unusually close to the slide. This size is so thin that it fractures easily. Thickness no. 1 (about 0.15 mm) usually is adequate for oil immersion work. Thickness no. 2 (about 0.20 mm) can be used for whole mounts or when oil immersion is unnecessary. Thickness no. 3 (about 0.30–0.35 mm) is occasionally used when a tougher glass is required and for large whole mounts. Circles and rectangles may be purchased in sizes up to $\frac{7}{8}$ inch in diameter. Circles are designed for mounts for which ringing the cover is necessary in order to create a seal against evaporation, as for whole mounts in glycerol jelly or any other volatile mountant. (A ringing table can be used to make the supporting ring for the cover glass and the final seal around the edge of the cover glass.) The rectangles come in 22 and 24 mm widths, and in 30, 40, 50 and 60 mm lengths. Cover glasses will mount more efficiently if they are cleaned in absolute alcohol and wiped dry, and fewer bubbles will cling to the glass when it is pressed into place over the mounting medium.

Mounting Techniques

Before sections are mounted, a glass-marking pencil should be used to make an identifying mark on one end of the slide, such as a tissue number, a student locker number, or a research project number, to prevent confusion with other slides. In addition, this mark enables the worker to distinguish on which side of the slides the sections are mounted and thus to avoid failure during the staining process.

Types of permanent glass-marking pencils include diamond points (a real commercial diamond mounted in a handle) and steel pencils with tungsten-carbide tips.[6] A vibro-engraver or vibro-tool can be used for engraved marking on the glass. Wax china-marking pencils make only temporary inscriptions, dissolving in staining solutions. Some inks are reasonably satisfactory; one of the best is Gold Seal Laboratory Ink.[7] Stafford (1960) uses a waterproof pencil (Venus Unique Blue Waterproof Pencil[8] on slides with frosted ends, but this may be removed by xylene or alcohol.

The customary means of affixing sections to slides is with egg albumen and water. (Gelatin and blood serum can also be used, page 537.) Tissue-tac,[9] a special solution for adhering sections and cellular substances to slides, is clear and never picks up stain. (See also subbed slides, page 538.) It is essential that slides be absolutely clean to insure adherence of the sections throughout the staining procedure. With one finger, smear a thin film of albumen fixative (adhesive, page 536) on the slide, and with a second finger, wipe off excess albumen. This should keep the film thin enough. Thick albumen picks up stain, makes a messy looking slide, and can prevent a clear image of sharp, uncluttered tissue elements. If there is a continued tendency to get too much albumen on the slide, try a "floating solution"; dilute a couple of drops of albumen fixative with about 10 ml of distilled water and float the sections on the solution on the slide (see also Schleicher 1951). Lillie (1965) suggests that, since water is used to float the sections on a thin coat of fixative and since the fixative constituents are water-soluble, it is doubtful that enough albumen remains on the slide to be an actual adhesive agent. He considers it probable that the albumen acts as a surface tension depressant and aids in closer attraction of sections to slide.

If a water bath is used for spreading the sections, have the temperature 5°–10°C below the melting point of the paraffin being used. Excessive heat for flattening may cause shrinkage of cell structure and tearing and displacement of tissue parts. When removing the sections from the microtome knife or from the box, stretch them as flat as possible as they are placed on the water surface. After they have warmed a bit, they can be pulled to more nearly their original size with a couple of dissecting

---

[6] Fisher Scientific Co., no. 13–378.
[8] No. 1206, American Pencil Co.

[7] No. A-1408 Clay-Adams, Inc., N.Y.
[9] Dade Reagents, Miami, Florida.

needles. With a hot needle or spatula separate the required sections from the rest of the ribbon. Dip an albumenized slide under them and with a needle hold them against the slide while removing it from the bath. When the sections are spread smoothly, the excess water can be drained off; this is facilitated by touching a piece of cleansing tissue or filter paper to the edge of the slide and around the sections. Permit slides to dry overnight on the warm plate or in an oven of comparable temperature. If they must be stained on the same day that they are mounted, dry them in one of the mechanical hot-air dryers.[10] If agar (Cordova and Polanco 1962) or gelatin is added to the water bath, unalbumenized slides can be used.

Frandsen (1964) describes mounting serial sections from a water bath, but I have always found direct mounting on slides quicker and simpler.

Sections collected (glossy side down) in a box can be separated with a spatula or razor and removed to an albumenized slide. Add enough distilled water to spread well beyond the edge of the sections. Place the slide on a slide warmer[11] to correct any compression acquired during sectioning and keep it on the warmer until the sections are stretched flat. Dense tissues will compress less than the paraffin surrounding them and often will develop folds because the paraffin does not expand enough on the warm plate to permit the tissues to flatten. Pieces with a lumen will invariably demonstrate this fault. One of the simplest ways to correct it is to break away the paraffin from the outside edge of the tissue. Do this when the ribbon is floating on the water on the slide, but not while it is warm, or the paraffin will stick to the dissecting needles. Allow the slide to cool and then split away the paraffin with care so that the tissue is not injured (Figure 10). If the paraffin tends to stick to the needles during this process, wipe off the needle tips with the fingers. Clean needles will minimize this problem. A rather violent method for straightening stubborn ribbons is the addition of some acetone or alcohol to the water. Lin and Corlett (1969) use a xylene spray to relieve compression; this must be applied with great care to avoid overcompensation.

Certain tissues frequently develop cracks through the sections while drying—a tendency common to spleen, liver, lymph nodes, and nervous tissue. To prevent this, as soon as the sections are spread, drain and blot them free of as much water as possible and dry them at a higher temperature (60°–65°), in the mechanical hot-air dryer, or run them rapidly through a Bunsen burner flame until the paraffin melts, but no longer (Figure 12).

Properly mounted sections will have a smooth, almost clear appearance.

---

[10] Technicon or Lipshaw companies.
[11] Fisher's Slide Warmer, no. 12–594–5, or Lipshaw's Slide Warmer, no. 371 or 374.

FIGURE 10. *Paraffin sections:* (a) *compressed or folded sections;* (b) *breaking paraffin with dissecting needles to permit ribbon expansion without folds;* (c) *expanded and flattened sections.*

If they have a creamy, opaque texture, and if they reflect light when examined from the undersurface, air is caught between the glass and the sections. Tissues so mounted will float off in aqueous solutions. The cause usually lies in poorly cleaned slides or in drying the slides at too low a temperature or not drying long enough. If, when mounting the sections, bubbles appear under them, use distilled water that has been boiled for some time in order to release trapped air.

a   b

FIGURE 11. *Thyroid sectioning:* (a) *crumbling colloid in paraf-*
fin section not soaked before sectioning; (b) *same*
thyroid sectioned after soaking in water.

Several other conditions may cause tissues to loosen from slides. These
include brittle sections, a large yolk or osmic acid content, and the use of
strong acids or alkalis during staining. In these situations the sections will
require special treatment as described below. In addition, poorly infiltrated
sections and nervous tissue that have been soaked in water will usually
fail to adhere, but can sometimes be saved by the following method.

After drying, and when ready to stain, dissolve out the paraffin with
toluene or xylene. Place the preparations for 1–2 minutes in absolute ethyl
alcohol, followed by 1–2 minutes in a dilute solution (about 0.5%) of
nitrocellulose in ether-absolute alcohol (50 : 50). Drain off excess nitrocellulose, but do not allow it to dry completely; wave the slide in air for a
second and place it in 70% or 80% alcohol. Continue with the planned
staining method. Also try mounting this type of tissue on subbed slides
(page 538) instead of on albumenized slides. This is much easier than
coating the mounted sections, especially for nervous tissue. I always use
subbed slides in preference to nitrocellulose-coated ones. When water used

a   b

FIGURE 12. *Adrenal sectioning:* (a) *cracks in an adrenal section caused by poor paraffin infiltration;* (b)
*cracking of same tissue has been prevented as sug-
gested on page 65.*

in mounting sections does not puddle uniformly on subbed slides, but
forms tight-up pools, sections are difficult to spread; to correct this, apply
a thin coat of egg albumen over the subbed surface before adding the
sections and water.

Serial Sections

With a few exceptions, the usual paraffin method is used for serial sections,
and in most cases, every section is mounted and in correct order. While
sectioning, arrange the ribbons in a cardboard box from left to right and
top to bottom (like a printed page). When the sectioning is finished,
remove the sections to slides in the above order. Serially number each
slide with a glass-marking pencil (do not use a grease pencil, which may
wash off in staining solutions). This system affords simple sorting of slides
when all processing is complete.

CHAPTER **6**

Methods for
Frozen Specimens

The freezing technique for preparing specimens is unsurpassed in certain situations. It is rapid, making it the best choice for diagnostic purposes, where speed is essential. It can be used for the preparation of sections containing such elements as fats, enzymes, and some radioisotopes that would be lost in alcohol, in paraffin solvents, or by heat. One disadvantage is that some distortion may be caused by the freezing and cutting.

Freezing methods formerly used ether, ethyl chloride, or other volatile liquids. Present methods employ compressed carbon dioxide gas blown into a chamber. There the gas expands, cooling the chamber, and freezing the tissue mounted on it. For this procedure most laboratories use the clinical microtome with a freezing attachment. An attachment can also be used on sliding microtomes, but not efficiently on rotary models. However, Strike (1962) and Stagg (1963) have adapted rotary microtomes for the freezing technique. There are also dry ice holders available, such as the Dry Ice Freezing Chamber.[1] Dry ice is held inside the chamber, and

[1] Fisher Scientific Co.

the specimen is frozen simply by pushing the dry ice up against the metal top holding the specimen. Nickerson (1944) fitted the object clamp of a rotary microtome with a small metal box holding dry ice. A new freezing system, which uses Freon, is on the market—the Histo-Freeze,[2] a portable freezing unit, 12 × 18 inches, with Freon 12 circulating through the unit. Model 67027A fits American Optical, 67027B fits Bausch and Lomb, and 67027C fits the Sartorius freezing microtomes. Lipshaw has a tissue-freeze refrigerating unit, counter or desk height, fitting American Optical, Bausch and Lomb, Sartorius, or Reichert microtomes. Tissu-Freez[3] and Bailey Freezing Stage[4] are solid state transistorized cooling devices using the Peltier effect to convert electricity into cold; they can be used with any freezing microtome. National Tissue Freeze[5] is an aerosol-type coolant for quick freezing of tissue. It uses a pistol grip with fingertip control and an extended nozzle.

FIXATION, BLOCKING, AND SECTIONING

Fresh tissues may be used; but it is preferable to fix the material, usually in formalin, which gives the tissue an ideal consistency for the freezing technique. Formalin is a water solution and requires no washing. Hartz (1945), however, recommends Bouin as better than formalin, but says that the tissue must be washed for a short time before freezing. If the fixative contains mercuric chloride, the crystals may be removed immediately after sectioning with an iodine solution, such as Lugol's, which contains no alcohol. If the tissues are extremely friable, it may be advantageous to embed them in gelatin (page 74), or in alginate gel (Lewis and Shute 1963), or to immerse them in a thick aqueous gum syrup, such as gum arabic. O.C.T. Compound[6] or Cryoform[7] facilitates cutting and handling of frozen sections. It is applied directly on the tissue immediately before freezing is started. O.C.T. Compound comes in three types for three ranges of temperature, $-10°$ to $-20°$, $-20°$ to $-35°$, and $-35°$ to $-50°$.

Purists among technicians may object to the following method of fixation, and it is recommended only for very limited use. It does, however, preserve sufficient cell structure for immediate diagnosis on surgical biopsies from patients being held under anesthesia. A small piece of the tissue

---

[2] Scientific Products, Evanston, Ill.   [3] Eric Sobotka Co., Farmingdale, N.Y.
[4] Bailey Instruments Co., Saddle Brook, N.J.
[5] Allied Chemical Corp., Morristown, N.J.
[6] Lab-Tek Instruments Co., Westmont, Ill.   [7] Fisher Scientific Co.

can be boiled for a couple of minutes in 10% formalin. Rinse in water and freeze. Reiner (1953) adds tissue to undiluted formalin at room temperature, then heats the solution and tissue to 56°–60°C in a water bath for 2 minutes. Agitate the tissue during the heating. Reiner claims inferior results without agitation and says that placing tissues in preheated formalin is disadvantageous. Additional pieces of tissue can be fixed in a more conventional manner for later study.

Friedland Method, Using Agar (1951)

1. Boil tissue in formalin.
2. Replace formalin with 2% sterile agar (previously heated to its melting point) in water. Boil gently 1 minute.
3. Pour off and freeze tissue immediately.

The agar can be kept ready in Pyrex tubes and melted quickly over an open flame.

Lev and Thomas Method (1955)

This method fixes tissue at 80°C and produces excellent results. The following solution is placed in a baby's bottle warmer and comes to a boil in 1 minute. Add the tissue and allow it to boil for 1 more minute.

formalin	10.0 ml
glacial acetic acid	1.0 ml
95% ethyl alcohol	80.0 ml

The tissue must be washed to remove the alcohol before freezing can be attempted. This can be a disadvantage if speed is important.

The pieces to be cut should be no more than 2–4 mm thick for rapid and uniform freezing. One surface should be flat to provide sufficient contact with the tissue carrier. If, during freezing, there is a tendency for the tissue to break loose from the carrier, cut a piece of filter paper to fit the top of the carrier and place it, wet, under the tissue to hold the tissue in place while freezing and sectioning. Young (1962) suggests, for extremely small tissues, the use of a small piece of sponge as a base to raise them high enough to prevent the knife from striking the metal of the object holder.

Consider the shape and type of tissue when it is oriented on the carrier. If one side is narrower than the other, place that side, or any straight side, parallel to the knife. If there is a tough membrane on one side of the tissue, place that side toward the knife to prevent the membrane from breaking away from the rest of the block during sectioning. Usually the amount of water carried over with the tissue is sufficient for firm attachment to the carrier. (Use normal saline for fresh tissues.) Avoid too much water and do not allow it to settle around the tissue in a wall of ice that can deflect the knife or tear through the tissue during sectioning. Uneven or torn sections result. If, however, gum or O.C.T. Compound is being used, it is necessary to build the compound around the tissue while freezing it. In this case the tissue must be completely encased by the medium.

With a finger, gently press the tissue onto the tissue carrier. Slowly freeze the tissue by turning on CO_2 for a moment or two, and then turn it off. A series of successive gas jets freezes better and wastes less gas than does a continuous stream. Other freezing methods, such as Histo-Freeze and Tissu-Freez use a continuous action. When the tissue is frozen fast and the finger can be removed, move the knife over the block so it too is in the path of the cooling gas. This aids in uniform freezing of the tissue by deflecting some of the gas onto the top of it, and at the same time cools the knife.

Good sections depend on proper tissue temperature. Material frozen too hard forms white brittle fragments on the knife; if too soft it forms a mushy mass. In both cases, the sections break up when placed in water. These difficulties can be avoided by slightly overfreezing the block and then, as it warms and reaches the correct temperature, a number of sections can be cut in close succession. When these sections fall on the tissue carrier, allow several to accumulate and pick up the group with one sweep of the finger. Or, as soon as the section is cut, remove it from the knife with a finger tip and place it in a dish of water. Do not allow water to collect on the knife. Keep both knife and fingers dry. Some tissues, such as adipose, tend to stick to the finger and will not shake loose in water. Transfer them to 70% or 80% alcohol.

Occasionally, when there is no time for gelatin embedding, one encounters a tissue difficult to section because it tends to fall apart. Bush and Hewitt (1952) support the tissue with Kodak Frozen Section Film. The film is softened in water, excess moisture is sponged off, and the film is pressed down on the freshly cut surface of the frozen tissue. The film freezes and adheres to the tissue while a section is cut; this film is carefully raised as the knife passes under it. The film with the section is laid face down on a slide. As the section thaws, it will cling to the slide, and a thin layer of moisture develops by capillarity between it and the film. When this has dried, the section can be stained. (See also use of cellulose tape, page 514.)

MOUNTING

If the sections are collected in a glass dish, place the dish on a black surface, such as a tabletop or a piece of black paper. The sections are more easily manipulated over the black background, and the best ones can be selected. Dip one end of an albumenized slide into the water. With a needle, or preferably a small round-tipped glass rod, gently bring a section against the slide. Hold the section at one corner; by maneuvering under water, creases can be unfolded and the section unrolled as the slide is drawn out of the solution. If the wrinkles refuse to straighten, drop the section on 70% alcohol and then remove it to a slide. Drain off excess water or pull it off with filter paper. Press section in place with filter paper moistened with 50% alcohol. With a pipette add absolute alcohol directly on top of the section; let stand for approximately 30 seconds. Replace with more absolute alcohol (must be ethyl alcohol). Drain thoroughly, and flood slide with 1% nitrocellulose (dissolved in absolute ethyl alcohol–ether, 1:1) for 1 second. Drain and wave slide in air until almost dry. Immerse in 70% or 80% alcohol, 5–10 minutes or longer; it can be left here for several hours or days. (*Caution:* Use an old solution of alcohol, either one from the staining series, or one which has been mixed for at least a day. A freshly prepared solution forms bubbles between section and slide.)

Fresh tissue sections may be removed from normal saline to the slide and thoroughly drained. Add 95% alcohol to remove the water; let stand 30–60 seconds. Drain off and allow the section to almost dry. Place in 95% alcohol for 1–2 minutes and proceed to stain.

A gelatin fixative can be used to affix the sections. Spread fixative on the slides and dry them in an incubator. (They can be stored for some time in a dustfree box.) Float the sections on the gelatin, drain off excess water, and place the slides in a covered dish over formalin. This converts the gelatin into an irreversible gel, holding the sections in place. After the slides have been over the formalin for 30 minutes, wash them in running water for 10 minutes and proceed to stain. Subbed slides can be used; I prefer them over other methods. Mount the sections out of the water onto the subbed slides, tilt the slides to drain off excess water, and allow to dry in a vertical position.

Snodgrass and Dorsey (1963) use an egg albumen method for embedding. It appears to be simple and practical.

Gelatin Embedding

Pearse Method (1953)

1. Fix in cold formalin (15%) at 4°C if enzymes are to be preserved: 10–16 hours.
2. Wash in running water: 30 minutes.
3. Infitrate with gelatin, 37°C: 1 hour.

gelatin	15.0 g
glycerine	15.0 ml
distilled water	70.0 ml
small crystal of thymol	

4. Cool and harden in formalin (40%), 17–22°C: 1 hour. Wash.
5. Store at 4°C or below until sectioned.

STAINING

Nitrocellulose-mounted sections are transferred from the 70% alcohol to water and then into any desired stain. After staining they probably will require special care. If after dehydration the sections are placed in absolute ethyl alcohol and show a tendency to fall off because the nitrocellulose dissolves, follow the 95% alcohol with isopropyl alcohol, 2 changes, and then xylene. Another method is to follow the 95% alcohol with carbol-xylol (page 532) or terpineol-xylol (1:9), several changes, until the sections look clear. Fogginess or colored droplets on the slides indicate traces of water. If carbol-xylol is used, rinse it off by dipping the slide for 2–3 seconds in xylene. Carbol-xylol must be removed; any of it left in the sections will fade stains in a short time, almost overnight. (*Caution:* Keep fingers out of carbol-xylol; it contains carbolic acid. Wash it off immediately if any spills on the skin, and use lanolin for skin irritation.)

Irregular staining usually is caused by an uneven coat of nitrocellulose; this can be prevented by tilting the slide immediately after the solution is applied. If the nitrocellulose is thicker than 1%, uneven staining may result. A colored background indicates a thick coat of nitrocellulose. Sometimes irregular staining is due to loosening of parts of the section. Sections will adhere to subbed slides better than to nitrocellulose-coated ones.

After the sections are cut, a 5-second immersion in Detec (Frozen Section Clarifier, Schain, Merck and Co.), then back into water for a few seconds before staining tends to make frozen sections clearer. (Jimenez and Schneider 1962)

See Albrecht (1956), Jacobson (1963), Peters (1961*a*), and Thompson (1957) for staining in quantity.

Rapid Staining Methods

Pinacyanole Method (Proescher 1933; Humason and Lushbaugh 1961)

Staining and mounting time: 3–4 minutes.

PROCEDURE

1. Mount frozen section (fresh or fixed) on subbed slide; drain off excess water and blot with two sheets of filter paper.
2. Cover section with several drops of stain: 3–5 seconds. (Pinacyanole,[8] 0.5% in either 70% methyl or ethyl alcohol; keep stock solution in refrigerator.)
3. Wash gently in tap water: 5–10 seconds or until free of excess stain (longer washing will not alter intensity). Blot.
4. Dehydrate in isopropyl alcohol, 2 changes: 1–2 minutes.
5. Clear in 2 changes xylene and mount.

Because differential staining is lost in paraffin sections, this method is recommended only for fresh or frozen sections. In order to preserve cytoplasmic staining, do not dehydrate the sections in ethyl or methyl alcohol; they decolorize cytoplasmic structures, and only the nucleus will retain appreciable amounts of stain. Steps 4 and 5 can be omitted and the sections mounted with an aqueous mountant or glycerine.

RESULTS

chromatin—well-differentiated blue to reddish blue
connective tissue—pink
elastic tissue—dark violet
muscle—violet to purple
plasma cells—red granuloplasm
hemosiderin—orange
hemoglobin; neutrophilic and eosinophilic granules—unstained
neutral fat—colorless to faint bluish violet
lipoids—bluish violet to purple
amyloid—carmine red

---

[8] Eastman Organic Chemicals.

Humphrey Method (1936)

This is perhaps the fastest method.

1. Place section on a slide and wipe off excess water.
2. Add 1 drop of 0.5% brilliant cresyl blue in saline.
3. Cover with cover glass and examine.
   This is never a permanent mount.

Thionin or Toluidine Blue Method

1. Remove section from water to a solution of either 0.5% thionin or toluidine blue O in 20% ethyl alcohol plus a few drops of acetic acid: 30 seconds.
2. Rinse in water and float on slide.
3. Drain off excess water, blot around edges of section, add drop of glycerine, and cover glass.
   This is not a permanent slide.

The above two methods stain the tissue elements in shades of blue and purple.

CHAPTER 7

Nitrocellulose Method

This form of embedding is often classified as the celloidin technique, and celloidin has become something of a generic term including the various cellulose compounds, such as nitrocellulose and soluble guncotton or collodion. These are solutions of pyroxylin consisting chiefly of cellulose tetranitrate. Obviously a purified nonexplosive form of pyroxylin is necessary, and there are several: Parloidin, Celloidin, and Photoxylin. LVN is an excellent embedding medium:[1] a solution of 30% low-viscosity nitrocellulose in 35% ether and 35% absolute ethyl alcohol. The special low-viscosity nitrocellulose allows more rapid infiltration with solutions of higher concentration than is possible with many of the other so-called celloidins. This permits formation of harder blocks in a shorter time, and cutting of thinner sections.

The chief advantages in using the nitrocellulose method are that larger and harder pieces can be cut in nitrocellulose than in paraffin, and the consistency of nitrocellulose allows mixed (hard and soft) tissues to be cut. This quality is useful for organs such as eyes, teeth, and bones and

---

[1] Tissue Embedding Solution no. 4700, Randolph Products Co., Carlstadt, N.J.

their surrounding tissues, and for problems of shrinkage and the formation of artificial spaces. But a slow dehydration and prolonged infiltration are preferable in these cases, and the method is costly.

In this method toluene cannot be used as a solvent, but there are many other solvents for nitrocellulose. One of the most common is a combination of two so-called latent solvents, not in themselves efficient solvents, but possessing excellent solvent qualities when mixed with other compounds: diethyl ether (generally referred to as ether) and ethyl alcohol. The solution is made up of equal parts of ether and absolute ethyl alcohol.

DEHYDRATING AND INFILTRATING

The tissue is fixed, washed, and can be stored in 70% alcohol as usual. The next steps are as follows:

1. 95% ethyl alcohol: 4–24 hours, or longer.
2. absolute ethyl alcohol, 2 changes: 4–24 hours or longer in each.
3. ether-alcohol (equal parts of absolute ethyl alcohol and anhydrous ether): 4–24 hours or longer.
4. 10% nitrocellulose (dissolved in absolute ethyl alcohol–ether, 1:1): 2 days or longer, in a screw-cap jar to keep evaporation to a minimum.
5. 33–35% nitrocellulose (in absolute ethyl alcohol–ether, 1:1): 2 days or longer, in tight jar.

Nitrocellulose solutions should be stored in the dark; light causes the solution to deteriorate. Ferreira and Combs (1951) warn that old light-affected solutions cause fading of nervous-tissue blocks.

EMBEDDING

Slow Method

Place the tissue in a small glass-covered dish (stender dish), and over it pour 33–35% nitrocellulose until it is $\frac{1}{4}$ to $\frac{1}{2}$ inch above the tissue. Over this pour a thin layer of ether-alcohol (1:1). Cover the dish tightly, and allow the solvent to evaporate slowly until the embedding medium reaches a proper consistency. If bubbles appear in the medium, evaporation is proceeding too rapidly. A little Vaseline on the ground-glass edge of the cover will help to seal it more tightly. In dry warm weather it may be necessary to enclose the evaporation dish in another dish or jar to slow down the evaporation of solvent. The process should take several days to a week or more, the slower the better. Do not allow the medium to become too hard; it should reach the consistency of hard rubber and should feel dry and not show fingerprints. If a tough film forms quickly on the surface of the solution and sticks to the side of the dish, carefully loosen the film from the glass to allow more efficient evaporation.

When the nitrocellulose is properly formed, trim it down to cutting size, and mount it on a fiber or wooden block (Figure 13). Wrap a band of hard bond paper tightly around the block, and secure the band with string or stick it down with a paper label. (Do not use a rubber band.) The paper must project high enough above the mounting block to enclose the tissue block. Moisten the mounting block and paper with ether-alcohol, and then place a small amount of 33–35% nitrocellulose in the container formed by the two. Roughen the underside of the tissue block with a needle, and cover it with a drop or two of ether-alcohol. Press the roughened surface tightly into the nitrocellulose on the paper-wrapped block. After a few minutes in the air, place the block in a closely capped jar with a small amount of chloroform; leave it there 20–30 minutes. Add more chloroform to immerse the entire block, and if possible allow it to "set" in chloroform overnight. There are other and more haphazard ways of mounting nitrocellulose blocks, but the above method is relatively sure.

Romeis (1948) claims that if wooden blocks are used they should be preseasoned with nitrocellulose. Cook the wooden blocks in distilled water and allow them to dry for several days. Extract with equal parts of 70% alcohol and glycerine for a day and wash the blocks in distilled water until shaking no longer causes a foam. Cover the surface with 8% nitrocellulose in ether-alcohol, dry, and put the blocks in a glass bottle.

Rapid Method

This method will not form as perfect a block as the slow method; the nitrocellulose is not always uniform and may section unevenly.

Wrap paper around the fiber block as for the slow method. Fill the cavity with some nitrocellulose from the infiltrating jar and transfer the tissue into the mass of nitrocellulose. Orient the piece of tissue so that it can be sectioned along the upper surface. Set the fiber block, nitrocellulose, and tissue in a tightly capped jar with a small amount of chloroform. Leave

FIGURE 13. *Preparation of a nitrocellulose tissue block for sectioning.*

them in this situation until the block is firm, but check frequently to see that the chloroform does not evaporate completely. When the nitrocellulose has hardened, add more chloroform to completely immerse the block and tissue until it is to be sectioned.

Mounted or unmounted tissue blocks can be stored in either chloroform, or 70%, 80%, or 95% alcohol, but for long storage add a little glycerine to alcohol solutions. (See Sallmen and Sherman 1961 for another rapid method.)

Hot Celloidin Technique

Koneff and Lyons Method (1937)

This is a rapid method and can be used when materials are not injured by heat. The transfers are handled in screw-cap jars in a paraffin oven, about 56°C.

PROCEDURE
1. 70% alcohol, 2 changes: $\frac{1}{2}$ hour each.
2. 80% alcohol, 2 changes: $\frac{1}{2}$ hour each.
3. 95% alcohol, 2 changes: $\frac{1}{2}$ hour each.
4. Absolute alcohol, 2 changes: $\frac{1}{2}$ hour each.
5. Absolute alcohol–ether (equal parts): 1 hour.
6. 10% nitrocellulose: 1 hour.
7. 25–33% nitrocellulose: overnight.
8. Embed as for rapid cold method.

COMMENTS

Walls (1932) uses a bottle with thick lips and corks and attaches the corks with a wire wound under the lip and over the cork. High pressure builds up in the bottle and, together with the high viscosity of the heated celloidin, produces rapid penetration. He maintains that heat is of no concern in this method; in the paraffin method it is the hot paraffin, not the heat itself, which produces a brittle block if used too long.

Do not hurry the cooling process. If the warm bottles are left on a wooden tabletop until they have reached room temperature, they will not break.

SECTIONING

Clamp the jaws of the tissue carrier tightly against the lower two-thirds of the mounting block. If the jaws are clamped against the upper portion, the resulting compression in the block can loosen the nitrocellulose.

If the knife has a concave surface, clamp that surface uppermost in the knife holder. The knife holder must be adjusted so the knife is pushed back on the instrument far enough to clear the tissue before the tissue is elevated in preparation for the next section. An automatic lever on a small slide (sliding surface) on the back of the knife block can be moved until the correct knife position is determined. If the knife does not clear the tissue before the tissue is elevated, the pressing of the knife can injure the tissue and may alter the thickness of the next section.

The maximum thickness of a section as controlled by the automatic feed lever will be only 30 or 40μ on most microtomes. Thicker sections can be cut by moving the hand feed counterclockwise and then releasing it. The hand feed is a large round knob (American Optical sliding microtome) mounted at the base of the tissue carrier and near the micrometer screw. Each movement of the hand feed will equal the number of microns indicated by the automatic feed.

Keep the surface of the slide for the knife block well oiled with the oil provided with the microtome (or Bear Brand Household Oil). A dry slide produces a jerky knife movement and rough, irregular sections.

Always maintain a pool of 70% alcohol on the edge of the knife and the top of the tissue block. Otherwise the sections may shrivel instead of slicing in a smooth sheet. The following methods are suggested for sectioning.

Method I

Wet knife and block with alcohol. Draw the knife through the block until a section is almost, but not quite, cut to completion, making certain that the cut has cleared the tissue in the block. If the cut does not clear the tissue, a striation will appear in the tissue at the point where the knife stopped or hesitated. The section rolls as it is cut. While the section is still attached at one corner, use the thumb or a camel's hair brush wet with 70% alcohol to unroll the section against the knife, and then finish the cut. Squares of paper cut to a size a little larger than the section are moistened with 70% alcohol. Hold one of these against the undersurface of the knife and slip the section off onto the paper (Figure 14). Invert the section and paper in a dish of 70% alcohol and press them down against

FIGURE 14. *A suggested method for removing sections from knife; thumb on top of section and other fingers holding paper under knife.*

the bottom of the dish. The sections can be stacked one on another in this fashion and stored until they are stained. (*Caution:* Use plain white paper, filter paper, or paper towels. Lined paper may stain the nitrocellulose and produce poor staining.)

Method II

Flood the knife with alcohol and moisten a camel's hair brush. Hold the brush close to the nitrocellulose block, so the alcohol is held between the block and brush by capillary action. Do not rest the brush too firmly on the block, lest it be cut by the knife. Pull the knife across the block; raise the brush slightly, and it will guide the section onto the knife, smoothly and without rolling. Remove as above. (Walls 1932.)

Method III

Walls (1936) developed a dry method that eliminates the constant soaking with alcohol during sectioning. After the block is hardened, soak it in a mixture of cedarwood oil and chloroform (50:50) for 24 hours. Blot the block and allow it to stand in the air for 10–20 minutes. Transfer to cedar-wood oil; this lubricates the block and knife during sectioning. The sections, as removed from the knife, are placed in the same kind of oil, then rinsed in 95% alcohol, then 80% alcohol, and transferred to water for staining. This type of oil-soaked block can be sectioned on a rotary microtome as well as a sliding one.

Difficulties in Nitrocellulose Sectioning[2]

 1. Scratches in sections. Caused by:
    a. Nick in knife.
    b. Hard material in tissue.
    c. Position where knife was stopped while unrolling section.
 2. Tissue soft and mushy; it crumbles or falls out. Caused by:
    a. Imperfect infiltration due to either incomplete dehydration or too short a time in nitrocellulose; re-infiltrate.
    b. Embedding too quickly, as in rapid method.
 3. Sections vary in thickness. Caused by:
    a. Worn parts in microtome.
    b. Pressing too hard on knife block.
    c. Insufficient hardening of nitrocellulose, which is too soft and compresses under knife.
    d. Tissue block rising before return stroke of knife has cleared it.
    e. Insufficient tilting of knife; shoulder of facets compresses block instead of cutting it and next section is thick.
    f. Embedding too quickly, as in rapid method.

STAINING AND MOUNTING

Staining Before Mounting

In the methods that follow sections are stained before mounting. Staining may be handled in syracuse watch glasses or similar flat dishes. With forceps or spatula, carry the sections through successive solutions in watch glasses and follow each change by draining the section against a paper towel or filter paper. In this way, contamination can be held to a minimum.

---

[2] Modified from Richards 1949.

If, however, the solutions do pick up considerable stain, change them as frequently as seems necessary, or even more often. Also, if they have remained uncovered for some time it is probable that the alcoholic content has been reduced by evaporation. Never, at any time during the transferring of sections, allow them to become dry. They may be carried directly from the 70% alcohol to the stain, if it is an alcoholic stain, or first into water, then into the stain, if it is an aqueous stain. (Remember to remove undesirable pigments or crystals before staining.)

When ready for dehydration, transfer through 50% and 70% alcohol into 95% alcohol. Absolute ethyl alcohol cannot be used with nitrocellulose sections unless about 5–10% of chloroform is added to it to nullify the softening effect of the alcohol. Absolute isopropyl alcohol, however, is perfectly safe to use and can be recommended before clearing action is applied. Do not attempt to dissolve or remove nitrocellulose from the sections; they will become too difficult to manipulate.

Method I

Transfer sections into carbol-xylol (page 532) or terpineol-xylol (1:9): 2–3 minutes. Keep carbol-xylol covered at all times to prevent evaporation of xylene. Dip the edge of a slide into the carbol-xylol dish and slip the section in place with a needle or forceps. Drain thoroughly and flush off excess carbol-xylol with toluene or xylene, 2 or 3 times. Add several drops of mountant and cover glass. Press cover firmly in place. Several hours later, after the resin has thickened, press down again. If the cover glass still tilts or lifts, place a small weight on it or clamp it with a clothespin overnight. Betram (1958) describes a weight that will not stick to the mountant and cover glass. *Caution:* when using carbol-xylol, keep fingers out of the solution. Wash off any that does get on the fingers and rub lanolin on the skin.

Method II

Transfer the sections to absolute alcohol containing 5% chloroform to prevent the alcohol from dissolving the nitrocellulose. Clear with benzene or toluene and mount as above, omitting the carbol-xylol.

A method particularly good for transferring thin sections to slides is as follows: slip a piece of cigarette paper under a section and spread it smoothly on the paper with a brush or needle. Place the paper plus the section on a slide with the section between the slide and paper. Blot firmly with a piece of filter paper. Peel off the cigarette paper and rinse the section with toluene or xylene. Add mountant and cover glass.

Mounting Before Staining

In the methods below sections are affixed to slides for staining.

Method I

1. Albumenize slide.
2. Press section on with filter paper.
3. Pour clove oil over it: 5–20 minutes.
4. 95% alcohol, 3 changes: 5–10 minutes total.
5. Absolute alcohol–ether, 2 changes, to dissolve nitrocellulose: 20 minutes total.
6. 70% alcohol, etc., to stain.

Method II (Lewis 1945)

1. Rub one drop Haupt fixative (page 537) on slide.
2. Transfer section on cigarette paper to slide.
3. Blot and firmly press with filter paper.
4. Roll off cigarette paper rapidly; section must not dry.
5. Place immediately in absolute alcohol–ether until nitrocellulose is dissolved: 10–20 minutes.
6. Remove to 70% alcohol, etc., to stain.

Method III

1. Float sections onto slide and blot firmly with filter paper.
2. Dip slide into 0.5% nitrocellulose for few seconds; drain and wipe back of slide clean.
3. Harden film in chloroform: 5–10 minutes.
4. 95%, 70% alcohol, etc., to stain.

Method IV (Culling 1957)

1. Transfer section to 95% alcohol: 1–2 minutes.
2. Float onto slide, drain for few minutes, blot lightly to flatten section, but do not dry.

3. Pour ether vapor over sections—only vapor, not liquid. This partially dissolves nitrocellulose and causes it to adhere to slide.
4. Place slide in 80% alcohol: 5 minutes, to harden nitrocellulose.
5. Running water: 10 minutes; stain.

Method V—Serial Sections

1. Arrange sections on knife from left to right.
2. Lay cigarette paper on sections and saturate with 70% alcohol.
3. With quick sweeping movement, pull paper off knife and carry to slide; smooth with brush.
4. Place filter paper on top and press flat, until much of alcohol is absorbed.
5. Jerk off paper and cover with clove oil: 3–5 minutes.
6. Submerge in 95% alcohol.
7. 70% alcohol, etc., to stain.

Method VI—Serial Sections (Williams 1957)

1. Coat clean slides by dipping in warm solution of 1% gelatin; allow to drain, and dry in vertical position.
2. Lay sections on blotting paper moistened with 70% alcohol; keep covered to prevent drying.
3. Pick up sections, one at a time, dip in 70% alcohol, and arrange in order on coated slides.
4. Blot sections with dry filter paper, flattening and pressing them into contact with gelatin.
5. Place in coplin jar containing 2–3 cc formalin; do not allow fluid to come in contact with sections, because it may cause them to float off slides.
6. Allow slides to remain tightly covered: 2–3 hours, room temperature.
7. 70% alcohol, etc., to stain.

Peters (1961*b*) after firmly pressing the sections onto albumenized slides, allows them to air dry overnight before staining.

I find the staining of unmounted nitrocellulose sections less traumatic than working with mounted ones. For preparing mounted sections, adopt the method that is most consistently effective.

CHAPTER 8

Specialized Embedding Techniques

WATER-SOLUBLE WAX EMBEDDING AND SECTIONING

The so-called Carbowax compounds and water-soluble polyethylene glycol waxes are used when it is necessary to go directly from fixative or water to the embedding medium, and no alcohols or clearing agents are required. These waxes are polymers, compounds composed of the same kind of atoms in the same percentage composition but in different numbers and therefore having different molecular weights; a polymer is designated by its average molecular weight.

The carbowaxes are these: Compound 4000, which is hard and dry and in crystalline flakes; Compound 1540, which is not so firm, and liquefies within a week; Compound 1000, which is in slippery lumps and liquefies within 24 hours; Compound 1500, a blend of equal parts of polyethylene glycol 300 (a fluid) and wax 1540.[1]

---

[1] Union Carbide and Shell Oil companies.

Other Carbowax compounds are the polyglycols E9000, E6000, E4000, E2000, and E1000,[2] and HEM (Harleco Embedding Media).[3]

Wade (1952) finds 4000 too hard for his use and suggests a mixture of 1 : 9 or 2 : 8 of 1540 and 4000, depending upon the weather; he recommends the 2 : 8 mixture except in the hottest and most humid climate.

Fixation

Any fixative may be used, but after a potassium dichromate fixative wash tissue for 12 hours before embedding.

Infiltration and Embedding

Use Carbowax 1–3 hours, 50°–56°C; agitate occasionally. Rinehart and Abul-Haj (1951*b*) preceded infiltration with pure Carbowax by placing the tissue for 30 minutes in 70% aqueous Carbowax and 45 minutes in 90% Carbowax. It is wise to include these steps if a tissue shows any tendency to infiltrate poorly under the short method.

Prepare the Carbowax ahead of time by placing the mixture in a beaker and allowing it to melt slowly in an oven. If it is melted rapidly over a flame, a block formed from it will not cut well. The mixture will keep for an indefinite time in the oven.

Embed by placing the tissue in a second batch of Carbowax in a small container or paper box in the refrigerator until it is hard, approximately 30 minutes. Blocks solidified at room temperature will not section as well as those made in the refrigerator. Do not chill blocks in water, which dissolves Carbowax. When completely hard, the blocks may be removed from the refrigerator. They will turn opaque as they warm to room temperature. This is of no disadvantage, but keep them in polyethylene bags or containers if storing them for some time. They must not pick up water, even from the atmosphere.

Reid and Sarantakos (1966) have improved their water-soluble embedding by adding a water-insoluble polyvinylacetate resin to the embedding compound. Their method is as follows:

1. Infiltrate with Carbowax 200,[4] 2 changes: 4 hours each.
2. Infiltrate with Carbowax 1540[4] and 4000[4] (3:1) mixed with an equal amount of Carbowax 200, 2 changes: 6 hours each.

---

[2] Dow Chemical Co.   [3] Hartman-Leddon Co.   [4] Union Carbide Co.

3. Infiltrate with Carbowax 1540 and 4000 (3:1) plus 7.5 g polyvinyl-acetate resin AYAF[4] to 100 ml of Carbowax mixture, 2 changes: 12 hours each.
4. Cast blocks; keep in a desiccator until used. Section at 24°C or less. Sections can be mounted on water.

Sectioning

A cool, dry room is recommended. The edges of the block must be parallel in order to obtain ribbons. Wade (1952) suggests that if the sections do not adhere, break up on handling, or do not ribbon at all, the block should be exposed to air for a day or two. For immediate use, paint a 25% solution of beeswax in chloroform on the upper and lower surfaces of the block. Make the layer uniform and allow it to become dull dry. Such surfaces should help the sections adhere to each other. Plain water may do as well in some cases.

Hale (1952) found that variable and high humidity produces erratic results, because with the absorption of water on the surfaces of the block, sectioning becomes impossible. He found that at 24°–25°C sections ceased to ribbon at 40% humidity; at 17°–18°C humidity had no effect. He therefore concluded that, as the temperature drops, greater humidity is permissible. At higher temperatures, a higher percentage of hard wax may counteract the effect of humidity but care must be taken that the block does not become too brittle for good sectioning.

Mounting Sections

Water mounting dissolves Carbowax, and will result in distorted sections and surface tension problems when trying to affix sections to slides. Various solutions for this problem follow.

Blank and McCarthy Method (1950)

potassium dichromate	0.2 g
gelatin	0.2 g
distilled water	1000.0 ml

Boil for 5 minutes. Filter.

The sections are picked up on a slide while they are floating on this solution and then allowed to dry.

Wade Method (1952)

Wade prefers to albumenize slides with a thin coat of Mayer egg albumen (page 536). Air dry overnight, or three hours or longer in an oven. To float ribbons that are intact and without wrinkles remains a problem. He suggests the addition of a wetting agent, turgitol 7 (0.005% in distilled water), to reduce surface tension effects. Add 10% of Carbowax 1540 to reduce shrinkage while sections are drying.

Giovacchini Method (1958)

Giovacchini smears the slides with a thin covering of the following solution: Dissolve 15 g gelatin in 55.0 ml distilled water by heating. Add 50.0 ml glycerine and 0.5 g phenol. Place sections on slides and place on warming plate at 58°–60°C for 15 minutes. Transfer to drying oven, 58°C, for 24 hours. The slide is then ready for staining.

Jones et al. Method (1959)

Jones and his coworkers float their sections on the following solution:

diethylene glycol	100.0 ml
formalin	7.0 ml
Carbowax	1.0 ml
distilled water	400.0 ml

In this method more spreading results from increasing the proportion of water. If the tissues overexpand, add a small amount of Zephiran chloride concentrate (5–10 drops to 500.0 ml flotation solution); this reduces surface tension and prevents air bubbles from being trapped between tissue and solution surface. Mount on albumenized slides and dry.

Pearse Method (1953)

This method combines features of the Giovacchini and Jones methods. Pearse smears with Giovacchini's fluid and mounts the sections with the diethylene glycol mixture.

Zugibe et al. Method 1958)

In this method, a section is cut from the ribbon with a razor blade. One edge of the section adheres to the blade. Touch the loose section edge against a slide coated with a flotation fluid and draw the rest of the section onto the slide.

Goland et al. Method (1954)

Goland and coworkers follow Carbowax infiltration with the following:

1. xylene, 61°C: 10 minutes.
2. paraffin, 61°C: 30 minutes.
3. embed.

Chill in refrigerator; do not apply ice directly. This method overcomes some of the disadvantages of Carbowax and produces less shrinkage and distortion than a regular paraffin method. After sectioning, the tissues may be handled as is customary for paraffin sections.

Either albumen or Giovacchini's fluid smeared on slides are successful. It is easier to mount sections out of a water bath (of floating solution, Jones et al. 1959) onto gelatin-smeared slides than to lay the sections directly onto the floating solution on the slide. Albumen slides mount readily either way. The razor blade method of moving the sections, as described above, is tricky, but works.

Other Methods

Reid and Taylor (1964) double embed with nitrocellulose and Carbowax.

Ashley and Feder (1966) combine hydroxyethyl methacrylate with Carbowax 200. Ruddell (1967a, b) combines methacrylate with Carbowax 400.

Riopel (1962) describes a method for botanical material that can be adapted to animal tissues. Before placing the tissue in pure Carbowax, he places it in aqueous dilutions of Carbowax and outlines a method for mounting serial sections.

(*A Final Warning:* Humidity and temperature are not always readily controlled in these techniques. As previously noted, at the usual room temperature, 21°–22°C, humidity must not rise above 40% and preferably should be much lower.)

DOUBLE EMBEDDING

Fragile, small, hard objects often crumble when processed by the paraffin method, and some difficulties can be eliminated by using a double-embedding technique.

1. Fix and wash tissue as usual.
2. Dehydrate in 50%, 70%, and 90% ethyl alcohol: 2 hours each.

3. Absolute ethyl alcohol: 2–16 hours.
4. Methyl benzoate-celloidin solution (see below): 24 hours. Pour off and replace with fresh solution: 48 hours. If tissue is not clear, repeat for 72 hours.
5. Pure benzene, 3 changes: 4 hours, 8 hours, and 12 hours, respectively.
6. Mixture of equal parts of paraffin and benzene, in embedding oven: 1 hour.
7. Paraffin, 2 changes: ½ to 6 hours each, depending upon thickness and nature of tissue (3 hours for 5 mm thickness).
8. Embed and proceed as with ordinary paraffin sections.

Methyl Benzoate-Celloidin Solution

To prepare, add 1 g air-dried celloidin flakes to 100.0 ml methyl benzoate. Shake well and allow bottle to stand upright for an hour or longer. Invert for an hour. Lay bottle on side for an hour, and then turn it upright again. Repeat this process until solution is completed, probably the following day.

Mounting Sections and Preparing to Stain

If the sections do not flatten well on slide, but instead curl and separate from it, float them on 95% alcohol to soften the celloidin. If there still is a curling problem, soften the sections with ether vapor, blot with filter paper, and soak in 0.5–1.0% celloidin or nitrocellulose. (Lillie 1954 or 1965.)

If albumen continues to fail as an adhesive agent, use a fresh solution of Masson gelatin fixative (page 537) under the sections and warm slides on warming plate. As soon as the sections have spread, remove the slides and allow them to cool for a few minutes. Drain off excess gelatin solution, blot sections with filter paper, and place in formalin vapor overnight (40°–50°C).

When ready to stain, place the slides in chloroform before proceeding to 95% alcohol. The chloroform will remove the paraffin and harden the celloidin simultaneously.

Briggs Double-Embedding Method (1958)

1. Fix, wash, and dehydrate through 70%, 95%, and 2 changes of absolute alcohol: 2 hours each.
2. Absolute alcohol-ether (50:50): 12–24 hours.
3. 4% celloidin (or nitrocelulose): 48 hours.
4. Dip tissue in alcohol-ether to remove excess celloidin from surface.

Harden in chloroform vapor for few minutes. Drop into chloroform and leave until tissue sinks: 12–24 hours.

5. Infiltrate, 2 changes paraffin: 12–24 hours each.
6. Embed.

Briggs suggests that during staining, $\frac{1}{3}$ chloroform by volume be added to all hydrating and dehydrating solutions above 70% alcohol to prevent removal of celloidin.

Banny and Clark Double-Embedding Method (1949)

This is similar to the Briggs method with the following modifications:

For step 3, place tissue in 2 changes of nitrocellulose, 3% and 6%. After step 4, place tissue in benzene overnight.

Dioxane in Double Embedding

1. Fix, decalcify if necessary, and wash.
2. Dioxane: 2 hours.
3. Dioxane-nitrocellulose: 3 days.

dioxane	70.0 ml
2% nitrocellulose	30.0 ml

4. Dioxane: 2 hours.
5. Dioxane-paraffin (50:50): 2 hours.
6. Paraffin, 3 changes: 20 minutes, 30 minutes, 1 hour.
7. Embed and section immediately.

At UCLA, this method was used successfully on hard fish with scales intact. (Also see Brown 1948.)

Crabb (1949) combines rosin with celloidin, and then double embeds with paraffin. Salthouse (1958) double embeds with tetrahydrofuran, particularly for insects.

ESTER WAX EMBEDDING

The ester waxes were introduced by Steedman (1947) as embedding media that combine the advantages of both paraffin and celloidin, but reduce their disadvantages. Tissue structure is well preserved, and structural lipids are retained perhaps better than when either paraffin or celloidin are used. The ester wax method is based on diethylene glycol distearate, whose hardness compares favorably with that of paraffin. Ester wax melts at a lower temperature and does not require the use of hardening hydrocarbons. The wax is soluble in the usual paraffin solvents and is also miscible with *n*-butyl alcohol, isopropyl alcohol, Cellosolve, glycol monobutyl ether, acetone, and aniline. It supports soft tissues adjacent to hard ones and adheres to smooth surfaces, and it can be sectioned on a rotary microtome.

Steedman Method (1960)

SOLUTION

diethylene glycol stearate	60.0 g
glyceryl monostearate	30.0 g
300 polyethylene glycol distearate	10.0 g

Melt the diethylene glycol distearate and heat until clear. Add glyceryl monostearate. When this is dissolved add the 300 polyethylene glycol distearate. Filter.

PROCEDURE

1. Transfer fixed tissues to 70% alcohol: 1–2 hours. Time depends on size of tissues.
2. Transfer to 70% alcohol–Cellosolve (ethylene glycol–monoethyl ether), (50:50): 1–2 hours.
3. Transfer to pure Cellosolve: 1–3 hours.
4. Transfer to Cellosolve–ester wax (50:50): 1–3 hours.
5. Infiltrate in pure ester wax, 3 baths: 24 hours total.
6. Block.

ALTERNATE PROCEDURE

1. Transfer fixed tissues to 95% alcohol: 1–3 hours.
2. Transfer to 95% alcohol–ester wax (50:50) in oven (35°–40°C): 1–3 hours.
3. Transfer to pure wax: 24 hours.
4. Block.

Have the melted wax about 10°C above the melting point (48°C) for blocking. Use a cold mold or metal L's. Cool block in ice water but do not submerge it. The more quickly it cools, the better the final texture of the wax.

SECTIONING

Ester waxes must be cut very slowly with a sharp knife. The wax is so hard that the knife must be held firmly in the clamps, and all parts of the microtome must be in good repair and tight, or the sections will be of uneven thickness. Thick sections may be cut the same day that the tissue is blocked, but thin sections will cut more easily on the second or third day following the blocking.

Mount as for paraffin sections on albumenized slides. Drain and dry at room temperature.

Sidman et al. modification (1961)

After receiving some poor batches of wax, Sidman and his coworkers tested a number of products and found that polyethylene glycol 400 distearate (Kessler Chemical Co., Philadelphia) gave the most consistent results. This wax can be purchased in 8-lb. pails and is used unmixed. It melts at $35°C$ and begins to solidify at $30°C$. Melt it overnight (no longer) at $56°C$ and then transfer it to a $37°C$ oven for use.

1. Fix and dehydrate the tissue as desired.
2. Infiltrate in 3 changes of wax, $37°C$: 24 hours at least.
3. Embed. Pour wax in mold and set it on a cube of ice until a solid base of wax forms. Remove from ice and add tissue. Orient it and place mold and tissue in refrigerator: 1 hour.
4. Allow to warm to room temperature: $\frac{1}{2}$ hour. Trim and mount on object disc as for paraffin blocks. Section under cool conditions. If the room is warmer than $25°C$, solid CO_2 held in a kitchen strainer on a ring stand about 6 inches above the knife will provide adequate cooling. Cut slowly.
5. Float sections on 0.1% gelatin in water ($25°C$) and onto clean slides. If the sections are thicker than 10μ, use $30°-32°C$ water. Drain. Dry at room temperature: several hours or overnight. Sections should not float too long on water, because the wax is soluble and loses its cohesiveness. A wet brush is easier to use than needles, because of the stickiness of the wax. Blocks should be stored in a cool place, and unused sections in the refrigerator.
6. For staining, remove the wax with absolute alcohol, hydrate, and stain.

See also Kuhn and Lutz (1958) and Ueckert (1960) for methods of polyester embedding and sectioning. Menzies (1962) recommends a paraffin-beeswax-stearic acid mixture for thin sections. See Wigglesworth

(1959) for an excellent double-embedding method using agar and ester wax for sections that are 0.5 to 1.0μ thick.

METHACRYLATE PROCESSING FOR THIN SECTIONS

Methacrylate-Plexiglas Processing for Thin Sections for Light Microscopy (Zambernard et al. 1969)

FIXATION

Any routine fixation followed by appropriate washing. Formalin must be thoroughly removed.

SOLUTIONS

Solution I

methyl methacrylate[5]	27.0 ml
polyethylene glycol distearate 1540 [6]	6.0 ml
dibutylphthalate	4.0 ml

Heat to 56°C to dissolve distearate. Cool.

Solution II

solution I	90.0 ml
benzoyl peroxide[5]	0.6 ml
wetting agent, Igepal 630 [7]	3 drops
Plexiglas molding powder A-1 [8]	30.0 g

Add Plexiglas gradually while stirring with magnetic stirrer. Store in refrigerator. Bring to room temperature before use.

PROCEDURE

1. Wash out fixative and dehydrate through 70%, 95% and absolute ethanol: 20–40 minutes each.

---

[5] Matheson Coleman and Bell.
[6] Ruger Chemical Co., Irvington on Hudson, N.Y.
[7] Chemical Sales Co., Denver, Colo.　　[8] Rohm & Haas Co., Philadelphia, Penn.

2. Transfer into anhydrous acetone (anhydrous sodium sulfate: acetone, 2 g/500 ml): 10–15 minutes.
3. Transfer into methyl methacrylate, 2 changes: 20 minutes each.
4. Transfer into equal parts of methyl methacrylate and solution II: 60 minutes.
5. Place in solution II: 60 minutes or more, may remain up to 48 hours (see comment 1).
6. Fill Peel-A-Way plastic embedding molds[9] with solution II and add tissue (see comment 2).
7. Polymerize plastic under ultraviolet light[10] at room temperature: 4 hours. Final cure under ultraviolet light, 56°C: 4 hours (see comment 3).
8. Peel off mold, moisten block with 50% alcohol, and trim with a coping saw or sharp knife.
9. Section on sliding or rotary microtome (see comment 4). Keep block moistened with 50% alcohol during sectioning.
10. Transfer moist sections to albumenized or subbed slides. Flatten with a couple of drops of 95% alcohol and dry on 45°C warm plate.
11. For staining, flood dried slides with clove oil: 10–20 minutes; 95% alcohol, 3 changes; and proceed as with deparaffinized slides.

COMMENTS

1. Infiltration of solution II may be increased by placing the solution and tissue in a desiccator and applying negative pressure of 500–600 mm, or rotating the vials at $\frac{1}{3}$ rpm.
2. To keep the tissue from settling too close to the bottom of the mold, a previously polymerized layer (1–2 mm thick) of solution II may be formed in the molds; I have not found this necessary.
3. The final curing described in step 7 can be done overnight in an oven. The molds can be placed in a small box and the ultraviolet light allowed to rest on the box edges, 5–6 cm above the molds; or blocks can be used to support the light.
4. Zambernard and his coworkers section on a sliding microtome, but I have had problems with a lightweight sliding microtome. The hardness of the plastic block lifted the knife and its support and cut a wedge-shaped section, thicker at the point of the knife's entry and thinner at the place of its exit. These problems were overcome by sectioning on a rotary microtome. The block maintains adequate

---

[9] Lipshaw Co., Detroit, Mich.
[10] Long Wave, UVL-21, Blak-Ray, Ultra-Violet Products, Inc. San Gabriel, Calif., also Fisher Scientific.

FIGURE 15. *Removing a plastic section from a microtome:* (a) *the section is straight-ened on the knife with a moist* (50% *alcohol*) *brush and is ready to be moved to the left;* (b) *the section is slipped onto the slide.*

moisture in the vertical position, and the section stays uniform in thickness. With a brush moistened in 50% alcohol, slip the section on the wet knife to one side of the tissue block (raised or lowered, whichever facilitates the action). Carefully place a slide behind the knife (between the knife and the microtome chuck) and with the wet brush slip the section onto the slide (Figure 15). Several sections can be placed on a slide.

5. Clamping the tissue block tightly in the microtome chuck can cause the block to compress and make the top of the block bulge outward. This can be corrected by mounting the tissue block on a wooden block with epoxy cement. Then clamp the chuck on the wooden block instead of on the tissue block.

EPOXY RESIN PROCESSING

The epoxy resins, popular for electron microscopy, can be used as well for light microscopy. They produce uniformly hard blocks with very little shrinkage of the tissue. The resin is cured (polymerized) with anhydrides; DDSA (dodecenyl succinic anhydride) produces a soft block, while NMA (nadic methyl anhydride) forms a very hard one. Varying proportions of the anhydrides, therefore, can be used to adjust the resin block to the desired hardness. Dibutyl phthalate, a plasticizer, sometimes is added to impart flexibility to the casting and to reduce viscosity. At final embedding, an accelerator, DMP [2, 4, 6-tri (dimethylaminomethyl)-phenol] is added. Propylene oxide is commonly used as the clearing agent following dehydration through alcohol. Epon 812 has been used following washing in water (see Craig et al. 1962 and Idelman 1964). Other water-miscible embedding media are Aquon, Durcupan, 2-hydroxy-propyl methacrylate, and glycol methacrylate.

There are many recommendations on the use of these resins: Luft (1961) uses Epon 812, highly favored, particularly for electron microscopy; Grimley et al. (1965) prefer Araldite 502 over Epon 812 and use it for light microscopy; I prefer Mollenhauer's mixture, which follows.

Epoxy Resin Processing for Light Microscopy (Mollenhauer 1964)

FIXATION

Any routine fixative, but preferably one of the aldehydes.

SOLUTIONS

Resin mixture

Epon 812 [11]	62.0 ml
DDSA[11]	100.0 ml
Araldite 502 [11]	81.0 ml
dibutyl phthalate[11]	4–7.0 ml

Mix thoroughly; a Teflon stirrer is recommended. To eliminate the difficulties of cleaning resins from glassware, use disposable beakers for mixing and disposable syringes for measuring the reagents. Store for short periods in the refrigerator, but it can be kept for several months in a freezer.

Catalyzed resin

resin mixture	5.0 ml
DMP-30 [11]	3 drops

Mix just before use.

PROCEDURE

1. After washing out fixative, dehydrate through 70% and 95% alcohol, and 4 changes absolute alcohol: 30 minutes (or more) each. (Tissue blocks should be no larger than 4–5 mm in thickness.)
2. Transfer to propylene oxide,[11] 2 changes: 30 minutes (or more) each.
3. Transfer to propylene oxide and catalyzed resin, (2:1): 1 hour (or longer).
4. Transfer to propylene oxide and catalyzed resin (1:2): 1 hour (or longer).
5. Transfer to catalyzed resin in "Peel-A-Way" molds and polymerize overnight at 60°C, or leave overnight at 35°C, the next day at 45°C, and then overnight at 60°C. The latter forms a more easily sectioned block.
6. Remove mold and section at 1–3μ with a very sharp knife. The sections can be removed from the knife with fine forceps and placed in a blackened box as in paraffin sectioning (page 60).
7. With fine forceps and a dissecting needle pull out any folds in the sections, straighten them as much as possible and place them on

---

[11] Polysciences, Inc., Warrington, Penn.

albumen-covered slides. The albumen is not essential, but does make the mounting solution puddle more uniformly.

8. Place on 45°C warm plate and immediately add 10% aqueous acetone under the sections. Allow the solution to evaporate and continue to straighten the sections with needles, if necessary. As soon as the slide is dry, it can be stained.

COMMENTS

As with the Zambernard method, I prefer to mount the tissue blocks on wooden blocks so that the microtome clamps on the wood rather than on the plastic. Sliding or rotary microtomes can be used, but the sliding microtome must be a heavy-duty model to prevent the same problem as described for the Zambernard blocks. A hard-grade steel or tungsten-carbide steel knife with a 55° cutting bevel edge is recommended. Plastic dulls the average microtome knife; only three or four sections can be cut successfully in a single area of the knife.

If, after curing, the block feels "sticky" soft and will not cut smoothly, return it to the 60°C oven overnight. Quickly peel off the mold and trim off the sloping sides of the block. Do this immediately after removal from the oven and while the block is still warm. It trims more easily in this condition than after it has cooled. Cut the first sections with an old knife, as recommended for paraffin sectioning, until the best part of the tissue is reached. Change to the good knife and collect the desired sections. Keep the edge of the knife clean; wipe it off after every section.

Plastic sections usually adhere to the slides through all solutions, but should loose sections develop, run the slides through a Bunsen burner flame three or four times. Allow them to cool before staining.

Staining Epoxy Sections

Tissues embedded in plastic will not respond to all staining methods; most stains cannot penetrate the plastic. Generally speaking, these sections will stain satisfactorily in alcoholic solutions and in basic dyes in alkaline solutions in the pH 11.0 range. An alcoholic and a buffered aqueous method are described below. Other aids to staining are pretreatment with mordants, hydroxide, peroxide, or iodine. I have had excellent staining results with the hydroxide method below. So far, among the stains I have tried, only PAS has given less than satisfactory returns.

Pinacyanole Stain (Humason and Lushbaugh 1961)

1. Cover sections with several drops of Pinacyanole[12] (0.5 g/100 ml 70% ethanol): 10–15 minutes. Cover the slides to prevent evaporation of alcohol.
2. Wash off excess stain, blot, and allow to dry.
3. Clear in xylene and mount.

Toluidine Blue Stain (Trump et al. 1961)

1. Flood slide with freshly prepared and filtered toluidine blue 0, C.I. 52040 (0.1 g/100 ml 2.5% aqueous sodium carbonate, pH 11.1): 30–120 minutes, or until cells are dark reddish-purple.
2. Tip slide and draw off extra stain with filter paper.
3. Flood slide gently with tap water from a medicine dropper and draw off water with filter paper.
4. Wash briefly with 90% ethanol and then absolute ethanol. This differentiates the stain, so check it under microscope.
5. Clear in xylene and mount.

Sodium Hydroxide Treatment (Lane and Dominic 1965)

1. Prepare a saturated solution of sodium hydroxide in absolute ethanol. Allow to stand: 2–3 days.
2. Immerse slides in solution: 1 hour. Keep container tightly closed to prevent evaporation and deposit of sodium hydroxide (NaOH) on slides.
3. Drain well; do not blot.
4. Transfer through absolute ethanol, 4 changes: 5 minutes each. If the first change becomes milky with accumulated NaOH, discard it, move numbers 2, 3, and 4 up into 1, 2 and 3 positions and add a new number 4.
5. Treat with phosphate buffer, pH 7.0 (page 544): 5 minutes.
6. Wash in 3 changes of distilled water: few seconds in each.
7. Treat with phosphate buffer, pH 4.0 (page 544): 5 minutes.
8. Wash in running water: 5 minutes.
9. Stain.

---

[12] Eastman Kodak, Rochester, N.Y.

COMMENTS

Berkowitz et al. (1968) use a dilution of saturated NaOH 1:5 with absolute ethanol. I have obtained the best results with the method as described above. For the peroxide method, see Aparicio and Marsden (1969) or Pool (1969). For the iodine treatment, see Yensen (1968), and, for permanganate mordanting, see Grimley et al. (1965) and Shires et al. (1969).

For descriptions of other stains, see Flax and Caulfield (1962); Grimley (1964); Grimley et al. (1965); Houck and Dempsey (1954); Richardson, Jarett, and Finke (1960); Trump et al. (1961); Schantz and Schecter (1965); Spurlock et al. (1966); Hoefert (1968); Martin et al. (1966); and Hendrickson et al. (1968).

CHAPTER **9**

The Microscope

THE COMPOUND MICROSCOPE

The fundamental parts of a compound microscope are the objectives and the eyepieces or oculars. Both are systems of lenses; the oculars at the top and the objectives at the bottom of a tube. The other basic parts are the stage on which the object to be viewed is placed and a mirror or system of mirrors for directing light. The ordinary laboratory microscope uses bright-field illumination with direct light amplified by a substage condenser and mirror. The condenser is also a system of lenses, the purpose of which is to concentrate light on the object to be viewed so it is seen in a bright, well-lit field.

The objectives magnify the specimen a definite amount, forming an image, which is magnified again by the ocular eyepiece. The lenses of a good microscope should not only magnify, but also improve the visible detail. This is the resolving power, and it is a function of (1) the wave length of the light used, (2) the lowest refractive index (refractive index is a measurement of the refraction or bending of light rays as they pass through one medium to another at an oblique angle) between the objective and the substage condenser (page 107), and (3) the greatest angle between two rays of light entering the front lens of the objective. Blue light increases resolving power over white light; ultraviolet increases it even more, but it cannot be seen, and the results must be photographed.

Oculars are manufactured in a variety of powers of magnification (engraved on them); 5X and 10X are the customary lenses for student and laboratory microscopes—the 5X for low power over large areas, the 10X for general work. Most of these microscopes are equipped with three objectives: 50 to 32 mm (low power); 16 mm (middle or intermediate power); and 8 to 4 mm (high power). All are engraved with their magnifying power. An additional objective will be included on the four-objective nosepiece and can be used interchangeably with the low power (or any other power) on the three-objective nosepiece. As one shifts from low power to higher powers, the magnification increases, but the size and depth of the examining field decreases; also, the clearance of the objective above the specimen (working distance) decreases, and more light is required. The working distance is the distance between the front of the objective and the top of the cover glass when the specimen is in focus.

Achromatic objectives (producing an image essentially free of color fringes) are found on most laboratory microscopes. These objectives are corrected chromatically for the light of two wavelengths, the C (red) and F (blue) lines of the spectrum, and spherically for the light of one color, usually in the yellow greens. Achromats give a good image in the central portion of the field but not as sharp an image at the edges as can be obtained with apochromatic objectives. These are essential for critical microscopy and color photography because they are corrected for three colors, including the G (violet) lines of the spectrum, and thereby reduce color fringes even further. Lenses, if not corrected in this manner, exhibit chromatic aberration, and color fringes will confuse the border of the specimen image. For best results, apochromatic objectives should be accompanied by a corrected condenser of a numerical aperture at least equal to that of the objective aperture and compensating ocular eyepieces. A correction collar on the apochromatic objectives can be adjusted to compensate for the thickness of the cover glass on the slide. The collar can be adjusted with one hand while the fine adjustment is manipulated with the other hand.

The body tube supporting the lenses may have a fixed length, or it may be equipped with an adjustable drawtube. A scale on the drawtube is used to determine the length of the body tube for the various lenses used in it—usually 160 mm for American-made objectives. With high-power objectives this scale becomes increasingly useful in adjusting the length of the drawtube to achieve sharp detail through a thick cover glass.

At the bottom of the microscope, under the stage, there is a mirror that directs light through the diaphragm and into a condenser, which increases the illumination of the specimen. Usually there are two mirrored surfaces: one plane (flat) to direct the light, reflecting a moderate amount of light with parallel rays but no change in form; the other, concave to converge the light rays to form a cone, concentrating a large amount of light. The concave surface replaces a condenser and can be used for low magnifications, but it should not be used when a condenser is present, nor should it be used for high magnifications in place of the plane mirror.

Immediately below the microscope stage is the substage condenser, whose lenses serve to converge the parallel beam of light from the mirror so that a cone of light passes through the aperture of the stage. This cone of light is focused on the specimen under examination and then is extended to fill the back lens of the objective. Without a condenser the back lens of high-power objectives is not filled with the proper amount of light. By opening or closing the iris diaphragm mounted below the condenser, the diameter of the light entering the condenser can be controlled. A two-lens Abbe condenser is found on most laboratory and student microscopes. For critical research microscopy, a corrected condenser with a centering mount should be used. It should be carefully centered to the objective, and immersion oil should be used between it and the slide for finest detail.

The image seen in the compound microscope is inverted: it is upside down and turned right side to left. The movement of the slide will be reversed. For clarity, most images depend on color in the specimen or color added to it and/or differences in refractive indices in different parts of the specimen and the medium in which it is mounted.

Innumerable types and sources of light are used in laboratories. If a table lamp, such as that found in many student laboratories, is used, use a daylight bulb in it. Good, simple types of microscope lamps are equipped with blue ground-glass filters. Place the lamp 10 to 15 inches in front of the mirror and with all of the light directed below the stage of the microscope. The surface of the bulb or of the filter is focused on the specimen by way of the substage condenser. If the lamp is the small substage type of lamp, it may be used in front of the mirror, or, if the mirror is removed, it is placed under the condenser. A more critical illumination (Köhler method) can be obtained with better kinds of lamps provided with condensing lens and a coil or ribbon filament.

Controlled illumination is lighting by a cone of rays whose proportions are regulated by a stop at the illuminator and by the iris diaphragm of the condenser. The whole area of the aperture of the objective is used, and none of the refracted light should fall on the lens mounting or on the drawtube.

THE OPERATION OF A MICROSCOPE

Student Microscopes

Without a substage condenser

1. Place slide on stage and center the field to be examined over the opening in stage.
2. Turn low-power objective over specimen a short distance above it— one-half inch will do.
3. Adjust concave mirror, moving it back and forth, until light is as uniform as possible through the hole and onto the specimen.
4. Raise the body tube by coarse adjustment until the specimen is in focus. Check mirror, adjusting for better light.
5. Higher powers may be swung into position and light adjusted if necessary for greater magnification.

With substage condenser and lamp without lenses

1. Place slide on stage as above. Adjust objective, plane mirror, and body tube for focus, as above.
2. Remove ocular eyepiece and, while looking down the tube, open iris diaphragm of the substage condenser until it coincides with the margin of the back lens of the objective, just enough to fill it with light, no more. Focus the condenser up and down until the light is uniform through the objective.
3. Replace ocular eyepiece. Make similar adjustments for all objectives.

With substage condenser and lamp with condenser lenses

1. Place lamp 8–10 inches from microscope and direct light onto center of plane mirror.
2. Adjust lamp bulb or lamp condenser until the filament image focuses on the microscope condenser. The visibility of the filament image can be increased by closing the lamp iris.
3. Place slide on stage and focus on it. With the lamp iris closed, raise

or lower the microscope condenser until the lamp iris appears in sharp focus with the slide specimen. Open the lamp iris until it disappears at the edge of the field.

4. Remove the ocular eyepiece and adjust the microscope iris by opening it until it coincides with the margin of the back lens of the objective.
5. Do not change diaphragm positions when proper adjustments have been made. Intense light can be reduced by the use of filters.

To adjust both parts of a binocular microscope, close the left eye and focus with the fine adjustment until the image is sharp for the right eye. Leave the fine adjustment alone and focus for the left eye with the focusing adjustment on the tube outside the ocular eyepiece. Between the two eyepieces is an adjustment for changing the interpupillary distance between the eyes. Turn the adjustment slowly until the right and left images blend and a single field is seen with both eyes.

Research Microscopes

Critical illumination (Köhler method) using a lamp with condensing lenses and an iris diaphragm

1. Place a slide with fine-detailed specimen in position on microscope stage. Center the condenser in relation to the stage aperture.
2. Place the lamp so the diaphragm is about 10 inches from the microscope mirror and line it up with the microscope so the light filament centers on the plane surface of the mirror. Insert a neutral filter.
3. Using a 10X ocular and 16 mm objective, adjust the mirror until the light passes through the substage condenser to the slide. Focus microscope on the specimen and center the light.
4. Rack up the substage condenser until it almost touches the slide and close the substage diaphragm. Close the lamp diaphragm about halfway and adjust the lamp condenser until an image of the filament is focused on the closed substage condenser diaphragm. This can be observed in the microscope mirror or with a small hand mirror. Partly open the substage iris diaphragm.
5. Focus with the substage condenser until a sharp image of the lamp diaphragm coincides with the specimen on the slide. With the mirror, center the diaphragm image.
6. Take out the ocular eyepiece, and, looking down the tube, observe the back lens of the objective. Open the substage diaphragm until its

image edge almost disappears in the back lens, but remains just visible. The back lens should be filled with light—full cone illumination. Because the light source is imaged on the diaphragm of the substage condenser, the light source is in focus at the back focal plane of the objective, and the filament of the lamp can be seen.

7. Replace the ocular eyepiece. If the tube length has to be adjusted, do so (usually 160 mm for American objectives).

Certain precautions are imperative for good illumination and bear repetition. The back lens should always be filled with even light, and the aperture of the substage diaphragm should be closed to a minimum to furnish as wide a cone of light as the specimen will take. Closing down the diaphragm to reduce light intensity, and thereby increasing contrast in transparent objects, is poor technique. Light intensity should be controlled by the kind of bulb, the use of filters, or a rheostat or variable transformer inserted in the lamp system.

Unless you have had previous experience with a certain microscope, when changing from one objective to another of higher power (high dry or oil-immersion) do not take it for granted that the microscope is parfocal (focal planes of all objectives lie in same position on the tube). The higher power may crush a valuable specimen. Proceed as follows: examine slide with the naked eye for approximate part to be examined. Center the area over the stage aperture. Check with low power, then middle power for the particular area of interest, then change to high power, but raise the tube $\frac{1}{4}$ to $\frac{1}{2}$ inch at least before revolving the nosepiece. Then, while watching the bottom of the objective from the side, lower the tube until the objective almost touches the cover glass. Look into the microscope and rack up the tube until the specimen is in focus. Remember, when an objective of a different numerical aperture is used, the condenser diaphragm must be adjusted (step 6). The numerical aperture (NA) number is used to compare the resolving power of the lenses: the higher the numerical aperture, the greater the resolving power and the wider the condenser diaphragm must be opened.

Use of Oil-Immersion Objective

The working distance with this objective is short, and great care must be taken to prevent crushing a cover glass. Thin cover glasses must be used, or the specimen mounted on the cover glass itself. Greater illumination is required than with the dry objectives, and a wider cone of light is needed. The space between the objective and the cover glass, also frequently between the slide and the condenser, must be filled with a suitable medium of a refractive index and dispersion like that of the glass of the lenses. This medium is provided by immersion oil (cedarwood oil, Cargille,[1] crown oil).

1. Using a dry objective, locate and center the area to be examined.
2. Raise microscope tube, swing oil-immersion objective into position.
3. Place a drop of immersion oil on the cover glass over the area to be examined. Watching (from the side) the lower edge of the objective, lower it slowly until it makes contact with the oil.
4. Observing through the microscope, very slowly lower the objective with the coarse adjustment until the specimen is in focus. Make more critical focus with the fine adjustment.
5. Remove ocular eyepiece and adjust the light on the back lens of the objective; open the condenser diaphragm.

After use, clean oil-immersion objective; never leave oil on it. Wipe it with clean lens paper, then with a little xylene or chloroform on lens paper, finally drying it with clean dry lens paper. Do not leave xylene or chloroform on the lens.

Other Hints for Efficient Microscopy

Always keep all surfaces clean and free from dust and grease. This includes the lamp condenser, filters, mirror, lenses of the substage condenser, lenses of objectives, slides, cover glasses, binocular prisms, and lens of ocular eyepieces. Use a good grade of lens paper and a fine camel's hair brush. Omit,[2] a lint and dust remover, assists in cleaning lenses and other parts of a microscope.

If glasses are worn, protect them from scratching by mounting a rubber band around the top of the ocular. (Fingers of old rubber gloves can be cut to fit the oculars.)

If, instead of a mechanical stage, clips are present on a fixed stage, place one clip over one end of the slide. Then, with two fingers of one hand controlling the loose end of the slide, move the slide for examining, while manipulating the fine adjustment with the other hand. Individual preferences will determine which hand to use for which motion. With a mechanical stage, the right hand has to control the knobs, and the left hand, the fine adjustment.

---

[1] Cargille Laboratories, Cedar Grove, N.J.
[2] Century Laboratories, Chicago, Ill.

If it is convenient to have a pointer in the eyepiece, unscrew and remove the top lens of the ocular eyepiece. Inside, about halfway down its tube is a circular shelf. Place a small drop of Permount (or other quick-drying mountant) on this shelf and, with small forceps, place an eyelash in this drop. Arrange the eyelash to lie flat and project slightly short of center of the hole formed by the circular shelf. Screw lens back in place.

When carrying a microscope, carry it upright, preferably with both hands. Never tilt it; the ocular eyepiece can fall out. When using a microscope to check staining effects or with any wet mounts, place a thin piece of glass on the stage. A lantern-slide cover glass ($3\frac{1}{4} \times 4$ inches) is usually available and is an ideal size for most stages. Clean it after use to prevent contamination of future slides or material.

MEASURING DEVICES USED ON A MICROSCOPE

Reading a Vernier

Most mechanical stages are equipped with a vernier—a device for the purpose of relocating the same spot previously examined on a slide. Two numerical scales run side by side, a long one constructed on a millimeter scale, and a short one constructed on a scale of nine millimeters divided into ten equal divisions. When the area of reference on the slide is centered in the objective, read the vernier. Check the point of coincidence of the zero of the short scale against the long scale. Read the lower whole figure on the long scale, to the left of the zero on the short scale. This determines the whole number, the number to the left of the decimal point in the final reading (21 on the long scale in Figure 16). *Exception:* If the zero of the short scale coincides exactly with a whole number on the long scale, then the number is recorded as a whole number with no decimals following it. The decimal is determined by the point of coincidence of the line of the short scale that perfectly coincides with any line on the long scale (.5 of the short scale in Figure 16). The reading, therefore, in Figure 16 is 21.5.

Measuring Objects with an Ocular Micrometer

The measurement of slide specimens usually is done with a micrometer disc placed in the ocular, but first the disc's actual value with respect to

FIGURE 16. *A vernier scale.*

the magnification at which it is being used has to be calibrated against a stage micrometer. Stage micrometers usually have a 2-mm scale divided into .01-mm divisions, or a 0.2-inch scale divided into .001-inch divisions. The ocular micrometer has a 5-mm scale divided into 50 (0.1-mm divisions or 100 (.05-mm) divisions.

Unscrew the top lens of the ocular, place the ocular micrometer on the circular shelf inside, and replace the lens. Focus on the stage micrometer, moving it until the zero lines of both micrometers coincide. Then a definite distance on the stage micrometer will correspond to a certain number of divisions on the ocular micrometer.

The rest is simple; for example:

The ocular micrometer has 100 divisions.

Suppose that 30 divisions (each measuring .01 mm) of the stage micrometer equal the 100 ocular micrometer divisions. Then $30 \times .01$ mm = .30 mm, the length of the 100 divisions on the ocular micrometer.

Next, $.30 \div 100 = .003$ mm (3.0μ), the value of each single division.

Therefore, when the ocular micrometer (each division measuring 3.0μ) is focused on a specimen, and the specimen requires 9 divisions to meet its length, multiply 9 by 3.0μ to equal 27.0μ (.027 mm), the total length of the specimen.

Obviously, each combination of oculars and objectives, and each different tube length must be calibrated in this fashion.

SPECIALIZED MICROSCOPY

The image of a biological specimen can be formed and used in several ways. The most common form, known in all laboratories, is the use of color images, either natural color in the specimen or differential color applied to it. But unless one of a few vital stains are employed, the color image method cannot be used on live material. Most living material is transparent to light; that is, the light wave passes through it with very little loss of intensity, so other means for examining it must be employed.

Dark-field Microscopy

In dark-field microscopy, the objects themselves turn the light into the microscope by reflecting it or scattering it, and the object appears luminous on a dark background. No light from an outside source reaches the eye. A stop in the substage condenser blocks out the central part of the solid cone of light formed in the condenser, and only oblique rays in a hollow cone of light, striking from the sides, illuminate the object.

The simplest form of dark field can be developed with a black central patch stop inserted in the carrier under the condenser. Dark-field elements in a threaded mount can be used in place of the upper lens of an Abbe condenser. This is practical with all objectives if the numerical aperture does not exceed 0.85, and immersion oil is used with the slide.

Dark-field illumination can be modified with colored stops and outer rings below the substage condenser, instead of black stops, to produce optical staining. The central stop determines the background color and the outer ring the color of the object, giving optical coloring to unstained objects and helping to reveal detail and structure.

Dark-field and optical staining can be useful for examining the following: (1) bacteria, yeasts, and molds; (2) body fluids, plant or animal; (3) colloids; (4) living organisms in water; (5) foods, fibers, and pigments; (6) insects and scales; (7) crystals; (8) bone, plant, and rock sections.

It is a common means of studying the results of microincineration, the investigation of minerals present in different parts of tissues. Paraffin sections of tissue fixed in a solution of formalin and alcohol are mounted on slides and placed in an electric furnace. The temperature is then slowly raised until all organic matter is burned away and only the mineral skeleton of the tissue is left. This appears white under dark-field observation.

Phase and Interference Microscopy

Although most living material is transparent to visible light, different components of tissue do alter to a different extent the phase of light waves passing through them. That is, the light velocity is altered, advanced, or retarded, and its vibration is said to be changed in phase. When two waves come together and are in phase, brightness increases, but, if they are out of phase and are of equal amplitude, interference occurs and the eye sees black, not light. Waves of intermediate difference in phase produce a series of grays. It is the aim of the phase microscope to change slight phase differences into amplitude differences and to produce a variation in intensity from light to dark contrast observable in the specimen under the microscope.

Both phase and interference microscopes follow this principle of interference phenomenon—combining light waves that are out of phase with each other to produce combined waves of greater and lesser amplitude. The means of applying the principle is different.

Phase microscopy

The cost of phase microscopy is not excessive, and the parts can be adapted to any bright-field microscope. The system requires phase contrast objectives (achromatic objectives with fixed-phase plates and a two-lens Abbe condenser with an iris diaphragm and a revolving ring below the condenser to carry four annular ring diaphragms, which are stops that produce different-sized cones of light). The correct annular ring diaphragm must be centered to the phase ring of the correct objective and must match its numerical aperture. The annular diaphragm causes the light to strike the object in the shape of a hollow cone and gives rise to two types of waves: one type passes straight through the object (undiffracted), and the other type are diffracted into a different course. Different components in the tissue diffract differently, producing the differences in intensity.

In phase microscopy, images exhibit a "halo" as a result of the diffraction of light at the phase-changing annulus (annular ring). The light that has passed straight through the specimen is made to interfere with the light diffracted sideways by it. Only the refracting structures are observed, and these edges and abrupt changes of refractive index produce the halo around the images.

Interference microscopy

A more sensitive and accurate instrument is the interference microscope, which is better adapted to measuring the refractive indices of a specimen. In this microscope, the light splitting and recombining is carried out externally to the specimen. Birefringent plates (doubly refracting) are cemented to the top lens of the condenser and to the front lens of the objective. One set of rays passes through the object, the other set passes through a clear region at one side, and they are then recombined. Any phase differences between them remain constant and can interfere to give light or dark. A refractile object in one beam causes a change in light intensity.

Phase versus interference

Both operate on the same principle—the interference phenomenon, which changes phase difference into amplitude difference. However, the special features of each should be considered in choosing between the two.

Phase: (1) Light passing through the specimen is made to interfere with light diffracted sideways by it. (2) It only shows up diffracting structures in the specimen; produces a halo. (3) Apparatus is simple, reasonable in cost, easy to operate, and can be added to any conventional microscope. (4) It can be used to study living material, cytoplasm, cell inclusions, nucleus, action of physical and chemical agents on living cells. (5) It is adequate for routine examination.

Interference: (1) Light splitting and recombining is carried out on outside of specimen and under control of experimenter. (2) No halo; variations in the optical path through the object are easily interpreted. Phase change can be measured. (3) Apparatus is expensive and complicated; requires constant checking and adjusting. (4) It can be used on living material to determine dry mass: for example, changes in mass during cell activity; protein distribution, both in cytoplasm and nucleus.

Polarizing Microscopy

Closely related to the above types of observation is the use of polarizing attachments. These may be used on most types of microscopes, but for continued use a polarizing microscope is preferable. When a ray of plane-polarized light (vibrating in one plane) falls on the object, it is split into two rays: one obeys laws of refraction, and the other passes through the object with a different velocity. After emerging from the object, the two rays are recombined, but because their velocity is different, they will be out of phase. This phase difference is the quantity measured in a polarizing microscope.

A polarizing prism in the fork substage (instead of a condenser) or a polaroid disc in the condenser slot polarizes the light so that it vibrates in one plane only. Part of the light (ordinary ray) is reflected to the side of the prism or disc and does not illuminate the object; the other part (extraordinary ray) continues straight through the prism to emerge as polarized light. (There is obviously a loss of light, so plenty of it must be used.) If the polarizer is a prism, it usually can be rotated 360° and the amount of rotation in the field checked.

In or above the ocular is fitted another polarizing prism, the analyzer, whose vibration direction is set at 90° to that of the polarizer. The extraordinary ray from the polarizer becomes the ordinary ray in the analyzer and is reflected out of the field, unless an anisotropic substance (doubly refracting when placed between the analyzer and polarizer) rotates the plane of polarization and interferes with the path of light. Such a substance divides the light from the polarizer into two beams, one of which passes through the analyzer, making the object appear to glow against a dark background. Isotropic substances (singly refracting) do not polarize and therefore do not divide the beam of light and do not glow.

Some uses of polarizing microscopy

1. To determine whether an object is isotropic or anisotropic (if it rotates the plane of polarization).
2. To determine differences in physical properties in different directions, such as those occurring in the study of mitotic spindles.
3. To reveal molecular orientation of structures, chemical constitution, chemical and physical intervention in the cell, pressures, or tensions; all can produce anisotropic effects.
4. To determine refractive indices.
5. Can be used on natural and artificial fibers; cellulose fibrils; lamellar plasma differentiation; pseudopodia; spindles and asters; nerve fibers, muscle fibers, chromosomes; chemical and mineral crystals; crystallized hormones and vitamins; dust counts; starch grains; and horn, claw, and bone sections.
6. Can be used on fresh unfixed material.

X-Ray Diffraction

All forms of matter scatter X-rays and form diffraction patterns. From these patterns information concerning the materials can be ascertained. Crystalline proteins show an X-ray diffraction of great complexity; fibrous proteins show less detail. Information about molecular orientation, particle and molecular size, and sometimes detailed molecular structure can be obtained. The image recorded can be related to simple absorption processes and thus be interpreted chemically.

The simplest X-ray method is to place a tissue section in contact with a fine-grained photographic emulsion and expose it. Magnification is obtained in an X-ray projection microscope by separating the sample and film. This requires very fine focusing. One of the principal disadvantages of the method is that vacuum is required; therefore material has to be dry, with absolutely no water present. Freeze-dried tissue can be used.

Uses of X-ray

Tissues with relatively high X-ray absorption are bone and tissues impregnated with heavy metals. The distribution of mineral salts in undecalcified sections of bone can be clearly shown and quantitative information obtained. Substances containing high percentage of elements of high atomic numbers can be injected into blood vessels, or the lymphatic system, and photographed with fine-grain film. Recorded images can be related to simple absorption processes and then be interpreted chemically. By using the oblique incidence of X-rays, it is possible to determine the thickness of bone tissue or nerves.

The determination of dry weight (mass) of cellular structures can be made with X-rays, but the interference microscope can also be employed. If the interference microscope is used, the material can be alive and in fluid. Sometimes it is advantageous to compare both methods as counterchecks on artifacts: either or both may be producing artifacts—the interference microscope because of fluid being present, or the X-ray product because the material is fixed or dried.

Ultraviolet Microscopy

Short waves beyond the visible spectrum have a profound effect upon living matter and are useful for the physical analysis of such matter. Some kinds of living material, under the influence of ultraviolet radiation, visibly radiate, glow, and fluoresce: some kinds remain dark. A microscope for this type of observation has been designed. It has optical glass in the objectives but is equipped with a quartz substage condenser. A quartz lamp condenser is part of the system to help concentrate the maximum amount of near-violet on the specimen. The object itself should not fluoresce but will absorb, partially absorb, or transmit the ultraviolet and thereby reveal structural differences in a photographic image without the use of stains. The specimen must be mounted in a nonfluorescent medium: water, glycerol, or mineral oil. Because glass fluoresces and will reduce contrast and possibly obscure some detail in the specimen, quartz cover glasses and slides should be used for best results, as should fluorescent-free immersion oil.

Resolution is increased over that of the conventional microscope, and differences in structure in the specimen are enhanced by the ultraviolet absorption. The technique is neither complicated nor expensive.

For greater resolution and higher selective absorption, a quartz ultraviolet microscope is better than the conventional ultraviolet microscope. It is, however, more difficult to handle and to focus, and it is more costly. The quartz microscope is equipped with fused quartz objectives, crystalline quartz eyepieces, a quartz substage condenser, and a quartz right angle prism. Quartz slides and cover glasses should be used. The light source must be ultraviolet.

Some cells are immune to ultraviolet, and others are only mildly affected; these can be examined with excellent results under this microscope. But in those that are killed by ultraviolet, formation of artifacts should be expected. Because of the possibility of damage to these cells, focusing can be done with visible radiation and then ultraviolet used only during photography or photoelectric measuring. Tissue cannot be fixed for ultraviolet (this decreases absorption), but ice-solvent or freeze-dry techniques can be used. Ultraviolet photomicrographs are difficult to interpret at times and their interpretation requires considerable experience.

The ultraviolet microscope has become a useful instrument for measuring the selective absorption of cellular components (measured at specific wavelengths). Its chief uses have been for the measurement of cellular concentration and localization of nucleic acids, RNA (ribonucleic acid), and DNA (desoxyribonucleic acid) content, and for observations on normal and neoplastic (tumoring) tissue (in which the nuclear RNA content is high). Some substances can interfere with such observations. Perhaps among the more interesting of these substances are the barbiturates, which absorb heavily and should not be used for sedation or anesthesia in animals to be prepared for this technique.

Fluorescent Microscopy

An object fluoresces when it absorbs ultraviolet light reflected on it or transmitted through it and then emits the energy as visible light of a specific violet, blue, green, yellow, orange, or red color. Secondary fluorescence can be induced by the use of fluorochromes (strongly fluorescent dyes or chemicals) applied to the specimen. Fluorochromes can induce fluorescence in objects that do not fluoresce naturally, and will intensify fluorescence in these that do. As with all ultraviolet preparations, nonfluorescent solutions, mounting media, and oil immersion must be used. Quartz slides and cover glasses may be used but are not essential for the techniques included in this book.

Apparatus required for fluorescent microscopy

Fluorescent microscopes can be purchased from several of the optical companies (Reichert, Zeiss-Winkel, American Optical). In addition to a compound microscope with a focusable condenser, minimum equipment includes a mercury lamp and filter and yellow check filters for each ocular eyepiece. Although some technicians feel that a front-surfaced aluminized mirror should be used, it is not essential. A monocular microscope gives a brighter image than a binocular microscope.

Keep all equipment free from grease, which is opaque to ultraviolet light.

Fluorescent microscopy differs from ultraviolet microscopy in that the fluorescent microscopy observes the specimen directly by innate autofluorescence or by secondary fluorescence induced by fluorochromes, and ultraviolet microscopy records specific absorption of the ultraviolet light by certain structures in the specimen.

Uses for fluorochromes

Alkaline phosphatase (Burstone 1960b)
Amyloid and connective tissue (Vassar and Culling 1959, 1962, Vasquez and Dixon 1956)
Bacteria (Braunstein and Adriano 1961, Carter and Leise 1958, Moody et al. 1958, Yamaguchi and Braunstein 1965)
Blood vessels and lymphatics (Schlegel 1949)
Bone marrow (Werth 1953)
Cancer (Vinegar 1961, von Bertalanffy and von Bertalanffy 1960, Umiker and Pickle 1960)
Chromosomes (Schiffer and Tiju 1962)
Counterstaining (Smith et al. 1959)
Elastin (Shelley 1969)
Fat (Metcalf and Paton 1944)
Fungi (Clark and Hench 1962)
Glycogen (Burns and Neame 1966)

Granules of islets of Langerhans (Hartroft, 1951)
Malarial parasites and other blood parasites (Ingram et al. 1961, Rothstein 1958, Rothstein and Brown 1960)
Motor nerve endings (Cole and Mielcarek 1962)
Mucin (Lev and Stoddard 1969, Sidney 1961)
Mucopolysaccharides (Kuyper 1957, Saunders 1962, 1964, Takeuchi 1962)
Nerve cells (Zeiger et al. 1951)
Nucleic acids and nucleoproteins (Armstrong 1956, Armstrong et al. 1957, de Bruyn et al. 1953)
Rickettsiae (Anderson and Greiff 1964)
Zinc (Sternberg et al. 1964)

See text for other uses.

Electron Microscopy

Among the methods used by investigators in biological and medical fields, one of the newest and fastest growing is electron microscopy. Work on electron microscopes began in the 1930s, and by 1940s electron microscopes had been developed in different parts of the world. Knoel and Ruska in Germany had a practical marketable research scope by 1939. England began work in 1936, and RCA in the United States began the development of one in 1938 and had one on the market in 1941. The electron microscope has a higher resolving power than the light microscope and reveals submicroscopic structures that were previously invisible. It uses electrons as the illuminating beam, and focuses the beam on the object by the use of magnets. As the beam passes through the object under observation, some of the electrons are scattered by the object, causing shadows of the scattered electrons on photographic film.

Electron beams must be handled in vacuum, therefore no wet or living tissue ordinarily can be observed. Sections must be thin (0.01–0.05μ) to prevent appreciable electron absorption, because the image depends on the differential scattering rather than absorption. The specimen is supported on a metal grid and is placed in the column at the electron-gun side of the object-lens aperture.

Special fixation, embedding, and sectioning are required to prepare ultrathin sections for use in the electron microscope. To facilitate the scattering of electrons, the density of the specimen must be increased. Because the usual fixing solutions do not meet this requirement, the so-called "electron stains" must be used in their place. These are compounds of heavy metals (tungsten and osmium), such as phosphotungstic acid and osmic acid. Potassium permanganate (Luft 1956), acrolein (Luft 1959), and more recently, aldehydes (Sabatini et al. 1963, 1964) have been advocated. Only plastic embedding permits ultrathin sectioning. Conventional microtomes have been modified for thin sections, but specially designed microtomes are available. Glass or diamond knives are essential for ultrathin sectioning; steel knives are rapidly dulled by cutting plastic.

Three-dimensional images may be obtained by metal shadowing, and this is finding favor with many researchers. Metals, such as silver, chromium, palladium, platinum, and others are evaporated on the specimen at a small angle. A thin film of the metal covers the specimen except for a "shadow" caused by the specimen being in the way of passage of the metal. From the shape and dimension of the shadow, the height and contour of the specimen can be observed. Of the many references, important ones are Kay (1965), Mercer and Birbeck (1961), Parsons (1970), Pease (1964), and Schultze (1969).

Further references for microscopy are as follows: Barer (1956, 1959); Beck (1938), Belling (1930), Bennett (1950), Cosslett and Nixon (1960), Davies (1958), Dempster (1944a and b), Engström (1956, 1959), Gage (1943), Ham (1957), Munz and Charipper (1943), Needham (1958), Nurnberger (1955), Oster (1955), Popper and Szanto (1950), Richards (1954), Ruch (1955), Scott (1955), Shillaber (1944), Vickers (1956), and Walker (1958).

CHAPTER 10

Stains and Staining Action

Because most tissues do not retain enough color after processing to make their components visible under a bright-field microscope, it is expedient to add colors to tissues by staining them. Correctly chosen stains aid in identifying tissues and their elements, and in diagnosing pathological conditions. Knowledge about the structure and action of the chemicals and stains used for tissue identification must be reviewed and understood. The subject is extensive, and the student is advised to consult specialized references for detail that cannot be included here. One of the best is H. J. Conn's *Biological Stains* (1969).

NATURAL DYES

Cochineal and Carmine

These are members of a group of dyes called "natural" dyes. Unlike other natural dyes, cochineal and carmine are derived from an animal source

—a minute insect, *Coccus cacti,* which lives on spineless cacti. The dye is present as a purple sap in the females, which are harvested, dried, and pulverized to produce cochineal. This dye by itself has little affinity for tissue unless iron, aluminum, or some other metal is present. With the salt of one of these metals as a mordant (see page 125), staining will result. Alum cochineal, a commonly used form of this dye with mordant, is an efficient nuclear stain.

The dye carmine is derived from cochineal by boiling the cochineal with a salt, usually alum, to produce a precipitate. This precipitate is insoluble in water, and before it can be used as a stain it must be converted into a soluble compound such as ammoniacal carmine or aceto-carmine, in a process that will be described in the section on mordants (p. 125).

Hematoxylin

In many respects, hematoxylin can be regarded as the most important of the natural dyes. It was one of the first histological dyes and remains one of the most widely known and used dyes. Hematoxylin is extracted from the heartwood of logwood trees from South and Central America and the West Indies. The tree is *Caesalpina campechianum,* one of the legumes similar to acacia or cassia trees. The crude material is exported as logs, as chips, or as dried aqueous extract of the heartwood. It is then extracted with ether in a continuous extraction apparatus evaporated to dryness, dissolved in water, filtered, and crystallized out of solution. All of these steps are slow and difficult to handle and require costly apparatus, thus making hematoxylin one of our most expensive dyes.

In the crystallized condition it is not yet a dye, and its color must be allowed to develop by oxidation into hematein (color acid—no relation to hematin, the colored constituent of red blood cells). Oxidation may be accomplished in either of two ways: "naturally" (a slow process of exposure to air for 3 to 6 weeks, as in Heidenhain hematoxylin) or artificially by the use of mercuric oxide, hydrogen peroxide, or another oxidizing agent (a more rapid process, as in Harris hematoxylin). Used alone, hematein is only a weak and diffuse dye with little affinity for tissues. A weak acid will not combine with nuclear elements in sufficient quantity to produce efficient staining. Some form of mordanting is required to form a base from this dye, which will then stain the acidic nuclear elements. The most commonly used mordants are alum salts of aluminum, potassium, or iron.

When properly oxidized hematoxylin is an exceedingly powerful dye with various shades of staining from purples, through blues, and into blue-blacks. The iron-mordanted form is one of the most valuable dyes for mitotic study, and it gives to the chromatin a precise black or blue-black color. This black color is the result of the presence in hematoxylin of some tannin, and the latter in combination with iron salts produces a lasting black color.

Hematoxylin staining is discussed further in chapter 12.

Other Natural Dyes

Other natural dyes are saffron from stigmas of *Crocus;* indigo from plants of the genus *Indigofera;* berberine from barberry; orcein and litmus from the lichens *Lecanora* and *Rocella;* and brazilin, chiefly from a few species of *Caesalpina,* a tropicopolitan genus of leguminous trees (Mohr 1950). Orcein, a specific dye for elastin (present in elastic fibers), is prepared by boiling the plants in water. The lecanoric acid in them splits to produce orcinol—a resorcinol with a methyl group attached to it. Orcinol with NH_3 (ammonia) and atmospheric oxygen forms orcein.

MORDANTS

Metallic mordants have been used for over one hundred years, but the exact nature of their action still is somewhat mysterious.

A mordant is a metallic salt or hydroxide that combines with a dye radicle to form an insoluble compound. Such a compound is called a lake. For example, carmine, dissolved in a solution of aluminum sulfate, becomes positively charged and acts as a highly basic dye. The term *mordant* should not be used for all substances that increase staining action, but only for salts and hydroxides of di- and trivalent metals.

The use of mordants in staining tissues has many advantages. Once the mordant-dye compound has formed in the tissue, the dye is relatively permanent, is insoluble in neutral solutions, and can be followed by many other forms of staining. Dehydration will not decolorize the tissue. There is considerable flexibility in mordant-dye procedures, as the mordant can be used before the dye, together with the dye, or following the dye (rare).

Carmine and hematoxylin are two dyes often used with mordants for staining tissue. For these dyes, the mordants commonly used are aluminum, ferric and chromium salts, and alums (potassium alum, ammonium alum, iron alum, and chrome alum). A solution of ferric chloride can be used in place of iron alum, and Cole (1933) recommends the use of a phosphate-ripened hematoxylin with it. The ferric chloride causes the tissues to stain more rapidly and more intensely than does the iron alum. Iron alum solutions are quite acid: the greater the amount of iron alum present, the more acid the solution. Adding even a weak acid such as acetic acid to an alum hematoxylin makes the solution more selective for nuclei.

For long-lasting solutions of combined mordant and dye, mordants with little or no oxidizing action must be used: ammonium alum, potassium alum, phosphotungstic acid, phosphomolybdic acid, and iron alum-ferrous sulfate. If a long life is not required it is possible to use mordants with a vigorous oxidizing action: ferric chloride, ferric acetate, and ferric alum (Cole 1943). Since the usefulness of these solutions lasts only a matter of hours, they must be prepared immediately before use. In iron-hematoxylin solutions that are premixed, a high mordant-to-dye ratio tends to produce nuclear staining, while a high dye-to-mordant ratio favors myelin staining. Though the term iron hematoxylin is widely used, it is actually chemically incorrect; hematoxylin cannot form a lake, but first must be oxidized to hematein (Puchtler and Sweat 1964).

Hematein forms blue-green lakes with copper, lilac with nickel, red with tin, dark brown with lead, and purple with bismuth.

When using two separate solutions, a mordant of any kind can be used if followed by a well-ripened hematoxylin solution. If the solution is unripened, the mordant should include a substance of considerable oxidizing power, a ferric or chromium salt. The value of two solutions lies in the fact that the dye can be preceded by a salt, which cannot be used in combination with the dye in a single solution. For example, ferric chloride, when added to an ammonia-ripened hematoxylin, will throw down a precipitate of ferric hydroxide. Double mordanting can be profitable; a mordant followed by a solution of hematoxylin containing a mordant gives excellent results. The mordants for separate use are ammonium or potassium alum, ferric ammonium alum (2–3 drops of HCl increases contrasts), and ferric chloride (HCl increases contrast). They yield the following colors with hematoxylin: ammonium alum, bright blue nuclei; potassium alum, lilac or violet; chrome alum, cold gray blue; iron alum, blue to black (Cole 1943).

When used with hematoxylin, some mordants are more effective than others for staining particular types of material. The following is a list of tissue elements and the metals effectively used on them (Mallory 1938).

Nuclei: aluminum, iron, tungsten
Myelin sheaths: chromium, iron, copper

Elastic fibers: iron
Collagen: molybdenum
Fibroglia, myoglia, neuroglia, epithelial fibers: tungsten
Axis cylinders: lead
Mucin: iron
Fibrin: tungsten

Progressive and Regressive Staining

Lakes, formed by mordant and dye, can be used progressively, but when mordant and dye are used separately, regressive staining usually is more effective. In progressive staining, the stain is added to the tissue until the correct depth of color is reached. In regressive staining, the sections are overstained, and excessive amounts are removed by one of the following methods.

Method I: Excess Mordant

With an excess of free mordant present outside of the tissue, the mordant-dye complex in the tissue is broken up, and, since the amount of mordant in the tissue is smaller than that in the differentiating fluid, the dye moves out of the tissue into the fluid. (If the tissue is left long enough in the fluid, it is conceivable that most of the dye could move out of it into the excess mordant, causing the sections to become colorless.) Because the nuclei hold considerably more dye than the cytoplasm, most of the dye is lost from the cytoplasm, but some still remains in the nuclei. When enough dye has been extracted from the cytoplasm, and ·when the correct intensity is left in the nuclei, remove the slides from the mordant and wash them thoroughly, usually in running water, to remove excess mordant. Remaining traces can cause the stain to fade in time.

Method II: Acids

Acids are effective differentiators for some dyes, but there is no completely adequate explanation for their action.

Method III: Oxidizers

Oxidizers furnish a third method of regressive staining, and by this method the dye can be oxidized to a colorless condition. Oxidizers are slow in action; the parts of the cell that hold only a small amount of dye will be bleached before those containing greater quantities of it. A complete expl...ation for this is not available. Picric acid, a commonly used oxidizer, has both a moderate oxidizing and a weak acidic action.

Accentuators and Accelerators

Not to be confused with mordants are accentuators and accelerators. Accentuators are substances that, contrary to the action of mordants, do not become a part of the dye complex or lake. Instead they increase the selectivity or stainability of the dye e.g., phenol in carbolfuchsine. Accelerators, as their name implies, accelerate the action of the dye— usually of importance in silver impregnation—e.g., chloral hydrate.

SYNTHETIC DYES

Natural dyes had no competition until the middle of the nineteenth century when William Perkin worked out the processes for making aniline or coal-tar dyes. These were the first synthetic dyes.

Synthetic dyes, like natural dyes, can be used either progressively or regressively. An acid solution is often used to remove excess basic dye; an alkaline solution is used to remove excess acidic dye. In some cases, alcohol can be used as a differentiator, particularly for basic dyes; but in general a sharper differentiation is achieved by using an acid.

The real importance of synthetic dyes lies in their use for double and triple staining—the use of two or more stains on the same slide. The dyes, if properly chosen, will stain histologically; that is, each dye, because of a known specificity, will stain only specific parts of the cells. This type of staining has obvious advantages. Owing to their chemical nature, synthetic dyes make this kind of staining possible. If the dyes are synthesized, their formula can be controlled, and the significant part of the dye is either anionic (acid) or cationic (basic) in action. Actually, dye powder as purchased is a salt, but the salts of the so-called basic dyes give up OH^- ions and act as cations, and the acidic dyes give up H^+ ions and act as anions. Therefore an acid, or anionic, dye is the salt of a color acid, usually a sodium salt; a basic, or cationic, dye is the salt of a color base, usually a chloride. (See structure of synthetic dyes, page 130.) The terms "anionic" and "cationic" are more appropriate than "acidic" and "basic" and are becoming more widely used. Basic dyes have an affinity for nuclei, which are basophilic (readily stained by basic dyes), and acidic dyes have an affinity for the cytoplasm, which is acidophilic (readily stained by acidic dyes).

When the sodium salt of an acid dye and the chloride of a basic dye are mixed, there is a tendency for ions to interchange. The so-called neutral dyes are formed and give results differing from those obtained with ordinary double staining using separate acid and base dyes. Polychroming is a process in which a dye forms other dyes spontaneously—the basis of the development of modern blood staining (Romanovsky stains). Methylene blue in solution is oxidized into one or more compounds of lower methylation. Therefore, all methylene blue solutions, on standing, contain lower homologs, primarily azure A and B, and the methylene blue is now called polychrome methylene blue. The azures are more violet in color and more selective in their action than unpolychromed methylene blue and account for the differential action of blood stains in the differentiation of white blood cells. When eosin is added, it enters into chemical combination with the above basic dyes to form an insoluble precipitate.

The dyes are not stable in aqueous solutions and precipitate out; the dye powder must therefore be dissolved in alcohol. Since alcoholic solutions do not stain well, different methods of using them have been employed in the Romanovsky stains. Sometimes they are used immediately after mixing (Giemsa) or, after applying the alcoholic solution, the stain is diluted with water (Wright). How the stains act is uncertain. Probably certain parts of the cells have an affinity for the neutral stain, others have an affinity for the basic dyes and break up the stain to acquire the basic part, and other cell parts do the same with the acidic portion. These actions by the cell parts are termed, respectively, neutrophilic, basophilic, and oxyphilic. The cells are stained in a solution of the proper *p*H, or they are overstained in a neutral solution and differentiated to the proper coloration of their parts (the Maximow method uses alcohol, the Giemsa method dilute acetic acid, and the Wolbach method rosin).

Polychroming must not be confused with another type of staining, metachroming, in which certain substances are stained in one color and others in another color by the same dye. The definition of metachromasia given by Paddy (1970) is "a characteristic reversible color change that any dye may undergo by virtue of a change in its environment not involving chemical reaction of the dye."

The explanation for the reaction to the dye Thionin may be as follows: Thionin stains chromatin blue; it stains mucus, ground substance of cartilage, and granules of mast cells red. The dye seems to exist in aqueous solution in two forms: (1) the normal color, blue; and (2) the metachromatic color, red. Both forms are always present, but the red is in a polymerized form of the blue, and color change may be due to some interaction between the chromophores present. The blue form is favored by an increase of temperature, a lowering of *p*H, a decrease of dye concentration, or the addition of salts, alcohol, or acetone. The red form is favored by a decrease of temperature, a raising of *p*H, or an increase in concentration of dye (Bergeron and Singer 1958). Also, sulfuric esters of high molecular weight and their salts increase the production of the red form. Mucin, the ground substance of cartilage, and the granules of mast cells contain substances of this nature and therefore take up the metachromatic form.

Some other metachromatic stains are methyl violet, brilliant cresyl blue, azure A and B, and toluidine blue.

The majority of dyes do not stain metachromatically, but are orthochromatic in action. This means that their action is direct and predictable under normal conditions; if it is a blue dye, it stains blue; if it is a green dye, it stains green; and so forth.

Some dyes are not stable in solution and will gradually give rise to other colored agents, which should be considered impurities. The presence of such agents is called allochromasia. Orthochromatic dyes may give rise to metachromatic impurities by allochromasy. Nile blue solutions, an example of allochromasia, contain cation of the dye (blue), anion of the dye (sulfate), Nile red (red or rose), and imino base of Nile blue (orange yellow).

A chromotrope is a substance that can alter the color of a metachromatic dye. Examples of chromotropic tissue substances are the matrix of cartilage, granules of mast cells, and secretions of some mucus glands. Bathochromic (Baker 1958) means a shift in the peak of the absorption curve of a dye in solution toward longer wavelengths. Hypsochromic means a shift toward the shorter wavelengths.

Structure of Synthetic Dyes

The synthetic (coal tar or aniline) dyes are derivatives of benzene, all built on the benzene ring.

There are an infinite number of ways in which this ring can combine with other radicals to produce compounds of various degrees of complexity. Certain chemical groups, called chromophores ($C=C$, $C=O$, $C=S$, $C=N$, $N=N$, $N=O$, and NO_2) are associated with color. When these groups are attached to a benzene derivative, the compound acquires the property of color and is known as a chromogen. A chromogen, how-ever, is not yet a dye. It has no affinity for tissues, will coat them only mechanically, and can be easily removed by mechanical means. The compound must also contain a group that gives it the ability to form a union with some tissue end group, either directly or through chelate ac-tion of a mordant metal. The commonest type group is one which permits electrolytic dissociation (the formation of anions and cations). This auxiliary group, known as an auxochrome, furnishes combining properties to the compound.

A simple demonstration of the principle of synthetic dye formation is picric acid. The nitro group ($-NO_2$) is a chromophore, and three of these can displace three hydrogen atoms on the benzene ring to produce tri-nitrobenzene.

$$O_2N \overset{NO_2}{\underset{NO_2}{\bigcirc}}$$

This is yellow, a chromogen, but not yet a dye. Replace one or more H^+ with an OH^-, the auxochrome, and produce dissociable picric acid.

$$O_2N \overset{NO_2}{\underset{NO_2}{\bigcirc}} OH$$

Since the auxochromes are the salt-forming groups, they usually deter-mine the action of the dye; the more basic or acidic groups present, the more basic or acidic is the action of the dye. If both a basic and an acidic group are present, the basic group predominates, but is weakened in action by the acidic group.

Most commercial dyes are sold as salt. An acid dye is the salt of a color acid, usually sodium salt; the basic dye is the salt of a color base, usually a chloride. The basic chromophores are: azo group ($-N=N-$), azin group $\left(\begin{smallmatrix} N \\ \bigcirc \\ N \end{smallmatrix} \right)$ and indamine group ($-N=$). The acid chromo-phores are: nitro group ($-NO_2$) and quinoid benzene ring,

Simpler compounds can be converted into more complex ones by substituting radicals for the H^+ atoms present. As more H^+ atoms are replaced by radicals, the deeper the color becomes. The simplest dyes tend to be yellow. As methyls or other groups are added, the colors pass through reds, violets, blues, and greens. Using ethyl groups instead of methyl groups deepens the color.

Nomenclature of stains has no absolute conformity. For the names of some stains, the color is used (orange G, Martius yellow); for others, a chemical term (methyl green). If the term is followed by a letter or numeral (Sudan III, IV, ponceau 2R, 4R, one dye is being distinguished from a related one. B indicates a more bluish color, Y or G a more yellowish color; WS means water-soluble; A, B, and C distinguish among azures.

NATURE OF STAINING ACTION

Biologists and biochemists disagree about the nature of staining action—is it chemical, physical, or a combination of both? If chemical, this can mean that some parts of the cells are acid and others alkaline, the acid parts tending to combine with cations and the alkaline with anions. Absorption and diffusion of the dye occur, penetrating the cellular elements and combining with them. This action can be combined with physical action, where there is an adsorption of the dye, an attraction of plus and minus charges for each other, and a condensation of the dye on the surface of the cell parts. Minute particles of the dye are deposited on the surface of the tissue by selective adsorption and then enter into combination with the tissue. The proteins, nucleic acids, and other components of the protoplasm proceed along lines of chemical laws by exchanging ions. Stearn and Stearn (1929, 1930) maintain that the confusion lies in the term "adsorption," that it may be either a chemical or a physical force, and that the adsorbent can form ions and then proceed along chemical lines. The development of physical chemistry has helped to resolve the question. There appears to be no doubt that both physical and chemical factors take part in the mechanism of salt formation—the basic action of stains.

In any case, the staining properties depend on three factors:

1. Strength of dye.
2. Rate of ionization of tissue proteins and dyes.
3. pH value of dye solution and tissue proteins.

In addition, staining can be affected by other conditions:

1. Alcoholic or aqueous solution of dye.
2. Low or high temperature during reaction.
3. Simple or multiple combinations of dyes.
4. Strong or weak concentration of dye in solution.
5. Permeability of tissues and dyes.

STANDARDIZATION OF STAINS

In the early days of tissue staining, it was difficult to secure reliable dyes. The textile dye industry was the sole source of dyes, and products received were often unsatisfactory. There was no adequate standardization of color, and no standardization of chemical content was undertaken. There were variations in color and quality and many impurities in the dyes. Grübler in Germany, the first to try to standardize dyes, built a highly specialized business in this field. He did not actually manufacture dyes, but bought up batches from other firms and tested them for technical use. After the beginning of the twentieth century, certain events changed Grübler's hold on the business of standardization. Perhaps of greatest influence were the two world wars, causing Germany to lose its monopoly in the dye industry. No country could afford to remain dependent on another country for its source of dye.

The lack of German dyes led to a new form of standardization in the United States. A body called the Commission on Standardization of Biological Stains was organized, and it later became the Biological Stain Commission. The object of this commission is to work with the manufacturers, showing them what the biologists require, testing their products, and permitting approved batches to be put on the market with the stamp of the commission on them. Specifications of the most important dyes have been drawn up by the commission. The specifications are partly chemical and partly spectrophotometric and contain detailed instructions for testing the dyes and the results to be expected from the tests.

Batches of dyes approved by the commission bear a special label furnished by the commission and known as a certification label. On it is a C.I. number (Color Index number) indicating the certification number of that particular batch. This number means that (1) a sample of that batch was submitted to the commission for testing, and a portion of it is on file; (2) the sample proved true to type by spectrophotometric tests; (3)

its dye content met specifications and is correctly indicated on the label; (4) it was tested by experts in the procedure named on the label and found satisfactory; (5) no other batch can be sold under the same certification number. Any description of the use of the stain should be followed by its C.I. number or the letters C.C., indicating that it is Commission Certified. C.I. numbers used in this text are from the new revised list.

The actual dye content of commercial dyes may vary from 32% to 99% and some batches may contain colored compounds with staining properties different from those of the actual dyes. Even different batches of a dye can vary. Rosenthal et al. (1965) developed a simple paper chromatography method for testing dyes that requires only simple laboratory equipment.

Unless otherwise specified, dyes can be purchased from most of the scientific supply houses with chemical outlets: Matheson Coleman and Bell Co., Hartman-Leddon Co., Fisher Scientific Co., Allied Chemical Corporation (National Aniline Division), and others.

Foreign stains can be obtained from the following sources:

Esbe Laboratory Supplies, Toronto, Canada, outlet for Gurr Stains.
Edward Gurr, London, England.
G. T. Gurr, London, England.
Pfatz and Bauer, Inc., Empire State Building, N.Y., N.Y.
Roboz Surgical Instrument Company, Washington, D.C. [American agent for Chroma-Gesellschaft (Dr. K. Holborn and Söhne, G. Grübler)].

For greater detail concerning stains and their actions, read: Baker (1945, 1958), Cole (1933, 1943), Conn (1946, 1948, 1961, 1969), Gurr (1960), Holmes (1929), Singer (1952), and Stearn and Stearn (1929, 1930).

CHAPTER **11**

Mounting and Staining Procedures

Certain standard principles usually apply to the staining of tissue on slides, but there are many exceptions and variations.

Paraffin sections first must be deparaffinized, because most stains are applied in either aqueous or alcoholic solutions and will not penetrate efficiently through paraffin-infiltrated tissues. The customary solvent for paraffin is xylene. This is followed by the removal of the xylene with absolute alcohol because stains rarely can be applied successfully in a xylene medium. After the removal of the xylene, the general rule is to transfer the slides to a medium comparable to the solvent of the dye being used. That is, if the dye is a water solution, the slides are hydrated through a series of decreasing alcoholic and increasing aqueous dilutions, such as 95%, 80%, 70%, and 50% alcohol, and finally they go into water. If the dye is dissolved in a 50% alcoholic solution, then the slides are carried only to 50% alcohol before going into the staining solution.

During the hydration process, undesirable pigments or other materials (mercuric chloride crystals, formalin pigment, etc.) are removed and the slides are washed well to remove the reagent. Counterstains (background color) or other special treatments are applied in their proper sequence,

FIGURE 17. *A staining coil carrying multiple slides.*

to allow each dye or chemical to maintain its specific effect on the tissue elements. Improper sequences of staining, decolorizing, or other solutions can result in a poorly stained slide. Hematoxylin-eosin staining (page 156) is a simple example: if the eosin (counterstain) is applied before the hematoxylin (nuclear) stain, the eosin stain will be completely removed by the action of the hematoxylin stain.

MECHANICAL AIDS

If slides are to be transported in quantity, rather than individually, several types of holders on the market are useful. One type holds 50 slides;[1] one holds six slides.[2] Another model [3] holds five slides and fits as a cap on coplin jars. Some are baskets; others are clips that fit into special staining dishes. The tissue-processing machines are equipped with slide carriers to fit the instrument. Phosphor-bronze spring wire, 0.05 inch diameter, can be fashioned into coils of ¾ inch in diameter and cut into any length to hold from 3 to 20 slides (Figure 17). This combination of slides and coil can be used in rectangular staining dishes with the slides resting on their long edges, or standing upright in tall stender dishes.

Cover glass mounts can be stained by using Chen's (1942) staining

FIGURE 18. *T. T. Chen's cover glass staining rack in a tall slender dish. Handle for rack lies beside dish.*

rack[4] (Figure 18), which holds 12 22-mm cover glasses, or racks[4] that hold either 5 or 30 cover glasses. Cover-glass staining dishes are far less convenient to use than racks and lead to breakage and wasted time in handling; each cover glass has to be processed individually with forceps.

PROCESSING SLIDES FOR MOUNTING

A final processing of slides is necessary to make permanent preparations for examination and storage without deterioration. All alcohol and water must be extracted (with certain exceptions) and a medium applied that maintains the tissues in a clear and transparent condition, does not alter the color or intensity of the stains, and holds a cover glass in place. The water is removed through increasing concentration of alcohol until absolute alcohol is reached. The final reagent is xylene (or a similar solvent), which removes the alcohol and makes the sections lose their opacity and become clear. Finally a mounting medium (mountant) is applied and the cover glass lowered into place, completely covering the sections. (The solvent of the mounting medium is usually either toluene or xylene.)

---

[1] Arthur H. Thomas Co.   [2] Ward's Natural Science Establishment.
[3] Peel-A-Way Scientific Co., Long Beach, Calif.   [4] Lipshaw Co.

COVER GLASS MOUNTING

The tidiest method for mounting a cover glass is shown in Figure 19: apply a thin streak of medium on the cover glass; turn cover glass over and rest it on edge on the slide, beside the sections; ease cover glass into place slowly to allow air to escape from under cover glass; press firmly from center outward to distribute the medium evenly. An alternate method is to apply mounting medium along one edge of sections on slide; rest one edge of cover glass adjacent to mounting medium; lower gradually to ease out air without bubble formation; press gently in place.

If, during microscopic examination of stained and mounted slides, dull black spots replace nuclear detail (Figure 21, b) the clearing solution was partly evaporated out of the sections before the cover glass was in place. Return such slides to xylene, dissolve all mounting medium to

FIGURE 19. *Mounting a cover glass:* (a) *placing mountant on cover glass,* (b) *turning cover over, and* (c) *lowering it onto slide.*

a
b

FIGURE 20. *Mercuric chloride crystals: (a) in paraffin sections as seen under microscope before deparaffinization; (b) in stained and cleared sections.*

a
b

FIGURE 21. *Common imperfections in sections: (a) knife scratch in paraffin section as seen under microscope before deparaffinization; (b) black areas caused by drying of sections before cover glass was applied.*

allow the air to leave the sections, and remount. If the slides appear dull, almost milky, instead of crystal clear, water is present. Remove all mounting medium in xylene and return slides to absolute alcohol, preferably a fresh solution, clear, and remount.

MOUNTING MEDIA (MOUNTANTS)

Natural Resins

Formerly, natural resins were used as mounting media: Canada balsam, composed of terpenes, carboxylic acid, and esters; gum damar, composed of unsaturated resin acids and a little ester; and gum sandarac, an unsaturated acid resin. These dried slowly, were variable in composition, and unpredictable in behavior. Some developed acidity, faded stains, and would turn yellow and crack after a few years. One still used, particularly for blood smears, is Euparal (refractive index 1.483), a eutectic (melts at low temperature) mixture of oil of eucalyptus, gum sandarac, salol, paraldehyde, menthol, and camphor. Euparal Vert (green because of copper salt content) is claimed to intensify hematoxylin stains. Dried smears may be mounted directly in Euparal, but sections must be carried into 95% alcohol and then into Euparal Essence before mounting. (If alcohol is carried into the Euparal, sections will fade.) Drain off excess essence and apply Euparal.

Synthetic Resins

The synthetic resins now available are superior to natural resins in most respects; their composition can be controlled; they are stable and inert; they dissolve readily in xylene or toluene, do not require long drying, and adhere tightly to glass; and they have the correct refractive indices, are pale in color, and do not yellow with age. The resins should have refractive indices of 1.53 to 1.54 or better (Lillie et al. 1953).

The most widely used synthetics are the β-pinene polymers (terpene resins), such as Permount; Piccolyte, refractive index 1.51 to 1.52; HRS (Harleco Synthetic Resin), refractive index 1.5202; Biloid Synthetic Resin, refractive index 1.5396; Kleermount; Histoclad; Pro-Texx; Technicon Mounting Medium; Namount; and Preservaslide.[5] There is no reason

---

[5] These products are manufactured by the following companies, respectively: Fisher Scientific Co., General Biological Supply Co., Hartman-Leddon Co., Will Corp., Carolina Biological Supply Co., Clay Adams Co., Scientific Products, Technicon Co., Allied Chemical Corp., and Matheson Coleman and Bell.

to recommend one of the above products over the others; all are equally efficient. They are soluble in xylene, toluene, aromatic hydrocarbon solvents, and in chlorinated hydrocarbons such as chloroform, but not in dioxane.

Lillie et al. (1950) recommend Lustron 2020 [6] for good preservation of Prussian blue mounts, which tend to fade in some resins. Lustron is a polystyrene, refractive index 1.59, and is water resistant and soluble in varying degrees in aromatic and chlorinated hydrocarbons. Lillie et al. use it in diethylbenzene as solvent. I have had excellent results using toluene or xylene without the formation of air bays as the xylene evaporated. If air bays do form, diethylbenzene is recommended; Lillie et al. suggest adding 5 ml dibutylphthalate to 70 ml xylene and 25 g polystyrene. This does not delay hardening.

There are also the German synthetics, Caedax, Eukitt and Rhenohistol, and the British synthetics, Xam, Cristalite and Clearmount. These are similar in most respects to American products.

Most synthetic mountants are allowed to air dry, but, if quick drying is a must, try this method of Manikas and Umiker (1959): After slides are covered, place them on metal trays in a drying oven, 160°–170°C: 3 minutes. Remove and chill in freezing compartment of refrigerator: 2 minutes. Slides can be cleaned, marked with ink, and stored without dislodging cover glasses. Leaving synthetic resin mounts on a slide warmer overnight will make them relatively firm.

Aqueous Mounting Media

Aqueous mounting media are indispensable for the preservation of tissue elements that are soluble in alcohol or hydrocarbons or are demonstrated by the use of dyes soluble in these fluids. Nearly all media that have been proposed use (1) gelatin and gum arabic as soldifying agents with water, (2) sugars and salts for increasing the refractive index, and (3) glycerols and glycine as plasticizing agents. Gum arabic slowly hardens by drying, gelatin sets by cooling, and the glycerine keeps the tissues from cracking or overdrying. The addition of phenol, thymol, merthiolate, or Zephiran prevents mold growth.

Kaiser Glycerine Jelly

water	52.0 ml
gelatin	8.0 g

---

[6] Monsanto Chemical Corp.

glycerine .... 50.0 ml

Add as a preservative either (1) carbolic acid
(phenol), 0.1 g; (2) merthiolate, 0.01 g; or
(3) Zephiran 0.1 ml.

Allow gelatin to soak for 1–2 hours in water, and add glycerine and
preservative. Warm for 10–15 minutes (not above 75°C) and stir until
mixture is homogeneous. This keeps well in covered jar in refrigerator.
If heated above 75°C, the gelatin may be transformed into metagelatin,
which will not harden at room temperature.

Von Apathy Gum Syrup

gum arabic (acacia) .... 50.0 g
sucrose (cane sugar) .... 50.0 g
distilled water .... 50.0 ml
formalin .... 1.0 ml

Dissolve lumps of gum arabic in warm water; add sucrose. When dissolved, filter and allow to cool. Add formalin. Do not use powdered
form of gum; it makes a milky solution, whereas the lump form produces
a clear medium. The cover glass does not have to be sealed as it does
with Kaiser glycerine jelly above.

Farrants Medium

distilled water .... 40.0 ml
gum arabic (acacia) .... 40.0 g
glycerine .... 20.0 ml
carbolic acid (phenol) .... 0.1 g
   (or merthiolate or Zephiran)

This solution can be purchased from Amend Drug and Chemical Company, New York, N.Y.

Gray and Wess Medium (PVA) (1952)

PVA (polyvinyl alcohol) 71–24 powder[7] ...... 2.0 g
70% acetone .... 7.0 ml
glycerine .... 5.0 ml
lactic acid .... 5.0 ml
distilled water .... 10.0 ml

---

[7] E. I. du Pont de Nemours and Co.

Make a paste of the dry alcohol with acetone. Mix half of water with glycerine and lactic acid; stir into paste. Add rest of water drop by drop while stirring. Solution will be cloudy but become transparent if warmed in water bath about 10 minutes.

Hoyer Medium (Beek 1955)

gum arabic	30.0 g
glycerine	16.0 ml
chloral hydrate	200.0 g
distilled water	50.0 ml

Dissolve gum in water; a little heat helps. Add chloral hydrate, then glycerine.

PVA Mounting Medium (polyvinyl alcohol)

PVA 71–24 powder	15.0 g
water	100.0 ml

Add PVA powder slowly to cold water. Heat in water bath (80°C) and stir until the solution becomes as viscous as thick molasses. Filter out undissolved lumps through two layers of cheesecloth. Solution will appear milky, but after standing for several hours, it will clear. Spread a thin film over stained blood smears when cover glasses are undesirable.

The following solutions are recommended for small whole mounts, insects, worms, tiny invertebrates, etc.

Berlese Mounting Medium

Gray formula (1952)

water	10.0 ml
glacial acetic acid	3.0 ml
dextrose syrup	5.0 ml
gum arabic (acacia)	8.0 g
chloral hydrate	75.0 g

Mix water with acid and syrup; dissolve gum in this. This requires a week or more. Stir at intervals. When solution is complete, add chloral hydrate. This is one of the best media for insects.

Thick formula

distilled water	40.0 ml
saturated aqueous solution of chloral hydrate	30.0 ml
gum arabic (acacia)	25.0 g
glycerine	5.0 ml

Dissolve gum arabic in water, add chloral hydrate and then glycerine. Filter.

Thin formula

distilled water	100.0 ml
gum arabic (acacia)	60.0 g
glycerine	40.0 ml
saturated aqueous solution of chloral hydrate	100.0 ml

Monk Karo Medium (1938)

Karo syrup, white	5.0 ml
Certo (fruit pectin)	5.0 ml
water	3.0 ml
thymol for preservative	

This is recommended for mounting a group of small parts that are easily disarranged. Spread a thin layer on a slide; remove parts to be mounted from glycerine and immediately arrange them in the medium. The mixture begins to set in about 2 minutes and holds parts in place. Dry to hardness over heat. Cover glass can be added with a drop of mixture or with Euparal. If using mixture, use enough to prevent air pockets from forming when the medium dries.

Lactophenol Mounting Medium

melted phenol (carbolic acid)	3 parts
lactic acid	1 part
glycerine	2 parts
distilled water	1 part

Turtox Mounting Media

General Biological Supply House offers nonresinous mounting media that permit the mounting of either living or preserved organisms from water or alcohol. Small specimens containing little air can be mounted directly, but larger ones must be mounted from water or alcohol. The CMC series (9, 9AB, 9AF and 10), refractive index 1.38 to 1.39, are fast-drying media used primarily for mounting small arthropods and diptera, such as fleas, ticks, mites, lice, and mosquito adults, larvae, and pupae. The CMCP series (9, 9AB, 9AF and 10) has a higher refractive index, 1.40 to 1.41, a more rapid killing and relaxing action than the CMC series, and can be used for mounting parasites and free-living worms.

Yetwin Mounting Medium for Nematodes and Ova (1944)

10% bacto-gelatin, granular, Difco, aqueous ...	150.0 ml
glycerine	50.0 ml
1% chromium potassium sulfate aqueous (chrome alum)	100.0 ml
phenol (carbolic acid), melted	1.0 ml

Dissolve gelatin in boiling water, add glycerine. After mixing, add chrome alum solution and phenol. The medium liquefies in 15 minutes at 65°C. Organisms may be transferred from glycerine or formalin directly into medium. The gelatin hardens to form a permanent mount.

Abopon

Abopon[8] is water-miscible and can replace glycerine jelly in several special staining techniques. It can be purchased in the form of lumps or solution. Dissolve the lumps approximately 50 g, in 25 ml distilled water. The fluid form is usually too thick; dilute with distilled water.

Aquamount

Aquamount[9] is an excellent, water-miscible, quick-drying, neutral mountant. It does not require ringing. I use it in preference to glycerine jelly for mounting acid phosphatase preparations, and for fat stains on frozen sections.

Diaphane

Diaphane[10] is available in two forms, colorless (refractive index, 1.4777) and green (refractive index, 1.4792). Slides can be mounted directly from absolute, 95%, or 70% alcohol.

---

[8] Glyco Chemicals, Williamsport, Penn.   [9] Edward Gurr, London.
[10] Will Corp.

Fluorescent Mounting Medium (Rodriquez and Deinhardt 1960)

Elvanol 51-05 (Du Pont)	20.0 g
0.14 sodium chloride buffered with 0.01 M	
$KH_2PO_4 - Na_2HPO_4 \cdot 12H_2O$ (pH 7.2)	80.0 ml
glycerine	40.0 ml

Agitate frequently during a period of 16 hours. Remove undissolved particles of Elvanol by centrifuging at 12,000 rpm for 15 minutes. The pH should be between 6 and 7.

AQUEOUS MOUNTING TECHNIQUES

Aqueous mounts—sections or whole mounts—are removed from the water, placed on a slide, and covered with a drop of mounting medium. The cover glass, held in a horizontal position, is placed directly on the medium. Do not drop it from a slanted position as this may push the object to one side. In many cases it is not necessary to press the cover glass into place; its own weight is sufficient. Since the sections or objects are not attached to the slides, too much pressure may result in disarranged and broken material.

Using Two Cover Glasses

The material is mounted out of glycerine into glycerine jelly between two cover glasses, one of which is smaller than the other. Clean away excess jelly and air dry overnight. Invert the pair of cover glasses on a drop of resinous medium, such as Permount or Piccolyte, on a slide. This will permanently seal the mount, and it can be treated like resin mount.

Ringing Cover Glasses

If the mountant contains a volatile substance like water and the slides are to be relatively permanent, the cover glass must be sealed with a ringing material. Ringing cements, such as Turtox Slide Ringing Cement[11] are sold by supply houses. Others easily obtainable are Duco cement,

---

[11] General Biological Supply House.

colorless nail polish, gold size, and asphaltum. Lustron 2020 [12] dissolved in xylene can be used.

Conger (1960) recommends Dentists' Sticky Wax.[13] It is solid and slightly tacky at room temperature, but flows easily when melted. It adheres well and does not leak, but will crack off cleanly if frozen with dry ice or liquid air. Good for acetocarmine and aceto-orcein preparations.

A firm and reasonably permanent ringing cement is cover glass cement Kroenig from Riedel-de-Häen AG, Seelze-Hannover, Germany.[14]

For temporary mounts, melted paraffin can be used to ring a cover glass. Because it is susceptible to temperature damage and will crack away from the cover, it is not recommended for slides subject to hard usage or for a long time.

If the cover glasses are round, a turntable rotating on a steel pin or ball bearing facilitates the ringing operation.[15] The turntable spins rapidly, and, with the ringing cement on a brush, a neat seal can be made by following concentric guidelines.

---

[12] Monsanto Chemical Co.   [13] Kerr Co., Detroit, Mich.
[14] Roboz Surgical Instrument Co., Washington, D.C.
[15] Watson & Sons, London.

CHAPTER 12

Hematoxylin Staining

The strength of hematoxylin as a staining agent depends on its proper oxidation and the use of a mordant, as discussed in chapter 10. A number of methods have been developed for oxidizing hematoxylin.

Baker (1958) recommends oxidation with sodium iodate for preparing hematoxylin solutions. Start with a wholly unoxidized hematoxylin dye powder, not hematein powder. Solutions started with the hematein powder tend to lose their strength by flocculation (sedimentation) of the products of oxidation. When in solution, the hematoxylin dye should be only partly oxidized by the sodium iodate. Use less chemical than would be required for complete oxidation; about one-fourth to one-half of the full amount of oxidizer is adequate. The solution will then continue to combine with atmospheric oxygen and thereby maintain its strength. Such solutions are allowed to ripen slowly for six weeks or more. (Mayer hematoxylin is an example of this type.) They will produce brilliant staining for many months.

The rate of oxidation is also affected by the type of solvent used. A neutral aqueous solution forms hematein in a few hours; an acid solution does this more slowly and an alkaline solution more rapidly. Alcoholic solutions are slow, and the addition of glycerine retards them even more. Color changes that take place in a stock solution indicate its efficiency. The changes, with no mordant present, are from water white through lilac, bright purple, deep purple, red, orange red, orange brown, to brown. At the purple stage, the solution is most vigorous; at the red stages less so; at the brown stage it is no longer useful. The lifetime of alcoholic solutions is five times greater than that of aqueous ones (Cole 1943).

SINGLE SOLUTIONS

Delafield Hematoxylin (Carlton 1947)

Dissolve 4 g hematoxylin in 25 ml absolute ethyl alcohol. Mix gradually into 400 ml ammonia alum ($NH_4Al(SO_4)_2 \cdot 12H_2O$), saturated aqueous (approximately 1 part alum to 11 parts distilled water). Leave exposed to light in a flask with a cotton plug for 3–5 days. Filter. To the filtrate add 100 ml glycerine and 100 ml methyl alcohol. Allow to ripen for at least 6 weeks. The ripened solution will keep for years in a stoppered bottle.

Hance and Green (1961) bubbled pure oxygen for 4 hours into 3 liters of freshly prepared Delafield solution. This reduced the oxidation time from weeks to hours. The solution thus oxidized stained more rapidly than solutions oxidized by atmospheric oxygen and could be diluted to half strength for use.

Ehrlich Hematoxylin (Gurr 1956)

hematoxylin	2.0 g
ammonia alum ($NH_4Al(SO_4)_2 \cdot 12H_2O$)	3.0 g
alcohol, methyl or ethyl	100.0 ml
glycerine	100.0 ml
distilled water	100.0 ml

Ripens in 6–8 weeks or may be ripened for immediate use with 0.24 gm sodium iodate.

Add 10 ml glacial acetic acid. Keeps for years.

Harris Hematoxylin (Mallory 1944)

Dissolve 1.0 g hematoxylin in 10 ml ethyl alcohol. Dissolve 20 g potassium alum ($KAl(SO_4)_2 \cdot 12H_2O$) or ammonia alum ($NH_4Al(SO_4)_2 \cdot 12H_2O$) in 200 ml water and boil. Add hematoxylin and boil $\frac{1}{2}$ minute. Add 0.5 g mercuric oxide. Cool rapidly. Add a few drops of acetic acid to keep away metallic luster and brighten nuclear structure. Does not keep longer than a month or two. (*Warning:* Mix with care; the reaction can be explosive.)

Mayer Hematoxylin (Mallory 1944)

Add 1 g hematoxylin to 1 liter distilled water. Heat gently and add 0.2 g sodium iodate and 50 g potassium alum ($KAl(SO_4)_2 \cdot 12H_2O$). Heat until dissolved; add 1 g citric acid and 50 g chloral hydrate. Allow to ripen, preferably for 6–8 weeks although it can be used within 1–2 weeks.

Papamiltiades Hematoxylin (1953)

hematoxylin, 1% aqueous	100.0 ml
aluminum sulfate, 5% aqueous	50.0 ml
zinc sulfate, 5% aqueous	25.0 ml
potassium iodide, 4% aqueous	25.0 ml
glacial acetic acid	8.0 ml
glycerine	25.0 ml

Ready for immediate use; keeps approximately 2 months.

Phosphotungstic Acid–Hematoxylin (Lieb 1948)

phosphotungstic acid	10.0 g
hematoxylin	0.05 g
red mercuric oxide	0.025–0.05 g
hydrogen peroxide	2.0 ml
distilled water	500.0 ml

Dissolve hematoxylin in a little water with heat. Dissolve phosphotungstic acid in rest of water with heat; add hematoxylin solution. Bring to boil. Cautiously add mercuric oxide and remove from flame. Cool. Add hydrogen peroxide. Ready for use in 5–7 days; should be brownish red in color.

DOUBLE SOLUTIONS

Weigert Iron Hematoxylin (Lillie Modification 1968)

Solution A

ferric chloride ($FeCl_3 \cdot 6H_2O$)	2.5 g
ferrous sulfate ($FeSO_4 \cdot 7H_2O$)	4.5 g
hydrochloric acid	2.0 ml
distilled water	298.0 ml

Solution B

hematoxylin	1.0 g
95% ethyl alcohol	100.0 ml

Mix A and B. The solution turns black at once and is usable for 2–3 weeks or until it turns brown.

Groat Variation of Weigert Hematoxylin (1949)

In this variation a single solution is used.

distilled water	50.0 ml
sulfuric acid (spgr 1.84, 94% H_2SO_4)	0.8 ml
ferric alum ($FeNH_4(SO_4)_2 \cdot 12H_2O$)	1.0 g
95% ethyl alcohol	50.0 ml
hematoxylin	0.5 g

Mix in order given, at room temperature. Filter.

Janssen Hematoxylin (Lillie and Earle 1939*a*)

Solution A

ferric alum ($FeNH_4(SO_4)_2 \cdot 12H_2O$)	15.0 g
ferrous sulfate ($FeSO_4 \cdot H_2O$)	15.0 g
distilled water	100.0 ml

Solution B

hematoxylin	1.0 g
95% ethyl alcohol	50.0 ml
glycerine	50.0 ml

Mix A and B in equal amounts. Can be kept unchanged for several months, but, if solution turns brown, it is no longer usable. Keeps better than Weigert's.

Hansen Iron Trioxyhaematin (Pantin 1946)

Solution A

ferric alum ($FeNH_4(SO_4)_2 \cdot 12H_2O$)	10.0 g
ammonium sulfate (($NH_4)_2SO_4$)	1.4 g
distilled water	150.0 ml

Solution B

hematoxylin	1.6 g
distilled water	75.0 ml

Dissolve both solutions with gentle heat. Cool. Pour solution B in a porcelain evaporating dish. Add A, stirring constantly. Heat slowly without stirring just to boiling point. Cool rapidly by floating dish on cold water. A deep violet color turns brown. Filter into stoppered bottle with little air space above solution. Keeps 6–8 months, but, if it develops a green sheen, it is unsatisfactory. It can be used progressively or regressively.

Krutsay Iron Hematoxylin (1962*b*)

hematoxylin	1.0 g
potassium alum ($KAl(SO_4)_2 \cdot 12H_2O$)	50.0 g
potassium iodate (KIO_3)	0.2 g
hydrochloric acid	5.0 ml
distilled water	1000.0 ml

This can be used like Mayer hematoxylin or converted to an iron hematoxylin like Weigert's by adding 8 volumes of the solution below to 100 of above hematoxylin just prior to use.

ferric alum ($FeNH_4(SO_4)_2 \cdot 12H_2O$)	2.0 g
hydrochloric acid	0.5 ml
distilled water	100.0 ml

A fresh solution is brownish violet; if it turns orange, it is no longer usable. Staining time: 5 min.

MacConaill Lead Hematoxylin (Solcia et al. Modification 1969)

Stock stabilized lead solution

Mix equal volumes of 5% aqueous lead nitrate and saturated aqueous ammonium acetate. Filter. Add 2 ml of 40% formalin for every 100 ml of filtrate. Store at room temperature. Keeps for several weeks.

Working solution

Add 0.2 g hematoxylin dissolved in 1.5 ml 95% ethanol to 10 ml of lead stock solution. Dilute with 10 ml of distilled water. Stir repeatedly. After 30 minutes, filter. Makes up to 75 ml with distilled water. Use immediately.

American hematoxylin solutions may ripen in 2–3 weeks. Rawlins and Takahashi (1947) say that bubbling air through hematoxylin solutions ripens them more rapidly. Hance and Green (1959, 1961) ripen solutions even more rapidly by bubbling oxygen into the bottom of a container of hematoxylin.

Double Solutions That Are Not Mixed Before Use

Heidenhain Iron Hematoxylin

Solution A

ferric alum $(FeNH_4(SO_4)_2 \cdot 12H_2O)$	4.0 g
distilled water	100.0 ml

Keep in refrigerator to prevent precipitation on sides of bottle.

Solution B

hematoxylin	10.0 g
95% ethyl alcohol	100.0 ml

Let stand until a deep wine-red color; 4–5 months is not too long. Add 4–5 ml of this stock solution to 100 ml distilled water; this gives a practically aqueous solution and is already ripe. Saturated aqueous lithium carbonate—3 drops—added to the working solution improves color.

A and B are not usually mixed. A is a mordant solution, and precedes B. If they are mixed for use, the solution deteriorates rapidly.

Mallory Iron Chloride Hematoxylin (1944)

Solution A

iron chloride ($FeCl_2$)	5.0 g
distilled water	100.0 ml

Solution B

hematoxylin	0.5 g
distilled water	100.0 ml

Prepare fresh each time. Never mix these solutions.

Testing Hematoxylin Solutions

Add several drops of the solution to tap (not distilled) water. If it turns bluish purple immediately, it is still satisfactory, but if it changes slowly, stays reddish or brownish, it has weakened or broken down and should be discarded.

SUBSTITUTES FOR HEMATOXYLIN SOLUTIONS

Gallocyanin (Berube et al. 1966)

gallocyanin	0.15 g
chrome alum ($CrK(SO_4)_2 \cdot 12H_2O$) 15% aqueous	100.0 ml

Boil 10–20 minutes. Cool. Filter. Restore volume to 100 ml by washing the precipitate. Keeps about a week, then deteriorates slowly. An iron lake may be prepared by substituting 5% aqueous iron alum for the chrome alum (Proescher and Arkush 1928; also see Terner and Clark 1960a,b).

Hematein (Hemalum) (Kornhauser 1930)

hematein	0.5 g
95% ethyl alcohol	10.0 ml

potassium alum $(KAl(SO_4)_2 \cdot 12H_2O)$
5% aqueous .... 500.0 ml

Grind hematin with alcohol in glass mortar, and add to the aqueous potassium alum. Immediately ready for use.

COUNTERSTAINS (PLASMA STAINS) FOR HEMATOXYLIN, GALLOCYANIN, AND HEMATEIN

Eosin

eosin Y (C.I. 45380) .... 1.0 g
70% ethyl alcohol .... 1000.0 ml
glacial acetic acid .... 5.0 ml

Dilute with equal volume of 70% alcohol for use and add 2–3 drops of acetic acid.

Eosin (Putt 1948)

eosin Y (C.I. 45380) .... 1.0 g
potassium dichromate .... 0.5 g
saturated aqueous picric acid .... 10.0 ml
absolute ethyl alcohol .... 10.0 ml
distilled water .... 80.0 ml
acetic acid (optional) .... 1 drop

Eosin–Orange G

1% eosin Y (C.I. 45380) in 95% ethyl alcohol ... 10.0 ml
orange G (C.I. 16230), saturated solution in 95% ethyl alcohol (approximately 0.5 g per 100 ml) .... 5.0 ml
95% ethyl alcohol .... 45.0 ml

Other Acceptable Counterstains

1. Acid fuchsin (C.I. 42685), 5% aqueous (slight acidity improves stain). If overstained, rinse with tap water.
2. Orange G (C.I. 16230), saturated in 95% ethyl alcohol.
3. Van Gieson or substitute (page 183).

4. Bordeaux red (C.I. 16180), 1% aqueous.
5. Biebrich scarlet (C.I. 26905), 1% aqueous; a good counterstain.
6. Additional "eosins"
   a. Eosin Y (C.I. 45380) or eosin B (C.I. 45400), 0.1–0.5% in 95% ethyl alcohol.
   b. Erythrosin B (C.I. 45430), 0.1–0.5% in 95% ethyl alcohol.
   c. Phloxine B (C.I. 45410), 0.5% aqueous, plus a few drops of acetic acid (page 265).
7. Congo red (C.I. 22120), 0.5% aqueous.
8. Light green SF, yellowish (C.I. 42095), 0.2–0.3% in 95% ethyl alcohol.
9. Aniline blue WS (C.I. 42780) (pages 173, 266).
10. Fast green FCF (C.I. 42053), similar to light green or aniline blue.
11. Wool green (C.I. 44090), 0.5% aqueous.

HEMATOXYLIN STAINING PROCEDURES

Delafield (or Harris) Hematoxylin, I (Progressive Method)

FIXATION

Any general fixative or one specific for nuclear detail.

SOLUTIONS

Hematoxylin, see page 149.

Counterstain, see page 155.

PROCEDURE

The slides are passed through a "down" series of jars, a process often termed running down slides to water (or hydration), because a series of alcohols of decreasing strength is used. (From left to right, top row of Figure 22.) Never at any time during this procedure allow the slides to dry.

1. Xylene (or toluene): 2–3 minutes or longer.

   Two changes may be necessary to insure removal of paraffin.

2. Absolute alcohol: 2–3 minutes or longer.
3. 95% alcohol: 2–3 minutes or longer.
4. 70% alcohol: 2–3 minutes or longer.

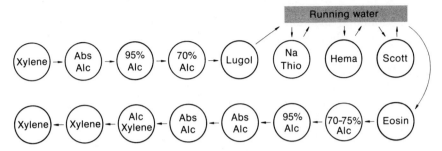

FIGURE 22. *A suggested arrangement of staining jars. An alternate arrangement can include two jars of xylene in the "down series" (left to right) when many slides are being stained. An absolute alcohol-xylene in "up series" (right to left) is optional.*

If mercuric chloride was absent from fixative, skip steps 5 through 7 and proceed to step 8.

5. Lugol solution (page 532): 3 minutes.
6. Running water: 3 minutes.
7. 5% sodium thiosulfate, $Na_2S_2O_3$: 2–3 minutes.
8. Running water: 3–5 minutes.
9. Delafield hematoxylin: 2–5 minutes, check after 1 minute for stain intensity. Fresh solutions stain faster than old ones. If not dark enough, return slides to stain. Rinse off stain in tap water before checking under microscope. If slide becomes too dark, convert to regressive type of stain.
10. Running water: 3–5 minutes.
11. Scott solution (page 533): 3 minutes.
12. Running water: 3–5 minutes.
13. Counterstain: 1 or more minutes, depending on stain used.

Transfer slides through "up" series, termed running up slides (or dehydration) using a series of alcohols of increasing strength. Steps 14 and 15 can control intensity of the counterstain. Watch timing carefully in these solutions. (From right to left, bottom row, Figure 22.)

14. 70% alcohol: 1 or more dips.
15. 95% alcohol: few dips.
16. Absolute alcohol: 3 minutes or longer.
17. Absolute alcohol: 2–3 minutes or longer.
18. Absolute alcohol–xylene (1:1): optional: 2–3 minutes or longer.
19. Xylene: 2–3 minutes or longer.

20. Xylene: 2–3 minutes or longer.
21. Mounting medium: keep sections moist with xylene during this process; they must not dry. Add cover glass.

RESULTS

nuclei—deep blue
cytoplasmic structure—pink, rose, etc., depending on counterstain

COMMENTS

1. Slides may be left in stronger alcohols and xylene longer than scheduled (indicated by "or longer"), but do not leave them much longer in any solutions weaker than 80% or 95% alcohol. Weaker alcohols and water can loosen the sections from the albumen.
2. Isopropyl alcohol can be substituted for ethyl alcohol in this schedule, but cannot be used as a stain solvent.
3. Two to five minutes for Delafield's (step 9) is an approximate time only. This may have to be varied according to the tissue used. Tissue sections are individualistic, often due to the type of fixative employed on them, and a certain amount of trial and error may be required to develop the exact timing schedule.
4. Poor staining can result from improper fixation, leaving gross tissues too long in alcohol or iodine, faulty processing during preparation for embedding, or careless handling of the slides in preparation for staining. In many instances no amount of trial and error will produce a perfect stain. The only correctable faults are those made during deparaffinization and hydration of the slides. Then the slides should be transferred in a reversed process through the solutions, back to xylene. Change to fresh solutions and try again.
5. Sometimes too many slides have been taken through the fluids and the fluids become contaminated. If more than 20 slides are being stained, use two changes of xylene in step 1 and follow step 20 with a third change. Prevent further contamination by draining slides properly. On removing them from a solution, touch the edges against the inner surface of the staining jar, and then against paper towels. But do not be too thorough and let the slides dry; merely drain off excess liquid.
6. Hematoxylin staining must be left in a "blued" condition. The original pinkish color of the hematoxylin, like Delafield's, must be converted to a blue color by alkalinity. In the progressive method this is accomplished in running water to remove excess dye and start the bluing action. Then use Scott solution (step 11), a couple of drops of ammonia in a final wash, or a weak solution of sodium bicarbo-

nate to insure the alkaline condition. In the regressive method (below) the same applies, and, as a precaution against neutralization caused by carrying one solution over into the other (step 5), after several exchanges of slides add a drop or two of acid and base to them. If there is any pinkness left in the nuclei, they are not adequately blued.

Delafield (or Harris) Hematoxylin, II (Regressive Method)

PROCEDURE

1. Deparaffinize and run slides down (hydrate) to water, removing $HgCl_2$ if present.
2. Stain in Delafield hematoxylin, until slides are well overstained: 15–20 minutes.
3. Wash in running water: 3–5 minutes.
4. Transfer to 70% alcohol.
5. Destain in ±0.5% HCl (2–3 drops in 60.0 ml of 70% alcohol) for a few seconds. Remove to 70% alcohol, with drop of ammonia added, until sections turn blue. Examine under microscope. If nuclei are too dark, repeat above procedure. When nuclei stand out as a sharp blue against colorless background, blue slides thoroughly in alkaline solution: 3–5 minutes. 70% alcohol, rather than water, is recommended for these procedures, because the water tends to loosen the sections from the slides.
6. Transfer to 70% alcohol.
7. Counterstain.
8. Dehydrate as for progressive staining.
9. Clear and mount.

RESULTS

nuclei—deep blue, sharper than in the progressive method
other elements—colors of counterstain

Mayer Hematoxylin

FIXATION

Any general fixative or one specific for nuclear detail.

SOLUTIONS

Hematoxylin, see page 150.

Counterstains, see page 155.

PROCEDURE

1. Deparaffinize and run slides down (hydrate) to water, removing $HgCl_2$ if present.
2. Stain in Mayer hematoxylin: 11 minutes.
3. Wash in running water: 3 minutes.
4. Blue in Scott solution: 3 minutes.
5. Wash in running water: 3–5 minutes.
6. Counterstain.
7. Dehydrate quickly through 70% and 95% alcohols.
8. Finish dehydration, absolute alcohol.
9. Clear and mount.

RESULTS

nuclei—deep blue
other elements—colors of counterstain

COMMENTS

Mayer hematoxylin is my favorite alum hematoxylin. It keeps well, and a single solution will stain several hundred slides over a period of two weeks without losing its intensity.

Heidenhain Iron Hematoxylin

FIXATION

Any general fixative, preferably one containing $HgCl_2$.

SOLUTIONS

Hematoxylin and iron alum, see page 153.

Counterstains, see page 155.

PROCEDURE

1. Deparaffinize and run slides down (hydrate) to water, removing $HgCl_2$.
2. Mordant in 4% iron alum: 15–30 minutes (or overnight, see comments below).
3. Wash in running water: 5 minutes.
4. Stain in hematoxylin: 15–30 minutes (or overnight).
5. Wash in running water: 5 minutes.
6. Destain in 2% aqueous iron alum or saturated aqueous picric acid

(see comments below) until nuclei stand out sharp and clear against colorless background. Before examining the slides, dip them in water containing a couple of drops of ammonia. *Warning:* too much ammonia will loosen sections.

7. Wash in running water: 15–30 minutes or longer.
8. Counterstain, if desired.
9. Run slides up through alcohols (dehydrate), clear and mount.

RESULTS

nuclei—blue black to black
muscle striations—blue-gray
other elements—colors of counterstain, or very light gray to colorless if
no counterstain used

COMMENTS

The shorter staining schedule produces blue-black nuclei and the overnight staining a truer black. The overnight staining is recommended for mitochondria. Picric acid usually perfects a more sharply differentiated nucleus than destaining with iron alum. Prolonged destaining with iron alum can leave a tan or yellow tinge in the cytoplasm. Yellow color left in the tissue after picric acid is due to insufficient washing following the destaining. Striations of muscle and some protozoan structures are better differentiated by iron alum. Thorough washing is essential after either method lest the sections continue to destain slowly and fade (picric acid destaining, Tuan 1930).

Destaining agents act as both an acid and an oxidizer, reacting in two ways: (1) as an acid they extract stain faster from the cytoplasm than from the nuclei; (2) as an oxidizer they bleach the stain uniformly. If an acid alone is used, the staining of the nuclei is favored, and if an oxidizer alone is used, the staining of the cytoplasm is enhanced.

Iron alum mordanting can precede a hematoxylin such as Mayer's and produce a blacker (iron type) nuclear staining. Destain, if necessary, in the same manner as for Heidenhain's.

Hutner (1934) recommendation for nuclei

1. Mordant in 4% iron alum: 1 hour.
2. Stain hematoxylin: 1 hour.
3. Destain in saturated aqueous picric acid.
4. Wash and blue by adding 1–2 drops ammonia to 70% alcohol.

Hutner (1934) recommendation for cytoplasm

1. Mordant in iron alum: 30 minutes.
2. Stain in hematoxylin: 30 minutes.
3. Destain in freshly made 95% alcohol 2 parts to Merk's superoxol (30% H_2O_2) 1 part.
4. Rinse in 1 change 70% alcohol: 10 minutes.

For brevity, intermediate steps have been omitted from the above procedures. Oxidizer and acid can be combined to approximate iron alum destaining effects:

0.25% HCl in 95% alcohol, 2 parts; superoxol, 1 part.
Wash in 70% alcohol and blue.

Cell membranes and various cytoplasmic structures stain well with Heidenhain hematoxylin, probably because of the use of iron alum. The metal is first bound to the tissues and then combines with the hematoxylin. It is likely that proteins are most frequently the bearers of iron-binding residues and are responsible for the precise staining action of Heidenhain hematoxylin (Puchtler 1958).

Acidified Hematein (Puchtler and Sweat 1964)

Dissolve 0.1 g hematoxylin in 2 ml 95% ethanol. Add 18 ml distilled water and 2 ml of a 1% aqueous solution of sodium iodate. Ripen for 15 minutes or longer. Add 60 ml of 10% acetic acid. This solution is not stable and should be prepared shortly before use.

Acidified hematein may be used in step 4 of Heidenhain method. Mordant (step 2) for 1 hour, wash thoroughly, and stain with acidified hematein for 1 hour. Wash thoroughly. Counterstain if desired. Dehydrate, clear, and mount. No differentiation is required.

Iron Hematoxylin (Groat Single Solution, 1949)

FIXATION

Any general fixative, preferably one containing $HgCl_2$.

SOLUTIONS

Hematoxylin, see page 151.

Counterstains, see page 155.

PROCEDURE

1. Deparaffinize and run slides down (hydrate) to water, removing $HgCl_2$.
2. Stain in Groat iron hematoxylin: approximately 10 minutes.
3. Rinse off excess stain with running water: 10 minutes.
4. Counterstain, if desired.
5. Dehydrate, clear, and mount.

RESULTS

nuclei—black
other elements—color of counterstain

COMMENTS

The 10-minute timing is approximate: check under the microscope for depth of color. If the background is too gray it may be decolorized in:

distilled water	50.0 ml
95% alcohol	50.0 ml
sulfuric acid (spgr 1.84, 95.5%–96.5% H_2SO_4)	0.18 ml

Follow by neutralizing in:

distilled water	50.0 ml
95% alcohol	50.0 ml
sodium bicarbonate	0.5 g

Phosphotungstic Acid–Hematoxylin (Lieb Modification 1948)

FIXATION

Any general fixative, preferably one containing $HgCl_2$. Lieb suggests that if tissue is formalin-fixed, mordant sections before staining in saturated aqueous mercuric chloride, rinse, and treat with Lugol's.

SOLUTIONS

Hematoxylin, see page 150.

Potassium permanganate, 0.5%

potassium permanganate	0.5 g
distilled water	100.0 ml

Oxalic acid, 2%

oxalic acid	2.0 g
distilled water	100.0 ml

Iron alum, 4%, see page 153.

PROCEDURE

1. Deparaffinize and run slides down (hydrate) to water, removing $HgCl_2$.
2. Oxidize in potassium permanganate, freshly prepared: 5 minutes.
3. Wash in running water: 5 minutes.
4. Bleach in oxalic acid: 5 minutes.
5. Wash in running water: 5 minutes; follow with rinse in distilled water.
6. Mordant in iron alum: 1 hour.
7. Wash out excess alum in running water 10 minutes or longer, rinse in distilled water.
8. Stain in phosphotungstic acid–hematoxylin: 2–24 hours.
   Usually complete in 2–3 hours, rarely requires overnight.
9. Transfer directly to 95% alcohol: brief rinse.
10. Dehydrate, absolute alcohol; watch that elements stained reddish brown are not destained.
11. Finish dehydration in 99% isopropyl alcohol, 2 changes: 3 minutes each.
12. Clear and mount.

RESULTS

nuclei, mitochondria, centrioles, fibrin, fibroglia, myoglia and neuroglia fibrils, and contractile elements of striated muscle—blue collagen, reticulum, elastin, cartilage, and bone—yellowish to brown red

COMMENTS

Mullen and McCarter (1941) use a chromium-chloride mordant after formalin fixation: Deparaffinize and hydrate to water; mordant for 2 hours or longer in:

chromium chloride ($CrCl_3$)	5.0 g
distilled water	100.0 ml
glacial acetic acid	50.0 ml

Good for several weeks. After this mordanting, the staining will be as brilliant as that following Zenker fixative.

(See Puchtler et al. 1963, for a 1-hour phosphotungstic acid–hematoxylin method.)

Sweat et al. (1968) have determined that this stain is relatively nonspecific for plasma proteins, including fibrin. There is a relation between the configuration of the proteins and the binding of the red and blue components of the stain. The affinity of the proteins for these components is affected by the fixative used, and by other experimental and pathological conditions.

Weigert Iron Hematoxylin (Lillie 1968)

FIXATION

Any general fixative or one specific for nuclei.

SOLUTIONS

Hematoxylin, see page 151.

Counterstains, see page 155.

PROCEDURE

1. Deparaffinize and hydrate slides to water, remove $HgCl_2$.
2. Stain in Weigert hematoxylin: 3–5 minutes or longer.
3. Wash in running water: 5 minutes.
4. Counterstain.
5. Dehydrate, clear, and mount.

RESULTS

nuclei—black
other elements—color of counterstain

COMMENTS

This is progressive; check results under microscope.

Mallory Iron Chloride–Hematoxylin (1944)

FIXATION

Any general fixative, or one specific for nuclei.

SOLUTIONS

Ferric chloride, see page 154.

Hematoxylin, see page 154.

Counterstains, see page 155.

Ferric chloride for destaining (0.25%)

ferric chloride 5% stock solution . . . .   5 ml
distilled water . . . .   95 ml

PROCEDURE

1. Deparaffinize and hydrate slides to water, remove $HgCl_2$.
2. Mordant in ferric chloride: 1 hour or longer.
3. Wash in running water: 5 minutes.
4. Stain in hematoxylin: 1 hour or longer.
5. Wash in running water: 3–5 minutes.
6. Destain in ferric chloride, 0.25%; agitate slides, check under microscope.
7. Wash thoroughly, running water: 10–15 minutes.
8. Counterstain, if desired.
9. Dehydrate, clear, and mount.

RESULTS

nuclei—deep blue
other elements—color of counterstain

HEMATOXYLIN SUBSTITUTE PROCEDURES

Gallocyanin (Einarson 1951, Berube et al. 1966)

Gallocyanin is an excellent substitute for hematoxylin and can be used in most procedures designed for hematoxylin. The solution is made in a few minutes and ready for immedate use; also an iron lake can be prepared. No differentiation is required, and it is better than hematoxylin for tissues of the central nervous system. It can replace thionin for ganglia and glia cells, but not for myelin sheaths. It is good for Negri bodies. The preferred fixatives are acetoformol, Zenker (acetic or formol), or formalin—fixatives that preserve chromophilic substance.

SOLUTION

Gallocyanin, see page 154.

PROCEDURE

1. Deparaffinize and hydrate to water.
2. Leave in stain overnight, or 3 hours at 56°C.
3. Wash thoroughly and proceed to counterstain.
4. Dehydrate, clear, and mount.

RESULT

nuclei—blue

Hematein (Hemalum) (Kornhauser 1930)

FIXATION

Any general fixative, preferably one containing $HgCl_2$.

SOLUTIONS

Hemalum, see page 154.

Counterstains, see page 155.

PROCEDURE

1. Deparaffinize and hydrate slides down to water, remove $HgCl_2$.
2. Stain hemalum, progressively: about 5 minutes.
3. Wash in running water: 5–10 minutes.
4. Counterstain.
5. Dehydrate, clear, and mount.

RESULTS

nuclei—blue
other elements—color of counterstain

RED NUCLEAR STAINING

Darrow Red (Powers et al. 1960, Powers and Clark 1963)

FIXATION

Any general fixative.

SOLUTION

> Darrow red [1] .... 50.0 mg
> 0.2 *M* glacial acetic acid (*p*H 2.7) .... 200.0 ml

Boil briefly. Cool and filter. Stable for 1 month.

PROCEDURE

1. Use frozen sections in water, or deparaffinize paraffin sections and hydrate to water.
2. Stain in Darrow red: 20–30 minutes.
3. Rinse in distilled water.
4. Differentiate and dehydrate through 50%, 70%, and 95% alcohol, not so slowly as to remove too much dye from nuclei, but slowly enough to decolorize background.
5. Completely dehydrate in *n-butyl* alcohol, clear, and mount.

RESULTS

nuclei—red
other elements—depending on other stains applied

COMMENTS

This is a good nuclear stain for contrast with blue cytoplasmic stains—following Luxol fast blue for example. It is a good Nissl stain at *p*H 2.7. For glial nuclei, use acetic acid–acetate buffer at *p*H 3.5 to prepare the staining solution. Stain 10–15 min. For celloidin sections, use 25 mg of dye in 200 ml of acetic acid-acetate buffer, 20 minutes. Dehydrate overnight in terpineol-xylene, 1:3.

Scarba Red (Slidders et al. 1958)

FIXATION

Any general fixative.

SOLUTIONS

Scarba red

> melted phenol (carbolic acid) .... 2.0 g
> neutral red .... 1.0 g
> Mix thoroughly, allow to cool, dissolve in:

---

[1] Matheson Coleman & Bell.

95% ethyl alcohol	15.0 ml
Add:	
2% aniline in water	85.0 ml
glacial acetic acid	1.0–3.0 ml

Mix well and filter. Keeps for at least 6 months.

Differentiator

70% ethyl alcohol	85.0 ml
formalin	15.0 ml
glacial acetic acid	15.0 drops

PROCEDURE

1. Deparaffinize and hydrate slides to water (remove $HgCl_2$ if present).
2. Stain in scarba red: 5 minutes.
3. Rinse in distilled water.
4. Transfer to 70% alcohol: 2–3 minutes.
5. Differentiate until nuclei are clear and sharp against colorless background.
6. Dehydrate in 95% and absolute alcohol.
7. Clear in xylene and mount.

RESULT

chromatin and calcium—red

PART II

SPECIFIC STAINING METHODS

13

Staining Connective Tissue and Muscle

Stains for connective tissue and muscle are used to demonstrate various components and features, such as collagen, reticulum, elastin, hyalin, fibrin, and striations in muscle cells. (Hemopoietic tissues are covered in chapter 16, page 245.)

MALLORY STAINING

Mallory Triple Connective Tissue Stain

Innumerable modifications of this method have appeared in the literature, particularly those using phosphomolybdic and/or phosphotungstic acid. Holde and Isler (1958) support the use of phosphomolybdic acid if selective staining of connective tissue is desired; the acid diminishes background and nuclear staining and stains connective tissue blue. They feel that the staining of connective tissue is due to the action of the acid on the fibers. Baker (1958) suggests that phosphomolybdic acid acts as a "colorless acid dye" in the tissue, chiefly on collagen. It acts as a dye excluder with acid dyes such as acid fuchsin, excluding them from collagen. Then the aniline blue stains the collagen selectively, but is excluded from other tissues. Since phosphomolybdic acid opposes the action of aniline blue as well as aiding it, perhaps it should be called a mordant.

It is true that various dyes (acid and basic triphenylmethane) form lakes with phosphomolybdic acid; such compounds are used in textile dying (Puchtler and Sweat 1963b). Basic dyes and some amphoteric dyes are bound by interaction between the acid groups of phosphomolybdic acid and accessible basic groups of the dye (Smith et al. 1966). If no phosphomolybdic acid is present, aniline blue stains more strongly and with very little selectivity.

In regard to the use of phosphotungstic acid, it may be of interest to note that certain tissue elements react more specifically to certain metals than to others: collagen to molybdenum; and fibroglia, myoglia, neuroglia, and epithelial fibers to tungsten. Lillie (1952) writes that phosphotungstic acid intensifies plasma staining, and phosphomolybdic acid, fiber staining. The acids may be used separately or together, depending on the desired final effect.

Oxalic acid is often used. Mallory (1944) claims that it makes the aniline blue stain more rapidly and intensely. Baker (1958) seems to agree when he suggests that oxalic acid lowers the pH, seeming thereby to aid staining with aniline blue and also with orange G.

No polychroming or metachroming takes place in the combination of dyes; they react orthochromatically in varying intensities of their own colors. In Mallory's combination of stains, aniline blue (acidic) stains connective tissue and cartilage; orange G (acidic) stains blood cells, myelin, and muscle; acid fuchsin (acidic) stains the rest of the tissue (including the nuclei) in shades of pink and red. This is not, however, an efficient method for staining nuclei, and they will fade in a few years. Modifications of Mallory's using basic fuchsin, carmine, or azocarmine result in more permanent nuclear staining, since the basic stain reacts more reliably with the nuclei than the acidic one. This method may also be preceded by hematoxylin staining for a brilliant permanent nuclear stain.

The final staining may be followed by an acetic acid wash (0.5–1%): 3–5 minutes or longer, which produces more transparent sections without altering the color.

Suggestion: Try Sirius supra blue in place of aniline blue (Sweat et al. 1968). Aniline blue WS was formerly a mixture of water blue 1 and methyl blue; that mixture now appears to be obsolete. Aniline blue ordered from some manufacturers is water blue; from others, methyl blue. The dyes are apparently considered interchangeable with each other as well as with the old mixture. However, I find the new dyes inferior in color to the old mixture and in most cases substitute Sirius supra blue. Treat sections for 5 minutes with 1% phosphomolybdic acid; wash with 3 changes of distilled water: 10 seconds each. Stain for 5 minutes in the following solution:

Sirius supra blue FGL-CF[1]	2.0 g
distilled water	100.0 ml
glacial acetic acid	2.0 ml

Wash in 2–3 changes of distilled water: 10 seconds each. Dehydrate, clear, and mount.

Note: Do not combine Sirius supra blue and phosphomolybdic acid as can be done with aniline blue; the combination is not successful with Sirius supra blue.

Pantin Method (1946)

FIXATION

Any general fixative, preferably one containing $HgCl_2$.

SOLUTIONS

Mallory I

acid fuchsin (C.I. 42685)	1.0 g
distilled water	100.0 ml

Phosphomolybdic acid

phosphomolybdic acid	1.0 g
distilled water	100.0 ml

Mallory II

aniline blue WS (C.I. 42780)	0.5 g
orange G (C.I. 16230)	2.0 g
distilled water	100.0 ml

---

[1] Roboz Surgical Instrument Co., Washington, D.C.

PROCEDURE

1. Deparaffinize and hydrate slides to water; remove $HgCl_2$. If $HgCl_2$ is absent from fixative, mordant in saturated aqueous $HgCl_2$, plus 5% acetic acid: 10 minutes. Wash, treat with Lugol's and sodium thiosulfate; wash and rinse in distilled water.
2. Stain in Mallory I: 15 seconds.
3. Rinse in distilled water to differentiate reds: 10 or more seconds.
4. Treat with phosphomolybdic acid: 1–5 minutes.
5. Rinse briefly in distilled water.
6. Stain in Mallory II: 2 minutes.
7. Rinse in distilled water.
8. Differentiate aniline blue in 90% ethyl alcohol.
9. Dehydrate in absolute alcohol, clear, and mount.

RESULTS

nuclei—red
muscle and some cytoplasmic elements—red to orange
nervous system—lilac
collagen—dark blue
mucus, connective tissue, and hyaline substance—blue
chitin—red
yolk—yellow to orange
myelin and red blood cells—yellows and orange
dense cellular tissue (liver)—pink with red nuclei
bone matrix—red

COMMENTS

From distilled water (step 7) the slides can be run through the "up" series of alcohols to control the blue color. Aqueous-alcoholic solutions differentiate the amount of Mallory II left in the various parts of the tissue. After the slides have remained in the absolute alcohol for a couple of minutes they should have changed from a muddy purple to a clear blue and red. An acetic acid rinse following step 7 contributes to the transparency of the sections.

For better nuclear detail, Mallory I can be preceded by alum hematoxylin staining (*results:* nuclei—blue). Lendrum and McFarlane (1940) recommend the addition of celestin blue. Krichesky (1931) proposes the following for Mallory II.

Solution A

aniline blue WS (C.I. 42780)	. . . .	2.0 g
distilled water	. . . .	100.0 ml

Solution B

orange G (C.I. 16230)	1.0 g
distilled water	100.0 ml

Solution C

phosphomolybdic acid	1.0 g
distilled water	100.0 ml

Keep solutions A, B, and C in separate bottles because the mixture deteriorates on standing. When ready to use, mix equal parts of each.

This modification is useful for instructors of microtechnique classes, because, if large stocks of Mallory II components are kept as individual solutions and freshly mixed before use, a more brilliant stain is achieved than from a single solution that has been stocked with all the components already combined.

Azan[2] Stain, Mallory-Heidenhain (Koneff Modification 1938)

FIXATION

Zenker formol; other general fixatives fair, but fixation is improved by mordanting slides overnight in 3% potassium dichromate.

SOLUTIONS

Azocarmine

azocarmine G (C.I. 50085)	0.2–1.0 g
distilled water	100.0 ml
glacial acetic acid	1.0 ml

Boil azocarmine in water 5 minutes, cool, filter, and add acetic acid.

Aniline alcohol

aniline	1.0 ml
85–90% ethyl alcohol	1000.0 ml

Acid alcohol

glacial acetic acid	1.0 ml
90–95% ethyl alcohol	100.0 ml

---

[2] Azan from first syllables of azokarmin B and anilinblau W.

Phosphotungstic acid

phosphotungstic acid	5.0 g
distilled water	100.0 ml

Aniline blue stain

aniline blue WS (C.I. 42780)	0.5 g
orange G (C.I. 16230)	2.0 g
oxalic acid	2.0 g
distilled water	100.0 ml
5% phosphotungstic acid (above)	1.0 ml

Acidified water

glacial acetic acid	1.0 ml
distilled water	100.0 ml

PROCEDURE

For strikingly beautiful results, particularly for pituitary, Koneff's longer method is recommended. Extra steps and changes in timing are indicated in parentheses.

1. Deparaffinize and hydrate slides down to water; remove $HgCl_2$.
2. (Treat with aniline alcohol: 45 minutes.)
3. (Treat with acid alcohol: 1–2 minutes.)
4. Stain in azocarmine, 56°C: 1 hour (or 2 hours). Check temperature carefully; if too high, azocarmine differentiates poorly.
5. Rinse in distilled water.
6. Differentiate in aniline alcohol. Check under microscope for brilliant red nuclei and very light red cytoplasm.
7. Treat with acid alcohol: 1–2 minutes.
8. Transfer to phosphotungstic acid: 2–3 hours. (4 hours.)
9. Rinse in distilled water.
10. Stain in aniline blue solution: 1–2 hours. (4 hours.)
11. Rinse in distilled water.
12. Treat with phosphotungstic acid: 3–5 minutes.
13. Rinse in distilled water.
14. Rinse in acidulated water: 1–2 minutes or longer.
15. Rinse in 70% alcohol: briefly dip.
16. Dehydrate in 95% alcohol, several dips in each of 2 changes.
17. Dehydrate in absolute alcohol, clear, and mount.

nuclei—brilliant red
collagen and reticulum—blue
muscle—red and yellow
basophilic cytoplasm—light blue
acidophilic cytoplasm—orange-red
chromophobes—colorless or light gray

COMMENTS

The azan method is particularly recommended for pituitary and pancreas but is also outstanding for connective tissue. If a tissue does not "take" the azocarmine, it may be because of formalin in the fixative. To correct this condition, mordant the sections overnight in 3% aqueous potassium dichromate.

Romeis (1948) suggests the substitution of acid alizarine blue for azocarmine; this shortens the time and does not require heat.

PROCEDURE

1. Stain in acid alizarine blue: 5 minutes.

aluminum sulfate ($Al_2(SO_4)_3$)	10.0 g
acid alizarine blue (C.I. 58610)	0.5 g
distilled water	100.0 ml

Bring to boil: 5–10 minutes. Let cool. Fill to 100 ml with distilled water. Filter. The solution will be a red-violet color.

2. Wash in running water.
3. Treat with phosphotungstic or phosphomolybdic acid, 5% aqueous: 30 minutes.
4. Wash in distilled water.
5. Proceed to aniline blue stain.

RESULTS

After phosphotungstic acid, the alizarine is red; it stays in nuclei and muscles but is removed from collagen. After phosphomolybdic acid, the nuclei and muscle are blue.

Mallory-Heidenhain Stain: Rapid One-Step Method (Cason 1950)

FIXATION

Any good general fixative.

SOLUTION

distilled water .... 200.0 ml

Dissolve each of below before adding next stain:

phosphotungstic acid .... 1.0 g
orange G (C.I. 16230) .... 2.0 g
aniline blue, WS (C.I. 42780) .... 1.0 g
acid fuchsin (C.I. 42685) .... 3.0 g

This solution keeps for several months.

PROCEDURE

1. Deparaffinize and hydrate slides to water, remove $HgCl_2$.
2. Stain: 5 minutes.
3. Wash in running water: 3–5 seconds.
4. Dehydrate rapidly, clear, and mount.

RESULTS

collagen—blue
ground substance of cartilage and bones—shades of blue
mucus, amyloid, other hyaline substances—shades of blue
nuclei, fibroglia, myoglia, neuroglia fibrils, axis cylinders, fibrin—red
erythrocytes, myelin—yellow
elastin—pale pink or yellow

TRICHROME STAINING

Masson Trichrome Stain (Gurr Modification 1956)

Masson (1928), when he developed this method, called it a trichrome
stain, although it includes four dyes. Since then the trichrome name has
been applied to many modifications of Masson's method that use many
combinations of different dyes.

FIXATION

Any general fixative.

SOLUTIONS

Iron alum

ferric ammonium sulfate .... 4.0 g
distilled water .... 100.0 ml

Hematoxylins, see page 149.

Acid fuchsin

acid fuchsin (C.I. 42685)	1.0 g
distilled water	100.0 ml
glacial acetic acid	1.0 ml

Ponceau de xylidine

ponceau de xylidine (C.I. 16150)	1.0 ml
distilled water	100.0 ml
glacial acetic acid	1.0 ml

Lillie (1940) suggests alternate stains for ponceau de xylidine: ponceau 2R, nitazine yellow, Biebrich scarlet, azofuchsin 3B, G, and 4G, Bordeaux red, chromotrope 2R, chrysoidin, eosin Y, orange G, and crocein.

Fast green

fast green FCF (C.I. 42053)	2.0 g
distilled water	100.0 ml
glacial acetic acid	2.0 ml

Phosphomolybdic acid

phosphomolybdic acid	1.0 g
distilled water	100.0 ml

Acidified water

glacial acetic acid	1.0 ml
distilled water	100.0 ml

PROCEDURE

1. Deparaffinize and hydrate slides to water, remove $HgCl_2$. Mordant formalin-fixed tissue in saturated aqueous $HgCl_2$: 5 minutes. Wash in running water: 5 minutes. Treat with Lugol's and sodium thiosulfate.
2. Mordant in iron alum: ½ hour.
3. Wash in running water: 5 minutes.
4. Stain in hematoxylin (Delafield or like): ½ hour.
5. Wash in running water: 5 minutes.

6. Differentiate in saturated aqueous picric acid.
7. Wash thoroughly in running water: 10 minutes or longer.
8. Stain in acid fuchsin: 5 minutes.
9. Rinse in distilled water until excess stain is removed. Check under microscope.
10. Stain in ponceau de xylidine: 1–5 minutes.
11. Rinse in tap water, control under microscope for proper intensity of acid fuchsin and ponceau de xylidine.
12. Differentiate in phosphomolybdic acid: 5 minutes.
13. No rinse; transfer directly into fast green: 2 minutes.
14. Differentiate fast green in acidified water and dehydrating alcohols.
15. Dehydrate in absolute alcohol, 2 changes: 3 minutes each.
16. Clear and mount.

RESULTS

nuclei—deep mauve blue to black
cytoplasmic elements—varying shades of red and mauve
muscle—red
collagen, mucus—green

COMMENTS

Among the many modifications of Masson stain, this is one of the best, because it offers good control of both red dyes by the use of two separate solutions. In most cases, the full 5 minutes in the acid fuchsin is advisable, but time in the ponceau de xylidine is not quite so critical.

Pollak Rapid Method (1944)

FIXATION

Any general fixative. Formalin-fixed tissues are improved by treating slides overnight with saturated aqueous $HgCl_2$.

SOLUTIONS

Hematoxylins, see page 149.

Trichrome stain

acid fuchsin (C.I. 42685)	0.5 g
ponceau 2R (C.I. 16150)	1.0 g
light green SF, yellowish (C.I. 42095)	0.45 g
orange G (C.I. 16230)	0.75 g
phosphotungstic acid	1.5 g

phosphomolybdic acid	1.5 g
glacial acetic acid	3.0 ml
50% ethyl alcohol, up to	300.0 ml

Add acetic acid to alcohol; split solution in 4 parts. Dissolve acid fuchsin and ponceau in one part; light green in the second part; orange G and phosphotungstic acid in the third part; and phosphomolybdic acid in the fourth part. Mix four solutions and filter. Keeps well.

Acidified water

| glacial acetic acid | 0.2 ml |
| distilled water | 100.0 ml |

PROCEDURE

1. Deparaffinize and hydrate slides to water; remove $HgCl_2$.
2. Stain in Mayer hematoxylin: 10 minutes. (Pollak uses Weigert: 10–20 minutes.)
3. Wash in running water: 10 minutes.
4. Stain in trichrome stain: 7 minutes.
5. Rinse briefly in distilled water.
6. Differentiate in acidified water: only a few seconds. Check under microscope if necessary.
7. Dip a few times in 70% alcohol.
8. Dehydrate in 95% alcohol, 2 changes: several seconds each.
9. Dehydrate in absolute alcohol, clear, and mount.

RESULTS

nuclei—dark blue
muscle, elastin—red
fibrin, calcium—purple
hyalin—pale blue
collagen, mucus—blue green

Toren (1963) combines this stain with Giemsa (page 255) for mast cell staining.

Gomori Method (1950*b*) (Sweat et al. Modification, 1968)

FIXATION

Any general fixative. Formalin-fixed tissues are improved by treating slides overnight with saturated aqueous $HgCl_2$.

SOLUTIONS

Bouin fixative

picric acid, saturated aqueous	1 part
formalin, concentrated	5 parts
glacial acetic acid	1 part

Mix shortly before use. Picric acid in solution is not dangerous, but is explosive when dry. Remove used Bouin from the oven and discard.

Trichrome solution

chromotrope 2R (C.I. 16570)	0.6 g
aniline blue WS (C.I. 42780)	0.6 g
phosphomolybdic acid	1.0 g
dissolve in distilled water	100.0 ml
add hydrochloric acid	1.0 ml

Allow to stand 24 hours in refrigerator before use. Store in refrigerator and use cold. Do not filter. The solution is stable until the reds begin to fade.

Acidified water

glacial acetic acid	0.2 ml
distilled water	100.0 ml

PROCEDURE

1. Deparaffinize and hydrate slides to water; remove $HgCl_2$.
2. Treat with Bouin solution, 56°C: 1 hour.
3. Wash in running water: 5 minutes, or until decolorized.
4. Stain in trichrome mixture: 1 minute.
5. Rinse briefly in distilled water to remove some of excess stain.
6. Rinse in acidified water: 30 seconds.
7. Dehydrate in 95% alcohol, 2 changes. Be careful not to remove too much green.
8. Dehydrate in absolute alcohol, clear, and mount.

RESULTS AND COMMENT

There have been a number of attempts to combine the Azan and Masson methods into a quicker method. Gomori considered Pollak's to be a disappointingly dull color scheme, lacking real red shades. Almost any combination of an acid triphenylmethane dye with a sulfonated azo dye in the presence of phosphomolybdic or phosphotungstic acid will give good results. The phosphomolybdic acid favors the green and blue shades; phosphotungstic acid, the reds. A short staining time produces more red; prolonged staining, more green and blue. Rinsing in tap water weakens the reds.

The Sweat et al. modification improves the colors of the fine connective tissue fibers, so that they closely resemble those obtained with the Mallory and Mallory-Heidenhain procedures. Pretreatment with Bouin improves the affinity for stains, offers more uniform results, and reduces variations due to different fixatives. Staining with hematoxylin has been eliminated; very little of its color remains after staining in the acid trichrome mixture. If an iron hematoxylin is used, it turns the clear red of the chromotrope to a murky color.

The staining of connective tissue fibers is affected by the pH of the solution, best around pH 1.3. The Gomori solution has a pH of 2.5–2.7; by replacing the acetic acid with hydrochloric acid the pH is lowered to approximately 1.3.

The rinse in acetic acid does not change the colors, but makes them more delicate and transparent.

I have had good results with most of the trichrome stains and can only recommend that the selection of a routine stain be made by each individual. All of the short trichrome methods offer clearer and sharper returns on thin $(5-7\mu)$ sections than on thicker ones. For the thick sections, the Mallory, Masson, or Azan stain should be used.

All of the many trichrome staining methods cannot be included, but following are references for some excellent ones.

1. Buffalo black staining, Conn et al. (1960), Lillie (1945)
2. Churg and Prado trichrome stain (1956)
3. Greenstein simplified five-dye stain (1961)
4. Koneff aniline blue stain (1936)
5. Kornhauser quad stain (1943, 1945)
6. Movat pentachrome stain (1955)
7. Menzies modification of picro-Gomori (1959)
8. Chlorazol black E, Levine and Morrill (1941)
9. PAS–phosphomolybdic acid–Sirius supra blue FGL–CF, Sweat, Puchtler, and Sesta. (1968)
10. Allochrome procedure, Lillie (1951c)
11. Sweat, Puchtler, and Woo, modification of no. 10 (1964)

COLLAGEN STAINING

Picro-Ponceau with Hematoxylin (Gurr 1956)

(Van Gieson Substitute, nonfading)

FIXATION

Any general fixative.

SOLUTIONS

Hematoxylins, see page 149.

Picro-ponceau

ponceau S (C.I. 27195), 1% aqueous	10.0 ml
picric acid, saturated aqueous	86.0 ml
acetic acid, 1% aqueous	4.0 ml

PROCEDURE

1. Deparaffinize and hydrate slides down to water, remove HgCl$_2$.
2. Overstain in hematoxylin: 5–15 minutes.
3. Wash thoroughly in running water until slides are deep blue: 10 minutes or longer.
4. Stain in picro-ponceau: 3–5 minutes. (This may be too long for some tissues; in addition to staining, the stain acts as a destaining agent on the hematoxylin.) Rinse for a few seconds in distilled water and check under microscope. Continue to stain and destain or differentiate in water until nuclei are sharp.
5. Dip several times in 70% alcohol.
6. Dehydrate in 95% alcohol, 2 changes, to insure complete removal of excess picric acid. Only that which has acted as a dye must be left in the tissue.
7. Dehydrate in absolute alcohol, clear, and mount.

RESULTS

nuclei—brown to brownish or bluish black
collagenous and reticular fibers—red
elastic fibers, muscle fibers, erythrocytes, epithelia—yellow

COMMENTS

A Weigert type of hematoxylin is excellent for this method, but it is not necessary to precede it with iron mordanting. The iron mordant is present in the staining solution. However, an additional step can be added after step 1: mordant with iron alum before Delafield (or similar) hematoxylin; follow by washing in running water. Then proceed to step 2.

This method is so superior to so-called Van Gieson stain, which uses acid fuchsin instead of ponceau S, that Van Gieson method has been omitted. The colors are identical but Van Gieson is unsatisfactory because it fades rapidly.

Katline (1962) recommends the following procedure if the picric solution tends to fade the hematoxylin: treat the sections prior to the picric staining with a solution of equal parts of 2.5% aqueous phosphotungstic and phosphomolybdic acid, 0.5–1 minute; wash briefly; and proceed to the picro-ponceau solution.

Sweat, Puchtler, and Rosenthal (1964) substitute 0.1% Sirius red F3BA[3] in aqueous picric acid for the acid fuchsin in Van Gieson method.

ELASTIN TISSUE STAINING

Verhoeff Elastin Stain (Mallory 1944)

FIXATION

Any general fixative.

SOLUTIONS

Verhoeff stain

On electric hot plate, dissolve 3 g hematoxylin in 66 ml absolute ethyl alcohol. Cool, filter, and add 24 ml of 10% aqueous ferric chloride and 24 ml Verhoeff iodine solution. Usefulness is limited to 1–2 weeks.

Verhoeff iodine solution

potassium iodide (KI)	4.0 g
distilled water	100.0 ml
Dissolve and add:	
iodine	2.0 g

Ferric chloride solution, 10%

ferric chloride ($FeCl_2$)	10.0 g
distilled water	100.0 ml

---

[3] Verona Dyestuffs.

Ferric chloride solution, 2%

10% ferric chloride	20.0 ml
distilled water	100.0 ml

Pico-ponceau solution

ponceau S (C.I. 27195), 1% aqueous	10.0 ml
picric acid, saturated aqueous	86.0 ml
acetic acid, 1% aqueous	4.0 ml

PROCEDURE

1. Deparaffinize and run slides down to 70% alcohol. Removal of $HgCl_2$ not necessary.
2. Stain in Verhoeff stain: 15 minutes.
3. Rinse in distilled water.
4. Differentiate in 2% ferric chloride: a few minutes. Elastic fibers should be sharp black; nuclei brown. If destained too far, return slides to Verhoeff for another 5–10 minutes.
5. Transfer to sodium thiosulfate, 5% aqueous: 1 minute.
6. Wash in running water: 5–10 minutes.
7. Counterstain in picro-ponceau: 1 minute.
8. Differentiate in 95% ethyl alcohol, 2 changes: a few seconds each.
9. Dehydrate in absolute alcohol, clear, and mount.

RESULTS

elastic fibers—brilliant blue-black
nuclei—blue to brownish black
collagen—red
other tissue elements—yellow

COMMENTS

If elastic fibers stain unevenly or not at all, Verhoeff solution is too old.
Verhoeff can be combined with Perl staining (page 269) to show both iron and elastin (Pickett and Klavins 1961).

Orcinol–New Fuchsin (Fullmer and Lillie 1956)

FIXATION

Any general fixative.

SOLUTION

new fuchsin (C.I. 42520)	2.0 g
orcinol	4.0 g
distilled water	200.0 ml
Boil for 5 minutes; add:	
ferric chloride (FeCl$_3$), 29.1% aqueous	25.0 ml

Boil for 5 minutes. Cool. Collect precipitate on filter paper and dissolve in 100 ml of 95% ethyl alcohol.

PROCEDURE

1. Deparaffinize and transfer slides into absolute alcohol.
2. Stain in orcinol–new fuchsin, 37°C: 15 minutes.
3. Differentiate in 70% alcohol, 3 changes: 5 minutes each.
4. Dehydrate in 95% and absolute alcohol, clear, and mount.

RESULTS

elastic fibers—deep violet
collagen—unstained by orcinol–new fuchsin

COMMENTS

Slides can be counterstained with hematoxylin, safranin, or picro-ponceau; of the three, safranin offers best color contrast.

Aldehyde-Fuchsin (Gomori 1950c) (Cameron and Steele Modification 1959)

FIXATION

Any good fixative, preferably one without dichromate. Bouin and formalin produce a colorless background.

SOLUTIONS

Potassium permanganate, 0.3%

potassium permanganate	0.3 g
distilled water	100.0 ml
sulfuric acid, concentrated	0.3 ml

Sodium bisulfite, 2.5%

sodium bisulfite	2.5 g
distilled water	100.0 ml

Aldehyde-fuchsin

Add 1 g basic fuchsin (C.I. 42500) to 200 ml boiling water: boil 1 minute. Cool and filter. Add 2 ml concentrated hydrochloric acid and 2 ml paraldehyde. Leave stoppered at room temperature. When mixture has lost reddish fuchsin color and is deep purple (3–4 days), filter it and discard filtrate. Dry precipitate on filter paper in an oven. Remove and store in bottle. Makes about 1.9 g. To make staining solution, dissolve 0.25 g in 50 ml of 70% alcohol. Keeps at least six months.

Paraldehyde decomposes readily. Freshly opened paraldehyde will give excellent results, but a bottle that has been open for a long time may give negative results (Gairdner 1969).

PROCEDURE

1. Deparaffinize and hydrate slides to water; remove $HgCl_2$.
2. Oxidize in potassium permanganate: 1 minute.
3. Rinse in distilled water.
4. Bleach in sodium bisulfite until permanganate color is removed.
5. Wash in running water: 5 minutes.
6. Transfer to 70% alcohol: 2 minutes.
7. Stain in aldehyde-fuchsin: 2–10 minutes.
8. Wipe off back of slide and rinse in 95% alcohol.
9. Differentiate in 95% alcohol until no more aldehyde-fuchsin comes out of sections.
10. Dehydrate in absolute alcohol, clear, and mount.

RESULTS

elastin—deep purple
mast cells, chief cells of gastric mucosa, beta cells of pancreas, basophils of pituitary, and some kinds of mucin—purple.

COMMENTS

This method can be combined with Perl (page 269) to show both iron and elastin (Pickett and Klavins 1961).

See also Krutsay (1962a) about the use of permanganate-resorcin-fuchsin for elastin.

See Puchtler and Sweat (1960 and 1964) concerning resorcin-fuchsin staining.

Orcein (Romeis 1948)

FIXATION

Any general fixative.

SOLUTION

orcein	1.0 g
70% ethyl alcohol	100.0 ml
hydrochloric acid (assay 37–38%)	1.0 ml

PROCEDURE

1. Deparaffinize and hydrate slides to water; remove $HgCl_2$.
2. Stain in orcein: 30–60 minutes.
3. Wash briefly in distilled water.
4. Dehydrate in 95% alcohol: 2 minutes.
5. Differentiate in absolute alcohol until background is almost colorless and elastin fibers are isolated.
6. Rinse in fresh absolute alcohol, clear, and mount.

RESULT

elastin—red

COMMENTS

Lillie et al. (1968) stain in orcein, 4 hours, and differentiate in ferric chloride hexahydrate (0.1 ml of 10% $FeCl_3 \cdot 6H_2O$ in 50 ml of 70% ethanol). Dehydrate and clear. This changes the elastin color to black or reddish black.

Roman et al. (1967) combine orcein, hematoxylin, ferric chloride, and iodine in a single solution and stain for two hours. Differentiation is controlled in a ferric chloride solution, and the elastic fibers are stained a deep purple.

Humason and Lushbaugh (1969) combine silver nitrate, orcein, and Sirius supra blue stains to show reticulum, elastin, and collagen in the same section.

Some disagreement appears in the literature concerning the pH at which orcein is most effective. Weiss (1954) writes that orcein stains only from an acid alcoholic solution between pH 3 and 8 and that this puts orcein in a category between basic and acidic dyes, which generally operate at the extremes of alkaline and basic pH. He suggests that there is a formation of hydrogen bond between orcein and elastin. Since this reaction takes place in acid alcohol, it is probably due to a uniquely low positive charge of elastin in such solutions. The use of alcoholic solutions is necessary to stabilize the positively charged orcein fractions. Dempsey and Lansing (1954) agree with Weiss.

Darrow (1952), experimenting with orcein, found that pH 1–2.4 is best for elastin staining, and that above pH 2.6 collagen stains as well. A dye content of 0.4% is adequate for elastin staining; a higher concentration adds to collagen staining. For a specific elastin reaction when collagen is present, therefore, it is advisable to check the *p*H of the acid alcohol used for the solution and to reduce the dye content to 0.4 g per 100 ml.

SUBCUTANEOUS TISSUE STAINING

Subcutaneous tissue, areolar tissue, and omentum (membranes) can be easily fixed for staining. A simple way is to spread a piece on a slide with dissecting needles, allow to dry slightly, and immerse in fixative. The tissue, in most cases, will adhere to the slide throughout staining procedures.

If histologic rings are not available for spreading tissue, cut thick filter paper into circles of a desired diameter. Cut a hole in the center of the circle, leaving an outer edge of 6–7 mm in width. Place the circle of filter paper under the omentum or other tissue to be spread and press against the tissue. Cut out a ring of the tissue, just beyond the outer edge of the paper circle, and place in fixative. Carry filter paper and tissue through staining procedures; connective tissue stains can be applied as usual. Finally remove the paper by peeling it off when the tissue is being placed on a slide preparatory to mounting with cover glass.

Bits of loose connective tissue that are difficult to handle may be dehydrated and infiltrated with nitrocellulose. Carefully spread the tissue in a film of nitrocellulose on a slide, allow to dry slightly, and harden in 70% alcohol. For more uniform staining, stain the tissue in place on the slide and carry it through solutions like a nitrocellulose section. This method gives more uniform staining than the dry method above.

BONE STAINING

Decalcified Sections (Romeis 1948)

FIXATION

Formalin, followed by decalcification (pages 28–31).

SOLUTIONS

Thionin

thionin (C.I. 52000) saturated in 50% ethyl alcohol	10.0 ml
distilled water	100.0 ml

If the color does not set correctly, add 1–2 drops of ammonia to staining solution.

Carbol-xylol, see page 532.

PROCEDURE
1. Deparaffinize and hydrate slides to water. If sections tend to loosen, treat with 0.5–1% nitrocellulose (page 67) or try subbed slides (page 538).
2. Stain in thionin solution: 10 minutes.
3. Wash, distilled water: 20 minutes, change several times.
4. Treat with picric acid, saturated aqueous: 1 minute.
5. Rinse in distilled water.
6. Differentiate in 70% alcohol: 5–10 minutes or more, until no more color comes off.
7. Dehydrate in 95% alcohol, 2 changes: 3 minutes each.
8. Dehydrate and clear in carbol-xylol: 5 minutes.
9. Clear in xylene and mount.

RESULTS
lacunae and canaliculi—bordered with bluish black
background—yellow

COMMENTS
Since paraffin sections occasionally loosen from the slides, some technicians prefer nitrocellulose embedding, but this takes more time.

Hand-Ground, Undecalcified Sections (Enlow 1954)

PROCEDURE
1. If it is necessary to remove organic materials and fat, treat in following manner:
   a. Boil in soap solution: 3–4 hours; and wash in running water: 3–4 hours.
   b. Suspend over ether or chloroform: 36–48 hours.
   c. Allow to dry thoroughly.
2. Cut slices as thin as possible without cracking them. A jeweler's saw is recommended.
3. Grind on sharpening hones with finely powdered Carborundum or household cleansing powder. Keep surface wet with water. Optical or metallurgical grinding and polishing equipment, if available,

can be used. The grinding can be done between two stones, or the section can be held on a wet rubber or cork stopper and ground against the hone. If this becomes difficult to manage, glue the section on a slide with Duco cement, and continue to grind. When the section is almost thin enough, loosen it from the slide with acetone, turn it over, glue it down, and grind the opposite side.

4. Polish off both sides with a fine leather strop.
5. Place in 95% alcohol: 5 minutes.
6. Air dry.
7. Place in plastic solution; agitate to liberate air bubbles.

parloidin	28.0 g
butyl (or amyl) acetate	250.0 ml

Let stand until parloidin is completely dissolved. Stir thoroughly.

8. Transfer to slide with a drop of solution.
9. Dry thoroughly; do not add more solution.
10. When completely dry, add mounting medium and cover glass.

COMMENTS

To bring out the density and distribution of the mineral in undecalcified bone, Frost (1959) recommends methods with basic fuchsin, silver nitrate, alizarine red S, and others.

Dowding (1959) uses methyl methacrylate as a plastic embedding medium and grinds the bone down as thin as 30μ.

Hause (1959) describes a block for holding the bone while sawing it and recommends a razor saw blade no. 35-ST[4] as a good cutting instrument.

Kropp (1954) describes a plastic-embedding method using heat and pressure.

Yaeger (1958) uses freeze drying and vacuum infiltration with butyl methacrylate–ethyl methacrylate.

Norris and Jenkins (1960) describe a method using epoxy resin for preparation of bone for radioautography. The resolution is good; there are no chemicals in resin to produce artifacts in nuclear emulsions; the medium does not warp or chip when machined or abraded. (Polyesters and methacrylate do distort when machined.) They include a design for a metal and lucite mold for embedding. Sections can be made with a microtome (6–10μ), with a circular saw (50–100μ), or ground to give sections as thin as 10–50μ. Kwan (1970) uses sticky wax and a circular diamond blade. Enlow (1961) describes staining of ground and de-

---

[4] Lipshaw Company.

calcified sections. Villanueva et al. (1964) stain ground sections with fast green FCF, orange G, celestin blue, and basic fuchsin.

Alizarine Red S Method for Embryos (Bone Formation)

FIXATION

A hardening action. Hollister (1934) suggests 70% alcohol for several days for fish; removal of scales desirable. Hood and Neill (1948) use 95% alcohol: 3 days. Organism preferably should be free of hair or feathers. Richmond and Bennett (1938) use 95% alcohol: 2 weeks. Some specimens require decolorization. The best method is to lay specimen in 95% alcohol in white tray. Place in direct sunlight for 24 hours each side. A sunlamp was used by Hollister on sunless days. Youngpeter (1964) bleaches specimens in 30% hydrogen peroxide: 1–2 hours, after step 2 in procedure below. Then he immerses in KOH until specimen is transparent.

SOLUTIONS

Potassium hydroxide, 2%

potassium hydroxide, white sticks	20.0 g
distilled water	1000.0 ml

Alizarine stock solution (Hollister 1934)

alizarine red S (C.I. 58005), saturated in 50% acetic acid	5.0 ml
glycerine	10.0 ml
chloral hydrate, 1% aqueous	60.0 ml

Alizarine working solution

alizarine stock solution	1.0 ml
1–2% potassium hydroxide in distilled water	1000.0 ml

Make up at least 500 ml; use at room temperature.

Clearing Solutions (Youngpeter 1967)

Solution A

glycerine	20 parts
4% potassium hydroxide	3 parts
distilled water	77 parts

Solution B

glycerine	50 parts
4% potassium hydroxide	3 parts
distilled water	47 parts

Solution C

glycerine	75 parts
distilled water	25 parts

PROCEDURE

1. From hardening solution, rinse few minutes in distilled water.
2. Leave in 2% potassium hydroxide until skeleton shows through musculature: 2–4 hours for small embryos, 48 hours or longer for larger forms. St. Amand and St. Amand (1951) warmed solution to 38°C for quicker action.
3. When clear, transfer to alizarine working solution: 6–12 hours or longer depending on specimen. Skeleton should be deep red. Large specimens may require fresh changes of dye solution.
4. Transfer directly to 2% potassium hydroxide: 1 day or longer until soft tissues are destained. The KOH may be 1.0–0.5% for small specimens. Sunlight or lamp speeds up process.
5. Transfer through the three clearing solutions: 24 hours each.
6. Transfer to pure glycerine with thymol added as preservative.
7. Store in sealed tubes or bottles. Mount on glass rods and seal in museum jars, or embed in plastic.

COMMENTS

Cumley et al. (1939) gradually replaces the glycerine with 95% alcohol, absolute alcohol, and finally toluene. Then the specimens are transferred to toluene saturated with naphthalene and stored in anise oil saturated with naphthalene. This method is supposed to produce greater clarity than the glycerine storage.

Crary (1962) adds alizarine red S to the KOH solution and clears in a mixture of glycerine, 70% ethanol, and benzyl alcohol.

Alizarine is the most specific stain for calcium. The red coloring depends on the presence of a Ca base; the only other substance staining red is strontium.

Taylor (1967) uses pancreatic enzymes to clear small vertebrates.

Ojeda et al. (1970) stain the cartilage of whole chick embryos with Alcian blue.

Plastic mounts can be made of these preparations. Excellent directions are supplied in: *Embedding Specimens in Transparent Plastic,* Turtox

Service Leaflet #33 (Chicago, Illinois, General Biological Supply House) and *How to Embed in Bio-Plastic* (Rochester, N.Y., Ward's Natural Science Establishment).

MUSCLE STAINING

Milligan Trichrome Stain (1946)

FIXATION

10% formalin in normal saline (I have used Gomori 1-2-3 fixative with excellent staining results).

SOLUTIONS

Mordant

Solution A

potassium dichromate	3.0	g
distilled water	100.0	ml

Solution B

hydrochloric acid, concentrated	10.0	ml
95% ethyl alcohol	100.0	ml

Mix 3 parts A with 1 part B; use within 4 hours.

Acid fuchsin

acid fuchsin (C.I. 42685)	0.1	g
distilled water	100.0	ml

Phosphomolybdic acid

phosphomolybdic acid	2.0	g
distilled water	200.0	ml

Use half of solution for orange G solution below.

Orange G

orange G (C.I. 16230)	2.0	g
1% phosphomolybdic acid	100.0	ml

Fast green

Stock solution

fast green FCF (C.I. 42053)	1.0 g
2% acetic acid (2 ml/98 ml distilled water)	100.0 ml

Working solution

fast green stock solution	10.0 ml
distilled water	90.0 ml

Aniline blue may be substituted for fast green.

PROCEDURE

1. Deparaffinize and transfer slides through absolute alcohol into 95% alcohol (remove $HgCl_2$ if present).
2. Mordant in potassium dichromate–hydrochloric acid solution: 5–7 minutes.
3. Rinse in distilled water.
4. Stain in acid fuchsin: 5–8 minutes.
5. Rinse in distilled water.
6. Fix stain in phosphomolybdic acid solution: 1–5 minutes.
7. Stain in orange G: 5–10 minutes.
8. Rinse in distilled water.
9. Treat with 1% aqueous acetic acid (1 ml/99 ml distilled water): 2 minutes.
10. Stain in fast green: 5–10 minutes.
11. Treat with 1% acetic acid: 3 minutes.
12. Rinse in 95% alcohol, transfer to second 95% alcohol: 5 minutes.
13. Finish dehydration in absolute alcohol, 2 changes: 3 minutes each.
14. Clear and mount.

RESULTS

nuclei, muscle—magenta
collagen—green (blue with aniline blue)
red blood cells—orange to orange red

COMMENTS

This is a beautiful precise stain showing strong contrast between muscle and connective tissue. Smooth muscle cells stand out sharply and clearly. Milligan reports that this stain's weak point is the nuclear stain, which is improved by using Gomori 1-2-3 fixative.

Galigher[5] recommended pinning out strips of muscle on a smooth piece

---

[5] Personal communication.

of wood. Allow just enough tension to hold the muscle straight without stretching it. Fix in Susa fixative (page 17): 3–6 hours. Stain the sections by the Azan method (page 177). The bands of the striations appear in red, blue, or yellow. Galigher also found phosphotungstic acid–hematoxylin to be a useful stain for muscle.

Iron hematoxylin staining (page 160) can produce beautiful striation staining if it is followed by careful destaining with iron alum instead of picric acid. Wash thoroughly (at least 30 minutes) in running water after destaining; do not counterstain.

Zenker fixative is probably the best fluid for the preservation of intercalated discs.

See Poley and Forbes (1964) for staining to show muscle infarction.

See Puchtler et al. (1969) for a PAS-myofibril stain that is excellent for muscle fiber bands.

CHAPTER **14**

Silver Impregnating I: Reticulum

SILVER IMPREGNATION

According to Baker (1958), methods under this heading depend upon the local formation within tissues of a colored substance that is not a dye. Impregnation applies to the condition which develops when an unreduced metal (silver, etc.) is taken up from a solution of salt or other compound and deposited in a colloidal state on a tissue element. Following impregnation, the tissue is removed to a reducing solution of a photographic type and the metal is reduced to the elementary state, probably in the form of a black deposit. Thus the tissue itself does not reduce the metal, but some extraneous reducer is required to perform the reaction.

Silver staining started in 1843 when Krause tried small pieces of fresh tissue in silver nitrate. Golgi, an Italian, in 1872 fixed nerve tissue for a long period of time in a dichromate solution. Then he tried soaking the same tissue in silver nitrate solution and found that silver dichromate was deposited selectively, leaving the impregnated components in sharp relief against an almost colorless background. Ramón y Cajal (1903, 1910) saw the possibilities of the silver method and experimented with it further, including ways of reducing it. In 1906, he and Golgi were jointly awarded the Nobel Prize in physiology and medicine. By continuing to improve the use of silver and gold impregnations, Ramón y Cajal and del Río-Hortega, one of his pupils, were able to systematically investigate the histology of the nervous system. In the early 1900s Protargol (Bodian method) and other organic silver compounds were introduced as substitutes for silver nitrate. The albumen fraction in the organic compounds is considered to act as a protective colloid that prevents too rapid reduction by form-aldehyde, thereby inducing the formation of finer-grained deposits of silver. Subsequent experimentation with silver on tissue sections led to many modifications of the methods of Ramón y Cajal, Bielschowsky, and others. Silver impregnation can be very complicated because of the many factors involved to make it specific for various tissue elements.

Ammoniacal silver is the familiar complex and, when reduced by form-aldehyde to metallic silver, forms a colloidal solution containing negatively charged particles. These may be precipitated out by oppositely charged surfaces, which can be changed to repel or to absorb the silver. Thus, the negatively charged silver (formed by reduction) is deposited on positively charged surfaces and allows selective impregnation of neurofibrils, retic-ulum, Golgi, and so on. It is probable that fixation and dehydration coagulate the proteins and leave them positively or negatively charged; this would account for the difference in charges of different tissue ele-ments. The pH of the silver solution is a strong factor in determining charges, and it depends on whether an ammonium or sodium hydroxide or sodium or lithium carbonate is used (Ramón y Cajal and del Río-Hortega methods).

As already mentioned, protective colloids, such as gum arabic and mas-tic, and the use of Protargol slow down the reduction of the silver to produce a finer grain (Liesegang method). Dilute reducing reagents, combined with the above, can have the same effect (Ramón y Cajal and Bielschowsky methods). Temperature is a factor, because it increases the kinetic energy of the particles and permits a greater number of collisions of the particles against the tissue surfaces. In some methods, copper is added to the silver solution, supposedly to speed up impregnation by initiating the reduction to metallic silver. Too heavy a deposit is prevented by removal of some of the silver from the solution. Thus, various applica-tions of the principles can be used to control the impregnation of different kinds of tissue elements (Silver 1942). Chemical properties of tissues and their responses to these conditions all help to determine the place and amount of deposition.

The types of silver impregnation (all ammoniacal) can be classified in the following manner: (1) ammoniacal silver nitrate, (2) ammoniacal silver hydroxide, and (3) ammoniacal silver carbonate. In all of these solutions the silver is present largely in the form of a complex silver ammonia cation $[Ag(NH_3)_2]^+$. In (1), ammonia alone is used to form the precipitate, the chief product in solution being silver diammino nitrate. In (2), the Bielschowsky method (also modified by Ramón y Cajal), sodium hydroxide is used to form the precipitate and ammonia to redissolve it, the chief product being silver diammino hydroxide. The difference between the Bielschowsky and the Ramón y Cajal methods lies in the way the silver is applied and the reduction performed. The Ramón y Cajal method uses a single ammoniated silver impregnation followed by reduction in Pyrogallol, hydroquinone, or one of the aminophenols. The Bielschowsky method uses double impregnation (silver nitrate followed by ammoniated silver) and reduction in formalin. In (3), the del Río-Hortega method, sodium carbonate (sometimes lithium carbonate) is used to form the precipitate and is followed by ammonia, the chief product being silver diammino carbonate. (Kubie and Davidson 1928.)

The reactions of these solutions to various conditions should be understood:

1. The ammoniacal silver nitrate is most stable, least sensitive to light, and least readily reduced, and it combines least easily with tissues. During formalin reduction, a cloud of finely divided gray dust slowly develops, and the staining is slow. This method is rarely used in preference to others.

2. Ammoniacal silver hydroxide is the least stable, the most sensitive to light, and the most readily reduced, and it combines most easily with tissues. Almost instantly a heavy black cloud appears. It combines almost at once with the tissue, the solution darkens quickly, and a precipitate begins to form. Since silver nitrate is not reduced in an acid solution, but reduces readily in an alkaline solution, it stands to reason that the ammoniacal silver hydroxide solution is the most sensitive of the three.

3. The properties of the ammoniacal silver carbonate solution are between those of the above two. Its precipitate forms more promptly than that of ammoniacal silver nitrate (1), but not as fast as that of ammoniacal silver hydroxide (2). Its precipitate is darker than that of 1 but not as dark as that of 2. The solution's reaction begins within 5 to 10 minutes and reaches optimum color before the solution begins to darken. Although solution 2 and this one are used interchangeably, this solution has the advantage in that its hydroxide $(OH)^-$ ion concentration is not high enough to render it as unstable and oversensitive as solution 2, and the presence of buffer salts makes the reduction proceed steadily and evenly.

As the acid (HNO_3) is formed during the reduction the buffer absorbs it and blocks its effect; thus the reduction is not lessened. In addition, the presence of CO_3 ion buffers the formalin and prevents formation of formic acid, also an effective stop to further reduction. Foot (1929) buffered his formalin and prolonged the reducing action, making his results darker and more intense. He warned, however, that the buffer must be kept to a minimum so the reaction would not become too intense. Equi-molar silver solutions produce the most uniform results. In most cases, it is wise not to dissolve completely the precipitate that is first formed since this can cause inferior results.

The use of pyridine, a fat solvent, precedes some methods, removes cephalin, lecithin, myelin, mitochondria, galactolipids, etc., and makes subsequent penetration of the silver easier. (This is particularly true for connective tissue.) Toning with gold chloride is optional in many methods, but it may yield a more desirable color and improve contrasts. The timing in gold chloride apparently is variable. It need be only long enough to make the desired change in color, usually a few seconds. If the reaction is slow, the solution has weakened. The final fixing in sodium thiosulfate (hypo) is necessary to remove all unreduced silver.

Corked bottles should be avoided, since cork extractives may disrupt selective impregnation of tissue elements (Deck and DeSouza 1959). In all silver methods, take care that metal instruments do not come in contact with the silver solution; a black precipitate may dribble down the surface of slides handled in such a manner. Coat forceps with paraffin, or use horn or wooden instruments. All glassware for metallic stains (silver, gold, etc.) must be acid-clean. Soak in cleaning solution (page 555), wash thoroughly in running water to remove cleaning solution, and rinse 4 or 5 times with distilled water.

The loosening of sections from slides is a common problem during silver impregnation. Davis and Harmon (1949) use a rinse (0.5 ml acetic acid, 2% aqueous, in 50.0 ml of water) before reduction in sulfite and hydroquinone in the Bodian method (page 227). Many technicians find that Masson gelatin fixative (page 537) is superior to Mayer albumen (page 536) for affixing sections for silver processes. Transferring unmounted paraffin sections through all solutions and mounting and deparaffinizing at the conclusion of the method is feasible. I use subbed slides (page 538).

Prepare ammoniacal silver solutions just before use. Smith (1943) warns that ammoniacal silver hydroxide solutions will become explosive if they stand for too long a time. Explosions have occurred in several laboratories as the result of the formation of explosive silver amide in ammoniacal silver hydroxide solutions that have been stored for some time.

For complete discussions concerning silver impregnation, see Baker (1958); Beech and Davenport (1933); Bensley (1959); Foot (1929); Kubie and Davidson (1928); Long (1948); and Silver (1942).

SILVER IMPREGNATION FOR RETICULUM

Among the numerous impregnation methods for reticulum, the following, when properly handled, have never failed to produce precise results for me, and usually with a minimal loss of sections. Subbed slides (page 538) can be used to reduce loss of sections.

del Río-Hortega Method (Mallory 1944)

FIXATION

Any good general fixative.

SOLUTIONS

Ammoniacal silver carbonate

To 10 ml of 10% silver nitrate (1 g/10 ml distilled water) add 10 ml of saturated aqueous lithium carbonate. Shake; allow precipitate to settle. Decant and wash precipitate with distilled water 5 times. Add 25 ml of distilled water. Almost dissolve precipitate with ammonia (28% reagent), drop by drop, with shaking. Allow a few grains of precipitate to remain. Add 95% alcohol up to 100 ml. Filter. Warm uncovered at 50°C for 20 minutes. Ready for use. Can be reused, but heat to 50°C and filter before doing so. Keep in brown bottle.

Formalin solution

formalin	10.0 ml
distilled water	50.0 ml

Gold chloride

gold chloride 1% stock solution (1 g/100 ml water)	12.5 ml
distilled water	50.0 ml

PROCEDURE

1. Deparaffinize and hydrate slides to water; remove $HgCl_2$.
2. Treat with potassium permanganate, 0.2% (0.2 g/100 ml water): 3 minutes.
3. Wash in distilled water: 2 minutes.
4. Bleach in oxalic acid, 5% (5 g/100 ml water): 3 minutes.
5. Wash thoroughly in distilled water, several changes: 10 minutes.
6. Impregnate with silver solution in 37°C oven: 15–30 minutes. Do not expose to bright light. Keep metal instruments out of solution.
7. Rinse quickly in distilled water.
8. Reduce in formalin solution: 3 minutes.
9. Wash in distilled water: 3 minutes.
10. Gold tone until yellow color turns to purplish gray.
11. Rinse briefly in distilled water.
12. Fix in sodium thiosulfate, 5% (5 g/100 ml water): 3 minutes.
13. Wash in running water: 5 minutes.
14. Counterstain (if desired) in hematoxylin and picro-ponceau (Van Gieson), page 188.
15. Dehydrate, clear, and mount.

RESULTS

reticulum—black
collagen—red to rose
nuclei—black, blue or brownish
cytoplasm—grayish yellow
muscle fibers, elastin—light yellow

COMMENTS

To prevent loosening of sections, Mallory (1944) substituted an alcoholic for the aqueous silver solution of the original method.

Gridley Method (1951)

FIXATION

Any good general fixative.

SOLUTION

Ammoniacal silver hydroxide

To 20.0 ml of 5% silver nitrate (1 g/20 ml distilled water) add 20 drops of 10% sodium hydroxide. Add fresh 28% (reagent) ammonia drop by drop until precipitate which forms is almost redissolved. Add distilled water up to 60.0 ml.

PROCEDURE

1. Deparaffinize and hydrate slides to water. Remove $HgCl_2$ if present.
2. Treat with 0.5% periodic acid (0.5 g/100 ml water): 15 minutes.
3. Rinse in distilled water.
4. Treat with 2% silver nitrate (2 g/100 ml water): 30 minutes, room temperature.
5. Rinse in 2 changes of distilled water.
6. Impregnate in ammoniacal silver solution: 15 minutes, room temperature.
7. Rinse rapidly in distilled water.
8. Reduce in 30% formalin (30 ml/70 ml water): 3 minutes. Agitate gently.
9. Rinse in 3 or 4 changes of distilled water.
10. Tone in gold chloride (10 ml 1% stock solution/40 ml water) until yellow-brown color has changed to lavender-gray.
11. Rinse in distilled water.
12. Fix in 5% sodium thiosulfate (5 g/100 ml water): 3 minutes.
13. Wash in running water: 5 minutes.
14. Counterstain if desired.
15. Dehydrate, clear, and mount.

RESULTS

reticulum fibers—black
other tissue elements—depends on counterstain

COMMENTS

Gridley uses periodic acid for oxidation in preference to potassium permanganate and oxalic acid because the latter combination frequently causes the sections to detach from the slides.

Gomori Method (Mallory 1944)

FIXATION

10% neutral formalin.

SOLUTIONS

Ammoniacal silver solution

To 20 ml of 10% silver nitrate solution (3 g/30 ml distilled water) and 4–5 ml of a 10% potassium hydroxide solution (0.5 g/5 ml distilled water), add 28% ammonia water, drop by drop, shaking the flask continuously, until the precipitate that forms is completely dissolved. Carefully add silver nitrate solution, drop by drop, until the precipitate that forms disappears when the solution is shaken. Make up the solution with distilled water to twice its volume. Always use acid-clean glassware.

Potassium permanganate

potassium permanganate	0.5 g
distilled water	100.0 ml

Potassium metabisulfite

potassium metabisulfite	2.0 g
distilled water	100.0 ml

Ferric ammonium sulfate

ferric ammonium sulfate	2.0 g
distilled water	100.0 ml

Formalin

formaldehyde, concentrated	20.0 ml
distilled water	80.0 ml

Gold chloride solution

gold chloride stock solution (1 g/100 ml water)	20.0 ml
distilled water	80.0 ml

Sodium thiosulfate

sodium thiosulfate	2.0 gm
distilled water	100.0 ml

PROCEDURE

1. Deparaffinize and hydrate slides to water.
2. Oxidize in potassium permanganate solution: 1 minute.
3. Wash in tap water: 2 minutes.
4. Decolorize in potassium metabisulfite: 1 minute.
5. Wash in tap water: 2 minutes.

6. Sensitize in ferric ammonium sulfate: 1 minute.
7. Wash in tap water: 2 minutes. Rinse in 2 changes of distilled water: 30 seconds each.
8. Impregnate in silver solution: 1 minute.
9. Rinse in distilled water: 20 seconds.
10. Reduce in formalin solution: 3 minutes.
11. Wash in tap water: 3 minutes.
12. Tone in gold chloride: 10 minutes. Sections turn purplish gray.
13. Rinse in distilled water.
14. Reduce in potassium metabisulfite: 1 minute.
15. Fix in sodium thiosulfate: 1 minute.
16. Wash in tap water: 2 minutes.
17. Counterstain, if desired; dehydrate, clear, and mount.

RESULT

reticulum fibers—black

COMMENTS

This is a good reliable method; it is quick, and I have had no loosening of sections with this procedure.

Nassar and Shanklin Method (1961)

FIXATION

Any good general fixative.

SOLUTIONS

Potassium permanganate

Solution A

potassium permanganate	0.5	g
distilled water	100.0	ml

Solution B

sulfuric acid, concentrated	0.5	ml
distilled water	100.0	ml

Mix equal parts of A and B. Use freshly mixed.

Oxalic acid

oxalic acid	2.0	g
distilled water	100.0	ml

Silver nitrate

silver nitrate	1.0 g
distilled water	50.0 ml
pyridine	15.0 drops

Silver diammino nitrate

To 1 ml of 28% ammonia contained in a flask, add 7 ml of 10% silver nitrate (2.5 g/25 ml water). Continue to add 10% silver nitrate drop by drop, while shaking the ammonia solution, until only a faint permanent turbidity remains. Dilute with an equal amount of distilled water. Add 3 drops of pyridine per 10 ml of solution.

Reducing solution

2% neutral formalin (1 ml/49 ml water)	50.0 ml
absolute ethyl alcohol	50.0 ml

Gold chloride

gold chloride stock solution (1 g/100 ml water)	10.0 ml
distilled water	40.0 ml

PROCEDURE

1. Deparaffinize and hydrate slides to water. Remove $HgCl_2$ if present.
2. Oxidize in permanganate solution: 1–2 minutes.
3. Rinse in distilled water.
4. Decolorize in oxalic acid solution: 2 minutes.
5. Wash in running water: 5 minutes.
6. Transfer through 70% alcohol into 95% alcohol.
7. Treat with silver nitrate solution, 50°C: $\frac{1}{2}$–1 hour. At room temperature: $1\frac{1}{2}$–2 hours.
8. Rinse in 95% alcohol.
9. Impregnate in silver diammine solution, 50°C: 5 minutes.
10. Rinse quickly in 95% alcohol.
11. Reduce in formalin-alcohol: 2 minutes.
12. Wash in distilled water: 3–5 minutes.
13. Tone in gold chloride until color turns to purplish gray.
14. Rinse in distilled water.
15. Fix in sodium thiosulfate, 5% (5 g/100 ml water): 3–5 minutes.

16. Wash in running water: 5 minutes.
17. Counterstain with hematoxylin or other stain.
18. Dehydrate, clear, and mount.

RESULTS

reticulum fibers—black
other tissue elements—depends on counterstain

Wilder Method (1935)

FIXATION

Any good general fixative.

SOLUTIONS

Phosphomolybdic acid, 10%

phosphomolybdic acid	10.0	g
distilled water	100.0	ml

Uranium nitrate

uranium nitrate	1.0	g
distilled water	100.0	ml

Ammoniacal silver nitrate

Add ammonia (28% reagent), drop by drop to 5 ml of 10% silver nitrate (0.5 g/5 ml water) until precipitate that forms is almost dissolved. Add 5 ml of 3.1% sodium hydroxide (3.1 g/100 ml water). Barely dissolve the resulting precipitate with a few drops of ammonia. Make the solution up to 500 ml with distilled water. Use immediately. Glassware must be acid-clean.

Reducing solution

distilled water	50.0	ml
formalin	0.5	ml
uranium nitrate, 1% (above)	1.5	ml

Make up fresh each time.

Gold chloride

> gold chloride stock solution (1 g/100 ml water) . 10.0 ml
> distilled water .... 40.0–80.0 ml

PROCEDURE

1. Deparaffinize and hydrate slides to water; remove $HgCl_2$.
2. Wash thoroughly in distilled water.
3. Treat with phosphomolybdic acid: 1 minute.
4. Wash in running water: 5 minutes.
5. Treat with uranium nitrate: 5 seconds or less.
6. Rinse in distilled water.
7. Impregnate with ammoniacal silver nitrate solution: 1 minute.
8. Dip quickly in 95% alcohol and immediately into reducing solution: 1 minute.
9. Wash in distilled water: 2–3 minutes.
10. Tone in gold chloride until yellow colors turn purplish gray.
11. Brief rinse in distilled water.
12. Fix in sodium thiosulfate, 5% (5 g/100 ml water): 3–5 minutes.
13. Wash in running water: 5 minutes.
14. Counterstain if desired.
15. Dehydrate, clear, and mount.

RESULTS

reticulum fibers—black
other tissue elements—depends on counterstain

COMMENTS

Phosphomolybdic acid replaces potassium permanganate as an oxidizer; the phosphomolybdic acid shows less tendency to loosen sections. Sensitization with uranium nitrate reduces the time and eliminates the heat required by some reticulum methods. Lillie (1946) disagrees that uranium nitrate is a sensitizer and claims that it is an oxidizer. (For a combination reticulum, collagen, and elastin stain, see Humason and Lushbaugh 1967.) Fitzgerald and Pohlman (1969) adapted a method for reticulum using silver proteinate reduced by hydroquinone.

CHAPTER 15

Silver Impregnating II: Neurological Elements

For nervous tissue, frozen sections are best; alcohol and xylene embedding may remove lipids, which can be an essential part of the tissue. The loss of lipids may result in no impregnation of oligodendroglia and weakened microglia, since their impregnation depends on lipid complexes. Periodic and chromic acid oxidation also weaken the reaction of oligodendroglia and microglia. (The reverse is true for connective tissue. Then pyridine, periodic, or chromic acid treatment is desirable to suppress the impregnation of nervous tissue elements.)

Myelin, because of its high lipid content (cholesterol, cerebroside, and phospholipids) is soluble in fat solvents, and most of it dissolves away in dehydrating and clearing solutions. Empty spaces remain at the former sites of myelin. However, special fixatives can be used to preserve myelin through dehydration, clearing, and paraffin embedding; osmic acid fixes it efficiently and at the same time colors it black. Overnight chromation of tissue blocks in 3% aqueous potassium dichromate preserves formalin-fixed myelin for paraffin processing.

If tissues from the central nervous system are to be embedded in paraffin, they require long periods of fixation to harden them; they also require extended treatment with alcohol, clearing solutions, and paraffin. If the tissue is not sufficiently fixed and hardened, it will compress badly during sectioning and the sections may not adhere to the slides during staining. The softness and almost jelly-like consistency of nervous tissue is due to the absence of supporting tissues like those of ordinary connective tissue which has tough intercellular substances, such as elastin and collagen. Only a delicate and ectodermally derived cellular substance—neuroglial cells and fibers—supports the nervous tissue elements by lying between and binding together the nerve fibers and blood vessels. (Connective tissue wrappings—meninges—do cover the brain and spinal column.)

Ordinary stains, such as hematoxylin and eosin, do not demonstrate the innumerable processes of neuroglial cells, because only the nuclei stain. Silver nitrate methods impregnate the fibers and aid in classifying them. When impregnating tissues from the central nervous system, bear in mind these facts: if the hydroxide method is used, astrocytes and microglia are not specially impregnated; if the carbonate method is used, the opposite effect takes place. For impregnation, do not heat the solutions above normal body temperature.

Neurological techniques, as has perhaps become evident, so often necessitate highly specialized methods that many technicians prefer to avoid them. Precise attention to all details, however, can produce beautiful and exciting slides. Carefully follow directions for fixation, such as the solution composition, the duration of fixation, and whether fixation is or is not followed by washing. Always wash in distilled water unless tap water is specified. For making silver and other special solutions, use double (glass) distilled water if possible and glassware cleaned in cleaning solution (page 555), washed well in running water, and rinsed 4 or 5 times in distilled water. All chemicals should be at least reagent grade. Use no corks in containers and no metal instruments in silver solutions. If in doubt about the age of solutions make fresh ones. If while being toned with gold chloride the tissue retains a yellow or brownish hue, the gold chloride is weakened. Prepare a new solution.

Artifact precipitates are difficult to avoid, but strict adherence to procedure details will reduce them to a minimum.

The method of embedding and sectioning will depend on the impregnating or staining technique to follow. Paraffin and nitrocellulose methods are used at times, but, as mentioned above, frozen sections are usually more satisfactory. Sections of neurological tissue are frequently thicker than for other tissues—7 to 20μ, or even more. Mount sections on subbed slides (page 538).

When fixing an entire brain, do not allow it to rest on the bottom of the container. Carefully insert a cord under the circle of Willis on the underside of the brain and support the two ends of the cord on the sides of the container. The brain, hanging upside down, should be free of the bottom of the vessel but should remain completely submerged in fixative. The spinal cord can be supported in a graduated cylinder filled with fixative. Run a thread through one end of the spinal cord and tie the thread around an applicator stick supported on the edges of the cylinder. For a perfusion method for the brain, see page 26.

Note: In neurological techniques, some methods employ section staining, and some, block staining. The letters **S** and **B** will be used in this chapter to indicate which type of staining is being described.

NEUROLOGICAL STAINING AND IMPREGNATING

Heidelberger Victoria Blue (Proescher Modification 1934) **S**

FIXATION

10% formalin.

PROCEDURE

1. Cut frozen sections, 10–15μ.
2. Place sections in distilled water. If not stained at once they may be stored in 10% formalin for 24 hours, but no longer.
3. Stain in Victoria blue B (C.I. 44045), saturated aqueous solution: 12–24 hours.
4. Wash rapidly in distilled water and mount on albumen-coated slides. Blot with filter paper and dry in air.
5. Expose to ultraviolet light: 30 minutes.
6. Immerse in $N/20$ iodine solution (aqueous): few seconds.
7. Remove from iodine, blot, air dry for 10 minutes.
8. Differentiate in xylene-aniline (1:1), clear in 2 or 3 changes of xylene, and mount.

RESULTS

glia cell bodies and fibrils—light to deep blue
nerve cell bodies, dendrites, and axis cylinders—faintly stained

COMMENTS

1. Staining time can be shortened to 30–35 minutes at 56°C.
2. Proescher experimented with oxidizing agents—potassium chromate and dichromate and hydrogen peroxide—and found they could be

substituted for ultraviolet. Potassium dichromate was most successful and can be used as a 0.5% aqueous solution for 30 minutes.

3. Victoria blue may be replaced by methyl violet 2B, ethyl violet, or crystal violet.

Nassar and Shanklin Silver Impregnation (1951) S

FIXATION

4 days in:

formalin	70.0 ml
ammonium bromide	14.0 g
distilled water	680.0 ml

WASHING, EMBEDDING, AND SECTIONING

1. After fixation, wash tissue blocks in distilled water plus strong ammonia (50 drops ammonia to 100 ml water): 10 hours. Keep container covered to prevent escape of ammonia.
2. Wash in 2 changes of distilled water: 1–2 hours each.
3. Dehydrate, clear, and embed in paraffin. Cut sections at 10–15μ and prepare slides.

SOLUTION

To 1 ml of concentrated ammonia, rapidly add 7–8 ml of 10% aqueous silver nitrate. Continue to add 10% silver nitrate drop by drop, shaking each time, until only a faint turbidity remains (9–10 ml). Dilute with an equal amount of distilled water. (Lillie 1946.)

PROCEDURE

1. Deparaffinize and hydrate to 3 changes of distilled water: 1–2 minutes each.
2. Sensitize in 5% aqueous sodium sulfite: 2 hours.
3. Transfer quickly through 3 changes of distilled water.
4. Impregnate in ammoniated silver solution: 2–5 minutes, room temperature.
5. Dip in distilled water: 1–2 seconds.
6. Reduce in 2% formalin (2 ml/98 ml water): 1 minute. Agitate gently, and change formalin solution 3–4 times.
7. Wash in distilled water: 1 minute.
8. Tone in gold chloride (1 g/500 ml water): few seconds to 1 minute, or until section turns gray.

9. Fix in 5% aqueous sodium thiosulfate (5 g/100 ml water): 1–2 minutes.
10. Wash in running water: 5 minutes.
11. Counterstain if desired. Dehydrate, clear, and mount.

RESULT

neuroglia—dark gray to black

COMMENT

Del Río-Hortega silver carbonate solution also can be used in this method.

Penfield Modification of del Río-Hortega Silver Carbonate Method (McClung, 1950) S

FIXATION

10% formalin at least one week (longer fixation also will give excellent results); or fix in:

formalin	14.0 ml
ammonium bromide	2.0 g
distilled water	86.0 ml

SECTIONING

Frozen sections, 20μ.

SOLUTIONS

Globus hydrobromic acid

40% hydrobromic acid	5.0 ml
distilled water	95.0 ml

Silver carbonate solution

Combine 5.0 ml of 10% aqueous silver nitrate and 20.0 ml of 5% aqueous sodium carbonate. Add ammonium hydroxide, drop by drop, until precipitate is just dissolved. Add distilled water up to 75 ml. Filter. The solution keeps for long periods if stored in a dark bottle.

PROCEDURE

1. Place sections in 1% formalin or distilled water.
2. Transfer to distilled water plus 1% of ammonia. Cover to prevent escape of ammonia and leave overnight.

3. Transfer to hydrobromic acid, 38°C: 1 hour.
4. Wash in 3 changes of distilled water, 2 minutes in each.
5. Mordant in 5% aqueous sodium carbonate: 1 hour. Tissue may remain in this solution 5–6 hours with no ill effect.
6. Impregnate in silver solution: 3–5 minutes or until sections turn a smooth gray when transferred to reducer. Try single sections at 3 minutes, 5 minutes, or longer.
7. Reduce in 1% formalin. Agitate during reduction, 2 minutes.
8. Wash in distilled water, 1 minute.
9. Tone in gold chloride solution (1 g/500 ml water) until bluish gray.
10. Fix in 5% sodium thiosulfate: 3 minutes.
11. Wash in running water: 5 minutes. Dehydrate, clear, and mount.

RESULT

oligodendroglia and microglia—dark gray to black

COMMENTS

If only oligodendroglia are to be shown, shorten fixation time to 2 days. Long fixation tends to increase the staining of the microglia and astrocytes.

Procedure for oligodendroglia

1. Follow fixation by treatment with 95% alcohol: 36–48 hours.
2. Wash, freeze, and cut sections.
3. Stain in a stronger silver solution by diluting the silver carbonate solution to only 45 ml: 15 minutes or more, until the sections begin to turn brown.
4. Wash for a few seconds in distilled water.
5. Reduce with agitation, wash, and tone as above.
6. Dehydrate, clear, and mount.

The oligodendroglia will stain black.

ASTROCYTES

Ramón y Cajal Gold Chloride Sublimate Method (Mallory 1944) S

FIXATION

Fix thin slices of tissue in:

formalin	15.0 ml
ammonium bromide	2.0 g
distilled water	85.0 ml

Maintain at 37°C: 1 day.

Frozen sections, 15μ thick.

SOLUTIONS

Gold chloride sublimate solution

mercuric chloride crystals (*not* powder)	0.5 g
gold chloride, 1% aqueous	35.0 ml
distilled water	6.0 ml

Prepare fresh, using acid-clean glassware. Pulverize mercuric chloride; add to distilled water. Heat gently. When dissolved, add gold chloride, and filter.

Del Río-Hortega carbolxylol-creosote mixture

creosote	10.0 ml
phenol (carbolic acid), melted	10.0 ml
xylene	80.0 ml

PROCEDURE
1. Place sections in 1% formalin.
2. Wash quickly in distilled water, 2 changes.
3. Place in gold chloride. Flatten sections, keep in dark at room temperature: 4–6 hours. When sections appear purplish, examine them for astrocytes.
4. Wash in distilled water: 5–10 minutes.
5. Fix in 5% aqueous sodium thiosulfate: 5–10 minutes.
6. Wash in distilled water, several changes.
7. Float on subbed slide, blot gently, dehydrate with 95% alcohol, 3 minutes.
8. Clear with del Río-Hortega solution.
9. Blot and mount.

RESULTS
astrocytes and processes—black
background—unstained or light brownish purple
nerve cells—pale red
nerve fibers—unstained

COMMENTS
After being fixed in hypo, extra sections can be preserved in 1% formalin for long periods.

For reliable results staining should be done within three weeks of initial fixation. If tissues have been in formalin for a long period, treat the frozen sections as follows:

1. Place in ammonium hydroxide, 1 part to 9 parts water: 24 hours.
2. Rinse in 2 changes of distilled water.
3. Treat in 40% hydrobromic acid (HBr), 1 part to 9 parts water: 2–4 hours.
4. Rinse in 2 changes of distilled water with 0.5% of dilute ammonia (step 1) added.

NISSL SUBSTANCE

Cresyl Violet (Powers and Clark 1955) **S**

Nissl substance is characteristic of fixed nerve cells. It is found in a granular form distributed throughout the cytoplasm and stains brilliantly with basic aniline dyes.

FIXATION

Bouin recommended; others satisfactory; Zenker not recommended.

SECTIONING

Paraffin method, 10μ; thinner sections are of no advantage.

SOLUTIONS

Cresyl violet

cresyl violet–acetate	0.2 g
distilled water	150.0 ml

Buffer solution, pH 3.5

0.1 *M* (approx.) acetic acid (6 ml/1000 ml water)	94.0 ml
0.1 *M* (approx.) sodium acetate (13.6 g/1000 water)	6.0 ml

Working solution

buffer solution	100.0 ml
cresyl violet solution	6–12 ml

PROCEDURE

1. Deparaffinize and hydrate sections to distilled water.
2. Stain 20 minutes in working solution. Use solution only once.
3. Rinse quickly in 70% ethyl alcohol and 95% ethyl alcohol.
4. Dehydrate in isopropyl alcohol, 2 changes: 3–4 minutes each.
5. Clear and mount.

RESULT

Nissl substance—purple

COMMENTS

Steps 7, 8, and 9 may be repeated several times until desired differentiation is reached.

Banny and Clark (1950) recommend Matheson Coleman and Bell cresyl violet for Nissl staining.

Manns (1960) stains with lithium hematoxylin for myelin, and with cresyl fast violet for Nissl substance.

Thionin (Clark and Sperry 1945) S

FIXATION

Bouin recommended; others satisfactory; Zenker not recommended.

SECTIONING

10μ; thinner sections of no advantage.

SOLUTIONS

Lithium carbonate, 0.55%

lithium carbonate	5.5 g
distilled water	1000.0 ml

Thionin

thionin (C.I. 52000)	0.25 g
0.55% lithium carbonate	100.00 ml

PROCEDURE

1. Deparaffinize and hydrate slides to water; remove $HgCl_2$.
2. Treat with lithium carbonate solution: 5 minutes.
3. Overstain in thionin: 5–10 minutes.

4. Rinse in distilled water.
5. Dip in 70% alcohol: few seconds.
6. Dehydrate in butyl alcohol, 2 changes: 2–3 minutes in each.
7. Clear and mount.

RESULT

Nissl substance—bright blue

COMMENTS

If differentiation is necessary, briefly rinse slides in 95% ethyl alcohol (following step 5) and place in aniline, then in lithium carbonate saturated in 95% alcohol. Proceed to step 6.

Gallocyanin S

FIXATION

Any general fixative.

SECTIONING

10μ.

SOLUTION AND PROCEDURE

See pages 154 and 166.

RESULT

Nissl substance—blue

NEUROFIBRILS

Ramón y Cajal Method (Favorsky Modification 1930) B

FIXATION

Cut slices perpendicular to organ surface, about 5 cm thick. Place in 70% alcohol, plus 0.5% glacial acetic acid: 6 hours.

PROCEDURE

1. Transfer to 80% alcohol: 6 hours.
2. Treat with ammoniacal alcohol: 24–36 hours.
   a. For cerebrum, cerebellum, spinal cord, or ganglia, add 4 drops of ammonia to 50 ml of 95% alcohol.
   b. For medulla, add 9 drops of ammonia to 50 ml of 95% alcohol.

3. Wash in distilled water, several changes, until pieces sink.
4. Treat with pyridine: 1–2 days.
5. Wash in running water: overnight; then wash in distilled water, several changes.
6. Blot on filter paper and place in relatively large volume of 1.5% aqueous silver nitrate: 5 days in dark, 38°C.
7. Rinse in distilled water and place in following fluid: 24 hours. Ramón y Cajal's reducing fluid:

pyrogallic acid or hydroquinone	1.0 g
distilled water	100.0 ml
neutral formalin	15.0 ml

8. Rinse in distilled water, several changes over a period of at least 1 hour.
9. Dehydrate and embed in paraffin, celloidin, or double embed.
10. Make sections perpendicular to surface of organ and about 15 or more μ thick.
11. Affix to slides, dry, remove paraffin with xylene, and mount. Lay celloidin sections on slides and mount in resin.

RESULTS

neurofibrils—black
background—brownish yellow

NERVE CELLS, FIBERS, AND ENDINGS

Golgi Rapid Method (Mallory 1944) **B**

FIXATION

Fix small pieces in:

potassium dichromate, 2.5% aqueous	40.0 ml
osmic acid, 1% aqueous	10.0 ml

Place cotton in bottom of container so fixative has access to all sides: 1–7 days. Renew fixative if it becomes turbid.

STAINING

1. Dry pieces of tissue with filter paper, place in 0.75% aqueous silver nitrate: 24–48 hours. If solution turns yellow, renew it.
2. Transfer to 40% ethyl alcohol, several changes: 1–2 hours.
3. Transfer to 80% and 90% alcohol: 1 hour each.

EMBEDDING AND SECTIONING

1. Dehydrate in absolute ethyl alcohol: 12 hours.
2. Transfer to absolute alcohol–ether (1 : 1): 2–4 hours.
3. Infiltrate with 4% celloidin or nitrocellulose: 1–2 days.
4. Embed and cut thick sections, 20–25μ, or more if desirable.

MOUNTING

1. Wash sections carefully in 80% alcohol to remove excess silver.
2. Dehydrate in absolute alcohol: a few minutes. (Add a few ml chloroform to absolute alcohol to prevent loss of nitrocellulose.)
3. Clear in clove oil, terpinol, or the like.
4. Mount on slides, press with filter paper. Add a drop of thick mounting medium. Warm to drive off solvent and, just before resin cools, add cover glass.

RESULTS

nerve cells and processes—deep black
background—light yellow

COMMENTS

If tissue has been fixed in 10% formalin, blot off excess formalin and chromate for 1–2 days, room temperature, in one of the following fluids (Davenport and Combs 1954).

1. potassium dichromate, 5% aqueous	100.0	ml
glacial acetic acid	6.9	ml
2. zinc or cadmium chromate	1.5	g
distilled water	25.0	ml
glacial acetic acid	6.0	ml
when chromate has dissolved, add:		
potassium dichromate, 5% aqueous	65.0	ml
3. potassium dichromate	3.0	g
potassium acetate, anhydrous	1.8	g
distilled water	100.0	ml
4. A. cadmium nitrate (4H_2O) or zinc nitrate (6H_2O)	3.0	g
distilled water	25.0	ml
B. potassium dichromate, 5% aqueous	65.0	ml
glacial acetic acid	6.0	ml

Mix A and B.

Remove tissues from any of above fluids, blot, and place immediately in silver solution: 18–24 hours. Proceed into step 2 above (under Staining).

Moliner Modification (1957) of Golgi Rapid Method

FIXATION

Fix 3–5 mm pieces: 3 days; change daily.

potassium dichromate, 6% aqueous	40.0 ml
potassium chlorate, 5% aqueous	20.0 ml
chloral hydrate, 20% aqueous	30.0 ml
formalin	10.0 ml

STAINING

1. Transfer tissue blocks to 3% aqueous potassium dichromate: 3 days, change twice a day.
2. Transfer to 1% aqueous silver nitrate: 3 days, room temperature.

SECTIONING AND MOUNTING

1. Cut frozen sections, 50–100μ. Sections can be kept in absolute alcohol until mounted.
2. Mount on slides and clear. Synthetic mountants can be used with a cover glass. (Before synthetic mountants were developed, Golgi preparations had to be mounted in balsam without a cover glass.)

Ramón y Cajal Pyridine-Silver Method (Davenport et al. Modification 1934) B

FIXATION

As soon as possible in:

absolute ethyl alcohol	98.0 ml
ammonia, concentrated	2.0 ml

Fix for 1–6 days, preferably no longer.

PROCEDURE

1. Treat tissue blocks in 5% aqueous pyridine: 24 hours. This is recommended; but Davenport et al. say it can be optional.
2. Wash in distilled water: 2–6 hours; change every half hour.
3. Impregnate in 1.5–2.0% aqueous silver nitrate, 37°C: 2 to 3 days, or longer, depending on size of tissue blocks. A minimum time yields the best-differentiated tissue. The longer the tissue is left in silver nitrate, the greater the tendency for everything to stain. The time will have to be determined by trial and error.
4. Wash in distilled water: 20 minutes to 1 hour. This is a critical step. The amount of silver washed from the tissue depends on whether

the water is changed often or whether the tissue is shaken in the water; change every 10 minutes or use a large volume of water and shake every 10 minutes. Equal staining of the periphery as well as central parts of the tissue determines correct washing. A light periphery indicates too much washing.

5. Reduce in 4% aqueous Pyrogallol: 4 hours.
6. Dehydrate, clear, embed, and section.
7. Mount on slides, deparaffinize, and clear. Add mountant and cover glass.

RESULTS

nerve—yellow to brown
neurofibrils—brown to black
axis cylinders of myelinated fibers—yellow to brown
axis cylinders of nonmyelinated fibers—black

COMMENTS

If the preparation is too light, reduce the time in pyridine and washing.

If the preparation is too dark, omit the pyridine and wash 48 hours after fixation, or reduce the concentration of the silver nitrate solution and wash longer between impregnation and reduction.

Bielschowsky Method (Davenport et al. Modification 1934) B

FIXATION

10% formalin: 2 days. For embryos add 0.5% of trichloracetic acid to 10% formalin.

SOLUTION

Add 5.0 ml of concentrated ammonia to 40.0 ml of 2% aqueous sodium hydroxide. Mix well. Add slowly from a burette or pipette 8.5% aqueous silver nitrate until opalescence remains in the solution (about 40.0 ml). Shake the hydroxide solution while adding the silver nitrate. Add 0.5–1.0 ml of ammonia. Dilute with about 5 parts of distilled water: dilution is not critical.

PROCEDURE

1. After fixation, wash: 1 hour.
2. Transfer to 50% aqueous pyridine: 1–2 days.
3. Wash in distilled water: 2–6 hours, depending on size of block. Change every half hour.
4. Impregnate with 1–1.5% aqueous silver nitrate: 3 days, 37°C.

5. Wash in distilled water: 20 minutes to 1 hour, depending on size. Change every 10 minutes or use a large volume of water and shake every 10 minutes. Periphery and central portions must be equally stained.
6. Impregnate in ammoniated silver solution: 6–24 hours.
7. Wash in distilled water: 15 minutes to 1 or 2 hours, depending on size.
8. Reduce in 1% formalin: 6–12 hours.
9. Wash in running water: 10–15 minutes.
10. Dehydrate, clear, and embed. Section and mount.
11. Deparaffinize, clear, and cover.

Alternate method (if gold toning is desired):

a. Following step 9, rinse in distilled water.
b. Gold tone and fix.
c. Return to step 10.

RESULTS

nerve fibers, neurofibrils—brown to black (no gold toning)
—gray to black (with gold toning)

Fluorescent Method (Zeiger et al. 1951) S

FIXATION

95% ethyl alcohol.

SECTIONING

Paraffin method.

PROCEDURE

1. Deparaffinize and hydrate slides to water.
2. Stain in 0.1% aqueous acridine orange (C.I. 46005): 6 minutes.
3. Differentiate in 95% ethyl alcohol: 2 seconds.
4. Blot with filter paper and mount in fluorescent mountant.

RESULTS

nonmyelinated fibers—bluish gray
myelinated fibers—brownish orange

COMMENTS

Fresh tissue can be frozen, cut, and stained in acridine orange made up in physiological saline or in Ringer solution.

Bodian Method (1936, 1937) **S**

FIXATION

Williams (1962) recommends:

formalin	40.0 ml
glacial acetic acid	10.0 ml
80% ethyl alcohol	100.0 ml
picric acid	2.0 g

Not suitable are chromates, chromic, and osmic acid, or mercuric chloride. 10% formalin causes excessive staining of connective tissue.

SECTIONING

Paraffin method.

SOLUTIONS

Protargol solution

Protargol [1]	1.0 g
distilled water	100.0 ml

Sprinkle Protargol on surface of water in a wide dish or beaker. Do not stir—this is critical. Keep on hot plate at 37°C until Protargol is dissolved.

Reducing solution

hydroquinone	1.0 g
sodium sulfite	5.0 g
distilled water	100.0 ml

Gold chloride

gold chloride	1.0 g
distilled water	100.0 ml

Acidified water

glacial acetic acid	0.5 g
distilled water	100.0 ml

---

[1] Roboz Surgical Instrument Co., imported "Silver Protein"; or Winthrop Laboratories, American "Protargol S."

PROCEDURE

1. Deparaffinize and hydrate slides to water.
2. Impregnate in Protargol solution plus 5 g clean copper (wire, shot or sheet, cleaned with 1 part HNO_3 to 3 parts HCl) to 100 ml of solution, 37°C: 12–24 hours.
3. Wash in distilled water, several changes.
4. Reduce: 5–10 minutes.
5. Rinse in distilled water, 3 changes: 1 minute total.
6. Tone in gold chloride: 4 minutes.
7. Rinse in distilled water, 3 changes: 1 minute total.
8. Develop in 2% aqueous oxalic acid. Check under microscope until background is gray and fibers are sharply defined: approximately 3 minutes.
9. Wash in distilled water, 3 changes: 1 minute each.
10. Fix in 5% aqueous sodium thiosulfate: 5 minutes.
11. Wash in running water: 5–10 minutes.
12. Rinse in distilled water.
13. Counterstain if desired.
14. Treat with acidified water: 5 minutes.
15. Dehydrate, clear, and mount.

RESULTS

nerve fibers—black
background colors—depending on counterstain

COMMENTS

Foley (1943) considers counterstaining essential, claiming that it serves as contrast between nervous and nonnervous tissue and adds to transparency of the sections. To counterstain by his method follow step 11 above with the following:

12. Stain in gallocyanin (page 154): overnight.
13. Wash thoroughly in running water: 5–10 minutes.
14. Mordant in 5% aqueous phosphotungstic acid: 30 minutes. utes.
15. Transfer directly to dilution (20 ml/30 ml water) of following stock solution: 1 hour.

aniline blue WS, (C.I. 42780)	0.1 g
fast green FCF, (C.I. 42053)	0.5 g
orange G, (C.I. 16230)	2.0 g
distilled water	92.0 ml
glacial acetic acid	8.0 ml

16. Differentiate through 70% and 95% alcohols.
17. Dehydrate, clear, and mount.

Ungewitter Modification of Bodian Method (1943)

FIXATION

As in Bodian method.

SOLUTIONS

Protargol solution, see page 227.

Elon solution

Elon	0.2 g
sodium sulfate, anhydrous	10.0 g
hydroquinone	0.5 g
sodium borate	0.1 g
distilled water	100.0 ml

PROCEDURE

1. Deparaffinize and hydrate sections to water.
2. Treat with Protargol solution (plus copper), 37°C: 12–24 hours.
3. Rinse in distilled water, 2 changes.
4. Reduce in Elon solution: 5–10 minutes, agitating for first minute.
5. Wash in distilled water, 4–5 changes.
6. Treat with 1% aqueous silver nitrate: 10–20 minutes, room temperature or 37°C.
7. Wash in distilled water, 2–3 changes.
8. Reduce (step 4 above).
9. Wash in distilled water, 4 changes. Examine. If nerve fibers show distinct differentiation, dehydrate, clear, and mount. If they are not well differentiated, repeat steps 6, 7, 8, and 9 until desired results are reached.
10. Toning may be done in 0.2% aqueous gold chloride until the tissue is grayish. Fix, dehydrate, clear, and mount.

Chloral Hydrate–Silver Method (Nonidez 1939) B

FIXATION

24 hours or longer (up to 3 days):

chloral hydrate	25.0 g
50% ethyl alcohol	100.0 ml

SOLUTIONS

Alcohol-ammonia

95% ethyl alcohol	60.0 ml
ammonia, concentrated	4 drops

Reducing solution

Pyrogallol	2.5–3.0 g
formalin	8.0 ml
distilled water	100.0 ml

PROCEDURE

1. Wipe off excess fixative and put blocks in alcohol-ammonia. Leave for 24 hours, but change at least once, particularly if blocks are large or contain a large amount of fat.
2. Rinse in distilled water: 5 minutes.
3. Place in 2% aqueous silver nitrate in dark cupboard or box, 37–40°C: 5–6 days. Replace silver solution after 2 days, or as soon as it becomes brownish in color.
4. Rinse in distilled water: 2–3 minutes.
5. Reduce for 24 hours in Pyrogallol solution.
6. Wash in several changes of distilled water: 2–3 hours.
7. Transfer to 50% alcohol, change twice over a period of 24 hours.
8. Dehydrate, clear, and embed.
9. Section, mount on slides, deparaffinize, clear, and cover.

RESULT

nerve fibers and endings—brown to black

COMMENT

Gold toning does not improve impregnation.

Chloral Hydrate–Silver Method (Davenport 1933) B

FIXATION

Perfusion (page 26) with 10% chloral hydrate is a desirable procedure, followed by fixation in 10% chloral hydrate: 1–2 hours. (After perfusion cut whole organ into tissue blocks, 5–8 mm thick.)

PROCEDURE

1. Treat tissue blocks with 20–40% aqueous pyridine: 2 days.
2. Wash in distilled water: 1–2 hours, no more. Change water every half hour.
3. Impregnate with 1.5% aqueous silver nitrate: 3–7 days. (Time is not critical within these limits.)
4. Wash in distilled water: 20–30 minutes, change water once.
5. Reduce in 2% aqueous Pyrogallol: 5 hours.
6. Dehydrate, clear, and embed.
7. Section (15–20μ) and mount on slides.
8. Deparaffinize and gold tone if desired, or clear and cover.

RESULTS

nerve fibers and endings—black
cells—pale yellowish brown

COMMENTS

Davenport suggests that the chloral hydrate be perfused into the aorta and pumped through the system for about 1–3 minutes before the heart stops.

The washing after the silver impregnation is critical. If the tissue is not sufficiently washed, the periphery will be overstained. A 30-minute washing leaves the superficial as well as the central zones well stained. Five hours of washing produces pale staining and leaves the end bulbs unstained. Cat tissue gave Davenport excellent results.

Abadía-Fenoli (1968) uses a chromation-silver method on frozen sections.

Nauta and Gygax Method (1951) **B**

FIXATION

10% formalin: 2 weeks to 6 months.

SECTIONING

Frozen sections, 15–20μ.

SOLUTIONS

Silver solution A

silver nitrate .... 1.5 g
distilled water .... 100.0 ml
pyridine .... 5.0 ml

Silver solution B

Dissolve 0.45 g silver nitrate in 20.0 ml distilled water. Add 10.0 ml of 95% ethyl alcohol. From calibrated pipettes add 2.0 ml ammonia (concentrated) and 2.2 ml of 2.5% aqueous sodium hydroxide. Mix thoroughly and keep the container covered to prevent escape of ammonia.

Reducing solution

10% ethyl alcohol	45.0 ml
10% formalin	2.0 ml
1% citric acid	1.5 ml

PROCEDURE

1. Demyelinate sections in 50% ethyl alcohol plus 1.0 ml ammonia per 100 ml: 6–12 hours. A longer time has no ill effect.
2. Wash in distilled water, 3 changes: few seconds each.
3. Impregnate in silver solution A: 12–24 hours.
4. With no washing, transfer into silver solution B: 2–5 minutes.
5. Transfer directly into reducing solution until the sections turn gold in color.
6. Transfer to 2.5% aqueous sodium thiosulfate: 1–2 minutes.
7. Wash in distilled water, at least 3 changes.
8. Dehydrate rapidly, clear, and mount.

RESULTS

nerve fibers and endings—black
cells—pale yellowish brown

COMMENTS

This method is nonselective and stains normal as well as degenerating axons. For degenerating axons see page 234.

Cole Method (1946) (Modified: Whole Mounts)

FIXATION

None, carry fresh tissue directly into step 1.

PROCEDURE

1. Tease striated costal muscle into strips 1 mm in diameter and a few millimeters in length, and place in either of following solutions (Zinn

and Morin 1962): (1) 1 part commercial lemon juice and 1 part distilled water; or (2) 0.001 M citric acid. Minimum time is 10 minutes; maximum time, 30 minutes.

Use a separate clean glass container for each step that follows.

2. Wash in several changes of distilled water: 5 minutes.
3. Transfer to 1% aqueous gold chloride. Make up this solution the day before it is to be used. Keep in dark: 60 minutes, or until tissue turns dark yellow. (If pieces are wider than 4 mm, a longer time is required.)
4. Wash in several changes of distilled water: 5 minutes.
5. Transfer to 20% formic acid (20 ml/80 ml water): 10–20 hours in dark. Do not use metal forceps; coat forceps with paraffin or use glass rods.
6. Rinse in tap water.
7. Transfer to 95% methyl alcohol–glycerine (1 : 1): several hours. Then remove top of container and allow alcohol to evaporate.
8. Transfer to pure glycerine.
9. To mount: Place a piece of muscle in a very small drop of glycerine (usually amount carried over by the piece is sufficient) on a round cover glass. Lay a smaller size cover glass over it. Spread the muscle fiber to single fiber thickness by using gentle pressure and strokes at right angles to the fibers. Turn cover glasses over and mount in Permount (or the like) on a glass slide. (Double cover glass mounting, page 146.)

RESULTS

muscle fibers—red-blue
motor end plates and medullated axons—black
muscle fiber nuclei—unstained

COMMENTS

Carey (1941) uses undiluted fresh filtered lemon juice in place of the citric acid. He claims that if the tissue requires more than 12 hours in the formic acid (step 5), the gold chloride technique is faulty. The color of the tissue should be gold, not brown.

Boyd (1962) fixes in fresh filtered lemon juice and formic acid, 3 : 1 : 2–10 minutes.

Cole and Mielcarek (1962) outline a fluorescent method.

Cavanagh et al. (1964) stain with Sudan black B.

Pyridine–Silver Method (Gladden 1970)

PROCEDURE

1. Fix muscle in absolute ethanol, 4.5 ml; distilled water, 5 ml; and concentrated nitric acid, 0.1 ml: 24 hours.
2. Transfer into absolute ethanol, 10 ml, and ammonia, 0.1 ml: 24 hours.
3. Wash in distilled water: 30 minutes.
4. Treat in pyridine: 2 days.
5. Wash in distilled water, 5–8 changes: 24 hours.
6. Place in 2% aqueous silver nitrate in dark, 25°C: 3 days.
7. Reduce in 5% aqueous formic acid, 10.0 ml, and Pyrogallol, 0.4 g: 6–24 hours.
8. Wash in distilled water and store in glycerine.
9. Tease muscle to show nerve endings and mount in glycerine.

DEGENERATING AXONS

Nauta and Gygax Method (1954) S

FIXATION

Perfusion method seems to be preferred.

1. Perfuse with 0.9% sodium chloride (intracardiac and intra-aortic cannulation) until escaping fluid is clear.
2. Perfuse with 500 ml of 5% aqueous potassium dichromate and 2.5% aqueous potassium chlorate. (This step is peculiar to Anderson (1959); others perfuse with a formalin solution.)
3. Remove tissue and fix in 10% formalin neutralized with carbonate: 1 week or longer. Nauta and Gygax recommend 1–3 months as best.

EMBEDDING AND SECTIONING

1. If embedding is necessary, use gelatin. Wash slices (5–10 mm thick) in running water: 24 hours. Incubate in 25% gelatin, 37°C (cover tightly): 12–18 hours. Wipe off excess gelatin and immerse in cool formalin: 6 hours or longer.
2. Frozen sections: Cut sections 15–25μ and place in 10% formalin, room temperature or cooler. Can be stored in formalin, preferably in refrigerator. Process only a few sections (6–10) at one time.

Adey et al. (1958) prefer a dual-freezing block method, using dry ice evaporated in 70% alcohol, over CO_2 freezing. This produces more even freezing. Rapid freezing and thawing with CO_2 may rupture some of the fibers; the use of Freon equipment will sometimes prevent this.

SOLUTIONS

Ammoniacal silver nitrate

silver nitrate	0.9 g
distilled water	20.0 ml

When dissolved add:

absolute ethyl alcohol	10.0 ml
ammonia, concentrated	1.8 ml
sodium hydroxide, 2.5% aqueous	1.5 ml

Reducing fluid

distilled water	400.0 ml
absolute ethyl alcohol	45.0 ml
formalin, 10%	13.5 ml
citric acid, 1% aqueous	13.5 ml

PROCEDURE

1. Soak sections in 15% alcohol: 30 minutes.
2. Wash briefly in distilled water and soak in 0.5% aqueous phosphomolybdic acid: 15–20 minutes.
3. No wash; transfer to 0.5% aqueous potassium permanganate: 4–10 minutes. This step can be critical; make trial runs at 5, 7, and 9 minutes. Turn sections over occasionally.
4. Rinse in distilled water.
5. Decolorize in equal parts of 1% aqueous hydroquinone and 1% aqueous oxalic acid: 1–2 minutes.
6. Wash thoroughly in distilled water, 3 changes.
7. Treat with 1% aqueous silver nitrate: 20–30 minutes.
8. Transfer sections individually; wash briefly in distilled water and then transfer to freshly prepared ammoniacal silver: 1 minute.
9. Transfer to reducing solution, allowing sections to float on surface. The color will become brown within 1 minute. If it is too light, add a little sodium hydroxide to ammoniacal silver solution; if too dark, add some ammonia.
10. Wash briefly, fix in 1% aqueous sodium thiosulfate: 2 minutes.
11. Wash in distilled water, 3 changes.
12. Mount using Albrecht (1954) method:
    a. Transfer into equal parts of 1.5% gelatin (1.5 g/100 ml water) and 80% alcohol: 5 minutes or longer.

b. Tease section onto slide, withdraw from fluid with slide tilted to allow rapid drainage of excess fluid. Wipe around section, but do not allow it to dry.

c. Blot gently with smooth filter paper. If section sticks to paper, moisten paper with 95% alcohol.

d. Immerse rapidly in 95% alcohol to congeal gelatin.

13. Dehydrate, clear, and mount.

RESULTS

disintegrating axons—black
normal axons—various shades of brown

COMMENTS

See also Nauta and Gygax (1951), Nauta and Ryan (1952), Wall (1950), White (1960), and additional references included therein.

The Nauta-Gygax impregnation of degenerating fibers is related to unsaturated lipids. Since the Nauta reaction is abolished by prior bromination, this suggests that the basic mechanism of the impregnation may involve unsaturated bonds, the ethylene bonds of cholesterol esters. The presence of these esters is attributed principally to the breakdown of the myelin. It is possible that degeneration of the axis cylinders is also part of the reaction (Giolli 1965).

Hamlyn (1957) and Guillery et al. (1961) describe methods for paraffin-embedded sections that are stained after mounting.

Marchi Method B

FIXATION

10% formalin plus 1% of potassium chlorate: 24–48 hours, no longer (Swank and Davenport 1934*a*).

SOLUTION

Marchi fluid (Poirier et al. 1954)

osmic acid, 0.5% aqueous	11.0 ml
potassium chlorate, 1% aqueous	16.0 ml
formalin	3.0 ml
acetic acid, 10% aqueous	3.0 ml
distilled water	67.0 ml

PROCEDURE

1. Cut tissue into thin slices, about 3 mm for easier impregnation.
2. Chromate in 2.5% aqueous potassium dichromate, in a dark enclosure: 7–14 days, change twice.
3. Transfer directly to Marchi fluid, a volume 15–20 times that of tissue: 1–2 weeks depending on size of tissue. Turn tissue over every day to improve penetration.
4. Wash in running water: 24 hours.
5. Dehydrate and embed. If using celloidin, keep embedding steps to a minimum. If using paraffin, avoid xylene and its solvent action on osmic acid; chloroform is safer and keeps steps to a minimum.
6. Deparaffinize slides with chloroform and mount in chloroform-resin.

RESULTS

degenerating myelin—black
background—brownish yellow
neutral fats—black

COMMENTS

The principle behind Marchi method is that the myelin of medullated nerves oxidizes more easily than degenerating myelin. Normal myelin is oxidized by chromating and will not react with osmic acid. Degenerating myelin contains oleic acid, which does not oxidize during chromating, therefore reduces the osmic acid, and stains black.

For additional reading about neurological staining, see Culling (1957); Lillie (1954*b*); Mettler (1932); Mettler and Hanada (1942); Poirier et al. (1954); and Swank and Davenport (1934*a, b;* 1935*a, b*).

MYELIN

Luxol fast blue B was first used by Klüver and Barrera (1953) for staining myelin sheaths. Margolis and Pickett (1956) combined Luxol fast blue MBSN with other methods to differentiate various neurological elements. Salthouse (1962) stained myelin sheaths a deep blue with Luxol fast blue ARN, which he said had a greater affinity for phospholipids than Luxol fast blue MBSN. In 1964 he reported Luxol fast blue G, with which he had stained myelin a blue-black color, to be superior to ARN. If dissolved in isopropyl alcohol, the Luxol dyes bind to more phospholipids than if dissolved in ethyl alcohol. The Margolis and Pickett and the Salthouse solutions are listed below: they can be used interchangeably. All Luxol fast blue solutions keep indefinitely.

Luxol Fast Blue S

FIXATION

Salthouse (1962) recommends 10% formalin or calcium-formalin, but the following solution is best:

calcium chloride	10.0 g
distilled water	900.0 ml

Dissolve and add:

cetyltrimethylammonium bromide	5.0 g

Dissolve and add:

formalin, concentrated	100.0 ml

EMBEDDING

Paraffin method.

SOLUTIONS

Luxol fast blue MBSN (Margolis and Pickett 1956)

Luxol fast blue MBSN [2]	0.1 g
95% ethyl alcohol	100.0 ml
acetic acid, 10% aqueous	0.5 ml

Luxol fast blue ARN (Salthouse 1962)

Luxol fast blue ARN [2] (C.I. solvent blue 37)	1.0 g
95% ethyl alcohol	1000.0 ml
glacial acetic acid	0.2 ml

Luxol fast blue G (Salthouse 1964)

Luxol fast blue G [3]	1.0 g
95% isopropyl alcohol	1000.0 ml
glacial acetic acid	0.2 ml

---

[2] E. I. du Pont de Nemours and Company.   [3] Matheson Coleman and Bell.

Periodic acid, 0.5%

| periodic acid | 0.5 g |
| distilled water | 100.0 ml |

Lithium carbonate, 0.05%

| lithium carbonate· | 0.5 g |
| distilled water | 1000.0 ml |

Schiff reagent, see page 327.

Sulfurous acid

sodium metabisulfite, 10% aqueous	6.0 ml
distilled water	100.0 ml
1 N HCl (see page 531)	5.0 ml

Mayer hematoxylin, see page 150.

PROCEDURE

1. Deparaffinize and hydrate slides to water. (Remove HgCl₂ if present.) Transfer to 95% alcohol: 2 minutes.
2. Stain in Luxol fast blue MBSN, 60°C: overnight; or Luxol fast blue ARN or G, 35–40°C: 2–3 hours.
3. Rinse off excess stain in 95% alcohol.
4. Rinse in distilled water.
5. Dip in lithium carbonate: 15 seconds.
6. Differentiate in 70% alcohol: 20–30 seconds.
7. Rinse in distilled water.
8. Dip in lithium carbonate, second solution: 20–30 seconds.
9. Differentiate in 70% alcohol: 20–30 seconds.
10. Rinse in distilled water. If differentiation is not complete, repeat steps 8 and 9.
11. Oxidize with periodic acid: 5 minutes.
12. Wash in distilled water, 2 changes: 5 minutes total.
13. Treat with Schiff reagent: 15–30 minutes.
14. Treat with sulfurous acid, 3 changes: 2 minutes each.
15. Running water: 5 minutes.
16. Stain in Mayer hematoxylin (or the like): 3 minutes.
17. Wash in running water: 5 minutes.
18. Dehydrate, clear, and mount.

RESULTS
> myelin—blue-green
> PAS positive elements—rose to red
> nuclei—blue

COMMENTS

Luxol is a trade name of the DuPont Company. The Luxol fast blue used by Klüver and Barrera was the amine salt of a sulfonated copper phthalocyanin. Phthalocyanin-amine dyes may have one to four sulfonyl groups on each molecule and one of several kinds of bases. Because of the copper phthalocyanin, Luxol fast blue methods are sometimes called copper phthalocyanin methods. Luxol fast blue ARN, however, is a diaryl quanidine salt of a sulfonated azo dye with no copper present, and it has a greater affinity for phospholipids than the MBS salt. Klüver and Barrera thought the reaction was due to the affinity of the copper phthalocyanin for the porphyrin present in myelin, but the superior reaction of the ARN salt proves that it does not depend on the copper. Pearse (1960) suggests that Luxol fast blue staining of fixed tissues is due to the presence of lipoproteins rather than lipids, that it is an acid-base reaction to form a salt, with the lipo-protein base replacing the phthalocyanin. Lycette et al. (1970) consider the phosphate group of phospholipids to be essential for the reaction. Whatever the explanation, the sensitivity of the dye for myelin is excellent. Also, stores of phospholipids can be clearly shown by this method. Salthouse (1962) demonstrates that the ARN salt forms complexes with many phospholipids.

Hale et al. (1960), using Luxol fast blue MBSN, differentiate abnormal from normal collagen; abnormal collagen did not stain, but normal collagen stained an intense blue.

Snodgrass and Lacey (1961) modify the Luxol fast blue method to differentiate degenerating myelinated fibers.

Lockard and Reers (1962) use it with neutral red to include Nissl staining.

Margolis and Pickett also describe methods for following Luxol fast blue with phosphotungstic acid–hematoxylin to differentiate neuroglia from myelin, and with oil red O to distinguish degenerating myelin from normal myelin.

Dziabis (1958) describes a method for staining gross brain sections.

Luxol Fast Blue–Holmes Silver Nitrate (Margolis and Pickett 1956) S

FIXATION

10% formalin preferred.

EMBEDDING

Paraffin method.

SOLUTIONS

Silver nitrate, 20%:

silver nitrate	20.0 g
distilled water	100.0 ml

Boric acid

boric acid	12.4 g
distilled water	1000.0 ml

Borax

borax	19.0 g
distilled water	1000.0 ml

Impregnating fluid

In a 500-ml cylinder, mix 55 ml boric acid solution and 45 ml borax solution. Dilute to 494 ml with distilled water. With pipette, add 1 ml of 1% silver nitrate (1 g/100 ml water). With another pipette add 5 ml of 10% pyridine (10 ml/100 ml water). Mix thoroughly.

Reducer

hydroquinone	1.0 g
sodium sulfite (crystals)	10.0 g
distilled water	100.0 ml

Can be used repeatedly, but only for a few days.

Luxol fast blue, see page 238.

PROCEDURE

1. Deparaffinize and hydrate sections to water.
2. Treat with silver nitrate in dark, room temperature: 1 hour. Prepare impregnating fluid.
3. Wash in distilled water, 3 changes: 10 minutes.

4. Impregnate, 37°C: overnight.
5. Shake off superfluous fluid and place in reducer: 2 minutes.
6. Wash in running water: 3 minutes; rinse in distilled water.
7. Tone in 0.2% aqueous gold chloride: 3 minutes.
8. Rinse in distilled water.
9. Treat with 2% aqueous oxalic acid: 3–10 minutes; when axons are thoroughly black, remove.
10. Rinse in distilled water.
11. Fix in 5% aqueous sodium thiosulfate: 3 minutes.
12. Wash in running water: 10 minutes.
13. Rinse briefly in 95% alcohol.
14. Stain in Luxol fast blue: overnight.
15. Rinse in 95% alcohol; rinse in distilled water.
16. Treat with 0.05% aqueous lithium carbonate: 15 seconds.
17. Differentiate in 70% alcohol: 20–30 seconds.
18. Rinse in distilled water.
19. Repeat steps 16 and 17 if necessary. (Three repeats are usually necessary for sharp differentiation.)
20. Dehydrate, clear, and mount.

RESULTS

axis cylinders—black
myelin sheaths—green blue

COMMENTS

Margolis and Pickett write that slides may be left in the distilled water (step 6) until it is time to prepare them for Luxol fast blue staining. However, leaving the sections in water tends to detach them from the slides. I have lost no sections by carrying the slides through to step 13. Transfer them into absolute alcohol and coat them with nitrocellulose. Harden the nitrocellulose and store the slides in 70–80% alcohol until you are ready to stain them. Slides can be left in this condition for a weekend with completely satisfactory results. The nitrocellulose protection also permits freer agitation of slides in the differentiating fluids. Subbed slides can be used.

Chromic Acid–Hematoxylin (Lillie and Henderson 1968)

FIXATION

10% formalin.

SOLUTIONS

Chromic acid

chromic acid	0.2 g
distilled water	100.0 ml

Acetic hematoxylin

1% stock alcoholic hematoxylin	10.0 ml
distilled water	89.0 ml
glacial acetic acid	1.0 ml

or

1% aqueous acetic acid	100.0 ml
hematoxylin	100.0 mg

Borax-ferricyanide

borax	1.0 g
potassium ferricyanide	1.0 g
distilled water	100.0 ml

PROCEDURE

1. Cut frozen sections, $10–15\mu$, transfer to 1% aqueous gelatin and pick up on slides.
2. Drain and blot firmly with filter paper. Immerse in 1% aqueous gelatin: 1–2 minutes. Drain.
3. Fix gelatin in formalin vapor: 30 minutes. Transfer to 5% formalin until ready to use.
4. Wash in running water: 10 minutes.
5. Mordant in chromic acid: 4 hours.
6. Wash in running water: 10 minutes.
7. Stain in acetic hematoxylin: 2 hours.
8. Rinse in distilled water.
9. Differentiate in borax-ferricyanide: approximately 1 hour. Check under microscope.
10. Dehydrate, clear, and mount.

RESULTS

myelin—blue black
cells and gray substance—yellow to colorless

COMMENTS

This method is recommended over the Ora or Mahon methods because the myelin sheaths remain a more normal size in relation to other structures in this method, whereas paraffin embedding may cause shrinkage in thin sheaths.

CHAPTER **16**

Staining Hematologic Elements and Related Tissues

Hemopoietic (blood-forming) tissue is connective tissue specialized to produce blood cells and to remove worn-out blood cells from the bloodstream. There are two varieties of hemopoietic tissue: (1) myeloid, which produces erythrocytes, granular leukocytes, and platelets (thrombocytes); (2) lymphatic, which produces most of the nongranular leukocytes.

Blood cells and fluid may become parasitized in various ways: for example, with malaria, trypanosomes, inclusion bodies of various diseases, such as rickettsia and psittacosis. Most of these foreign elements are best demonstrated with a Romanovsky type of stain.

See chapter 24, page 431, for techniques for demonstrating enzymes found in hemopoietic tissue.

BLOOD SMEARS

Preparation for Thin Smears

Slides must be clean for a uniform smear. Handle slides at the edges, keeping fingers off the clean surface. Prick the finger and, when a small

FIGURE 23. *Preparation of thin smears: (a) drop of blood is placed near one end of slide and second slide pulled against it; (b) second slide pulls blood across first slide.*

drop of blood appears, wipe it away. Touch the next drop of blood to the clean surface of the right end of the slide. Place the narrow edge of another slide at a 20°–30° angle on the first slide and to the left of the drop of blood. Pull to the right until the slide touches the blood. As soon as the blood has spread along the line of contact, push the right hand toward the left. Push steadily until all the blood disappears or the other end of the slide is reached. Move the hand rapidly; if the smear is spread too slowly, the leukocytes concentrate along the edges and in the tail of the smear. (Christophers 1956.) This method drags the blood but does not run over it and crush some of the cells. The hand can be kept from shaking by resting it on the table. Also, do not use a slide with a rough edge, which will leave streaks in the smear. If the blood seems thick, reduce the 20° angle to feed it out at a slower rate. For thin blood, increase the angle (Figure 23). See Pequeno, 1960, for a clever method of using a pen to make thin smears.

Dry the slides rapidly in the air; waving them facilitates drying and prevents crenation (notching or scalloping of edges) of the red cells.

For cover-glass preparations, place a small drop of blood in center of cover glass; place a second cover glass directly over the top of the drop at a slight angle to the bottom glass. Allow the blood to spread to edges of cover glasses. Quickly slide cover glasses apart. Air dry (Figure 24). See page 432 for a similar method of spreading blood on slides.

Preferably, blood smears should be stained immediately or within 24 hours. If they must be stored, place them in a tight box away from dust and flies.

Blood smears are commonly stained with a Romanovsky type of stain, or neutral stain (page 129). Neutrality is essential, and therefore dilution

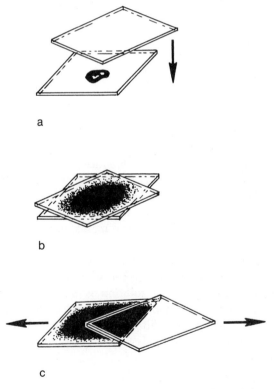

a

b

c

FIGURE 24. *Preparation of cover-glass films:* (a) *a square cover glass is placed on top of blood on another square cover glass;* (b) *blood is allowed to spread;* (c) *top cover glass is pulled across and off bottom cover glass.*

is usually made with a buffer solution of a known pH. Distilled water is often too acid, and tap water too alkaline. The smears require no fixation in the usual sense, since Wright stain (1902) includes both fixing agent (methyl alcohol) and stain.

A common practice is to leave stained blood smears uncovered, because some mounting media tend to fade Romanovsky stains after a period of time. The fading takes place slowly and is an inconvenience only if slides are to be stored for years. Plastic coatings have been recommended, but they tend to be uneven in thickness and are easily scratched. See Steinman (1955) for preparation of a plastic coating. Hollander (1963) adds 1% of 2,6-di-*tert*-butyl-*p*-cresol to the synthetic mountant Permount to prevent the fading of blood smears. Try plastic mountants, see page 451.

Thin Smear Method

SOLUTIONS

Wright stain

The stain may be purchased in three forms: (1) dry powder, by the bottle, 1, 10, 25, or 100 g; (2) dry powder in capsule form, 0.1 g per capsule; or (3) in solution, 1, 4, 8, or 16 ounces.[1]

The powder (bulk or capsule) is ground up with methyl alcohol (0.1 g to 60.0 ml). The alcohol must be labeled "neutral" and "acetone free." Grind thoroughly in a glass mortar, and pour off supernatant (surface-floating) liquid. If undissolved powder remains, pour back the liquid and grind again.

A slower but perhaps easier method is satisfactory. Add 0.3 g stain to 100.0 ml methyl alcohol and 3 ml glycerine in a stoppered bottle. Shake occasionally. After 24 hours, the solution is usually ready for use. To check whether a staining solution is satisfactory, with a pipette release a small drop of the solution onto filter paper. A good stain forms a deep blue spot, but a poor stain spreads into a broad pink halo around the central blue spot.[2]

---

[1] Scientific Products sells Camco Quik Stain (no. B4130) a single solution of Wright stain with built-in buffering agent. A dip of only 30 seconds is required.

[2] From "Lab Hint: A Quick Check for Wrong Wright's," *Clinical Reporter,* Fisher Scientific Co.

Buffer solution, pH 6.4

monobasic potassium phosphate (KH_2PO_4) .... 6.63 g

anhydrous dibasic sodium phosphate

(Na_2HPO_4) .... 2.56 g

distilled water .... 1000.0 ml

Bohorfoush diluent (1963)

sodium thiosulfate ($Na_2S_2O_3$) .... 0.1–0.2 g

distilled water .... 1000.0 ml

Allow to stand 1 hour. Bohorfoush claims good uniform staining with this solution.

PROCEDURE

1. With wax pencil draw 2 marks across slide delineating region to be stained (length of 40 mm cover glass).
2. Cover with 10–12 drops of Wright stain: 1–2 minutes.
3. Add an equal amount of buffer or diluent: 2–4 minutes.
4. Rinse in distilled water, 1 or 2 dips. Precipitate deposit on the slide can be avoided by flushing the slide with a pipette, or carry the stain with the slide into the wash water; do not drain it off first.
5. Blot with 2 sheets of filter paper; press but do not rub.
6. Allow slide to thoroughly dry before applying cover glass.

ALTERNATE PROCEDURE

1. Fix dried film in methyl alcohol: 1–2 minutes.
2. Dilute Wright stain with an equal amount of buffer. About 10 drops of each is sufficient for one slide. Use immediately.
3. Cover slide with diluted stain: 4 minutes.
4. Rinse and dry as in above procedure.

RESULTS

erythrocytes—pink

nuclei—deep blue or purple

basophilic granules—deep purple

eosinophilic granules—red to red-orange, bluish cytoplasm

neutrophilic granules—reddish brown to lilac, pale pink cytoplasm

granules of monocytes—azure

lymphocyte granules—larger and more reddish than monocyte granules, sky blue cytoplasm

platelets—violet to purple

COMMENTS

Precipitate formation can be troublesome; the dark granules obscure the blood cells and are confusing in malarial smears. In addition to poor washing in step 4, another cause of precipitate may be evaporation during exposure to undiluted stain. Methyl alcohol is highly volatile and readily lost by evaporation. In dry warm weather use more Wright stain or shorten the time a bit, or cover the slides with a petri dish. Rapid evaporation is easily detected, and more stain can always be added.

Proper rinsing in step 4 is one of the most important steps in this procedure. Never use tap water; it is too alkaline. Distilled water is usually satisfactory, but for best results rinse in the buffer, pH 6.4

The longer the washing with water, the more stain is removed from the white cells. Only a dip or two is usually sufficient, but, if the white cells are overstained, differentiate them by longer washing.

If the slides are overstained, the erythrocytes are too red, the white cells too pale, or the stain has precipitated, the slides can be recovered: (1) Cover the entire slide almost to excess with additional Wright stain: 15–30 minutes; (2) rinse with distilled water; (3) dry (Morrison and Samwick 1940).

Wright Stain for Cold-Blooded Vertebrates (Heady and Rogers 1962)

PROCEDURE

1. Cover with Wright stain: 2 minutes.
2. Add equal portion of phosphate buffer, pH 6.7: 3 minutes (page 542).
3. Wash in distilled water and blot as above.

Wright Stain for Birds (Santamarina 1964)

SOLUTIONS

Wright stain

Wright stain powder	3.3 g
absolute methyl alcohol	500.0 ml

This is stronger than the solution for human blood. Do not use one that is too dilute.

Formalin solution

formalin, concentrated	0.25 ml
distilled water	500.0 ml

Adjust to *p*H 6.8 by adding either 0.25% sodium carbonate or 0.25% hydrochloric acid as needed.

PROCEDURE

1. Cover dried smears with Wright solution: 8 minutes.
2. Slowly add formalin solution with a medicine dropper. Do not allow solution to overflow.
3. When a metallic sheen covers entire surface of fluid, pour off and flush immediately with distilled water adjusted to *p*H 6.8 as above.
4. Blot.
5. Differentiate in ether–absolute methyl alcohol (1 : 1). Dip slides up and down 6 to 10 times. Examine them under microscope; if they are too bluish, differentiate further.
6. Clear and mount.

RESULTS

erythrocytes—yellow to pinkish cytoplasm, purple chromatin
thrombocytes—gray-blue cytoplasm, purple chromatin
lymphocytes—blue granules, purplish-red chromatin
monocytes—light blue granular cytoplasm, purple chromatin
heterophils—yellowish to brownish-red rods, light purple chromatin
eosinophils—similar to heterophils
basophils—dark purple granules

Preparation of Thick Blood Films

With this type of film, a relatively large quantity of blood is concentrated in a small area, thereby increasing the possibilities of finding parasites. The concentration and timing of staining are adjusted so the action is stopped at the point when the leukocytes have stained, some hemoglobin has been dissolved, and the red cell membranes have not yet begun to stain. At this point only the leukocytes, platelets, and protozoa are stained and lie on an unstained or very lightly stained background, which is yellowish from the remaining hemoglobin. Freshly prepared films stain better than films one or more days old.

Puncture the skin deeply enough to form a large drop of blood. On a slide, cover a space the size of a dime with enough blood (about 3–4 average drops) to spread easily. Too much blood will crack and peel off when dry. Smear it by circling the slide under the finger without making actual contact. Some find it easier to swirl the blood with a dissecting needle or the corner of a slide. Burton (1958) taps the slide against the table top, so the drop of blood flows transversely across the slide for $\frac{5}{8}$ to $\frac{3}{4}$ inch. Practice will determine the best method and how much blood to use. Too thin a film has no advantage over a thin smear. An ideal film is several cell layers thick in the center, tapering off to one cell thickness at the periphery.

Allow the slide to dry in a horizontal position; if tilted the blood will ooze to one edge of the film. Protect from dust and flies, and do *not* fix by flaming or alcohol.

Field-Wartman Thick Film Method (Field 1941; Wartman 1943)

SOLUTIONS

Solution A

methylene blue (C.I. 52015)	0.8 g
azure A (C.I. 52005)	0.5 g
dibasic sodium phosphate, anhydrous	5.0 g
monobasic potassium phosphate	6.25 g
distilled water	500.0 ml

Solution B

eosin Y (C.I. 45380)	1.0 g
dibasic sodium phosphate, anhydrous	5.0 g
monobasic potassium phosphate	6.25 g
distilled water	500.0 ml

Dissolve phosphate salts first, then add stains. (The azure will go into solution by grinding in glass mortar with a small amount of phosphate solvent.) Set aside for 24 hours. Filter. Store in refrigerator.

PROCEDURE

1. Dip slides in solution A: 1 second.
2. Rinse gently in clean distilled water: few seconds, or until stain ceases to flow from film.
3. Dip in solution B: 1 second. Use this solution with care; it tends to decolorize the leukocytes stained with methylene blue-azure, and accelerates the dissolving of hemoglobin.
4. Rinse gently in distilled water: 2–3 dips.
5. Dry, do not blot. Stand slides on end and allow to air dry.
6. When completely dry, add mounting medium and cover glass.

RESULTS

There will be a thicker, central area of partially laked blood. This may not be well suited for examination. But the surrounding area, and especially at the edge toward which the hemoglobin has drained, will be creamy, sometimes mottled with pale blue. This is the best area for study.

Leukocytes

cytoplasm—pale blue, poorly defined
nuclei—dark blue, well defined
eosinophilic granules—bright red, large, well defined
neutrophilic granules—pale purple-pink, small, indistinct
basophilic granules—deep blue with reddish cast

Platelets—pale purple or lavender

Parasites

cytoplasm—blue
chromatin—purplish red or deep ruby red
pigment—unstained yellow granules of varying intensity

Wilcox Thick Film Method (1943)

SOLUTIONS

Wright-Giemsa stock solution

Giemsa powder	2.0 g
glycerine	100.0 ml

Heat in water bath, 55°–60°C: 2 hours, stirring at intervals. Avoid absorption of moisture by covering mouth of flask with double thickness of paper secured with rubber band.

Add:

aged Wright staining solution (2.0 g/1000.0 ml methyl alcohol)	100.0 ml

Let stand overnight; add in addition:

Wright staining solution	800.0 ml

Filter; ready for use.

Working solution

Wright-Giemsa stock solution	1 part
neutral distilled water (buffer)	9 parts

Buffer, pH 7.0

Solution A

dibasic sodium phosphate, anhydrous	9.5 g
distilled water	1000.0 ml

Solution B

monobasic potassium phosphate	9.7 g
distilled water	1000.0 ml

Working solution

Solution A	61.1 ml
Solution B	38.9 ml
distilled water	900.0 ml

PROCEDURE

Short method

1. Stain film: 10 minutes.
2. Flush scum from top of solution with buffer to avoid picking up precipitate on slides.
3. Remove slides to neutral distilled water: 1 minute.
4. Dry slides standing on end. Do not blot. Mount.

Long method

This method produces a more brilliant stain.

1. Stain: 45 minutes in dilution of 1 to 50 of buffer.
2. Flush off scum and transfer slides to buffer: 3–5 minutes.
3. Dry as above.

Thin smear and thick film can be stained on same slide. Fix thin portion in methyl alcohol for 1–2 minutes, taking care that thick film does not come in contact with alcohol. Stain as above (short method), but shorten washing time to 2 or 3 dips so thin smear is not too light; this shorter washing time will leave a deeper background in the thick film.

COMMENTS

Modifications of this method have been made by Fenton and Innes (1945), Manwell (1945), and Steil (1936).

Old smears will stain more brilliantly if treated 5–10 minutes in alcohol-acetic solution—10 drops glacial acetic acid to 60 ml absolute alcohol.

BLOOD TISSUE ELEMENTS AND INCLUSION BODIES [3]

Giemsa Stain (Wolbach Modification, Mallory 1944)

FIXATION

Any good general fixative.

SOLUTIONS

Acid alcohol

70% ethyl alcohol	100.0 ml
glacial acetic acid	0.5–1.0 ml

Giemsa stock solution

Giemsa powder	1.0 g
methyl alcohol, neutral, acetone-free	66.0 ml
glycerol	66.0 ml

Work Giemsa powder into glycerol and place in 60°C oven for 2 hours. Add methyl alcohol and stopper tightly.

Giemsa working solution

Giemsa stock solution	2.5 ml
methyl alcohol	3.0 ml
distilled water	100.0 ml

Rosin stock solution (colophonium)

white wood rosin	10.0 g
absolute alcohol	100.0 ml

---

[3] See page 401.

Rosin working solution

rosin stock solution	5.0 ml
95% ethyl alcohol	40.0 ml

PROCEDURE

1. Deparaffinize and hydrate slides to water. Remove $HgCl_2$.
2. Treat in acid alcohol: 5 minutes (not necessary if tissue has been subjected to acid decalcification).
3. Wash in running water: 5 minutes.
4. Rinse in distilled water.
5. Place in freshly mixed Giemsa working solution: 1 hour.
6. Transfer to second jar of Giemsa working solution: overnight.
7. Rinse in distilled water.
8. Differentiate in rosin working solution.
9. Rinse in 95% alcohol.
10. Dehydrate in absolute alcohol, 2 changes; clear and mount.

RESULTS

nuclei—reddish-purple
other tissue elements—similar to Wright stain
inclusion bodies—blue to purplish-blue

COMMENTS

Bouchard (1963) finds that heating Giemsa solution accelerates the staining. The staining solution can be heated to 80°–85°C on the slide: 10 minutes; or it can be heated to 80°–85°C and then poured on the slide and let stand: 10 minutes. The solubility of Giemsa dye in water is increased by heat, thus promoting faster staining.

For other stains for inclusion bodies, see page 401.

Jenner-Giemsa Stain for Malarial Parasites (McClung 1939)

FIXATION

Smears may be fixed during staining process. Fix tissue sections in any good general fixative, preferably one containing mercuric chloride and alcohol.

SOLUTIONS

Jenner solution

Jenner stain	0.2 g
methyl alcohol, neutral, acetone-free	100.0 ml

Giemsa stock solution

Giemsa powder	3.8 g
methyl alcohol, neutral, acetone-free	75.0 ml
glycerine	25.0 ml

Work stain into glycerine, warm for 2 hours, 60°C oven. Add methyl alcohol.

Working solution

Giemsa stock solution	10.0 ml
distilled water	100.0 ml

PROCEDURE

1. Deparafinize sections and run down to 50% alcohol. Remove $HgCl_2$, if present.
2. *a.* Sections—flood with ample amount of Jenner solution and add an equal amount of distilled water, or mix equal amounts of Jenner solution and water and flood slides or stain slide in coplin jar: 4 minutes.
   b. Smears—flood with Jenner solution: 3 minutes. Add equal amount of distilled water: 1 minute.
3. Pour off Jenner; no rinse.
4. Flood with diluted Giemsa solution or place in coplin jar of solution: 15–20 minutes.
5. Rinse off stain with distilled water and continue to differentiate with distilled water. If too blue, eosin color can be brought out by rinsing in 1% acetic acid in water.
6. Dehydrate sections in following sequence of acetone and xylene.

   acetone 95 parts: xylene 5 parts
   acetone 70 parts: xylene 30 parts
   acetone 50 parts: xylene 50 parts
   acetone 30 parts: xylene 70 parts
   acetone 5 parts: xylene 95 parts

   Smears may be blotted and air-dried before mounting.
   Suggestion: In place of the above solutions, use 3 changes isopropyl alcohol, xylene-isopropyl alcohol (1 : 1). Do not use ethyl or methyl alcohol, which extract the stain.

7. Clear in xylene and mount.

RESULTS

chromatin of parasite—red or purplish-red
cytoplasm of parasite—blue (other elements—color similar to Wright stain)
pigment—yellow-brown to black
Schüffner's stippling—red

COMMENTS

A beautiful clear stain for any blood picture, particularly blood parasites.

Brooke and Donaldson (1950) recommend the use of Triton X-30, a nonionic liquid detergent to prevent the transfer of malarial parasites between films when mass staining.

For a fluorescent antibody technique, see Ingram et al. (1961).

Giemsa Staining for Cold-Blooded Vertebrates (Pienaar 1962)

1. Fix dried smears in undiluted Wright (page 248) or Jenner (page 256) stock stains: 3–5 minutes. (Smears may be previously fixed in absolute methyl alcohol.)
2. Add equal volume of distilled water buffered to pH 6.2–6.8: 5–6 minutes. (pH is important.)
3. Pour off solution, but do not rinse. Stain in diluted Giemsa stock (page 257) (1 ml/10 ml distilled water): 20–30 minutes.
4. Flush with distilled water to prevent stain precipitate on slide.
5. Differentiate, if necessary, in distilled water, pH 6.2.
6. Rinse well in distilled water. Blot, dry, and mount.

Pienaar recommends the following modification for snake eosinophils, which are distorted in the above method.

1. Jenner solution: 1–3 minutes.
2. Buffered water dilution: 10 minutes.
3. Dilute Giemsa: 10–15 minutes.
4. Rinse briefly in distilled water. Dehydrate rapidly through acetone, xylene-absolute alcohol (1 : 1), xylene. Mount.

Giemsa Staining for Birds (Lucas and Jamroz 1961)

1. Fix dried smears in Wright stain: 4 minutes.
2. Add distilled water: 4–5 minutes.
3. Wash in running water: 2 minutes.

4. Stain in dilute Giemsa (15 drops/10 ml water): 15–20 minutes.
5. Wash in distilled water. If precipitate is present, run Wright stain over the slide. Wash off with running water.
6. Stand slides on edge to dry. Do not blot.

Reticulocyte Staining (Brecher 1949)

SOLUTION

new methylene blue N (C.I. 52030)	0.5 g
potassium oxalate	1.6 g
distilled water	100.0 ml

PROCEDURE

1. Mix equal amounts of stain and fresh or heparinized (or oxalated) blood on a slide. Draw mixture into a capillary pipette and allow to stand: 10 minutes.
2. Expel mixture on a slide and mix again.
3. Make thin smears. Air dry.

RESULTS

reticulum—deep blue
red cells—light greenish blue

COMMENTS

If slides fade, fix in methanol and restain with Wright stain. The slides can be stained with Wright immediately following the air drying in step 3, if desired.

Heinz bodies are demonstrated by this stain, pale to deep blue against the pale green background of the red cells (Thompson 1961). Simpson et al. (1970) recommend Rhodanile blue as a selective stain for Heinz bodies.

Lucas and Jamroz (1961) use the following for birds: Mix a drop of blood with a drop of 1% brilliant cresyl blue in 0.85% NaCl or Ringer avian solution (page 534). Let stand 2 minutes and smear. Dry, fix in methyl alcohol, and stain with Wright.

For cold bloods (Pienaar 1962), mix a few drops of blood and stain together in equal quantity in a small covered dish, and leave 5 minutes. Make up the stain, 1% in Ringer or Locke solution for cold bloods (page 534). Make smears on slightly warm slides. Dry, fix in methyl alcohol, and stain with Wright.

HEMOGLOBIN STAINING

See page 272 for other hemoglobin methods.

Buffalo Black for Hemoglobin (Puchtler et al. 1964)

FIXATION

Any general fixative.

SOLUTIONS

Tannic acid

tannic acid	5.0 g
distilled water	100.0 ml

Phosphomolybdic acid

phosphomolybdic acid	1.0 g
distilled water	100.0 ml

Buffalo black

Saturate a solution of buffalo black NBR (C.I.20470) in methanol–glacial acetic acid (9 : 1). Let stand 48 hours. Do not filter. Keeps several months. Filter the used portions back into the storage bottle.

PROCEDURE

1. Deparaffinize and hydrate to water.
2. Treat with tannic acid: 10 minutes.
3. Transfer through 3 changes of distilled water.
4. Treat with phosphomolybdic acid: 10 minutes.
5. Transfer through 3 changes of distilled water.
6. Stain in buffalo black: 5 minutes.
7. Differentiate in methanol–acetic acid (9 : 1), 2 changes: 5 minutes total.
8. Place in absolute ethanol or propanol, clear, and mount.

RESULTS

hemoglobin—dark blue to purplish-black
other tissues—yellow

COMMENTS

Do not use freshly prepared tannic acid. The tannic acid should be more than 48 hours old; it keeps indefinitely.

Fetal Hemoglobin (Modified from Kiossoglou et al. 1963)

FIXATION

Air dry blood smears for 1 hour. Fix in absolute methanol: 5 minutes; follow by 80% ethanol: 5 minutes.

SOLUTIONS

Buffer

McIlvaine citric acid–phosphate, pH 3.2–3.4, page 544.

Stain

eosin Y (C.I. 45380)	0.5 g
distilled water	100.0 ml

or

ponceau S (C.I. 27195)	0.5 g
1% aqueous acetic acid	100.0 ml

PROCEDURE

1. Rinse fixed slides in tap water. Follow by a rinse in distilled water.
2. Treat in buffer, 37°C: 5 minutes. Slides must be in vertical position.
3. Rinse in tap water.
4. Stain: 3 minutes.
5. Rinse in tap water: few seconds. Dry standing in vertical position.

RESULTS

cells containing fetal hemoglobin—deep pink to red
adult cells—colorless or pale-staining ghosts

COMMENTS

The smears can be stained with Mayer hematoxylin for 5 minutes before staining with eosin or ponceau S. Also they may be cleared and mounted. Count 500 cells and express the number of F-containing cells as a percentage of the total. Smears must be fresh; if left unstained over-night they can give false determinations.

BONE MARROW STAINING

Maximow Eosin-Azure Stain (Block et al. 1953)

FIXATION

Zenker-neutral formalin: 30 minutes. Neutralize formalin by adding 2 g lithium or magnesium carbonate to 500 ml of formalin. Excess of carbonate should be present. Wash in running water 1–24 hours before preparation for embedding. (See comment 1.)

SOLUTIONS

Solution A

eosin Y (C.I. 45380) [or eosin B (C.I. 45400)]	0.1 g
Wright buffer	100.0 ml

Solution B

azure II	0.1 g
Wright buffer	100.0 ml

Working solution

Wright buffer	2.0 ml
distilled water	40.0 ml
Solution A	8.0 ml

Stirring vigorously, add gradually

Solution B	4.0 ml

Fresh solutions are best; their action deteriorates after 3 or 4 weeks. Working solution should appear deep violet in color, and a precipitate should not form in it for an hour or more. If a precipitate forms on the slides, the stain mixture was improperly made; solution B was added to solution A too rapidly or without stirring. If the eosin loses its brilliance, solutions are old. Store solutions A and B in the refrigerator.

PROCEDURE

1. Deparaffinize and hydrate slides to water; remove $HgCl_2$. If tissue was fixed in a fixative without potassium dichromate, chromate slides overnight in 2.5%–3% aqueous potassium dichromate. Wash thoroughly in running water: 15 minutes. Proceed to step 2.
2. Stain in Mayer hematoxylin: 30–45 seconds, no more.

3. Wash in running water: 5–10 minutes.
4. Wash in distilled water: 5 or more minutes.
5. Stain in eosin-azure overnight.
6. Differentiate in 95% alcohol. In stubborn cases of differentiation (old solutions) a brief immersion in colophonium alcohol (page 255) may help to sharpen the colors. Differentiation must be done under the microscope.
7. Dehydrate in absolute alcohol, clear, and mount. If, after mounting, the stain still appears undifferentiated, remove the cover glass and mountant with xylene. Transfer to absolute alcohol: 1–2 minutes. Transfer for a brief period to 95% alcohol. Examine the slide under a microscope immediately, since the tissue destains rapidly after this treatment. Dehydrate, clear, and mount.

RESULTS

nuclei—dark purple-blue
erythrocytes—light pink
eosinophilic granules—brilliant red
cytoplasm—pale blue

COMMENTS

1. After 30-minute fixation, cells in center are not as well-fixed as those at periphery, but longer fixation produces granular cytoplasm in eosinophilic erythrocytes and erythroblasts. If 30-minute fixation is not adequate for fixing bone marrow contained in bone, follow Zenker-formol fixation by overnight treatment with 3% aqueous potassium dichromate. Also crack the bone and break away a section of the bone to permit penetration of the fixative.
2. Block[4] embeds in methacrylate. This is recommended for thin sections of 1–3μ with a minimum of shrinkage in the tissue and with precise cellular detail. (See page 97.)
3. Lillie (1965) stains 1 hour in the following eosin-azure solution:

azure A, 0.1%	4.0 ml
eosin B, 0.1%	4.0 ml
0.2 *M* acetic acid	1.25 ml
0.2 *M* sodium acetate	0.75 ml
acetone	5.0 ml
distilled water	25.0 ml

---

[4] Personal communication.

The solution should be made up immediately before use. Transfer to acetone, 2 changes; acetone-xylene (1 : 1) and xylene. Lillie alters the amount of acetic acid and sodium acetate in the solution. The above mixture is for material fixed in Zenker. The proportions of acetic acid and sodium acetate should be adjusted following other fixatives (Lillie).

4. Bone marrow when collected should have an anticoagulant added to it. Gardner (1958) and many other workers mix the bone marrow in a tube containing 0.5 mg heparin powder. Kniseley[5] wets his syringe with d-potassium EDTA as an anticoagulant. Smears can be made with some of the aspirated (drawn out by suction) marrow. The remaining material is poured into a small funnel lined with Filtrator Coffee Paper (Zbar and Winter 1959). Rinse marrow, while in funnel, several times with saline, or until the marrow has lost most of its color. This also helps to wash the material into one mass at the apex of the funnel. Partially clotted marrow will be broken up and washed free of blood, leaving excellent clear particles of marrow. Fold in the filter paper, place in tissue receptacle and carry through fixation, washing, dehydration, and infiltration without removing from filter paper. No marrow will be lost. Minute amounts of marrow may prove difficult to recover from the filter paper without loss of material. I allow the paraffin around the marrow (on the filter paper) to congeal slightly, then carefully scrape paraffin and marrow together, and remove to melted paraffin in mold. The congealed paraffin containing the marrow will sink to the bottom of the mold, or it can be pressed down into desired position. This method keeps the marrow concentrated reasonably well. For other methods for handling bone marrow, see Berthrong and Barhite (1964), Conrad and Crosby (1961), Leach (1960) and Raman (1955).

5. Gude and Odell (1955) recommended for dilution Vinisil (Abbott Laboratories), a 3.5% solution of polyvinylpyrrolidone (PVP) in normal saline. Endicott (1945) used plasma serum as a diluting fluid to thin smears. The serum can be kept on hand for several months if stored in refrigerator.

6. For a fluorescent method, see Werth (1953).

Phloxine–Methylene Blue (Thomas 1953)

FIXATION

Any general fixative.

---

[5] Ralph M. Kniseley, Oak Ridge Associated Universities. Personal communication.

SOLUTIONS

Phloxine solution

phloxine B (C.I. 45410)	0.5 g
distilled water	100.0 ml
glacial acetic acid	0.2 ml

A slight precipitate will form. Filter before use (I have found this unimportant; the precipitate settles at the bottom of the bottle and does not collect on the tissue).

Methylene blue–azure solution

methylene blue (C.I. 52015)	0.25 g
azure B (C.I. 52010)	0.25 g
borax	0.25 g
distilled water	100.0 ml

PROCEDURE

1. Deparaffinize and hydrate slides to water; remove $HgCl_2$.
2. Stain in phloxine: 1–2 minutes.
3. Rinse well in distilled water.
4. Stain methylene blue–azure: 0.5–1 minute.
5. Partially destain in 0.2% aqueous acetic acid.
6. Complete differentiation in 95% alcohol, 3 changes.
7. Dehydrate in absolute alcohol; clear and mount.

RESULTS

nuclei—blue
plasma cell cytoplasm—blue
other tissue elements—shades of rose and red

COMMENTS

Thomas substitutes azure B for azure II of other methods because of the uncertain composition of the latter. It is unnecessary to use colophonium alcohol as in other methods. I recommend Thomas' method over all others I have tried. It is practically foolproof. See Greenstein's (1957) modification. He uses an alcoholic phloxine solution and mixes equal parts of separate solutions of methylene blue and azure B.

Delez and Davis (1950) make up their phloxine with oxalic acid; 1% phloxine in 0.05% aqueous oxalic acid.

This staining method is frequently used for general staining in place of a hematoxylin-eosin method.

STAINING FOR FIBRIN

Ledrum Acid Picro-Mallory Method (Culling 1957)

FIXATION

Any good general fixative.

SOLUTIONS

Celestin blue

celestin blue B (C.I. 51050)	0.25 g
ferric alum, 5% (5 g/100 ml water)	50.0 ml

Boil 3 minutes. Cool and filter. Add:

glycerine	7.0 ml

Keeps for several months.

Mayer hematoxylin, see page 150.

Picro-orange

picric acid saturated in 80% alcohol	100.0 ml
orange G (C.I. 16230)	0.2 g

Acid fuchsin

acid fuchsin (C.I. 42685)	1.0 g
distilled water	100.0 ml
trichloracetic acid	3.0 g

Phosphotungstic acid

phosphotungstic acid	1.0 g
distilled water	100.0 ml

Aniline blue

aniline blue WS (C.I. 42780)	2.0 g
distilled water	100.0 ml
glacial acetic acid	2.0 ml

PROCEDURE

1. Deparaffinize and hydrate slides to water; remove $HgCl_2$.
2. Stain in celestin blue: 3–5 minutes.
3. Rinse in tap water.
4. Stain in Mayer hematoxylin: 5 minutes.
5. Wash in running water: 5 minutes.
6. Rinse in 95% alcohol.
7. Stain in picro-orange: 2 minutes.
8. Stain in acid fuchsin: 5 minutes.
9. Rinse in distilled water.
10. Dip in equal parts of picro-orange and 80% alcohol: few seconds.
11. Differentiate in phosphotungstic acid: 5–10 minutes, until colors are clear.
12. Rinse in distilled water.
13. Stain in aniline blue: 2–10 minutes.
14. Rinse in distilled water: 2–3 minutes.
15. Dehydrate, clear, and mount.

RESULTS

fibrin—clear red
erythrocytes—orange
collagen—blue
nuclei—blue-black

COMMENTS

This method is specific for fibrin, setting it off sharply from other tissue elements. See also phosphotungstic acid-hematoxylin (page 163), the Gram-Weigert method (page 381), and the Rosindole method, Puchtler et al. (1961). The latter is also considered specific for fibrin.

CHAPTER 17

Staining Pigments and Minerals

Pigments are a heterogeneous group of substances containing enough natural color to be visible without staining. Sometimes, however, color is added to give them more intense differential staining. Some pigments are artifacts, such as formalin pigment (page 8). Others, exogenous pigments, are foreign pigments that have been taken into the tissue in some manner. Carbon is a common pigment found in the lungs of city dwellers, particularly of people from coal-burning cities. Endogenous pigments are found within the organism and arise from nonpigmented materials. Iron-containing hemoglobin can be broken down into iron-containing pigment, hemosiderin, and a brown pigment containing no iron, called hematoidin, or bilirubin, which can be oxidized to biliverdin (green). Normal hemoglobin, when not broken down into hemosiderin, will not show a positive Prussian blue reaction (page 269). The iron is masked or occult, and the organic part of the hemoglobin molecule must be destroyed in order for the iron to be unmasked.

Melanin (brown or black pigment) is found normally in the skin, hair, and eye, but may occur pathologically anywhere in the body. The pigment is formed from tyrosine by a special enzyme, tyrosinase. Lipofuscin, sometimes known as the "wear and tear" pigment, can be found in the heart muscle, adrenals, ganglion cells, and liver. The lipofuscins are derived from lipid or lipo-protein sources, at least partly by oxidation. Several enzymes are found in them, nonspecific esterases and acid phosphatase in particular. Melanin and lipofuscin are brownish pigments stainable by fat dyes and some basic aniline dyes such as fuchsin. They are metachromatic (page 129) with methyl green and give a positive Schiff reaction after periodic acid treatment (page 328).

Calcium may be present in tissues in a number of different forms, but the methods here are applicable to the insoluble forms—deposits of calcium phosphate and calcium carbonate (Ham 1957).

STAINING FOR IRON

According to Baker (1958) the iron reaction is an example of local formation of a colored substance which is not a dye. It is a type of histochemical test wherein a tissue is soaked in a colorless substance. Certain tissue elements react with the substance and become colored. This well-known test is called the Berlin blue, Prussian blue, or Perl reaction: the iron is dissolved from hemosiderin by hydrochloric acid, and then reacts with potassium ferrocyanide to form the Berlin blue precipitate, ferric ferrocyanide.

Sometimes fading occurs in slides due to the reduction of the Berlin blue to colorless ferro ferrocyanide. The mounting resin probably takes up oxygen while drying and deprives the sections of oxygen, thereby reducing them to the colorless condition. If this takes place and it is essential to recover the slides, treat them with hydrogen peroxide. The newer synthetics are not as prone to reduce Berlin blue as former resins; also Lustron 2020 is recommended (page 141).

Iron Reaction (Hutchison 1953)

FIXATION

Hutchison recommends:

sodium sulfate	12.0	g
glacial acetic acid	33.0	ml
formalin	40.0	ml
distilled water to make	200.0	ml

Go directly into 70% alcohol from fixative.

SOLUTIONS

Ferrocyanide-hydrochloric acid solution

Solution A

potassium ferrocyanide	2.0 g
distilled water	50.0 ml

Solution B

hydrochloric acid	2.0 ml
distilled water	48.0 ml

Prepare the two parts of the solution separately, mix them together, warm slightly, and filter. Place in paraffin oven, 56°C, a short time before using.

Safranin

safranin O (C.I. 50240)	0.2 g
distilled water	100.0 ml
glacial acetic acid	1.0 ml

Kernechtrot

Kernechtrot (nuclear fast red)[1]	0.1 g
5% aluminum sulfate (5 g/100 ml water)	100.0 ml

Dissolve with heat. Cool and filter.
Add a crystal of thymol as preservative.

PROCEDURE

1. Deparaffinize and hydrate slides to water.
2. Wash in distilled water: 3 minutes.
3. Treat with ferrocyanide–hydrochloric acid, 56°C: 10 minutes.
4. Rinse, several changes distilled water: 5 minutes.
5. Counterstain with safranin: 2–5 minutes, or Kernechtrot: 5–8 minutes.
6. Rinse in 70% alcohol.
7. Dehydrate, clear, and mount.

---

[1] Edward Gurr, Roboz Surgical Instrument Co., Washington, D.C.

RESULTS

iron pigment—brilliant greenish blue
nuclei—red
other tissue elements—shades of red and rose

COMMENTS

Hutchison claims that warming the solution is the most important step in this method. Do not leave the slides in the solution longer than 10 minutes. If they have been well washed in distilled water, precipitate seldom forms on the sections. This method produces deep, brilliant colors.

For masked or occult iron (Glick 1949), pretreat slides with 30% hydrogen peroxide alkalized with dilute ammonia (1 drop/100 ml peroxide): 10–15 minutes. Wash well and proceed to step 3. If the hydrogen peroxide tends to form bubbles under the sections and loosen them, keep the solution and slides cool in the refrigerator during the treatment.

I prefer 3% nitric acid in 95% alcohol overnight. Wash in 90% alcohol and proceed as usual. This method does not unmask iron in hemoglobin.

Past (1961) decalcifies osseous tissue in 5% EDTA (ethylenediaminetetraacetic acid) adjusted to pH 12 N NaOH before staining for iron.

Turnbull Blue Method for Ferrous Iron (Pearse 1968)

FIXATION

10% buffered formalin; other general fixatives satisfactory if acid is absent.

SOLUTIONS

Ammonium sulfide

A saturated solution of $(NH_4)_2S$, 20%–30% content, analytical reagent.

Potassium ferricyanide

potassium ferricyanide ($K_3Fe(CN)_6$)	20.0 g
distilled water	100.0 ml

Hydrochloric acid

hydrochloric acid (sp gr. 1.188–1.192, 37%–38% HCl)	1.0 ml
distilled water	99.0 ml

Safranin or Kernechtrot (see page 270).

PROCEDURE

1. Deparaffinize and hydrate slides to water; remove $HgCl_2$.
2. Wash in distilled water.
3. Treat with yellow ammonium sulfide: 1–3 hours.
4. Rinse in distilled water.
5. Treat with equal parts of potassium ferricyanide and hydrochloric acid, freshly mixed: 10–20 minutes.
6. Rinse in distilled water.
7. Counterstain with safranin: 2–5 minutes, or Kernechtrot: 5–8 minutes.
8. Rinse in 70% alcohol.
9. Dehydrate, clear, and mount.

RESULTS

ferrous iron and ferric iron converted to ferrous iron—deep blue
other tissue elements—shades of rose and red

COMMENTS

The dinitroresorcinol used by Humphrey (1935) gives a disappointingly dull slide. Hukill and Putt's (1962) technique with 3,7-diphenyl-1, 10-phenanthroline (Bathophenanthroline) also gives disappointing results.

Kutlík (1970) stains for 2 hours at room temperature in chlorate hematoxylin (1% hematoxylin in 7.3% aqueous potassium chlorate, filtered). Wash well and counterstain. This forms a black iron hematein lake where iron compounds are present. I have not tried this method, but feel it has promise, particularly when photographs are desired. The solution will keep for 6 months.

HEMOGLOBIN STAINING

See page 260 for other hemoglobin methods.

Benzidine Method (Ralph 1941, Glick 1949)

FIXATION

Alcohol or formalin; also try frozen sections of unfixed tissue.

SOLUTIONS

Benzidine reagent

| benzidine | 1.5 g |
| absolute methyl alcohol | 100.0 ml |

Peroxide reagent

| peroxide (reagent 3%) | 25.0 ml |
| 70% ethyl alcohol | 75.0 ml |

PROCEDURE

1. Deparaffinize sections and hydrate to water. Not necessary for smears, which will be fixed by benzidine reagent.
2. Flood with benzidine reagent: 1–2 minutes.
3. Drain, and flood with peroxide reagent: 1.5 minutes.
4. Wash in distilled water: 15 seconds.
5. Counterstain if desired with a red nuclear stain or Wright's blood stain.
6. Dehydrate, clear, and mount.

RESULTS

hemoglobin—dark brown
hemosiderin—see iron reactions, page 269.

COMMENTS

Tissues containing heme and hematein compounds have weak peroxidase activity; most other tissues lack it. Since the peroxidases need the presence of H_2O_2 to catalyze the oxidation of peroxidase substrates such as benzidine, the tissue is first treated with benzidine and then exposed to peroxide. The peroxide activates the peroxide-peroxidase system to produce rapid oxidation of the benzidine to an unstable blue compound. This compound undergoes auto-oxidation to form a stable, brown-colored compound at the site of the activity.

Ericsson (1965) modified the Ralph method by staining 1 minute in a freshly prepared solution made by mixing 40 ml of the benzidine reagent with 60 ml of the peroxide reagent.

Putt (1951b) combined thionin with the benzidine reagent to produce an olive green instead of a brown substance that is not so easily confused with iron or formalin pigment.

Puchtler and Sweat (1962b) use Buffalo black for staining hemoglobin (see page 260).

Puchtler and Sweat (1963*a*) combine hemoglobin staining with a ferrocyanide staining for hemosiderin.

Cyanol Reaction (Dunn, 1946)

FIXATION

Formalin, preferably buffered to *p*H 7; fix no longer than 48 hours. Dried smears may be methyl-alcohol fixed and then taken to step 2 below.

SOLUTIONS

Cyanol reagent stock solution

cyanol	1.0 g
distilled water	100.0 ml
zinc powder, CP	10.0 g
glacial acetic acid	2.0 ml

Bring to boiling point. In a short time the solution is decolorized. Stable for several weeks.

Working reagent

Filter 10 ml of stock reagent, and add:

glacial acetic acid	2.0 ml
hydrogen peroxide, commercial, 3%	1.0 ml

PROCEDURE

1. Deparaffinize and hydrate slides down to water.
2. Treat with working reagent: 3–5 minutes.
3. Rinse in distilled water.
4. Counterstain in red nuclear stain.
5. Dehydrate, clear, and mount.

RESULT

hemoglobin—dark blue to bluish gray

COMMENTS

Gomori (1952) uses 1.0 g acid fuchsin instead of cyanol, and produces a red hemoglobin. Counterstain with hematoxylin or other blue nuclear stain.

BILE PIGMENT (BILIRUBIN) STAINING

Glenner Method (1957)

FIXATION

Use fresh tissue quickly frozen by the Adamstone-Taylor cold knife method (page 425), or fix tissue in formalin for 6 hours and prepare Carbowax sections.

SOLUTION

potassium dichromate, 3% aqueous . . . .   25.0 ml
buffer, pH 2.2 (0.1 N HCl, 8.0 ml; 0.1 N potas-
sium dihydrogen phosphate, 17.0 ml) . . . . . .   25.0 ml

PROCEDURE

1. Treat sections with dichromate solution, room temperature: 15 minutes.
2. Wash in running water: 5 minutes.
3. Counterstain with hematoxylin, safranin, or Kernechtrot.
4. Dehydrate rapidly through 95% alcohol: 5–10 seconds.
5. Finish dehydration in absolute alcohol: 1 minute.
6. Clear through absolute alcohol–xylene and xylene.
7. Mount in Permount or the like.

RESULT

bilirubin—emerald green

COMMENTS

The pH of 2.2 is crucial; it must be low enough to result in complete oxidation of all bilirubin, but not so acid that it causes loss of the pigment from the sections. Glenner also includes a method for the demonstration of bilirubin, hemosiderin, and lipofuscin in the same section.

This reaction is recommended in preference to Stein's iodine method (Glick 1949), which cannot be considered reliable at all times because the reactants are diffusible and the final color can spread from the original site. The test is based on oxidation of the bile pigment (bilirubin) to green biliverdin by the iodine solution. The Gmelin method using nitric acid is even less permanent.

Kutlík (1958) and Lillie and Pizzolato (1968) describe argentaffin methods (page 336). For definite blackening of the bile pigment, Lillie and Pizzolato recommend 5–16 hours exposure to diammine silver at 25°C.

The solution is made as follows: to 2.0 ml of 28% NH_4OH add 35–40 ml of 5% $AgNO_3$, rapidly at first with swirling to dissolve the precipitate that forms. Then add smaller amounts of $AgNO_3$ until only a faint turbidity persists.

See Lillie and Pizzolato (1968) for a more complete discussion of bile pigments.

Azo-Coupling Method.

FIXATION

Formalin.

SOLUTIONS

Buffer solution

Solution A

sodium acetate	1.17 g
sodium barbital	2.94 g
distilled water	100.0 ml

Solution B

hydrochloric acid	0.84 ml
distilled water	100.0 ml

Working solution, pH 9.2

solution A	5.0 ml
solution B	0.25 ml
distilled water	17.75 ml

Staining solution

fast red B salt[2]	50.0 mg
buffer, pH 9.2	50.0 ml

PROCEDURE

1. Deparaffinize and hydrate sections to water or cut frozen sections.
2. Stain: 30 seconds.
3. Wash off excess stain in tap water.
4. Stain in hematoxylin: 5 minutes.

---

[2] Sigma

5. Wash in tap water: 5 minutes.
6. Dehydrate, clear, and mount.

RESULTS

bile pigments—orange-red
nuclei—blue

MELANIN AND LIPOFUSCIN STAINING

Lillie (1956*a,b*) Nile Blue Method

FIXATION

Any good general fixative.

SOLUTION

Nile blue A (C.I. 51180) (Nile blue sulfate) . . . 0.05 g
sulfuric acid, 1% (1 ml conc. H_2SO_4/99 ml dis-
tilled water) . . . . 100.0 ml

PROCEDURE

1. Deparaffinize and hydrate slides to water: remove $HgCl_2$.
2. Stain in Nile blue A solution: 20 minutes.
3. Wash in running water: 10–20 minutes.
4. Mount in glycerine jelly.

RESULTS

lipofuscins—dark blue or blue-green
melanin—pale green
erythrocytes—greenish yellow to greenish blue
myelin—green to deep blue
nuclei—poorly stained

Lillie Alternate Nile Blue Method

PROCEDURE

Steps 1 and 2 as above.
3. Do not wash in water. Rinse quickly in 1% sulfuric acid to remove excess dye.
4. Dehydrate at once in acetone, 4 changes: 15 seconds each.
5. Clear in xylene and mount.

RESULTS

cutaneous, ocular, meningeal melanins—dark green
mast cells—purple-red
lipofuscins—unstained but appear yellow to brownish
muscle, myelin, erythrocytes—unstained
nuclei—greenish to unstained

COMMENTS

Nile blue sulfate stains basophilic tissue and acidic lipids, such as free fatty acids, phospholipids, and certain lipid components of some pigment deposits of a lipogenic character.

Lipofuscins stain with Nile blue by two mechanisms: a fat solubility mechanism operating at pH below 1.0; and an acid-base mechanism operating at levels above pH 3.0. When stained by the second mechanism, they retain a green stain after acetone or brief alcoholic extraction, but when the first mechanism is used, they are promptly decolorized by acetone or alcohol (Lillie 1956*b*). (See also Lillie 1965.)

Melanins stain with basic dyes at pH levels below 1.0 and retain the stain when dehydrated and mounted (Lillie 1955).

Ferro-Ferricyanide Method (Lillie 1965)

FIXATION

Avoid chromate fixatives; others are satisfactory.

SOLUTIONS

Ferrous sulfate

ferrous sulfate ($FeSO_4 \cdot 7H_2O$)	2.5	g
distilled water	100.0	ml

Potassium ferricyanide

potassium ferricyanide ($K_3Fe(CN)_6$)	1.0	g
distilled water	99.0	ml
glacial acetic acid	1.0	ml

PROCEDURE

1. Deparaffinize and hydrate slides to water.
2. Treat with ferrous sulfate: 1 hour.
3. Wash in distilled water, 4 changes: total 20 minutes.
4. Treat with potassium ferricyanide: 30 minutes.

5. Wash in 1% aqueous acetic acid: 1–2 minutes.
6. Counterstain if desired: picro-ponceau satisfactory; do not use hematoxylin.
7. Dehydrate, clear, and mount.

RESULTS

melanins—dark green
background—faint greenish or unstained; with picro-ponceau, collagen stains red, and muscle and cytoplasm yellow and brown

COMMENTS

Lillie says this method is highly selective. No other pigments react in this procedure, except occasionally hemosiderin.

Indophenol Method (Alpert et al. 1960)

FIXATION

Any general fixative, or use fresh-frozen sections.

SOLUTION

sodium-2,6-dichlorobenzenone-indophenol	0.1 g
50% ethyl alcohol	100.0 ml

Prepare just before use. Filter and add 1% aqueous hydrochloric acid until color of solution is red (approximately 1:5) or until pH is 2.0.

PROCEDURE

1. Deparaffinize and hydrate slides to water.
2. Treat with HCl-indophenol: 5 minutes.
3. Dip in tap water several times until background loses color.
4. Mount in glycerine jelly or Von Apathy (page 142).

RESULTS

lipofuscins—red
erythrocytes—blue
other tissue elements—colorless to blue or slightly pink

COMMENTS

Avoid alcohol; it removes color immediately. Stain fades in about 1 day.

STAINING FOR CALCIUM DEPOSITS

Von Kossa Method (Mallory Modification 1944)

FIXATION

10% formalin or alcohol, although other fixatives give reasonably good results.

SOLUTIONS

Silver nitrate

silver nitrate ($AgNO_3$)	5.0 g
distilled water	100.0 ml

Fresh solution always best; never use one more than 1 week old.

Safranin or Kernechtrot, see page 270.

PROCEDURE

1. Deparaffinize and hydrate slides to distilled water; remove $HgCl_2$.
2. Treat with silver nitrate in dark: 30 minutes.
3. Rinse thoroughly in distilled water.
4. Expose slides (in distilled water) to bright light (75–100W bulb satisfactory): 1 hour. Lay them over a white background to expedite reaction.
5. Wash thoroughly in distilled water.
6. Counterstain in safranin: 2–3 minutes, or Kernechtrot: 5–8 minutes.
7. Rinse in 70% alcohol.
8. Dehydrate, clear, and mount.

RESULTS

calcium deposits—dark brown to black
nuclei and other tissue elements—shades of red and rose

COMMENTS

If the exposure to light in step 4 produces too much brown or yellow background, try developing in a photographic developer: 5 minutes (step 4).

Step 5. Rinse in distilled water and fix in hypo: 5 minutes. Wash well and continue to step 5 above.

If iron blocks out the calcium, treat 10–15 minutes with .005 *M* sodium EDTA in normal saline. Wash in distilled water and proceed to step 2 (McGee-Russell 1958).

This reaction actually demonstrates phosphates and carbonates rather than calcium itself, but since soluble phosphates and carbonates are washed out, the calcium phosphates and calcium carbonates remain to react with the silver (Lillie, 1954*b* or 1965), and the test can be regarded as sufficiently specific.

Renaud (1959) demonstrates that a high percentage of alcohol (at least 80% content) in the fixing fluid is necessary in order to detect calcium in some tissues (heart and coronary vessels). Water and even buffered formalin can dissolve out some small deposits of calcium salts.

Pizzolato and McCrory (1962) demonstrated that neither chemical reduction nor exposure to illumination was necessary in the Von Kossa technique.

Pickett and Klavins (1961) followed step 5 with an elastin stain (preferably aldehyde-fuchsin) to demonstrate calcium and elastin simultaneously.

Carr et al. (1961) describe a method using chloranilic acid (red-brown) to differentiate calcium deposits from carbon deposits.

Alizarine Red S (Puchtler et al. 1969)

FIXATION

Alcohol or Carnoy fixative should be used for best preservation of calcium. Formalin (even buffered) or Zenker-formol remove calcium deposits.

SOLUTION

alizarine red S (C.I. 58005)	0.5 g
phosphate or barbital buffer, *p*H 9	100.0 ml

PROCEDURE

1. Deparaffinize and hydrate slides to water.
2. Stain in alizarine red: 30–60 minutes.
3. Rinse in buffer, *p*H 9: 5 seconds.
4. Dehydrate in 3 changes absolute ethanol: 2 minutes each.
5. Clear and mount.

RESULT

calcium—orange-red

COMMENTS

The recommendations made by Puchtler et al. (1969) must be observed to obtain reliable results with this method. Proper fixation is important; calcium staining is most intense after alcohol and Carnoy fixation. Formalin and Zenker produce weakly or moderately colored deposits and staining decreases after storage in formalin, disappearing completely after 2–3 weeks. Alizarine red S stains red more reliably in a buffer solution of pH 9 than at other pHs. It stains red without a yellow tinge within 5 minutes, but intensity of color increases with staining times up to an hour. The alizarine dyes in aqueous solution of pH 7 or below become only slightly dissociated with little tendency to form calcium salts. If the solutions are made at pH 9, then many of the 2-OH groups of alizarine become dissociated and the formation of monoalizarate yields the red color for calcium. At a higher pH of 12, almost all of the 2-OH and about half of the 1-OH groups are dissociated and a blue alizarate forms. The sulfonic acid group of alizarine red S also forms a salt with calcium. Alizarine, however, cannot be considered specific for calcium alone; other metals also react with the stain. It is sensitive toward ions of uranium, titanium, bismuth, thallium, zirconium, hafnium, and thorium. The Puchtler article should be read for a better understanding of this stain and its reaction.

REMOVAL OF PIGMENTS

Melanin Pigment

This will appear as brown, grayish, or almost black granules.

Permanganate Method of Removal (Lillie, 1965)

1. Hydrate slides to water.
2. Immerse slides in 0.5% aqueous potassium permanganate: 12–24 hours.
3. Wash in running water: 5 minutes.
4. Immerse slides in 1% aqueous oxalic acid: 1 minute.
5. Wash in running water: 10 minutes, and proceed to stain.

Performic or Peracetic Acid Method of Removal

Lillie (1954*b* or 1965) claims that the permanganate method can be unpredictable and that the best bleaching is done with performic or peracetic acid: 1–2 hours:

Performic acid

90% formic acid, aqueous	8 ml
30% H_2O_2 (undiluted reagent)	31 ml
H_2SO_4, concentrated	0.22 ml

Keep at or below 25°C. About 4.7% performic acid is formed within 2 hours, but it deteriorates after a few more hours.

Peracetic acid

glacial acetic acid	95.6 ml
30% H_2O_2 (undiluted reagent)	259 ml
H_2SO_4, concentrated	2.2 ml

Let stand 1–3 days. Add 40 mg disodium phosphate as stabilizer. Store at 0.5°C. Keeps for months in refrigerator.

Chlorate Method of Removal

Treat sections 24 hours in 50% alcohol plus small amount of potassium chlorate and a few drops of HCl (concentrated). Wash 10 minutes before staining.

Bromine Method of Removal

1% bromine in water: 12–24 hours. Wash well before staining.

Chromic Acid Method of Removal

Mixture of equal parts of 1% chromic acid and 5% calcium chloride, aqueous: 8–12 hours. Wash well.

Peroxide Method of Removal

10% H_2O_2: 24–48 hours. Wash well before staining.

The peroxide method is considered by some to be the best; it is specific for melanin; other pigments resist longer than 48 hours. The permanganate and chromic acid methods allow no differentiation between pigments.

Formalin Pigment

Brown and black crystalline granules and artifacts, produced by formalin, are found in and around blood, and are considered a hematein derivative.

The crystals are formed when the tissue is fixed with formalin at an acid *p*H. They usually will not form when the *p*H of the formalin fixative is above 6.0.

Barrett Method of Removal (1944)

Saturated solution of picric acid in alcohol: 10 minutes to 2 hours.

Malarial Pigment

This appears as amorphous brown-black granules. Its histochemical properties apparently are the same as those of formalin pigment.

Gridley Method of Removal (1957)

Method 1

1. Hydrate slides down to water.
2. Bleach for 5 minutes in:

acetone	50.0 ml
3% H_2O_2	50.0 ml
28–29% ammonia, concentrated	1.0 ml

   or overnight in 5% aqueous ammonium sulfide (diluting 20% analytical reagent 4 : 1).
3. Wash thoroughly in running water: 15 minutes or longer.

Method 2

Bleach in 3% H_2O_2: 2 hours.

Carbon

Carbon (opaque black) usually appears in the lungs and adjacent lymph nodes, sometimes in the spleen and liver. It is possible to distinguish it from malarial pigment, iron, or some other pigment or precipitate, since the carbon is black and is insoluble in concentrated sulfuric acid in which other pigments will dissolve.

Hemosiderin

This yellowish, brownish, or greenish brown pigment resists bleaching, does not dissolve in alkalis or acids, and is not argentaffin. It can be identified by Perl's test, page 269.

Bile Pigments

These are yellowish green, resist bleaching, and are argentaffin. They can be converted by H_2O_2, Lugol solution, nitrous acid, or dichromates into emerald green biliverdin.

CHAPTER 18

Staining Proteins and Nucleic Acids

PROTEIN STAINING

The proteins found in cells may be classified as either (1) simple proteins, which on hydrolysis yield *a*-amino acids and their derivatives, or (2) conjugated proteins, which yield, in addition to amino acids, nonprotein materials. The simple proteins may be fibrous or globular proteins. Collagens, reticulins, keratin, and fibrin are simple fibrous proteins and insoluble in most aqueous media. The globular proteins—albumins, globulins, globins, and histones—are soluble in aqueous media. Among the conjugated proteins are nucleoproteins, mucoproteins, lipoproteins, and glycoproteins.

Because of their omnipresence in cells, proteins can be disappointing subjects for precise and brilliant localization. Sometimes, however, it does become necessary to ascertain the protein or nonprotein nature of a granulation or another kind of locale in cells. "Protein" techniques exist, but few are specific for proteins as such. Most of these techniques demonstrate a component of the protein (such as one of the amino acid groups), or they are specific for the aromatic nuclei, the phenolic functions, or the guanidine grouping (Gomori 1952; Pearse 1968).

The Millon reaction, one of the classical methods, is not specific for proteins, but for aromatic groups found in the amino acid tyrosine. The method, therefore, is specific for tyrosine, which is in most proteins. Two Millon variations follow.

Millon Reaction (Gomori 1952)

FIXATION

Any good general fixative.

SOLUTIONS

Solution A

mercuric acetate	5.0 g
distilled water	100.0 ml
tricloracetic acid	15.0 g

Solution B

sodium nitrite, 1% aqueous	10.0 ml

PROCEDURE

1. Run sections down to water. (Cut sections at least 10μ thick.)
2. Incubate sections in solution A, 30°–37°C: 5–10 minutes.
3. Add solution B; incubate another 25 minutes.
4. Rinse in 70% alcohol.
5. Dehydrate, clear, and mount.

RESULT

tyrosine—pink to brick red

Millon Reaction (Romeis 1948)

SOLUTION

Dilute 400 ml of concentrated nitric acid with distilled water to make 1 liter. Let stand 48 hours. Dilute again, 1 part of this solution with 9 parts of distilled water. Add an excess of mercuric nitrate and allow to stand several days with frequent shaking to complete saturation. To 400 ml of filtrate, add 3 ml of the 40% original solution and 1.4 g sodium nitrite.

PROCEDURE

1. Fix, section, and hydrate to water as in the procedure for the first Million reaction above.
2. Immerse sections in reagent: 3 hours.
3. Give the slides a short dip in 1% nitric acid.
4. Transfer quickly through 70%, 95%, and absolute alcohol, and xylene; mount.

RESULT

tyrosine—rose or red

Mercuric Bromphenol Blue Method (Mazia et al. 1953)

FIXATION

Any general fixative.

SOLUTION

mercuric chloride	10.0 g
bromphenol blue	100.0 mg
95% ethyl alcohol (or distilled water)	100.0 ml

PROCEDURE

1. Deparaffinize sections. (Use thin sections, no more than 5μ, or stain will be too intense.)
2. Transfer to absolute alcohol: 1–2 minutes.
3. Stain in alcoholic solution: 15 minutes. (Note: if stain is aqueous, hydrate slides to water, then to stain.)
4. Wash in 0.5% aqueous acetic acid to remove excess dye: 20 minutes.
5. Immerse in tap water or buffer of pH 6–7 to convert dye to blue alkaline form: 3 minutes. Do not use ammonia, which distorts morphological detail.
6. Dehydrate rapidly or transfer directly to tertiary butyl alcohol, clear, and mount.

RESULT

This appears to be limited to proteins and peptides that are not removed by washing—blue.

COMMENTS

See Kaniwar (1960) about the uncertain specificity of this method. The identification of the individual amino acids in tissue sections involves a number of procedures, some of which are controversial and unreliable as to specificity, and some of which are complex. Available techniques include methods for amino groups, carboxyl groups, tyrosine, tryptophane, histidine, the SS and SH groups, and arginine. Only a small number of these can be included here.

Ninhydrin Reaction (Serra 1946)

Ninhydrin (triketo-hydrindene-hydrate) reacts with the *a*-amino groups, oxidizing them to carbon dioxide, ammonia, and an aldehyde. The reaction is not stable and its products are diffuse, but the aldehydes that are formed will react with Schiff reagent to produce a red color. Glenner (1963) questions the value of this method, since it reacts with both free and bound *a*-amino acids, but is not limited to them. See also Kasten (1962) and Puchtler and Sweat (1962).

FIXATION

10% formalin.

SOLUTION

phosphate buffer, *p*H 6.98 (page 543) ....   50 ml
ninhydrin, 0.4% aqueous (triketo-hydrindene-
hydrate) ....   50 ml

PROCEDURE

1. Deparaffinize and hydrate slides to water, or use frozen sections.
2. Place slides on a rack over boiling water, cover with stain, and steam for 1–2 minutes.
3. Drain stain off the slide and mount in glycerine jelly.

RESULT

Blue or violet color indicates the presence of amino acids, peptides, and proteins.

COMMENTS

Slides should be examined at once because the color diffuses readily and begins to fade within a day or two.

Ninhydrin-Schiff Reaction (Yasuma and Ichikawa 1953)

FIXATION

Any general fixative, but 10% formalin, Zenker or absolute alcohol are recommended.

SOLUTIONS

Ninhydrin

ninhydrin	0.5 g
absolute ethyl alcohol	100.0 ml

Schiff reagent, see page 327.

PROCEDURE

1. Deparaffinize sections. (Use 10μ sections.)
2. Transfer to absolute ethyl alcohol, 2 changes: 2 minutes each.
3. Incubate in ninhydrin solution, 37°C: 5–24 hours.
4. Wash in tap water: 3–5 minutes.
5. Proceed as in Feulgen reaction, page 332.

RESULT

protein or peptide—red or purple

COMMENTS

See Van Duijn (1961) for acrolein-Schiff reaction, which is based on the reaction of acrolein with tissue compounds. All proteins except arginine-rich protamines are positive. Alloxan also may be used (Yasuma and Ichikawa 1953).

Staining of Histones

Histones are the basic protein of chromatin.

Alfert and Geschwind Method (1953)

FIXATION

10% formalin: 3–6 hours. Zenker, Susa, and Carnoy do not permit specific staining. Wash overnight in running water.

SOLUTIONS

Trichloracetic acid (TCA)

trichloracetic acid	5.0 g
distilled water	100.0 ml

Fast green

fast green FCF (C.I. 42053)	0.1 g
distilled water	100.0 ml

Adjust *p*H to 8.0–8.1 with a minimum of NaOH. Prepare fresh.

PROCEDURE

1. Deparaffinize and hydrate to water.
2. Immerse in TCA solution in boiling water bath to remove nucleic acids: 15 minutes.
3. Wash out TCA with 2 changes of 70% alcohol: 10 minutes each. Then rinse in distilled water.
4. Stain in fast green: 30 minutes.
5. Wash in distilled water: 5 minutes.
6. Transfer directly to 95% alcohol, absolute alcohol, and xylene; mount.

RESULT

histone—green

COMMENTS

This is a specific chromosomal stain, and all experiments indicate that histones or protamines are responsible for the histological stain. The stain is not retained unless nucleic acids are removed. The histones occur in a constant quantitative ratio to the DNA in cell nuclei, and neither histones nor DNA vary in amount when cells undergo physiological changes.

Ammoniacal Silver Method (Black and Ansley 1964)

FIXATION

Buffered formalin (page 15).

SOLUTION

Add 10% aqueous silver nitrate, drop by drop, to 3–4 ml concentrated ammonium hydroxide until the first turbidity occurs. Stir constantly. Approximately 38 ml $AgNO_3$ is required for 4 ml of NH_4OH. Prepare fresh.

PROCEDURE

1. Deparaffinize and hydrate to water.
2. Treat with buffered formalin: 15 minutes.
3. Wash in 5 changes of distilled water: several seconds each.
4. Stain in silver solution with gentle agitation: 5–10 seconds.
5. Wash in 5 changes of distilled water: several seconds each.
6. Treat with 3% neutral formalin with agitation: 2 minutes.
7. Wash in 3 changes of distilled water: several seconds each.
8. Dehydrate, clear, and mount.

RESULT

histone—black

COMMENTS

Black et al. (1964) have written an excellent article concerning the cell specificity of histones. Also see Vidal et al. (1971) for a silver stain.

Arginine Reaction

The arginine reaction uses the reaction of naphthol with guanidine derivatives in the presence of hypochlorite or hypobromite to produce an orange-red color. The reaction takes place in guanidine derivatives when one hydrogen atom of one of the amino groups is replaced by alkyl, fatty acid, or cyano radical. Proteins containing arginine give the color. In tissue fixed by ordinary methods the reaction is specific for protein-bound arginine and other guanidine derivatives such as glycocyamine or agmatine. Guanidine, urea, creatine, creatinine, and amino acids other than arginine do not give the color.

Sakaguchi Reaction (Modified by Deitch 1961)

FIXATION

Absolute ethyl alcohol–glacial acetic acid (3 : 1). Formalin fixation may block amino and guanidyl groups and reduce the reaction. Smears may be air-dried and fixed in methyl alcohol.

SOLUTIONS

Barium hydroxide

barium hydroxide	4.0	g
distilled water	100.0	ml

Filter just before use.

Hypochlorite solution

Clorox, freshly obtained (5% solution)	1.0 ml
distilled water	4.0 ml

Naphthol solution

2,4-dichloro-*a*-naphthol [1]	0.15 g
tertiary butyl alcohol	10.0 ml

PROCEDURE

1. Deparaffinize and hydrate sections to water. Use sections of 4μ or thinner.
2. Blot and place slides in an empty coplin jar.
3. Pour 5 parts of barium hydroxide solution in a flask. Add 1 part hypochlorite solution and then 1 part of dichloronaphthol, shaking flask during each addition. Pour immediately on slides in jar. Room temperature: 10 minutes.
4. Transfer slides rapidly through 3 changes of tertiary butyl alcohol containing 5% of tri-*N*-butylamine.[2] Move each slide vigorously in first change for about 5 seconds, transfer through next 2 changes: 30–60 seconds each. Change first solution frequently.
5. Transfer rapidly through 2 changes of xylene containing 5% tri-*N*-butylamine: 30–60 seconds each.
6. Drain rapidly and mount in Cargille's[3] (formerly Shillaber's) oil containing 10% tri-*N*-butylamine.

RESULT

arginine—orange to red

COMMENTS

Always keep slides moist. Drying causes a crust of barium carbonate to form. The precipitate on the underside of the slide can be wiped off with dilute acetic or hydrochloric acid.

Deitch has substituted barium for sodium or potassium hydroxide for better tissue morphology, and the sections do not tend to fall off. The addition of an organic amine to the solutions and mountants reduces the fading problem of former methods. See also Baker (1947), Carver et al. (1953), Liebman (1951), and Thomas (1950).

---

[1] Eastman Organic Chemicals.   [2] Eastman Organic Chemicals.
[3] Cargille Laboratories, Cedar Grove, N.Y.

The following references are only a few of the many excellent papers concerning proteins and amino acids. For proteins: Deitch (1955) and Spicer and Lillie (1961). For tyrosine: Glenner and Lillie (1959) and Lillie (1957d). For histidine: Landing and Hall (1956), Pauly (1964), and Reaven and Cox (1963). For histamine: Lagunoff et al. (1961), Enerbäck (1969). For tryptophane: Adams (1957), Bruemmer et al. (1957), and Glenner and Lillie (1957b). For SH groups: Barrnett and Seligman (1952), Bennett (1951), Bennett and Watts (1958), and Chévremont and Fréderic (1943). For SS groups: Adams (1956) and Adams and Sloper (1955, 1956). For arginine: Notenboom et al. (1967). For fluorescent labeling of proteins: Rinderknecht (1960).

NUCLEIC ACID AND NUCLEOPROTEIN STAINING

The two types of nucleic acids are: (1) DNA (thymonucleic acid, deoxyribonucleic acid, and deoxypentose ribonucleic acid), the principal components of nuclei; and (2) RNA (ribonucleic acid, and plasmonucleic acid), which is found in the nucleolus and ribosomes of the cytoplasm. Nucleoproteins are combinations of basic proteins and nucleic acids in various highly polymerized forms. The nucleolus is composed largely of concentrated protein, with variable amounts of DNA. Its function is part of the synthetic processes of the cell (Love and Rabotti 1963; Vincent 1955).

On hydrolysis, the nucleic acids yield: (1) purine and pyrimidine bases, (2) carbohydrates, and (3) phosphoric acid. DNA contains deoxyribose (deoxypentose) sugar, and RNA contains ribose (pentose) sugar.

Because the nucleic acids exist in many states of polymerization, fixation will cause some redistribution. Organic solvents, formalin, and most acid fixatives, however, will not cause significant losses from the cell. Pearse (1968) prefers alcohol and acetic acid to formalin for nucleic acids and nucleoproteins, but Kurnick (1955a) says neutral (pH 7.0) 10% formalin appears to be satisfactory. Pearse warns against long exposure if Carnoy fixative is used, since it causes first the extraction of RNA and then DNA. He recommends Lillie's acetic-alcohol-formalin, 4°C: 24 hours.

40% formalin, concentrated	10.0 ml
glacial acetic acid	5.0 ml
absolute ethyl alcohol	85.0 ml

Pure nucleic acids can be stained with basic dyes, but nucleic acids in living cells are bound with protein and will not react with basic dyes. Inactivation of the protein-bound amino groups by formalin or deamination (page 301) will increase the basophilia of the nucleic acids by releasing increasing numbers of nucleic acid phosphoryl groups to bind the basic dye. By using different exposures to these reagents, Walsh and Love (1963) demonstrated several nucleoproteins. (See also Love 1957, 1962; Love and Liles 1959.)

For specific staining of nucleic acids, the Feulgen technique is considered specific for DNA (page 332). It is thought to depend on the uncovering of aldehyde groups of deoxyribose by the hydrolysis of purine-deoxyribose bonds by hydrochloric acid. The released aldehyde gives the Schiff reaction. Proper hydrolysis is critical. Two actions occur almost simultaneously: (1) the purine bases are rapidly removed and the aldehyde groups uncovered; (2) the histones and nucleic acids are progressively removed. At the beginning, the first action predominates; as hydrolysis proceeds, the second action begins to take over and staining with Schiff will decrease. Overhydrolysis can produce a negative Feulgen reaction. The optimum time is in the area of 8 minutes; the range is 6–12 minutes.

In methyl green–pyronin staining for nucleic acids, the two dyes are used to distinguish between polymerization of the nucleic acids, not between the acids as such. Methyl green is readily bound by highly polymerized DNA; the binding is acquired at two sites, two amino groups of the dye with two phosphoric acid groups of the DNA. Depolymerizing treatment, such as heat and certain fixatives (Zenker, picric acid), must be avoided. If formalin is used, it must be accurately adjusted to pH 7.0 and fixation must be brief, so that the solution does not become acid. Pyronin, for RNA staining, stains low polymers of both nucleic acids, but it does permit methyl green to compete effectively for the highly polymerized DNA to the exclusion of the pyronin. Nuclei will stain red with pyronin alone, but they will stain green when the two dyes are used together. Control slides should be used with all methyl green–pyronin staining. If one slide is exposed to ribonuclease and one is not, and then both are stained with methyl green–pyronin, the material stained red with pyronin and removable with ribonuclease is considered RNA, and that which is not removable is not RNA. (See Tepper and Gifford 1962, and Kasten et al. 1962.)

The pages of literature are full of techniques, theories, controversies, and rationalizations for the demonstration of the nucleic acids and related components of the cell. Obviously they cannot all be included here, but references to some of them are listed below.

Spicer (1961) reports that Bouin fixes RNA as well as or better than any of a number of fixatives and yields DNA without acid hydrolysis. By staining at controlled pH (RNA stains more strongly at low pH) with Schiff reagent for DNA and methylene blue for RNA, the chromatin stains red and the cytoplasmic RNA stains blue.

Kurnick (1952) uses 1% toluidine blue in 95% alcohol as a rough screen test to demonstrate areas of high concentration of nucleic acids. By comparing an unextracted slide with a preparation made after extraction of the nucleic acids, one may differentiate the basophilia caused by nucleic acids from that caused by other components of the cell.

Daoust (1964) uses 0.1% toluidine blue in veronal buffer, pH 5.0 (5 minutes); dehydrates in 3 changes of tertiary butyl alcohol (1 minute each), clears, and mounts.

Other stains used for nucleic acids and nucleoproteins are these: azure A (Anderson 1961; Flax and Pollister 1949); Biebrich scarlet (Spicer 1962); cresyl violet (Ritter et al. 1961) (page 477); methenamine silver (Korson 1964); methylene blue (Deitch 1964); fluorescent methods (Armstrong 1956; Armstrong et al. 1957; deBruyn et al. 1953); Feulgen-oxidized tannin-azo technique (Malinin 1961); trimethylthionin basic fuchsin (Menzies 1963a).

Aceto-orcein and acetocarmine are popular chromosome stains (page 465); the orcein probably depends on DNA and the carmine on protein. Hematoxylin used with an alum mordant probably stains chromatin specifically and may depend on the state of polymerization of DNA.

Tagging with ^{14}C and ^{32}P can be used for autoradiography.

Nucleoli can be "stained" with silver (Tandler 1955).

For general discussion about nucleic acids and nucleoproteins, see Gomori (1952), Kurnick (1955b), and Pearse (1968).

Pyronin–Methyl Green for Nucleic Acids (Kurnick 1952)

FIXATION

Absolute alcohol, Carnoy, or cold acetone. If formalin is used, it must be adjusted to pH 7 and used only briefly before the solution turns acid and produces a faint green staining of cytoplasm. Picric acid depolymerizes DNA and shows no green chromatin.

SOLUTIONS

Methyl green

methyl green (C.I. 42590) .... 0.2 g
0.1 M acetate buffer (pH 4.2) or distilled water   100.0 ml

Before making the solution, purify the methyl green by extraction with chloroform. Add approximately 10 g methyl green to 200 ml of chloroform in a 500-ml Erlenmeyer flask and shake. Filter with suction and repeat with smaller amounts of chloroform until the solution comes off blue-green instead of lavender (usually requires at least 3 extractions). Dry and store in stoppered bottle. Purified dye is stable.

To detect crystal violet contaminant, place a drop of methyl green solution on a piece of filter paper. Hold the stained area over ammonia solution. The green color disappears, and violet color, if present, will remain. (Kasten and Sandritter 1962.)

Pyronin

Pyronin Y saturated in acetone, or for lighter color in 10% acetone. The British Pyronine Y for RNA (G. T. Gurr or Edward Gurr) is recommended.

PROCEDURE

1. Deparaffinize and hydrate slides to water.
2. Stain in methyl green solution: 6 minutes.
3. Blot and immerse in *n*-butyl alcohol: several minutes in each of 2 changes.
4. Stain in pyronin: 30–90 seconds. Shorten to a few dips if stain is too dark.
5. Clear in cedarwood oil, xylene, and mount.

RESULTS

RNA containing cytoplasm—red
nucleoli—red
chromatin—bright green
erythrocytes—brown
eosinophilic granules—red
cartilage matrix—green
osseous matrix—pink with trace of violet

COMMENTS

These two dyes distinguish between states of polymerization of nucleic acids, not between the acids themselves. Highly polymerized nucleic acid stains with methyl green; low polymers of both DNA and RNA stain with the pyronin. The usual techniques result in too pale a methyl green, or, if excess staining is tried, water rinses remove the methyl green and acetone removes the pyonin. Dehydrating with ethyl alcohol removes most of the color; isopropyl and tertiary butyl are no improvement. Kurnick developed the above pyronin–methyl green method when he found that *n*-butyl alcohol differentiated the methyl green but removed the pyronin.

Since the pyronin was lost anyway, he therefore decided to leave it out of the first solution and tried following the methyl green with pyronin saturated in acetone. (Because the acetone must be free of water, always use a fresh solution.)

Combined Stain Method (Kurnick 1955*a*)

SOLUTION

pyronin Y, 2% (2 g/100 ml water)	12.5 ml
methyl green, 2% (2 g/100 ml water)	7.5 ml
distilled water	30.0 ml

PROCEDURE

1. Deparaffinize and hydrate slides to water.
2. Stain in methyl green–pyronin: 6 minutes.
3. Blot carefully with filter paper.
4. Dehydrate in *n*-butyl alcohol, 2 changes: 5 minutes each.
5. Clear in xylene: 5 minutes; transfer to cedarwood oil: 5 minutes.
6. Mount.

Pyronin–Methyl Green for Nucleic Acids (Elias 1969)

SOLUTION

methyl green GA[4]	0.5 g
Walpole acetate buffer, *p*H 4.1 (page 540) . . .	100.0 ml
pyronin GS or Y	0.2 g

PROCEDURE

1. Hydrate slides to water.
2. Stain, 37°C: 1 hour.
3. Rinse in ice-cold distilled water: 1–2 seconds.
4. Blot.
5. Rinse, agitating vigorously in tertiary butanol.
6. Dehydrate in 2 changes of tertiary butanol: 5 minutes each.
7. Clear and mount.

COMMENTS

The Elias method gives a consistent, qualitative, nucleic-acid differentiation. The ice water differentiation rinse is critical, and tertiary butanol is superior to *n*-butanol.

---

[4] Chroma Gesellschaft. Roboz Surgical Instrument Co.

Thionin–Methyl Green for Nucleic Acids (Roque et al. 1965)

FIXATION

Fix thin (2 mm) pieces in 4% formaldehyde in 1% sodium acetate: 3 hours. Dry smears of those fixed in methyl alcohol can be used.

SECTIONING

Cut 2–4μ paraffin sections.

SOLUTION

methyl green (C.I. 42590)	0.1 g
thionin, (C.I. 52000)	0.0165 g
citrate buffer, 0.01 M, pH 5.8 (0.01 M HCl, 42 ml; 0.01 M sodium citrate, 58 ml)	100.0 ml

Dissolve the thionin in a little water and add the buffer and methyl green. Shake well and filter. This solution is best when freshly mixed. The methyl green should be chloroform-extracted (page 296).

PROCEDURE

1. Deparaffinize and hydrate sections to distilled water. Smears can go directly into water.
2. Stain in methyl green–thionin, 40°C: 30 minutes.
3. Rinse in distilled water.
4. Dehydrate in 3 changes of alcohol mixture; 3-butyl alcohol, 80.0 ml; absolute ethyl alcohol, 20 ml: 30 seconds in first change, 3 minutes each in second and third changes.
5. Rinse in absolute ethyl alcohol, clear in xylene, and mount.

RESULTS

chromatin—green or blue-green
nucleolar and cytoplasmic basolphilic substance—red or reddish purple

COMMENTS

Objection to the specificity of the pyronin–methyl green method centers around the unsuitable staining properties of pyronin. Roque et al. tried various basic dyes in an effort to find a suitable substitute for pyronin. Thionin gave the best results; it did not obscure the green coloring of chromatin and was less readily extracted by aqueous and alcoholic rinses. Also it stained more selectively than pyronin.

Use a short fixation time, no more than 3 hours. Zenker, Helly, and Carnoy can be used, but the staining is less intense, although the color contrasts are satisfactory.

A range of pH of 6–4 is satisfactory, but the color varies with the type of buffer, becoming blue when acetate or phosphate buffers are used, and green when a citrate buffer is used.

Thin sections, 2–4μ should be cut for best color contrasts; if the RNA content is very low, sections of 4–6μ may be necessary.

Nucleic acids are stained deep blue with gallocyanin–chrome alum, page 166.

See Mundkur and Brauer (1966) for selective staining of nucleoli.

CONTROL SLIDE TECHNIQUES

In some techniques the reaction is so specific that controls are unnecessary. But in others, controls are essential in order to separate a genuine reaction from one similar in appearance but not specific in nature. An essential ingredient may be omitted from a solution used in the control. This is the common procedure in histochemistry, in which the specific substrate is omitted from the incubating mixture used for the control. In other methods, inactivators, inhibitors, or extractives are used.

Extraction Techniques

In the demonstration of the specificity of a reaction for nucleic acids, the extraction of one or both of the nucleic acids is frequently essential. The original preparation can then be compared with another preparation that has been subjected to extraction techniques.

Removal of Both Nucleic Acids

Hot 5% trichloracetic acid, 90–98°C: 30 minutes.

Removal of RNA

1. 5% trichloracetic acid, 60°C: 30 minutes.
2. 10% perchloric acid, 4°C: 12 hours. The cold temperature is critical (Aldridge and Watson 1963; Kasten 1965).
3. 1 N hydrochloric acid, 60°C: 6–12 minutes (Feulgen reaction, page 332).
4. Ribonuclease (RNAse)

RNAse (crystallized, saltfree)[5] . . . .   1.0 mg
distilled water . . . .   1.0 ml

Incubate sections, 40°C: 4 hours.,

RNAse usually removes nucleolar and cytoplasmic RNA, but chromosomal RNA may be resistant.

Removal of DNA (Deoxyribonuclease [DNAse] method, Daoust 1964)

DNAse (crystallized)[5] . . . .   0.05 mg
0.1 M tris maleate buffer, pH 6.5, containing
MgSO$_4$·7H$_2$O at a concentration of 0.2 M . . .   1.0 ml

Incubate, 37°C: 24 hours.

Amano (1962) claims the results are poor if the slides are incubated in a horizontal position with a few drops of the enzyme. He recommends putting the slides in a coplin jar with 40 ml of DNAse solution.

Blocking Methods

A blocking method may become an essential part of a technique, not because it stains a component, but because it prevents its staining and thereby verifies that a certain chemical group is responsible for the staining reaction.

Blocking Protein End Groups, Reactions for Amines

Deamination

In this reaction nitrous acid blocks amino groups, most rapidly the *a*-amine groups; it converts amino groups into hydroxyl groups. Van Slyke reagent (Lillie 1954*b*) is best for this.

concentrated sodium nitrite (6 gm/10 ml water)   10.0 ml
glacial acetic acid . . . .   5.0 ml
distilled water . . . .   25.0 ml

Use at room temperature: 1–12 hours (10 minutes for blood films).

Acetylation (Lillie 1954*b*, 1958, 1964*a,b*)

Acetylation produces acyl derivatives of primary and secondary amines and amine compounds which prevent 1,2 glycols from reacting with PAS.

---

[5] Worthington Biochemical Corporation.

Acetylation and deacetylation are most commonly used for controls with PAS for mucopolysaccharides, even if the validity of the control is not absolutely infallible.

1. Take sections to absolute alcohol.
2. Dip in pyridine.
3. Transfer to following solution, 25°C: 1–24 hours, or 58°C: 30 minutes–6 hours.

   acetic anhydride .... 16.0 ml
   anhydrous pyridine .... 24.0 ml

   (The acetylation of glycols is progressive but slow, and it may require 24–48 hours for a complete blockage of PAS. It can be somewhat accelerated by adding 0.4–0.5% by volume of concentrated sulfuric acid.)
4. Wash in 2 changes each of absolute, 95%, and 80% alcohols. Proceed to method for end groups under investigation.

Rapid acetylation and complete blockage of PAS can be accomplished by use of the following for 4 minutes:

   acetic anhydride .... 25.0 ml
   glacial acetic acid .... 75.0 ml
   sulfuric acid .... 0.25 ml

Alcohol acetylation of phenolic groups (Lillie, 1964b) can be done in a mixture of equal volumes of acetic anhydride and absolute ethyl alcohol. Use at 23°–25°C: 30 minutes.

Benzoylation (Lillie 1954b; Barnard 1961)

Benzoylation is similar to acetylation, and steps 1 and 2 are the same.

3. Transfer to following, 25°C: 1–24 hours; or 58°C: 30 minutes–6 hours.

   benzoyl chloride .... 2.0 ml
   anhydrous pyridine .... 38.0 ml

4. Wash in alcohols as in step 4 above and proceed to method for group under investigation.

Barnard's method requires anhydrous conditions throughout benzoylation, room temperature, in a desiccator: 3 hours.

   dry acetonitrile .... 50.0 ml
   benzoyl chloride .... 4.2 ml
   anhydrous pyridine .... 2.2 ml

Deacetylation (Saponification)

Acetylation blockage is almost completely reversed by saponification, which is effected by placing sections in the following for 20–30 minutes:

potassium hydroxide	1.0 g
70% ethyl alcohol	100.0 ml

and then treating for 16–24 hours in

absolute ethyl alcohol	70.0 ml
distilled water	10.0 ml
ammonia, 28% concentrated	20.0 ml

Keep solution in closed container during action. Wash in 80% alcohol and water and proceed to method under investigation.

Blocking Aldehydes

Aniline–acetic acid (Lillie and Glenner 1957), room temperature: 20–30 minutes.

glacial acetic acid	45.0 ml
aniline	5.0 ml

If this fails to block a Schiff reaction, aldehydes are excluded.

If aldehydes are made to combine with sodium sulfite or phenylhydrazine, the reaction with Schiff after hydrolysis (Feulgen) or oxidation (PAS) can be blocked (Pearse 1960).

Phenylhydrazine

phenylhydrazine	5.0 ml
glacial acetic acid	10.0 ml
distilled water to make	50.0 ml

Treat at 60°C: 2–3 hours.

Sodium bisulfite

absolute ethyl alcohol	10.0 ml
50% aqueous sodium bisulfite	40.0 ml

Treat at room temperature: 2–8 hours.

Blocking Carboxyl Groups or Other Acid Groups

This reverses metachromasia of most components and blocks basophilia of nucleic acids; it abolishes reactivity of phosphoryl, carboxyl, sulfate groups of polysaccharides, and carboxyl groups of proteins.

Methylation (Lillie 1954b, 1958)

0.1 N HCl (8 ml/1000 ml) made up in methyl alcohol, 25°C: 2–3 days. Use screw-cap jars for the procedure. A temperature of 58°C is required to block the PAS of glycogen, epithelial gland mucins, thyroid colloid, collagen, and reticulin.

1% HCl in methyl alcohol is required to block nuclear and cytoplasmic staining with azure A, and it destroys the reactivity of argentaffin cells in the methenamine silver procedure.

Geyer (1962) says that methanol-thionyl chloride is more effective than methanol-HCl under some conditions. With 100 ml of absolute methanol in an Erlenmeyer flask, add 4 ml thionyl chloride very slowly down a wall of the flask, held at a slant and away from the face. Use transparent thionyl chloride. Mix immediately before use. Use at 56°C: 30 minutes–2 hours.

Demethylation (Saponification)

Saponification is used to deblock tissue: that is, to return it to the original active form that was blocked by one of the blocking methods. However, it is not always successful in reversing the effect of methylation.

Cover sections with thin nitrocellulose and place in 1% potassium hydroxide in 70% ethyl alcohol, 25°C: 30 minutes. This restores reactivity of methylated sulfonic acid and of carboxyl and phosphoryl groups, but not that of sulfate groups.

Blocking Tryptophane and Cystine

Performic acid oxidation (Pearse 1960) is specific for the indole ring of tryptophane.

98% formic acid	40.0 ml
30% H_2O_2 (must be fresh)	4.0 ml
sulfuric acid	0.5 ml

Use within 8 hours and stir vigorously before using, room temperature: 15–60 minutes.

Peracetic acid reagent, see page 282.

Blocking Tyrosine

Treat with 1% 2,4-dinitrofluorobenzene (DNFB) in 90% ethyl alcohol containing 0.01 M NaOH, room temperature: 16–20 hours.

Lillie (1968) and Pearse (1968) include detailed discussions of blocking methods.

CHAPTER 19

Staining Lipids
and Carbohydrates

LIPIDS

Lipids (lipoids, fatty substances) include a large number of substances that have been grouped together because of their solubility properties. They are insoluble in water and soluble in fat solvents such as alcohol, ether, chloroform pyridine, benzene, and acetone. Classification of the lipids is not undertaken here, but a few familiar groups can be mentioned: carotenoids (vitamin A), fatty acids, triglycerids (acetone soluble) (neutral fats), phospholipids, cerebrosides, sterols, and lipid pigments (essentially insoluble in acetone) (lipofuscins—page 277).

For the fixation of fats, formalin is best, particularly if 1% of calcium chloride is added to make the phospholipids insoluble (Gomori 1952). Because of the use of fat solvents, the tissue cannot be embedded in paraffin or nitrocellulose, but it can be embedded in Carbowax (Rinehart and Abul-Haj 1951*b*), polyvinyl alcohol (Feder 1962, Masek and Birns 1961), polyethylene glycol (Hack 1952), or ester waxes (page 94). Frozen sections are simpler to make and are the most frequently used. During any processing, alcohol higher than 70% must be avoided. (For the fixation of lipoids, see Elftman 1958.)

The dying of lipids is one of the simplest forms of staining: the coloring agent merely dissolves in a fluid contained within the tissues. In addition, it should be emphasized that a dye solvent that does not dissolve the lipid itself should be used. The dye, therefore, must meet certain requirements:

1. The dye must be strongly colored.
2. It must be soluble in the substance that it is intended to show, but it must not be soluble in water, the major constituent of cells.
3. It must not attach itself to any tissue constituents except by solution.
4. It must be applied to tissues in a solvent that will not dissolve the substance to be dyed, and it must be less soluble in the solvent than in the substance.

Baker (1958) suggests the name "lysochromes" for these dyes that dissolve in the tissue elements to be colored. The name is derived from the Greek *lúsis* meaning solution. The Sudans were the first synthetic dyes of this sort, followed by the Nile blues and reds.

These dyes are used in saturated solutions and often create the problem of dye precipitate on the tissue. Vlachos (1959) makes the sensible proposal that the precipitate is probably formed by the solution becoming oversaturated and that perhaps a saturated solution is unnecessary. He suggests two alternatives:

1. Make up the concentration below the saturation point; for example, 0.25 g Sudan IV per 100 ml 60% alcohol.
2. Desaturate the solution either by dilution of a saturated solution with equal volumes of 60% alcohol, or by refrigeration. When the solution has been refrigerated long enough to have acquired the refrigerator temperature, filter. Neither method alters the staining quality of the solution, and the amounts of precipitate produced are negligible.

Lipid Staining

Oil Red O for Lipids

FIXATION

10% formalin or other aqueous general fixatives.

SOLUTION

oil red O (C.I. 26125)	0.7 g
absolute isopropanol	200.0 ml

Shake; leave overnight. Filter. Dilute 180 ml of oil red O solution with 120 ml of distilled water. Leave overnight at 4°C. Filter. Let stand 30 minutes and filter. Ready for use. Stable for 6–8 months.

PROCEDURE

1. Mount frozen sections (15μ) on subbed slides. Allow to dry: 5–10 minutes. Loose, unmounted sections also may be stained this way.
2. Rinse in 60% isopropanol: 30 seconds.
3. Stain in oil red O: 10 minutes.
4. Rinse in 60% isopropanol: few seconds.
5. Wash in running water: 2–3 minutes.
6. Stain in Mayer hematoxylin (or the like): 2–3 minutes.
7. Wash in tap water: 3 minutes.
8. Blue in Scott solution (page 533): 3 minutes.
9. Wash in tap water: 5 minutes. Mount in gum syrup or glycerine jelly. For permanency, ring cover glass.

RESULTS

lipids—orange-red or brilliant red
nuclei—blue

COMMENTS

Gurr Aquamont (page 145) can be used as a mountant for oil red O sections and does not require sealing.

This is by far the best of the oil red O methods that I have used. It never leaves a stain precipitate on the sections.

For a blue coloration, oil blue N (C.I. 61555) can be used, but the color is not as intense as that of oil red O.

Sudan IV or Sudan Black B (Chiffelle and Putt 1951)

FIXATION

10% formalin, no alcohol.

SOLUTION

Dissolve 0.7 g Sudan IV (C.I. 26105) or Sudan black B (C.I. 26150) in 100 ml propylene or ethylene glycol. Add small amounts at a time, and heat to 100°–110°C, stirring, for a few minutes. Do not exceed 110°C or a gelatinous suspension is formed. Filter hot through Whatman no. 2 paper and cool. Filter again through fritted glass filter with the aid of suction, or through glass wool by vacuum.

PROCEDURE

1. Frozen sections, 15μ, into water.
2. Place in pure propylene or ethylene glycol: 3–5 minutes, 2 changes. Agitate.
3. Stain, 2 changes: 5–7 minutes each, agitate occasionally.
4. Differentiate in glycol and water (85 : 15): 3–5 minutes. Agitate.
5. Wash in distilled water: 3–5 minutes.
6. Counterstain in hematoxylin if desired.
7. Mount in glycerine jelly.

RESULTS

lipids—Sudan IV, orange to red; Sudan black B, blue-black to black

COMMENTS

Sudan black B is considered the most sensitive of the lipid dyes.

Chiffelle and Putt recommend glycol as a perfect solvent for a fat stain because it does not extract lipids.

Zugibe et al. (1958, 1959) suggest Carbowax 400 as a solvent for oil red O and Sudan IV.

Gomori (1952) questions the use of glycols because they are solvents for so many water-insoluble substances, and he prefers triethylphosphate. It has a low volatility and is harmless to lipids. His method follows.

1. Frozen sections are rinsed in water and transferred into 50% alcohol: few minutes.
2. Stain in a saturated, filtered solution of any of the fat dyes in 60% triethylphosphate: 5–20 minutes.
3. Differentiate in 50% alcohol: 1 minute.
4. Counterstain in hematoxylin or any preferred stain.
5. Mount in glycerine jelly.

Osmic Acid (Mallory 1944)

Oleic acid or triolein react with the osmic acid by oxidation of their double bonds.

FIXATION

10% formalin, no alcohol.

SOLUTION

osmic acid (osmium tetroxide)	1.0 g ampule
distilled water	100.0 ml

With file, score a circle around ampule and drop it into bottle with the distilled water. Several sharp shakes will break the ampule and allow the water to dissolve the crystals. This method eliminates the possibility of breathing the fumes. However, new packaging methods have virtually eliminated this problem.

PROCEDURE

1. Frozen sections, 15μ, into water.
2. Transfer to osmic acid solution: 24 hours.
3. Wash in several changes distilled water: 6–12 hours total.
4. Treat in absolute alcohol: 4–5 hours, for secondary staining of fat.
5. Wash well in distilled water: 5 minutes or more.
6. Mount in glycerine jelly.

RESULTS

lipids—black
background—yellowish brown

Naphthalene Yellow (Sills and Marsh 1959)

Naphthalene yellow in 60% acetic acid was used to restore the yellow color to the fat of formalin-fixed gross specimens. This cannot be used for sections; the color is too pale.

Fluorescent Method (Metcalf and Patton 1944; Peltier 1954)

FIXATION

10% formalin (salts of heavy metals—$HgCl_2$—have a quenching effect on fluorescence), or use fresh tissue.

SOLUTION

phosphine 3R [1]	0.1–1.0 g
distilled water	100.0 ml

PROCEDURE

1. Cut frozen sections, 10μ.
2. Wash in distilled water.
3. Stain in phosphine solution: 5 minutes.
4. Rinse quickly in distilled water.
5. Mount in glycerine or examine as a water mount.

---

[1] Roboz Surgical Instrument Co., Washington, D.C.

RESULT

lipids (except fatty acids, cholesterol, and soaps) will fluoresce brilliant white

Sudan Black B Method for Phospholipids (Elftman 1957*b*)

FIXATION

mercuric chloride, 5% aqueous	100.0 ml
potassium dichromate	2.5 g

Adjust *p*H to 2.5 with HCl. Mix fresh. Fix for 3 days; oxidizes phospholipids and decreases their solubility in fat solvents such as are used in paraffin embedding. Nonetheless, leave tissues for short periods in all solvents.

SOLUTIONS

Sudan black B, see page 308.

Ethylene glycol solution

ethylene glycol	85.0 ml
distilled water	100.0 ml

PROCEDURE
1. Deparaffinize and transfer slides into absolute alcohol: 2 minutes.
2. Transfer to absolute ethylene glycol.
3. Stain in Sudan black solution: 30 minutes.
4. Differentiate in 85% ethylene glycol: 2–3 minutes.
5. Wash in distilled water.
6. Counterstain, if desired.
7. Wash well, mount in von Apathy's gum syrup or in glycerine jelly.

RESULT

phospholipids—black

COMMENT

Phospholipids are increasingly important in the study of mitochondria and Golgi apparatus, and are an important constituent of myelin.

Roozemond (1969) recommends fixing phospholipids in 4% formalin with 1% of calcium chloride added and then cutting frozen sections: fixation at 0°C reduces the breakdown of lipids. He suggests that a strong interaction between calcium and phospholipids may aid in the retention of the phospholipids when they are surrounded by a calcium-containing medium.

Acid Hematein for Phospholipids (Hori 1963)

FIXATION

Formalin-calcium (10% formalin/1% calcium chloride), formalin-calcium-cadmium (1% cadmium chloride added to above), or glutaraldehyde-formalin-calcium chloride (2.5 : 25 : 1) in 0.1 M sodium cacodylate, pH 7.0, (page 417): 18 hours.

SOLUTIONS

Acid hematein

hematoxylin	50.0 mg
0.01% sodium iodate	50.0 ml

Heat until it begins to boil, cool, and add:

glacial acetic acid	1.0 ml

Prepare fresh.

Borax-ferricyanide

borax $(Na_2B_4O_7 \cdot 10H_2O)$	250.0 mg
potassium ferricyanide $[K_3Fe(CN)_6]$	250.0 mg
distilled water	100.0 ml

Keeps well in refrigerator.

PROCEDURE

1. Cut frozen sections.
2. Chromatize sections in 5% aqueous potassium dichromate, 60°C: 4 hours. Transfer sections with a glass rod.
3. Wash in several changes of distilled water.
4. Stain in acid hematein, 37°C: 30 minutes.
5. Wash briefly in distilled water.
6. Differentiate in borax ferricyanide, 20°–25°C: 18 hours.
7. Dehydrate, clear, and mount.

RESULTS

phospholipids and some proteins—blue, blue-gray or blue-black.

COMMENTS

This is a simplification of Baker's (1946) original acid hematein test for phospholipids. If it is desired to distinguish between phospholipids and certain proteins that also give a positive reaction in this test, apply Baker's pyridine extraction method:

1. Fix small pieces of tissue in weak Bouin solution: 20 hours.

picric acid, saturated aqueous	50.0 ml
formalin, concentrated	10.0 ml
glacial acetic acid	5.0 ml
distilled water	35.0 ml

2. Transfer to 70% alcohol: 1 hour.
3. Transfer to 50% alcohol: 30 minutes.
4. Wash in running water: 30 minutes.
5. Transfer to pyridine, room temperature, 2 changes, 1 hour each.
6. Transfer to pyridine, 60°C: 24 hours.
7. Wash in running water: 2 hours.
8. Transfer to step 2 of acid hematein method.

RESULTS

phospholipids—unstained
nuclei—stain after extraction, not before extraction
erythrocytes—stain before and after extraction

COMMENTS

Luxol fast blue can be used for phospholipids, see page 238.

For unsaturated lipids, see Norton et al. (1962); for lipoproteins, "masked" lipids, see Carmichael (1963), Holczinger and Bálint (1961), Serra (1958).

CARBOHYDRATES (SACCHARIDES)

Polysaccharides are a complex group of substances that are found throughout animal tissues. Complete agreement in nomenclature is still lacking, but in general they are classified as follows. (Pearse 1968; Zugibe 1970).

Polysaccharides (glycans) include glycogen, starch and cellulose. Glycogen is the only naturally occurring member of this group that remains in animal tissue after aqueous fixation and embedding.

Polysaccharide-protein complexes include (1) acid mucopolysaccharides —hyaluronic acid (Wharton's jelly), chrondroitins (cartilage) and heparin; (2) neutral mucopolysaccharides—chitin; (3) mucoproteins—lens capsule and vitreous humor.

Glycoproteins and glycopolypeptides—ovomucoid and salivary gland mucoid.

Glycolipids—cerebrosides and gangliosides.

Glycolipid-protein complexes—ox brain mucolipid.

Certain specific stains and aldehyde reactions are commonly used to demonstrate carbohydrates. The aldehyde groups must first be liberated by some chemical agent, either oxidized (periodic or chromic acid) or hydrolyzed (dilute hydrochloric acid). Then the specific reagent for aldehydes can be applied. See the Feulgen and periodic acid–Schiff reactions, page 326.

Alcoholic mixtures are usually recommended as fixatives for saccharides. Most conventional fixatives cause glycogen to migrate to one side of the cell. Gomori (1952) theorized that, when glycogen is enclosed in a complex mixture of proteins and lipids, a good protein precipitant may coat the glycogen with a protein membrane. This will be impermeable to the large glycogen molecules and keep them in situ. The fixative should act and harden rapidly. Alcohol or acid fixatives form glycogen in coarse droplets, and fixatives containing formalin distribute the glycogen in fine granules. In some tissues, such as liver, unless the tissue and fixative are chilled, enzymes can cause a loss of glycogen. Although in most tissues glycogen is more stable, Gomori recommended placing the tissue in fixative in the refrigerator. Leske and von Mayersbach (1969) have demonstrated that block fixation does not safely preserve glycogen. Sections should be cut fresh frozen, and fixed in Carnoy fixative or 2.5–10.0% aqueous trichloracetic acid: 5–15 minutes. Murgatroyd (1971) reports that cold formal alcohol or cold Rossman fluid preserve the most glycogen.

After proper fixation paraffin embedding is satisfactory, but after deparaffinizing the mounted slides, rinse them in absolute ethyl alcohol and protect the sections with a coat of 1% nitrocellulose to prevent diffusion of the glycogen during staining. Freeze-drying or freeze substitution methods are satisfactory. Lillie (1962) recommends the use of 5% formic acid if a tissue must be decalcified. Check the tissue each day by the oxalate method (page 28). Stop decalcification as soon as a negative oxalate test is obtained. Avoid any use of heat. Lillie also suggests that 1–3 days of infiltration of the tissue with 1% nitrocellulose before the decalcification can improve the resistance of glycogen to acid extraction. Trott (1961a) uses EDTA, pH 7.5, for decalcification.

For mucoproteins and mucopolysaccharides, any protein precipitant fixative—Zenker, Helly, or Regaud—is satisfactory (Hale 1957; Gomori 1952; Thompson 1966).

Glycogen Staining

Best Carmine

FIXATION

Avoid aqueous media (McManus and Mowry 1958)

absolute alcohol	9 parts
formalin, concentrated	1 part

Use ice cold.

This method starts dehydration for embedding with 95% alcohol, but the loss of glycogen is about 20%–30% (Trott 1961a,b; Trott et al. 1962). Double embedding with nitrocellulose and paraffin can be used to reduce the loss of glycogen.

MOUNTING SLIDES

I have achieved better localization of glycogen by mounting paraffin sections with 95% alcohol in place of water. As soon as sections are spread, drain off excess fluid and continue to dry.

SOLUTIONS

Best carmine stock solution

carmine (C.I. 75470)	2.0 g
potassium carbonate	1.0 g
potassium chloride	5.0 g
distilled water	60.0 ml

Boil gently: 5 minutes. Cool. Filter. Add:

ammonium hydroxide	20.0 ml

Lasts 3 months at 0°–4°C.

Working solution

carmine stock solution	30.0 ml
ammonium hydroxide	25.0 ml
methyl alcohol	25.0 ml

Lasts 2–3 weeks.

Differentiating fluid

absolute ethyl alcohol	16.0 ml
methyl alcohol	8.0 ml
distilled water	20.0 ml

PROCEDURE

1. Deparaffinize, coat with nitrocellulose, and transfer into absolute alcohol: 3 minutes.
2. Coat with 1% celloidin; dry slightly in air.
3. Dip 2 or 3 times in 70% alcohol and then into water.
4. Stain in Mayer hematoxylin: 5 minutes.
5. Wash, blue in Scott, and wash.
6. Stain, Best carmine working solution: 15–30 minutes.
7. Treat with differentiating fluid: 5–15 minutes.
8. Rinse quickly in 80% alcohol.
9. Dehydrate, clear, and mount.

RESULTS

glycogen—red
mast cell granules, mucin, and fibrin—light red
nuclei—blue

COMMENTS

Pearse (1968) uses Lison "Gendre-fluid" method of fixation, claiming that it shows a localization of glycogen comparing favorably with that in tissue preserved by freeze-drying.

picric acid, saturated in 96% alcohol	85 parts
formalin	10 parts
glacial acetic acid	5 parts

Cool before use to $-73°C$ in an acetone–CO_2 snow mixture. Use small pieces; fix for 18 hours.

Trott et al. (1962) fix in 1% periodic acid in 10% formalin, which they claim shows greater amounts of glycogen than tissues fixed in acetic acid–formalin–alcohol. Schiff reagent methods are replacing the classical staining methods for glycogen.

Since it has been suggested that carmine staining is due to the hydrogen bonding between glycogen and carminic acid (Goldstein 1962), other dyes that carry groups capable of forming hydrogen bonds also should stain glycogen (Murgatroyd and Horobin 1969). Acid alizarine blue SWR, and alizarine brilliant blue BS stained glycogen red. Alizarine red S stained it red, gallein purple, hematein and hematoxylin red changing to brown. The staining solution was made as follows: 1 g dye, 1 g potassium carbonate, and 5 g potassium chloride are added to 60 ml of boiling distilled water. Let stand 1 hour. To 20 ml of dye solution add 15 ml of ammonia and 15 ml of absolute methanol in that order. The solution is stable for only 24 hours. Stain: 2–5 minutes. Dehydrate in 2–3 changes of absolute methanol, clear, and mount.

Staining of Acid Mucopolysaccharides

For metachromatic methods see page 348.

Alcian Blue Method, *p*H 2.5 (Luna 1968)

At this *p*H the Alcian blue stains both sulfated and nonsulfated mucosubstances.

FIXATION

Any general fixative but formalin-99% alcohol (1 : 9) or Rossman fluid are recommended.

SOLUTIONS

*Alcian blue, *p*H 2.5*

Alcian blue 8GX (C.I. 74240) . . . . 1.0 g
distilled water . . . . 100 ml
glacial acetic acid . . . . 3.0 ml

Filter; add thymol crystal to prevent mold.

Kernechtrot, see page 270.

PROCEDURE
1. Deparaffinize and hydrate slides to water.
2. Rinse in 3% acetic acid: 3 minutes.
3. Stain in Alcian blue: 30 minutes.
4. Wash in running water: 10 minutes.
5. Counterstain with Kernechtrot: 5 minutes.
6. Rinse in distilled water.
7. Dehydrate, clear, and mount.

RESULTS

acid mucopolysaccharides—blue green
nuclei—red to bluish red

COMMENTS

Can be followed by PAS (page 327) for demonstration of both acidic groups and 1,2 glycols, or the Feulgen reaction (page 332).

Lison (1954) counterstains with chlorantine fast red for efficient differentiation of mucin from collagen (mucin—bluish green; collagen—cherry red). Follow steps 1, 2, and 3 above, and then substitute the following steps.

4. Treat with phosphomolybdic acid, 1% aqueous: 10 minutes.
5. Rinse in distilled water.
6. Stain in chlorantine fast red: 10–15 minutes.

chlorantine fast red 5B [2] .... 0.5 g
distilled water .... 100.0 ml

7. Rinse in distilled water.
8. Dehydrate, clear, and mount.

Williams and Jackson (1956) notes possible diffusion of acid mucopolysaccharides under aqueous conditions. The fixative should form complexes insoluble in water and alcohol. They suggest two possible solutions containing organic chemicals that form such insoluble complexes with acid mucopolysaccharides:

a. 0.5% cetylpyridium chloride (CPC) in 4% aqueous formalin. (See also Conklin 1963 and Wolfe 1964).
b. 0.4% 5-aminoacridine hypochloride in 50% ethyl alcohol.

Spicer and Meyer (1960) precede staining with Alcian blue by immersing the tissue for 5 minutes in aldehyde-fuchsin (page 190), transferring it to 70% alcohol, and washing it in running water.

Mowry (1960) and Scott and Mowry (1970) report that lots of Alcian blue have differed in staining efficiency and fastness. Current stocks of Alcian blue 8GX are more soluble, less fast, and more stable; the 8GX prepared by ICI, Ltd., Blackly, Manchester, England, is recommended.

Alcian blue staining is attributed to the presence of carboxyl groups and in some reactions to sulfates. All acidic mucopolysaccharides contain either sulfates alone or sulfate and carboxyl groups, and the stainability should be consistent. Yamada (1963), however, thinks that the basis for staining remains to be determined precisely and that it is better to examine

---

[2] Edward Gurr, London, or Ciba, Ltd., Montreal, Canada.

acidic groups by using metachromatic dyes at different pH levels. See page 320 for staining at pH 1.0 for sulfated carbohydrates.

Mowry and Emmel (1966) successfully use national fast blue 8XM, a basic copper phthalocyanin dye, as a substitute for Alcian blue 8GX, using 0.1% in 3% acetic acid for 2 hours. Some slight nuclear coloration frequently occurs.

The Alcian green method (Putt and Hukill 1962) substitutes Alcian green[3] for Alcian blue. Since Alcian green stains more rapidly than Alcian blue, the time required for staining (step 2) can be reduced from 30 minutes to 10 minutes. The procedure and results are identical, except that acid mucopolysaccharides stain deep green rather than blue-green.

Carlo (1964) identifies sulfated groups with Alcian blue and carboxyl groups with Alcian yellow.

Wolfe (1964) stains extremely water-soluble mucopolysaccharides in a series of toluidine blue dissolved in decreasing concentrations of alcohol. Horn and Spicer (1964*a,b*) use azure A at a low pH of 4.0–5.0.

For fluorescent staining of mucopolysaccharides, see Saunders (1962).

Fluorescent Method (Hicks and Matthaei 1958)

FIXATION

10% formalin preferred.

SOLUTIONS

Iron alum

ferric ammonium sulfate	4.0 g
distilled water	100.0 ml

Acridine orange

acridine orange (C.I. 46005)	0.1 g
distilled water	100.0 ml

PROCEDURE

1. Deparaffinize and hydrate paraffin sections to water, or cut frozen sections and place in distilled water.
2. Treat with iron alum solution: 5–10 minutes.
3. Wash briefly in distilled water.
4. Stain in acridine orange: 1½ minutes.

---

[3] ESBE Laboratory Supplies, Toronto, Canada.

5. Rinse briefly in distilled water, blot, and mount in Harleco Fluorescent Mountant.[4]

RESULT

mucin—brilliant reddish orange fluorescence

COMMENTS

This is not a permanent slide. The iron inhibits the production of fluorescence with acridine orange in nearly all tissue components except mucin. Hicks and Matthaei also suggest that the mucins of acid polysaccharides (rather than mucoproteins) fluoresce.

For a fluorescent antibody technique, see Kent (1961).

Lev and Stoward (1969) use eosin as a fluorescent dye to demonstrate mucous cells. The concentration of eosin can be 0.1% to 10%; the pH should be 9.5 to 5.7 but no lower. Staining time is 5–10 minutes. The fluorescence is stronger after Carnoy than after formalin fixation. Isopropyl extracts less stain than ethyl alcohol, thereby producing a stronger fluorescence. Hematoxylin used with the eosin improves the contrast due to the orange-red color it gives to nuclei.

Alcian blue method, pH 1.0 (Luna 1968)

FIXATION

Formalin-95% alcohol (1 : 9) or Rossman fluid.

SOLUTION

Alcian blue 8GX	1.0 g
0.1N hydrochloric acid	100.0 ml

PROCEDURE

1. Deparaffinize and hydrate sections to distilled water.
2. Stain in Alcian blue: 30 minutes.
3. Do not rinse the sections in water; blot them dry with filter paper.
4. Dehydrate in 95% alcohol, absolute alcohol and clear and mount.

RESULTS

sulfated mucosubstances—greenish blue
nonsulfated mucosubstances—unstained

---

[4] Hartman Leddon Co., Philadelphia.

Colloidal Iron Method (Ritter and Oleson 1950)

FIXATION

10% formalin in 90% alcohol; do not use dichromate—it oxidizes polysaccharides.

SOLUTIONS

Colloidal iron solution

dialyzed iron, Merck's	50.0 ml
2 M acetic acid (11.5 ml/88.5 ml water)	50.0 ml

Potassium ferrocyanide in HCl

potassium ferrocyanide, 1% aqueous	86.0 ml
N HCl (page 531)	14.0 ml

Periodic acid

periodic acid	0.4 g
distilled water	100.0 ml
$M/5$ sodium acetate (135 mg hydrated salt/35 ml ethyl alcohol)	5.0 ml

Reducing rinse

potassium iodide	1.0 g
sodium thiosulfate	1.0 g
distilled water	20.0 ml

Add while stirring:

absolute ethyl alcohol	30.0 ml
2 N HCl (page 531)	0.5 ml

Sulfite rinse

water	50.0 ml
concentrated HCl	0.5 ml
potassium metabisulfite	0.2 g

Schiff reagent, see page 327.

PROCEDURE

1. Deparaffinize and hydrate slides to water.
2. Treat with colloidal iron: 10 minutes.
3. Wash well in distilled water.
4. Treat with potassium ferrocyanide: 10 minutes.
5. Wash well in distilled water.
6. Flood with 70% alcohol.
7. Treat with periodic acid: 5 minutes.
8. Rinse with 70% alcohol.
9. Transfer to reducing rinse: 5 minutes.
10. Rinse with 70% alcohol.
11. Transfer to Schiff reagent: 1 hour.
12. Use 3 sulfite rinses: 1.5 minutes each.
13. Wash well in running water: 5 minutes.
14. Stain with hematoxylin if desired.
15. Dehydrate, clear, and mount.

RESULTS

acid polysaccharides—blue
glycogen—magenta

COMMENTS

Nito and Stokes (1960) prepare a dialyzed iron as followers. Dissolve 100 g ferric chloride in 230 ml distilled water. Add 130 ml glycerol and mix well. Divide into 2 portions: $\frac{1}{4}$ and $\frac{3}{4}$. To the large volume add concentrated ammonia in small quantities with vigorous stirring. A precipitate forms and dissolves more and more slowly until the mixture becomes pasty (approximately 110 ml ammonia). Add the smaller portion drop by drop and with constant stirring until the solution becomes clear, dark reddish brown. It now contains a slight excess of ferric chloride.

Dialyze against distilled water, using a cellophane bag—capacity about four times that of the solution. Change the distilled water every day until the solution in the bag has reverted to its original volume (in about two weeks). Store at room temperature. For use, dilute with equal amount of water. (See also Nito 1961.)

Acid mucopolysaccharides combine with dialyzed iron and then become blue in potassium ferrocyanide (Prussian blue). Mowry (1958) considers this method more specific than others. It stains more strongly than Alcian blue.

Colloidal Iron Method (Mowry 1958)

SOLUTIONS

Colloidal iron stock solution

Ferric chloride 29% (USP XI)	4.4 ml
distilled water	250 ml

Bring water to boil. While the water is boiling, pour in the ferric chloride solution. Stir. When solution is dark red, remove from heat and allow to cool. The color must be dark red and clear. It is stable for months.

Working solution

glacial acetic acid	5.0 ml
distilled water	15.0 ml
stock colloidal iron solution	20.0 ml

pH should be 1.1–1.3. Nonspecific staining takes place at pH 1.4 or higher. Effective for 1 day.

Hydrochloric acid–ferrocyanide

2% HCl (2ml/98 ml water)	50.0 ml
2% potassium ferrocyanide (2 g/100 ml water) .	50.0 ml

Mix immediately before use.

PROCEDURE

1. Deparaffinize and hydrate slides to water: remove $HgCl_2$
2. Rinse in 12% acetic acid (12 ml/88 ml water): 30 seconds (to prevent dilution of reagent).
3. Transfer to freshly prepared colloidal iron (working solution): 60 minutes.
4. Rinse in 12% acetic acid, 4 changes: 3 minutes each.
5. Treat with hydrochloric acid–ferrocyanide, room temperature: 20 minutes. (Include a control slide, not treated with colloidal iron.)
6. Wash in running water: 5 minutes.
7. Counterstain if desired: Feulgen, PAS, or hematoxylin.
8. Optional: dip in aqueous picric acid to color cytoplasm and erythrocytes; rinse for few seconds in tap water. (Acetic–orange G can also be used.)
9. Dehydrate, clear, and mount.

RESULTS

acid mucopolysaccharides—bright blue; uncolored in control slide.

mucins of connective tissue, epithelium, mast cell granules, capsules of some microbial agents, pneumococci—bright blue.

COMMENTS

Hale was the first to suggest the use of colloidal iron for the demonstration of acid polysaccharides, claiming that acidic groups can react with metals and cations. In order that such acidic groups may be demonstrated, the acid moiety should be a firmly bound part of a fixed and insoluble tissue constituent and be free to react. Also the link between the metal and acid should be strong enough that it will not be disrupted when water is used to wash away excess unbound metal. Using a positively charged colloidal solution is the easiest way to prevent this disruption. The metal is held tightly, probably by the multiple valencies of both reacting particles. Since the range of reactivity of the colloidal iron reagent depends on the pH of the solution, it is possible to stain strongly acidic groups selectively by making the solution strongly acid. If strongly acidic groups are blocked by bound iron, it should be possible to stain the weakly acid groups with another colloidal solution. Wolman (1956, 1961) developed his Bi-Col staining technique by using two colloidal metal solutions—colloidal iron and colloidal gold (Wolman 1956).

CHAPTER 20

PAS and Feulgen Techniques, and Related Reactions

Feulgen and PAS techniques involve two chemical reactions: (1) the oxidation of a-amino alcohol and/or 1,2 glycol groups to aldehydes, and (2) the reaction of resulting aldehydes with Schiff reagent (page 326) to form a purple-red color. The Schiff reaction sequence for detecting the aldehydes is complicated, and two interpretations have been suggested: (1) the formation of an amino sulfinic acid followed by an additional reaction of the aldehyde (Wieland and Scheuing 1921); (2) the formation of an amino alkylsulfonic acid (Hörmann et al. 1958; Rumpf 1935). Hardonk and van Duijan (1964a,b,c) and Nauman et al. (1960) confirm the second interpretation. The two oxidizers most commonly used are chromic and periodic acids. The periodic acid breaks the carbon chains of the polysaccharides containing the 1,2 glycol groupings and oxidizes the broken ends into aldehyde groups. Chromic acid is a weaker oxidizer whose action is limited almost exclusively to glycogen and mucin (the principle of the Bauer method, page 330). If necessary, glycogen and starch can be demonstrated to the exclusion of other reactants, by iodine or Best carmine (page 315), and mucin by Alcian blue (page 317), or metachromatic methods (page 347). If it is desirable to prevent the reaction of glycogen or starch, the saliva or diastase treatment is simple and effective (page 329).

Among the nucleic acids, oxidation will not form aldehydes, and acid hydrolysis is required (see pages 300 and 332).

SCHIFF REAGENT

A century ago, Hugo Schiff carried out extensive research on the reactions of amines with aldehydes and reported that the addition of a few drops of aldehyde would restore a red-violet color to a rosanilin (fuchsin, magenta) dye solution that had been decolorized by SO_2. His name has been applied to the colorless derivative formed by the action of SO_2 on basic fuchsin, or any of its component dye moieties. Feulgen (1914) and Feulgen and Rossenbeck (1924) developed the use of Schiff reagent for DNA, and Bauer (1933) developed its use for glycogen.

In an acid solution and with an excess of SO_2 present, basic fuchsin (a mixture of several related phenyl methane dyes, rosanilins, and pararosanilins built on the quinoid ring) is reduced to form a colorless N-sulfinic acid (fuchsin sulfurous acid). This reagent, with the addition of aldehydes, forms a new phenyl methane dye, slightly different from basic fuchsin since the color is more purple-red than pure red. The chemical reaction is not wholly understood. Thus, areas rich in DNA show deep coloration after hydrolysis. Schiff is stable as long as an excess of SO_2 with high acidity is present. Anything removing these conditions restores the original dye and produces a pseudoreaction. But, when regenerated by an aldehyde, the dye becomes extremely resistant to such agents. Schiff reagent is not a dye; it lacks a chromophore and is, therefore, colorless. Baker (1958) considers this an example of localized synthesis of a dye; when Schiff reagent comes in contact with an aldehyde, the chromophore of the triarylmethane (quinoid ring structure) dye is reconstituted. The additive compound of the aldehyde with the Schiff could be called a dye.

Schiff reagent deteriorates rapidly at temperatures above 40°C, but at 0–5°C, if kept tightly stoppered, the deterioration is slow. Always keep it in the refrigerator when not in use. Under these conditions it will keep as long as six months.

Chung and Chen (1970) restore exhausted Schiff reagent by adding 0.5 g $NaHSO_3$ to 100 ml of the reagent. This can be repeated 2 or 3 times as the solution becomes exhausted, and is advisable even when the solution is kept in the refrigerator. If the solution shows a red tinge it is partially exhausted; if it shows a purple tinge it is completely exhausted. In either case it can be recovered by this method.

For references concerning the Schiff reaction, see Atkinson (1952); Baker (1958); Bensley (1959); Glegg, et al. (1952); Gomori (1952); Kasten (1960); Lhotka and Davenport (1949); Lillie (1951*a,b*, 1953*a,b*, 1954*a,b*); McManus (1961); and Stowell (1945, 1946); Lodin et al. (1967).

PERIODIC ACID–SCHIFF (PAS) REACTION

Two methods are outlined: one with alcoholic solutions; the other with aqueous solutions. Sometimes there is a need for alcoholic PAS, because complex carbohydrates are present that are soluble in water, but not in alcohol. The treatment time is longer in the alcoholic than in the aqueous solution.

PAS Technique, Alcoholic

FIXATION

Any general fixative, but if glycogen or other soluble polysaccharides are to be demonstrated, fixation and washing should be done in alcoholic fluids of no less than 70% alcoholic content.

SOLUTIONS (McManus and Mowry 1960)

Periodic acid

periodic acid (HIO_4)	1.0 g
90% ethyl alcohol	100.0 ml

Keep in dark; solution is unsatisfactory if it turns brown.

Schiff reagent (Lillie 1951*b*)

basic fuchsin (C.I. 42500)	0.5–1.0 g
distilled water	85.0 ml
sodium metabisulfite ($Na_2S_2O_5$)	1.9 g
N HCL	15.0 ml

Place in bottle with approximately 50–60 ml of free air space. Shake at intervals for at least 2 hours or overnight. Add 200 mg activated charcoal: 1 minute; shake occasionally. Filter. If solution is not water-white the charcoal is old. Try a fresh batch and refilter. Store Schiff reagent in a bottle with a minimum of air space above the solution and keep in refrigerator. This will decrease loss of SO_2 (Elftman 1959c). It can be stored frozen, but never at room temperature.

Kasten (1960) recommends the use of less than 50 mg of charcoal per 100 ml of solution, claiming that too much charcoal abstracts some of the discolored dye and reduces its sensitivity. Do not leave the charcoal in the solution too long, about 45 seconds, and filter.

Sodium bisulfite

sodium metabisulfite ($Na_2S_2O_5$)	0.5 g
distilled water	100.0 ml

PROCEDURE

1. Deparaffinize and run slides down to 70% alcohol; remove $HgCl_2$, using iodine in 70% alcohol. (If no $HgCl_2$ present, carry slides into absolute alcohol only; then proceed to step 2.)
2. Treat with alcoholic periodic acid: 2 hours.
3. Wash with 90% alcohol: 5 minutes.
4. Hydrate quickly to water.
5. Treat with Schiff reagent: 10 minutes.
6. Transfer through sulfite solutions, 3 changes: $1\frac{1}{2}$–2 minutes each (see comment 5).
7. Wash in running water: 5 minutes.
8. Counterstain if desired (see comment 4).
9. Dehydrate, clear, and mount.

COMMENTS

Mowry now uses 1% periodic acid in absolute methanol (personal communication).

See also comments under following technique.

PAS Technique, Aqueous

FIXATION

Same as for alcoholic technique.

SOLUTIONS

Periodic acid

periodic acid (HIO_4)	0.6 g
distilled water	100.0 ml
nitric acid, concentrated	0.3 ml

Schiff reagent, see page 327.

Sodium bisulfite, see alcoholic technique.

PROCEDURE

1. Deparaffinize and hydrate slides to water: remove $HgCl_2$.
2. Treat with periodic acid, aqueous: 5 minutes.
3. Wash in running water: 5 minutes.
4. Treat with Schiff reagent: 10 minutes.
5. Transfer through sulfite solutions, 3 changes: $1\frac{1}{2}$–2 minutes each (see comment 2).
6. Wash in running water: 5 minutes.
7. Counterstain, if desired (see comment 4).
8. Dehydrate, clear, and mount.

RESULTS

fungi—red
nuclei and other tissue elements—color of counterstain
Many tissues give positive PAS reactions: glycogen, starch, cellulose, mucins, colloid of thyroid, cartilage matrix, chitins, reticula, fibrin, collagen—rose to purplish red (see comment 6).

COMMENTS

1. When preparing slides, avoid excessive use of egg albumen, which contains sufficient carbohydrate to react with the Schiff.

2. Some technicians eliminate the sulfite rinses, but high chlorination of the water makes this a questionable practice. The reagent must be removed, and the sulfite solutions insure this. Renew them often.

3. *Control slides:* To remove glycogen, run the slides down to water and subject them to the saliva test. Saliva contains a diastatic enzyme which dissolves glycogen and starch. Human saliva can be negative for diastase and should be tested on a known starch before its use on control slides. A reliable substitute is a 1% solution of diastase of malt[1] in a phosphate buffer, (page 544) pH 6.0, 37°C: 1 hour.

To remove mucin, treat slides with lysozyme, 0.1 mg to 10.0 ml of Sorensen M/15 phosphate buffer (page 543), pH 6.24, room temperature: 40–60 minutes.

To remove RNA, treat slides with 0.01 mg RNAse in 10.0 ml 0.2 M acetate buffer (page 539), pH 5.0 at room temperature: 10–15 minutes (or tris-HCl buffer, 0.05 M, pH 7.5, 56°C: 2 hours).

---

[1] Fisher Scientific Co.

4. *Counterstains:* Nuclei—hematoxylin; for glycogen—fast green FCF as in Feulgen method; mucin or acid polysaccharides—an acid dye (fast green); other polysaccharides—a basic dye (malachite green).

5. If it is necessary to remove the PAS from the tissue, treat it with potassium permanganate until all the color is removed; then bleach the permanganate with oxalic acid.

6. A number of carbohydrate and carbohydrate-protein substances give PAS reaction: acid mucopolysaccharides, glycolipids, mucoproteins, glycoproteins, pituitary gonadotropins, neutral mucopolysaccharides, unsaturated lipids, phospholipids, and others. Among the polysaccharides that give PAS reaction are glycogen, starch, and cellulose; all of these have 1,2 glycol groups. Cartilage has a polysaccharide compound that makes this tissue react positively. Among the mucoproteins, the mucins are carbohydrates and give PAS reaction. Certain other structures that are of unknown chemical composition, but that contain polysaccharides, will show positive reactions: striated and brush borders and reticulin fibers are examples. Fats may become colored with Schiff's after application of periodic acid if they are glycolipids. Unsaturated fats color, probably because of the oxidation of C=C sites to 1,2 glycols. Simple sugars are always lost, no matter what technique is employed. To retain complex sugars, avoid use of water solutions (McManus 1961).

7. Horobin and Kevill-Davies (1971) substitute 0.5% basic fuchsin in ethanol–water–concentrated HCl (80 : 20 : 1): 20 minutes for Schiff reagent. They report their method as less troublesome to prepare and use than Schiff. The reaction is inhibited by aldehyde blocking, thus demonstrating that the reagent reacts with aldehyde groups.

8. The PAS technique can be combined with many other procedures, such as silver, aldehyde-fuchsin, Luxol fast blue, Sudan black, alkaline phosphatase (Elftman 1963; Himes and Moriber 1956; Lazarus 1958; and Moffat 1958). For black periodic and black Bauer methods, see Lillie et al. (1961). See page 333 for additional observations.

FEULGEN REACTION

Bauer-Feulgen Reaction (1933)

FIXATION

Avoid aqueous media (see McManus and Mowry 1958).

absolute ethyl alcohol	9 parts
formalin	1 part

Start dehydration for embedding in 95% alcohol. Mount slides with 95% alcohol.

SOLUTIONS

Chromic acid

chromic acid (CrO_3)	....	4.0 g
distilled water	....	100.0 ml

Schiff reagent, see page 327.

Sodium bisulfite

sodium meta bisulfite ($Na_2S_2O_5$)	....	1.0 g
distilled water	....	100.0 ml

PROCEDURE

1. Deparaffinize, coat with nitrocellulose, and hydrate slides to water; remove $HgCl_2$ if it is present.
2. Oxidize in chromic acid: 1 hour.
3. Wash in running water: 5 minutes.
4. Treat with Schiff reagent: 10–15 minutes.
5. Transfer through sodium bisulfite, 3 changes: 1, 2, and 2 minutes each. (See comment 2, page 329.)
6. Wash in running water: 5 minutes.
7. Counterstain in hematoxylin, if desired.
8. Wash and blue in Scott and wash.
9. Dehydrate, clear, and mount.

RESULTS

glycogen—red
nuclei—blue

See also Best carmine reaction for glycogen, page 315.

COMMENTS

This reaction may be of historical rather than practical interest. The PAS technique supersedes oxidation with chromic acid (Bauer) and with potassium permanganate (Casella), which further oxidize the aldehyde residues and do not give as intense a stain as PAS does.

Feulgen Reaction

FIXATION

A fixative containing $HgCl_2$ is preferred.

SOLUTIONS

5 N hydrochloric acid, see page 531.

Schiff reagent, see page 327.

Bleaching solution (sulfurous acid)

1 N HCl (page 531)	5.0 ml
potassium bisulfite $(K_2S_2O_5)$ or sodium bisulfite $(Na_2S_2O_5)$, 10% aqueous	5.0 ml
distilled water	100.0 ml

or

HCl, concentrated	1.0 ml
potassium or sodium bisulfite	0.4 g
distilled water	100.0 ml

For best results, make up bleach fresh each time.

Fast green

fast green FCF (C.I. 42053)	0.05 g
95% ethyl alcohol	100.0 ml

PROCEDURE

1. Deparaffinize and hydrate slides to water; remove $HgCl_2$. (Leaving slides in 95% alcohol overnight will remove lipids that might cause a plasmal reaction.)
2. Rinse at room temperature in 5 N HCl: 2 minutes.
3. Hydrolyze in 5 N HCl, room temperature: 1 hour (see comment 1).
4. Rinse at room temperature in 5 N HCl; rinse in distilled water.
5. Stain in Schiff reagent: 2 hours in dark.
6. Drain and transfer quickly into bleaching solution, 3 changes: 1½–2 minutes each.
7. Wash in running water: 10–15 minutes.
8. Rinse in distilled water.
9. Counterstain in fast green: 10 seconds.
10. Dehydrate, clear, and mount.

RESULTS

thymonucleic acid–containing substance—red-violet
other tissue elements—shades of green

COMMENTS

1. The specificity of the Feulgen reaction has been attacked periodi-
cally, but it seems evident that a properly applied Feulgen reaction can
be reasonably specific. Because the results of the reaction can vary with
conditions, optimum conditions are essential. For instance, excessive
hydrolysis will allow the degraded nucleic acids to diffuse from the tissue.
Depending on the fixative used, the following times of hydrolysis are recommended (Pearse 1968):

Carnoy	6 minutes	Formalin sublimate	8 minutes
Carnoy, formula B	8 minutes	Helly	8 minutes
Carnoy-Lebrun	6 minutes	Regaud	14 minutes
Champy	25 minutes	Susa	18 minutes
Chrome acetic	14 minutes	Zenker	5 minutes
Flemming	16 minutes	Zenker-formol	5 minutes
Formalin	8 minutes		

The conventional hydrolysis temperature has been 60°C, but Deitch
et al. (1968) and Decosse and Aiello (1966) agree that 5 N HCl at room
temperature is preferable. The Feulgen reaction time is less critical at
room temperature, and Feulgen values may be from 5 to 30% higher than
when hydrolysis is done at 60°C.

2. Kasten and Burton (1959) make the following Schiff reagent. It is
quickly prepared and colorless. It does not stain the hands and does not
require refrigeration. It can be made more sensitive by boiling it for 1
minute.

basic fuchsin (C.I. 42500)	0.05 g
distilled water	300.0 ml
sodium hyposulfite ($Na_2S_2O_4$)	6.0 g

Solution should decolorize immediately. Filter if necessary. Ready for
immediate use.

3. Pink solutions of Schiff reagent may have lost their potency. Test
by pouring a few drops into 10 ml of 40% formalin. A good solution
changes rapidly to reddish purple, but, if the color changes slowly and
becomes blue-purple, the solution is breaking down. Older batches of
reagent tend to fade more than new batches (de la Torre and Salisbury
1962).

4. Chen (1944*b*) uses the following weak Flemming fixative for avian

parasites. It gives beautiful results on any kind of smear preparation and on very small pieces of tissue, which should be fixed for 1–4 hours.

chromic acid, 1% in normal saline	25.0 ml
acetic acid, 1% in normal saline	10.0 ml
osmic acid, 2% in normal saline	5.0 ml
normal saline	60.0 ml

Wash smears for 1 hour in running water; proceed to step 2, hydrolysis. Wash pieces of tissue: overnight before embedding for sectioning.

5. The first of the series of three bleaching solutions (step 6) will begin to accumulate Schiff reagent and turn pink. Then it is advisable to remove that solution, shift steps 2 and 3 to steps 1 and 2, and add a new step 3 solution.

6. Swartz and Nagy (1963) found less fading of Feulgen stain in tissues that were mounted in Eukitt[2] than in those mounted in Clearmount or Permount.

7. Block and Godman (1955) hydrolyzed in 1 N trichloracetic acid and stained with a trichloracetic acid–Feulgen. The trichloracetic acid was substituted for hydrochloric acid, mole for mole in the hydrolyzing solution and the Feulgen reagent. In this way, the histone proteins were left intact.

Deoxyribonucleic Acid (DNA) Fluorescent Technique
(Culling and Vasser 1961)

FIXATION

10% formalin for sections; methyl alcohol for smears. (Other fixatives may require a different time for hydrolysis.)

SOLUTIONS

Fluorescent Schiff reagent

acriflavine dihydrochloride	1.0 g
potassium metabisulfite	2.0 g
distilled water	200.0 ml
N hydrochloric acid	20.0 ml

Dissolve acriflavine and potassium metabisulfite in distilled water; add hydrochloric acid. Keep overnight before using.

---

[2] O. Kindler, Freiburg, Germany.

Periodic acid

periodic acid	1.0	g
M/15 sodium acetate	10.0	ml
absolute ethyl alcohol	90.0	ml

Acid alcohol

70% ethyl alcohol	99.0	ml
hydrochloric acid, concentrated	1.0	ml

PROCEDURE

1. Deparaffinize and hydrate sections to water. Smears can be carried directly into next step.
2. Transfer to preheated N hydrochloric acid, 60°C: sections, 10 minutes; smears, 3–4 minutes (or 1% periodic acid: 10 minutes for PAS).
3. Wash briefly in distilled water.
4. Transfer to fluorescent Schiff reagent: 20 minutes.
5. Wash in acid alcohol: 5 minutes.
6. Wash in fresh acid alcohol: 10 minutes.
7. Dehydrate in absolute alcohol: 1 minute.
8. Clear in xylene: 1 minute, and mount.

RESULTS

DNA—bright golden fluorescence
other elements—dark green to black

COMMENTS

The advantage of this method over the conventional Feulgen reaction is that smaller amounts of dye molecules are more easily observed.

Culling and Vasser warn that previously heated hydrochloric acid is important; cold hydrochloric acid can produce negative results. Also timing is important; the reaction may weaken if slides are left in hydrolysis too long.

See Betts (1961) and Kasten and Glover (1959) for other fluorescent Schiff-type reagents. Also see Armstrong et al. (1957); Metcalf and Paton (1944); Keeble and Jay (1962); and Nash and Plaut (1964) for fluorescent method for DNA and RNA.

CHAPTER 21

Staining Cellular Elements

THE ARGENTAFFIN REACTION

The argentaffin reaction should not be confused with silver impregnation (page 200). In argentaffin reaction some substances (ascorbic acid, aldehydes, uric acid, polyphenols, and others) reduce silver solutions under specific conditions. This reaction with the tissue itself can therefore be used histochemically to identify these substances.

The only source of error in the method is found in calcification areas, but only if these are in large masses. Most silver phosphates and carbonates will be dissolved out during the process. If not, treat the slides with 0.2 to 0.5% nitric acid or hydrochloric acid in absolute alcohol: 2–3 minutes before step 3. Wash off the acid in absolute and 95% alcohol and proceed with step 3.

URIC ACID STAINING

Argentaffin Method (Gomori 1952)

FIXATION

95% or absolute alcohol.

SOLUTIONS

Methenamine silver solution

silver nitrate, 5% (5 g/100 ml water)	5.0 ml
methenamine, 3% (3 g/100 ml water)	100.0 ml

Mix and shake. A precipitate forms but will redissolve. Keep in cool, dark place. Good for several months.

Buffer solution, pH $\pm$ 9.0

$M/5$ boric acid (12.368 g/1000 ml water)	20.0 ml
$M/20$ borax (19.071 g/1000 ml water)	80.0 ml

Working solution

methenamine silver solution	30.0 ml
buffer solution	8.0 ml

PROCEDURE

1. When mounting sections on slides, float on water only a few seconds, or try 95% alcohol.
2. Deparaffinize and run slides down to 95% alcohol.
3. Treat in methenamine silver working solution, prewarmed to 37°C: 30 minutes.
4. Rinse in distilled water.
5. Fix in sodium thiosulfate, 5% aqueous: 3 minutes.
6. Wash in running water: 5 minutes. (Gold toning optional.)
7. Counterstain if desired.
8. Dehydrate, clear, and mount.

RESULT

uric acid crystals—black crystalline fine granules

COMMENTS

Uric acid occurs in the form of acid sodium urate. Heavy metal salts of uric acid are insoluble and can be converted into colored compounds, as can phosphates and carbonates. Uric acid is, however, one of the few argentaffin substances found in the body and can be identified by argentaffin methods.

ENTEROCHROMAFFIN (EC) CELL STAINING

Fontana Method (Culling 1957), **and Methenamine Silver** (Gomori 1952, 1954*b*)

FIXATION

Zenker formol preferred. Do not use alcohol—it dissolves argentaffin granules. During embedding, do not expose to paraffin longer than 30 minutes.

SOLUTIONS

Silver solution A (Fontana method)

silver nitrate, 10% (2.5 g/25 ml water)	25 ml
ammonia, 28%	as needed
distilled water	25 ml

To silver nitrate add ammonia drop by drop until the precipitate that forms is almost dissolved. Add distilled water. Store for 24 hours in brown bottle. Filter before use. Good for 2 weeks.

Methenamine silver solution B (Gomori method)

Solution I

methenamine, 3% (3 g/100 ml water)	100.0 ml
silver nitrate, 5% (5 g/100 ml water)	5.0 ml

Shake until white precipitate disappears. Keeps for several months.

Solution II

$M/5$ boric acid (12.368 g/1000 ml water)	80.0 ml
$M/5$ borax (19.071 g/1000 ml water)	20.0 ml

The *p*H should be 7.8 to 8.0.

Working solution

Solution I	30.0 ml
Solution II	8.0 ml

Gold chloride

gold chloride stock solution 1%	10.0 ml
distilled water	100.0 ml

Safranin (or Kernechtrot, page 270)

safranin O (C.I. 50240)	1.0 g
distilled water	100.0 ml

Add a few drops of glacial acetic acid.

PROCEDURE

1. Deparaffinize and hydrate slides to water.
2. Treat with Lugol solution (page 532): 30 minutes to 1 hour.
3. Wash in running water: 3 minutes.
4. Bleach in 5% aqueous sodium thiosulfate: 3 minutes.
5. Wash in running water: 5 minutes. (Steps 2–4 suppress background staining.)
6. Treat with silver solution A in the dark, room temperature: 18–48 hours. *Alternate method:* Use methenamine silver solution B at 60°C for 3.5 hours; or at 37°C for 12–24 hours—until EC cells stand out black.
7. Rinse in several changes of distilled water.
8. Tone in gold chloride: 10 minutes.
9. Rinse in distilled water.
10. Fix in 5% aqueous sodium thiosulfate: 3 minutes.
11. Wash in running water: 5 minutes.
12. Counterstain in safranin (or Kernechtrot).
13. Rinse in 70% alcohol: few seconds.
14. Dehydrate, clear, and mount.

RESULTS

argentaffin granules—black
melanin—black
other tissue elements—reds and pinks
background grayish to blackish—slide is overstained

COMMENTS

Chromaffin material is found only in the adrenal medulla and the gastrointestinal tract. It received its name because of its reaction with chromium salts and certain other metals to produce a yellowish to brown color, a brownish chromium oxide. To preserve this material, either use a dichromate fixative or post-chromate with potassium dichromate (page 27). Chromaffin material is also argentaffin and is blackened in the above method. Melanin, also brownish, is not a chromaffin substance.

Methenamine silver methods are also used for fungi and bacteria, see page 397.

Diazo-safranin Method (Lillie et al. 1953)

FIXATION

10% formalin buffered with 2% calcium acetate: 1–3 days.

SOLUTION

Diazotized safranin

The three parts of the diazotized safranin solution can be kept as stock solutions. To prepare the working solution they must be mixed *immediately before use.*

Acid safranin

safranin O (C.I. 50240)	3.6 g
distilled water	60.0 ml
N HCl	30.0 ml

This solution will remain stable for several weeks.

N sodium nitrite

sodium nitrite ($NaNO_2$)	6.9 g
distilled water	100.0 ml

Keep in refrigerator, stable for 3 months.

Disodium phosphate, $M/10$

disodium phosphate (Na_2HPO_4), anhydrous ...	14.2 g
distilled water	1000.0 ml

To prepare the working solution add 0.5 ml of sodium nitrite solution to 4.5 ml of ice-cold safranin solution. The resulting mixture turns deep blue and foams. Keep at 0° to 5°C for 15 minutes for diazotization. Dilute 1 ml of the solution with 40 ml of ice-cold disodium phosphate solution and use immediately (pH should be about 7.7).

PROCEDURE

1. Deparaffinize and hydrate slides to distilled water.
2. Place slides in a previously chilled coplin jar and pour over them the freshly prepared diazo-safranin solution: 5 minutes.

3. Decant stain and wash slides in 3 changes of N/10 aqueous hydrochloric acid or acid-alcohol (1 ml concentrated HCl/99 ml 70% alcohol): 10–13 seconds total, to remove most of adherent stain. Longer extraction lightens the colors but does not improve contrast.
4. Wash off acid with water (if N/10 HCl is used) or 95% alcohol (if acid-alcohol is used).
5. Dehydrate, clear, and mount.

RESULTS

enterochromaffin granules—black
gastric gland, chief cell granules—dark red
Paneth cell granules—red
mucin—unstained
nuclei and cytoplasm—pink to red

Ferric-Ferricyanide Method (Modified Schmorl Technique)
(Lasky and Greco 1948; Lillie and Burtner 1953*a*)

FIXATION

10% formalin buffered with 2% of calcium acetate: 1–3 days.

SOLUTION

Ferric-ferricyanide

potassium ferricyanide, 1% aqueous	10.0 ml
ferric chloride, 1% aqueous	75.0 ml
distilled water	15.0 ml

PROCEDURE

1. Deparaffinize and hydrate slides to water.
2. Stain in freshly prepared ferric-ferricyanide: 5 minutes.
3. Rinse in 3 changes of distilled water.
4. Counterstain in safranin, or the like.
5. Dehydrate, clear, and mount.

RESULTS

enterochromaffin granules—dark blue
nuclei—red

COMMENTS

Lillie and his coworkers devoted several years to research on the enterochromaffin granules along with melanins, etc. (Glenner and Lillie 1957*b*;

Lillie 1955, 1956*a*, *b*, *c*, 1957*a*, *b*, *c*, 1960, 1961; Lillie et al. 1953; Lillie et al. 1957; Lillie et al. 1960.)

Azo-Coupling Method (Gurr 1958)

FIXATION

Formalin or Bouin.

SOLUTION

Garnet reagent

garnet GBC salt[1]	0.5 g
distilled water	100.0 ml
borax, saturated aqueous	2.5 ml

If it is necessary to purify the garnet GBC see comments on page 440.

PROCEDURE

1. Deparaffinize and hydrate to water.
2. Transfer to garnet solution: 30–60 seconds.
3. Wash in running water: 30 seconds.
4. Stain in Mayer (or other) hematoxylin: 3 minutes.
5. Wash in running water: 5 minutes.
6. Dehydrate, clear, and mount.

RESULTS

argentaffin granules—red
nuclei—blue
background—light yellow

AMYLOID STAINING

Congo Red (Puchtler et al. 1962)

FIXATION

10% formalin or alcoholic fixative.

---

[1] Imperial Chemical Industries, Manchester, England, Dajac Laboratories, or American Cyanamid Company.

SOLUTIONS

Solution A

1% aqueous sodium hydroxide	0.5 ml
80% ethanol saturated with NaCl	50.0 ml

Use within 15 minutes.

Solution B

80% ethanol saturated with Congo red (C.I. 22120) and NaCl	50.0 ml
1% aqueous sodium hydroxide	0.5 ml

Filter. Use within 15 minutes.

PROCEDURE

1. Deparaffinize and hydrate slides to water.
2. Stain in Mayer hematoxylin: 10 minutes.
3. Rinse in 3 changes distilled water: few seconds each.
4. Treat with solution A: 20 minutes.
5. Stain in solution B: 20 minutes.
6. Dehydrate rapidly in 3 changes absolute ethanol, clear, and mount.

RESULTS

amyloid—red
nuclei—blue

COMMENTS

Amyloid is a hyaline substance that appears in connective tissue as a result of disease, and methods should be on hand for pathological study.

The working solutions must be freshly made, but their major components, 80% ethanol saturated with NaCl and 80% alcohol saturated with Congo red and NaCl, will keep for several months. The latter should stand 24 hours before use.

This method is used in preference to Bennhold's or Highman's because in it there is no need for differentiation. Bennhold's method lacks uniformity, and Highman's easily overdifferentiates.

Staining is more intense in alcohol-fixed tissues than in formalin-fixed tissues, but even in the latter small deposits of amyloid are visible.

Puchtler et al. suggest that the selectivity of Congo red for amyloid may be due to the polysaccharide moiety of amyloid. Puchtler et al.

also found that certain direct cotton dyes stain amyloid. See Puchtler and Sweat (1965) and Puchtler and Sweat (1966) for a review of the concepts of amyloid substance. See also metachromatic methods, page 346.

Fluorescent Method

Thioflavine T is recommended for amyloid by Vassar and Culling (1959 and 1962), but it colors amyloid only faintly. Puchtler and Sweat (1965) use their Congo red method (see above) with excellent fluorescence and pink to red coloring of amyloid, and with other tissues greenish gray. Mount in Harleco Fluorescent Mountant[2] for a black background. Synthetic mountants, such as Permount, have only moderate fluorescence, but slides mounted with them can be used for fluorescent observation; the faintness of the fluorescence does not interfere too seriously with slide study.

MAST CELL STAINING

Mast cells are of common occurrence in connective tissue. Because of their cytoplasmic granules, however, staining methods for these cells have been included here. The specific staining of these granules is the primary means of identification of mast cells. (See also metachromatic methods, page 349.)

Chrysoidin (Harada 1957)

FIXATION

Any general fixative.

SOLUTION

chrysoidin Y (C.I. 11270) .... 0.5 g
distilled water .... 100.0 ml

PROCEDURE

1. Deparaffinize and run slides down to 80% alcohol; remove $HgCl_2$ if present.
2. Stain in chrysoidin: 5–10 minutes.
3. Rinse in distilled water.
4. Dehydrate, clear, and mount.

---

[2] Hartman-Leddon Co.

RESULTS

mast cell granules—deep brown to black
other elements—yellowish

COMMENTS

Staining with chrysoidin can be preceded with hematoxylin or PAS. Toren (1963) combines Pollak (1944) trichrome stain with a Giemsa stain. Barlow (1957) uses a tribasic stain of acridine red, neutral red, and basic fuchsin. Menzies (1962a) combines methylene blue and basic fuchsin. Spatz (1960) uses Bismarck brown.

Fluorescent Method (Jagatic and Weiskopf 1966)

FIXATION

Any general fixative.

SOLUTIONS

Weigert hematoxylin, see page 151.

Acridine orange

acridine orange (C.I. 46005)	0.1 g	
distilled water	100.0 ml	

PROCEDURE

1. Deparaffinize and hydrate sections to water.
2. Stain in hematoxylin: 5 minutes.
3. Wash in running water: 3–5 minutes.
4. Stain in acridine orange: 5–6 minutes.
5. Rinse in distilled water: 1 minute.
6. Dehydrate in 95% ethanol or absolute isopropanol.
7. Clear and mount in Harleco Fluorescent Mountant.

RESULT

mast cell granules—bright red-orange

COMMENTS

Old H-and-E slides have been restained by this method, also sections from paraffin blocks 10 to 15 years old.

METACHROMASIA

A few tissue elements are stainable by a particular group of cationic dyes, changing in the tissues from blue (the usual orthochromatic form) to purplish red or reddish purple. Such a dye, called metachromatic, is of considerable value in the study of specific elements of connective tissue. Among the metachromatic dyes most commonly used are toluidine blue O, thionin, methylene blue, azures, crystal violet, cresyl violet, methyl violet, safranin O, celestin blue, gallocyanin, and Pinacyanole. Some of the tissues identified by this means are mast cells, amyloid, cartilage, and mucus materials (Schubert and Hamerman 1956).

The methods are tricky, and a technician must learn to distinguish a true metachromasia from a false one. The difficulty in preserving metachromasia lies in the dehydration of the tissue after staining. Increasing strengths of alcohol (ethyl) revert the dye back to the orthochromatic form. Sections can be examined in an aqueous condition, but this produces at best a semi-permanent preparation. Some workers use acetone, isopropyl, or tertiary butyl alcohol for dehydration, then follow with one of the clearing agents. Padawer (1959) uses ether 181 (tetramethylene glycol ether)[3] (page 350), and Levine (1928) uses oil of clove for dehydration. Always stain at room temperature; heat can destroy metachromasia.

Amyloid

Crystal Violet (Lieb 1947)

FIXATION

10% formalin or alcoholic fixative.

SOLUTION

Crystal violet stock solution

crystal violet (C.I. 42555), 14.0–15.0 g saturated
in 95% ethyl alcohol . . . .   100.0 ml

Working solution

stock solution . . . .   10.0 ml
distilled water . . . .   300.0 ml
hydrochloric acid, concentrated . . . .   1.0 ml

---

[3] Ansul Chemical Co., Marinette, Wisconsin.

PROCEDURE

1. Run frozen sections or deparaffinized sections down to water.
2. Stain in working solution: 5 minutes to 24 hours.
3. Rinse in water.
4. Mount in Abopon (page 145).

RESULTS

amyloid—purple
other tissue elements including hyalin—blue

COMMENTS

Acid in the staining solution makes it self-differentiating, and staining time is flexible. Lieb suggests that if there is only a small amount of amyloid in the tissue, thicker sections should be cut to make the color reaction more clearly visible. Slides mounted in Abopon have retained their color for 2 years. Abopon mounts must be sealed with ringing cement for permanency.

Gomori (1952) outlines a simple method for permanent mounts that preserve the metachromasia of the crystal violet.

1. Float paraffin sections on dye solution: 15–20 minutes.
2. Float sections on distilled water to remove excess dye.
3. Float on 1–2% aqueous acetic acid, to differentiate.
4. Mount on slides in usual fashion and dry.
5. Remove paraffin with xylene and mount.

See Bancroft (1963) for staining with methyl violet and methyl green. See also amyloid staining, page 342.

Mucin

Toluidine Blue (Lillie 1929)

FIXATION

Any general fixative, but an alcoholic one is preferred. Do not use a chromate fixative.

SOLUTION

toluidine blue O (C.I. 52040)	2.0 g
distilled water	100.0 ml

PROCEDURE

1. Deparaffinize and hydrate slides to water; remove $HgCl_2$.
2. Stain in toluidine blue: 1 minute.

3. Wash in water: 2–3 minutes.
4. Dehydrate in acetone, 2 changes: 3–5 minutes each.
5. Clear in xylene and mount.

RESULTS

mucin—reddish violet
nuclei and bacteria—deep blue
cytoplasm, fibrous tissue—bluish green
bone—bluish green
cartilage matrix—bluish violet
muscle—light blue
cell granules—blue violet
hyalin and amyloid—bluish green

COMMENTS

Gomori (1952) uses 0.02%–0.05% toluidine blue O in citrate buffer, pH 3.5–4.5: 10–15 minutes, or until the nuclei are blue and the mucin, intense pink.

For nonmetachromatic staining of mucin, see page 317.

Thionin (Mallory 1944)

FIXATION

Any general fixation, but an alcoholic one is preferred.

SOLUTION

thionin (C.I. 52000) ....   1.0 g
25% ethyl alcohol .... 100.0 ml

PROCEDURE

1. Deparaffinize and hydrate slides to water; remove $HgCl_2$.
2. Stain in thionin: 15 minutes to 1 hour.
3. Differentiate in 95% alcohol.
4. Dehydrate in absolute alcohol, clear, and mount.

RESULT

mucin—light to dark red or purple

Acid Mucopolysaccharides

Thionin (Gurr 1958)

FIXATION

10% formalin or other general fixative.

Thionin

thionin (C.I. 52000) saturated aqueous	0.5 ml
distilled water	100.0 ml

Molybdate-ferricyanide solution

ammonium molybdate, 5% aqueous	50.0 ml
potassium ferricyanide, 1% aqueous	50.0 ml

Make up solutions fresh each time.

PROCEDURE

1. Deparaffinize and hydrate slides to water.
2. Stain in thionin: 5–15 minutes.
3. Rinse in distilled water.
4. Treat with molybdate-ferricyanide: 2 minutes.
5. Wash in distilled water: 2–3 minutes.
6. Dehydrate, clear, and mount.

RESULTS

acid mucopolysaccharides—purple
other cell elements—bluish

COMMENTS

The molybdate-ferricyanide solution prevents loss of metachromasia. See Kuyper (1957) for suggestions concerning fluorescent methods for mucopolysaccharides. See also acid mucopolysaccharides, page 317.

Mast Cells

For nonmetachromatic staining of mast cells, see page 344.

Thionin Method (Lillie 1954 or 1965)

FIXATION

Any general fixative.

SOLUTION

thionin (C.I. 52000)	0.5 g
0.01 *M* acetate buffer (see results)	100.0 ml

PROCEDURE

1. Deparaffinize and hydrate slides to water; remove $HgCl_2$.
2. Stain in thionin: 30 minutes for light stain.
3. Rinse in water.
4. Dehydrate, clear thoroughly in xylene, and mount.

RESULTS

mast cell granules—red-purple
nuclei—faint blue-violet

COMMENTS

At pH 2, metachromasia stains only mast cells and cartilage; at pH 5, muscle and connective tissue stain light green; at pH 3–2, cytoplasm stains poorly.

Quick Toluidine Blue Method [4]

FIXATION

Any general fixative.

SOLUTION

toluidine blue O (C.I. 52040)	0.2 g
60% ethyl alcohol	100.0 ml

PROCEDURE

1. Deparaffinize and run slides down to 60% alcohol. Remove $HgCl_2$ if present.
2. Stain in toluidine blue: 1–2 minutes.
3. Rinse quickly in tap water.
4. Dehydrate in acetone, 2 changes: 2–3 minutes each.
5. Clear in xylene and mount.

RESULTS

mast cells—deep reddish purple
background—faint blue

Toluidine Blue (Padawer 1959)

FIXATION

Formalin-alcohol preferred, 10% formalin satisfactory.

---

[4] Personal communication from Miss Marlies Natzler, U.C.L.A.

formalin . . . . 10.0 ml
95% alcohol . . . . 90.0 ml

SOLUTION

toluidine blue O (C.I. 52040) . . . . 0.05 g
ether 181 . . . . 50.0 ml
distilled water . . . . 50.0 ml

Dissolve dye in water, add ether 181. Filter. Use within 24 hours.

PROCEDURE

1. Deparaffinize and transfer into absolute methyl alcohol, 2 changes: 10 minutes each.
2. Transfer to ether 181, 50% aqueous: 10 minutes.
3. Stain in toluidine blue: 20–60 minutes.
4. Dehydrate in ether 181, 50% aqueous: 10–15 minutes.
5. Dehydrate in ether 181 concentrated, 2 changes: 20 minutes each.
6. Clear in xylene and mount.

RESULTS

mast cell granules—reddish purple
background—faintly blue
cartilage—reddish purple (cartilage is the only tissue other than mast cells staining metachromatically)

COMMENTS

Because toluidine blue in many methods reverts to its orthochromatic (blue) color during dehydration, it is the usual practice to prepare only a semi-permanent aqueous or glycerine mount to preserve metachromasia. However, Padawer suggests that partial dehydration is required for metachromasis. The ether 181 used in his method does not completely dehydrate, and in the specific concentration used both mast cell polysaccharides and toluidine blue are poorly soluble.

Izard (1964) mounts toluidine blue preparations in prepolymerized butyl methacrylate to preserve metachromasia. Radden (1961) dissolves toluidine blue in McIlvaine citric acid–disodium hydrogen phosphate buffer; at pH 3 the granules stain rapidly and orthochromatic staining is retarded.

Bélanger and Hartnett (1960) use the following solution:

potassium acid phthalate–tartaric acid . . . . 1.0 g
distilled water . . . . 100.0 ml
toluidine blue O (C.I. 52040) . . . . 0.5 g

The dye requires 24 hours to dissolve. Filter before use. Good for one week. Stain for 10 minutes; rinse in fresh 1% phthalate–tartaric acid buffer (1 g/100 ml water): 2 minutes; dehydrate in tertiary butyl alcohol, 4 changes: 2 minutes each; clear and mount.

Cell Membranes

See Puchtler and Leblond (1958).

ENDOCRINE GLAND STAINING

Pituitary Cells

Staining the cytoplasm and its elements is used to characterize and differentiate the cells of the anterior pituitary, but the naming of cell types is still in a state of confusion. Different systems of nomenclature are offered by different investigators. The possibilities include names for cell types based on the hormones they secrete or the morphological characteristics they display during staining. The Greek terminology is still in contention. To add to the confusion—the glands of all species do not react with the same color specificity. It is, therefore, impossible to say "this is it" of any one stain. Some of the more useful stains, each valuable for some particular cell or group of cells, are Mallory types, PAS, aldehyde-fuchsin, aldehyde-thionin and Alcian blue. When followed with care, Herlant methods rate well among the best methods, particularly for basophilic staining, and I use these. An aldehyde-fuchsin method is included, and the Azan (Mallory type) stain can be found on page 177. For extensive study of the pituitary, see Harris and Donovan (1966).

In the anterior pituitary, the glandular cells are sometimes classed as either chromophils or chromophobes. Seventy-five percent of the chromophils are normally acidophilic (acidophils, sometimes called alpha cells) and twenty-five percent are basophilic (basophils). The basophilic group is made up of gonadotrophs (delta cells) and thyrotrophs (beta cells). Thyrotrophs and gonadotrophs have an affinity for Schiff reagent; thyrotrophs also stain with Gomori aldehyde-fuchsin or aldehyde-thionin. Acidophilic granules differentiate sharply in Elftman's (1960) method. Orange G also stains the acidophilic granules. Paget and Eccleston (1959, 1960) use Luxol fast blue to distinguish them. Separating the chromophobes from the chromophils is not always as simple as it may sound;

transitional stages, or the state of secretion of the cells, can make the two types of cells difficult to identify. The cells of the pars tuberalis contain no cytoplasmic granules. Those of the pars intermedia are pale, with a few of the cells containing basophilic granular cytoplasm. The pars nervosa (posterior lobe) does not show well-organized cell structure, but the neurosecretory substance (NSS) can be demonstrated. Gomori chrome hematoxylin is perhaps the most commonly used method for its demonstration, but Bargmann's modification should be used. The substance is easily identified, even though the nuclei, Nissl substance, lipofuscin, and the basophils of the anterior pituitary are also stained. Since the NSS is rich in cystine, the performic acid–Alcian blue method of Adams and Sloper (1955, 1956) and Sloper and Adams (1956) demonstrate the substance.

For additional information, some of the following may be of assistance: Elftman (1957a; 1959a, b; 1960); Glenner and Lillie (1957a); Gomori (1941b); Kerenyi (1959); Kerenyi and Taylor (1961); Landing (1954); Landing and Hall (1955); Lazarus (1958); Paget and Eccleston (1959, 1960); Pearse (1949, 1950, 1960); Shanklin et al. (1959); Thompson, 1966.

Herlant (1960) Pituitary Stain I

FIXATION

Any good general fixative, but Zenker formol preferred.

SOLUTIONS

Erythrosin

erythrosin B (C.I. 45430)	1.0 g
distilled water	100.0 ml

Mallory II

aniline blue WS (C.I. 42755)	0.5 g
orange G (C.I. 16230)	2.0 g
distilled water	100.0 ml
glacial acetic acid	1.0 ml

Acid alizarine blue

acid alizarine blue (C.I. 63015)	0.5 g
aluminum sulfate	10.0 g
distilled water	100.0 ml

Bring to boil, approximately 5 minutes. Cool. Adjust to 100.0 ml with distilled water and filter. Stable, but add thymol.

Phosphomolybdic acid

phosphomolybdic acid	5.0 g
distilled water	100.0 ml

PROCEDURE

1. Deparaffinize and hydrate to water; remove $HgCl_2$ if present.
2. Stain in erythrosin: 5 minutes.
3. Rinse briefly in distilled water.
4. Stain in Mallory II: 5–10 minutes.
5. Rinse briefly in distilled water.
6. Stain in acid alizarine blue: 5–10 minutes.
7. Rinse briefly in distilled water.
8. Treat with phosphomolybdic acid: 5–10 minutes.
9. Rinse briefly in distilled water.
10. Differentiate in 70% ethanol.
11. Clarify stain in 70% ethanol with a few drops of acetic acid added.
12. Dehydrate, clear, and mount.

RESULTS

alpha cells—yellow
gamma cells—violet
beta cells—pale blue
delta cells—dark blue
epsilon cells—red to rose
nucleus—dark blue to violet

Herlant (1960) Pituitary Stain II

FIXATION

Bouin or Hollande Bouin preferred.

SOLUTIONS

Permanganate solution

2.5% aqueous potassium permanganate	10.0 ml
5% aqueous sulfuric acid	10.0 ml
distilled water	60.0 ml

Mix just before use.

Alcian blue, pH 3

  Alcian blue 8GX (C.I. 74240)   1.0 g
  distilled water   100.0 ml
  glacial acetic acid   1.0 ml

Alcian blue, pH 0.2

  Alcian blue 8GX (C.I. 74240)   1.0 g
  10% sulfuric acid   100.0 ml

The Alcian blue dissolves less easily in the acid solution than in water. Warm the solution until stain dissolves; cool, and filter. Good for 1 month; filter occasionally.

Sodium metabisulfite

  sodium metabisulfite   5.0 g
  distilled water   100.0 ml

Periodic acid

  periodic acid   1.0 g
  distilled water   100.0 ml

Schiff reagent, see page 327.

PROCEDURE

1. Deparaffinize and hydrate slides to water.
2. Oxidize in permanganate solution: 1–2 minutes.
3. Rinse in distilled water: few seconds.
4. Bleach in metabisulfite solution: 1 minute.
5. Wash in running water: 5 minutes.
6. Rinse in distilled water.
7. Stain in Alcian blue: 15–30 minutes.
8. Wash in running water: 5–10 minutes.
9. Treat with periodic acid: 10 minutes.
10. Rinse in distilled water: 1–2 minutes.
11. Treat with Schiff reagent: 10–20 minutes.
12. Wash in 3 changes of metabisulfite solution: total of 5 minutes.
13. Wash in running water: 15 minutes.
14. Dehydrate, clear, and mount.

Step 13 can be followed by staining with hematoxylin if desired.

This method separates the two kinds of gonadotrophic cells. The potassium permanganate oxidation confers on the alpha cells only a faint affinity for the PAS reaction. They will stain a faint rose, much less marked than the color of the gamma cells. If step 13 is followed by staining with 1% orange G, the rose color of the alpha cells is masked, and they become yellow; beta cells, violet; delta cells, dull blue; gamma cells, brick red.

With Alcian blue, increased acidity improves staining specificity. At pH 3, the beta and delta cells show a similar affinity for the stain and react only feebly with PAS. At pH 0.2 the two cells are more easily distinguished; the beta cells are paler than the delta cells and PAS does not change the blue of the latter, but colors the beta cells violet. Apparently the beta cells are so weakly stained by the Alcian blue that they do not mask the PAS.

Cameron and Steele Method (1959)

FIXATION

Any good fixative.

SOLUTIONS

Potassium permanganate, 0.3%

potassium permanganate	0.3 g
distilled water	100.0 ml
sulfuric acid, concentrated	0.3 ml

Sodium bisulfite, 2.5%

sodium bisulfite	2.5 g
distilled water	100.0 ml

Aldehyde-fuchsin, see page 190.

Halmi mixture (1952)

distilled water	100.0 ml
light green SF, yellowish (C.I. 42095)	0.2 g
orange G (C.I. 16230)	1.0 g
chromotrope 2R (C.I. 16570)	0.5 g

| phosphotungstic acid | 0.5 g |
| glacial acetic acid | 1.0 ml |

Keeps indefinitely.

PROCEDURE

1. Deparaffinize and hydrate slides to water; remove $HgCl_2$.
2. Oxidize in potassium permanganate: 1 minute.
3. Rinse in distilled water.
4. Bleach in sodium bisulfite until permanganate color is removed.
5. Wash in running water: 5 minutes.
6. Transfer to 70% alcohol: 2 minutes.
7. Stain in aldehyde-fuchsin: 2–10 minutes.
8. Wipe off back of slide and rinse in 95% alcohol.
9. Differentiate in 95% alcohol until no more aldehyde-fuchsin comes out of sections.
10. Transfer to 70% alcohol: 2 minutes.
11. Rinse in distilled water: 1 minute.
12. Stain in Halmi mixture: 45 seconds.
13. Wipe off back of slide, differentiate in 95% alcohol plus 0.2% acetic acid: 2–3 minutes.
14. Rinse in fresh 95% alcohol.
15. Dehydrate in absolute alcohol, clear, and mount.

RESULTS

granulation of beta cells—dark purple
delta cells—green
acidophilic granules—orange
nucleoli—bright red
coagulated contents of cytoplasmic granules—orange

Ewen (1962) Modification of Cameron and Steele Method

This modification increases the method's specificity.

FIXATION

Fix in Bouin with 0.5%–1.0% trichloracetic acid instead of acetic acid. Helly also is good.

SOLUTION

Aldehyde-fuchsin stain

Stock solution

aldehyde-fuchsin stain (page 190) . . . .   0.75 g
70% ethyl alcohol . . . .   100.0 ml

Working solution

stock solution . . . .   0.75 g
70% ethyl alcohol . . . .   75.0 ml
glacial acetic acid . . . .   1.0 ml

PROCEDURE

Steps 1 through 8 as in Cameron and Steele method.

9. Differentiate in acid alcohol (0.5 ml HCl/100.0 ml absolute ethyl alcohol): 10–30 seconds.

10. Transfer through 70% alcohol and water.

11. Mordant in the following: 10 minutes.

phosphotungstic acid . . . .   4.0 g
phosphomolybdic acid . . . .   1.0 g
distilled water . . . .   100.0 ml

12. Rinse in water and stain for 1 hour in a modified Halmi's mixture: increase the light green to 0.4 g and omit the phosphotungstic acid.

Steps 13 through 15 as in Cameron and Steele method.

Chrome-Hematoxylin, Bargmann Modification for NSS (Pearse 1960)

FIXATION

Bouin preferred. Susa and Stieve satisfactory. If tissue is alcohol-fixed, float sections on Bouin solution instead of on water when mounting sections on slides.

SOLUTIONS

Bouin chrome alum

Bouin solution (page 14) . . . .   100.0 ml
chrome alum . . . .   3.0–4.0 g

Potassium permanganate–sulfuric acid

2.5% aqueous potassium permanganate . . . .   1 part
5% aqueous sulfuric acid . . . .   1 part
distilled water . . . .   6–8 parts

Oxalic acid

oxalic acid .... 1.0 g
distilled water .... 100.0 ml

Chromium hematoxylin, see page 361.

Acid alcohol

hydrochloric acid, concentrated .... 1.0 ml
70% ethyl alcohol .... 100.0 ml

Phloxin

phloxin B (C.I. 45410) .... 0.5 g
distilled water .... 100.0 ml

Phosphotungstic acid

phosphotungstic acid .... 5.0 g
distilled water .... 100.0 ml

PROCEDURE

1. Deparaffinize and hydrate sections to water.
2. Mordant in Bouin chrome alum solution, 37°C: 12–24 hours.
3. Wash in running water until sections are colorless.
4. Oxidize in permanganate solution: 2–3 minutes.
5. Wash in distilled water: 1 minute.
6. Bleach in oxalic acid: 1 minute.
7. Wash in running water: 5 minutes.
8. Stain in chrome hematoxylin: 10 minutes.
9. Differentiate in acid alcohol: 30 seconds.
10. Wash in running water: 2–3 minutes.
11. Stain in phloxin: 2–3 minutes.
12. Treat with phosphotungstic acid: 2 minutes.
13. Wash in running water: 5 minutes.
14. Dehydrate, clear, and mount.

RESULTS

NSS—deep purple
nuclei—lighter purple
backgrounds—pinkish red

Lead Hematoxylin Method (Solcia et al. 1969)

FIXATION

An aldehyde fixative is best.

SOLUTION

Lead hematoxylin, see page 153.

PROCEDURE

1. Deparaffinize and hydrate to water.
2. Stain in lead hematoxylin, 37°C: 2–3 hours.
3. Wash in running water: 10 minutes.
4. Dehydrate, clear, and mount.

RESULTS

endocrine cell types, pancreas A and D cells, thyroid C cells, hypophysis MSH (melanocyte-stimulating hormone) and ACTH (corticotrophin hormone) cells—blue-black
nucleoli and nuclear chromatin—dark blue
calcium deposits, nerve fibers, A and Z bands of striated muscle—blue

Pancreas Cells

Lane in 1908 established the difference between the zymogen granules and the islet granules, and showed that there were two types of islet cells. He named cells with alcohol-soluble granules, beta cells; those with water-soluble granules, alpha cells. The water-soluble granules are larger than those of the beta cells. Various techniques have been used to bring out the difference in the granules. Mallory's will stain alpha cells red and beta cells blue, but the Azan method is better, and good differentiation depends on proper fixation. Glenner and Lillie (1957*b*) specifically stain alpha cells by a post-coupled benzilidine reaction for indoles. The alpha cells are associated with a strong nonspecific esterase reaction, and the beta cells, with a strong acid phosphatase reaction. Gomori chrome-hematoxylin is a favorite method for staining the islets.

Recently, a D cell was added to the A and B cells. See Epple (1967) for methods of demonstrating D cells and the use of THF as an aid in overcoming difficulties encountered in embedding pancreas blocks. Also Solcia et al. (1968, 1969).

Chromium-Hematoxylin-Phloxin (Gomori 1941*b*)

FIXATION

Bouin preferred. Stieve satisfactory. Zenker, Carnoy, and formalin unsatisfactory.

SOLUTIONS

Bouin solution, see page 14.

Potassium dichromate–sulfuric acid

This can be made up as separate 0.3% solutions and mixed, or it can be made as follows:

potassium dichromate	0.15 g
distilled water	100.0 ml
sulfuric acid, concentrated	0.15 ml

Hematoxylin solution

hematoxylin	0.5 g
distilled water	50.0 ml

When dissolved, add:

potassium chromium sulfate (chrome alum) 3% aqueous	50.0 ml

Mix well and add:

potassium dichromate, 5% aqueous	2.0 ml
$N/2$ sulfuric acid (about 2.5 ml/100 ml water)	2.0 ml

Allow to ripen for 48 hours. Can be used until a film with a metallic luster does not form on its surface after 1 day's standing. Store at 0–4°C. Filter before use.

Phloxin

phloxin B (C.I. 45410)	0.5 g
distilled water	100.0 ml

PROCEDURE

1. Deparaffinize and hydrate slides to water; remove $HgCl_2$.
2. Refix in Bouin solution: 12–24 hours.

3. Wash in running water: 5 minutes.
4. Treat with potassium dichromate–sulfuric acid: 5 minutes.
5. Decolorize in 5% aqueous sodium bisulfite: 3–5 minutes.
6. Wash in running water: 5 minutes.
7. Stain in hematoxylin solution until beta cells are deep blue (check under microscope): 10–15 minutes.
8. Differentiate in hydrochloric acid, 1% (1 ml/99 ml water): about 1 minute.
9. Wash in running water until clear blue: 5 minutes.
10. Stain in phloxin: 5 minutes.
11. Rinse briefly in distilled water.
12. Treat with 5% aqueous phosphotungstic acid: 1 minute.
13. Wash in running water: 5 minutes. Sections turn red again.
14. Differentiate in 95% alcohol. If the sections are too red and the alpha cells do not stand out clearly, rinse 15–20 seconds in 80% alcohol.
15. Dehydrate in absolute alcohol, clear, and mount.

RESULTS

beta cells—blue
alpha cells—red
delta cells (not present in all animals—pink to red, actually indistinguishable from alphas

COMMENTS

See also Heidenhain Azan method, page 177.

If the zymogen granules (acidophilic) are to be preserved in the acinar cells of the pancreatic lobules, avoid fixatives containing acetic acid.

For fluorescent differentiation of the alpha and beta cells, see Hartroft (1951).

Herlant stain I (page 353) is also excellent for pancreas.

Monroe and Spector (1963) use tannic acid, hematoxylin, Alcian blue, and basic fuchsin.

Thyroid Cells

Outstanding features of the thyroid are the secretory epithelial cells and the extracellular amorphous colloid—the stored secretion. The colloid consists of proteins associated with carbohydrates. Protein reactions can be used; some of the proteins are of the basic type and contain arginine, which can be demonstrated by Sakaguchi modifications. The ferric ferricyanide reduction technique (Lillie 1965) reacts with the colloid. Since the colloid is strongly PAS positive, mucopolysaccharides are probably present, and a positive dialyzed iron reaction would indicate acid mucopolysaccharides. The colloid is basophilic. The thyroid uses iodine to make up its hormone, and since radioactive iodine was one of the first isotopes prepared for biologic purposes, the isotope tracer technique has been used for some time in thyroid studies. Bélanger and Bois (1964) recommended AFA fixation—1 part acetic acid, 5 parts formalin, and 15 parts absolute ethyl alcohol: 24 hours. This gives excellent fixation of the colloid, with less vacuolization than after formalin alone, and well-stained colloid and cell detail with any of the staining techniques. However, I find that preservation of cellular detail and colloid is just as good after fixation in cold (4°–10°C) calcium formalin (page 15). The C cells (assumed to produce polypeptide hormones) of the thyroid can be demonstrated by MacConaill lead hematoxylin as modified by Solcia et al. (1969), page 360.

CHAPTER 22

Staining Golgi Apparatus, Mitochondria, and Living Cells

GOLGI APPARATUS STAINING

The Golgi apparatus (Golgi bodies, Golgi substance, Golgi complex) is usually lost in routine fixation and requires special treatment. Since the methods are not always predictable under all conditions, it may be necessary to modify the fixing and/or staining time in order to attain precise results. In osmium and silver methods the Golgi appears as either a dark net, a granular mass, a cord, or even a more diffuse condition. These reactions seem to indicate the presence of lipids, principally phospholipids. The Golgi apparatus (or adjacent lysosomes) can have a high level of acid phosphatase activity and lesser levels of alkaline phosphatase and other enzymes. Nucleoside diphosphatase is restricted to the Golgi apparatus.

There is a great deal of controversy about the responses of the Golgi apparatus to different conditions, physiological changes, and techniques. Golgi composition appears to differ from cell to cell, and it can be related to the specific activity of a specific cell. Silver or osmic acid leaves the

Golgi in the form of globules that have been described as spheres with an osmophilic cap surrounding an osmophobic center. But a thin lamella also is found adjacent to the spherical elements, indicating a duplex structure. Most of the fixatives recommended for Golgi fixation contain heavy metal salts; yet it has been claimed that these are not necessary and the light metal or organic salts or acids do just as well (*Proceedings of the Histochemical Society* 1953; Pollister and Pollister 1957). (See histochemistry, acid phosphatase, page 435.)

Much of the current research on the Golgi apparatus is being carried out with electron microscopes.

Osmium Tetroxide, Ludford Method (Lillie 1954*b* or 1965; Cowdry 1952)

FIXATION

Mann osmic sublimate: 18 hours.

osmic acid, 1% aqueous	50.0 ml
mercuric chloride, saturate aqueous, plus 0.37 g sodium chloride	50.0 ml

PROCEDURE

1. Wash blocks of tissue in distilled water: 30 minutes.
2. Impregnate:
   2% osmic acid: 3 days, 30°C
   2% osmic acid: 1 day, 35°C
   1% osmic acid: 1 day, 35°C
   0.5% osmic acid: 1 day, 35°C.
3. Wash in distilled water: 1 day.
4. Dehydrate, clear, and embed.
5. Section 6–7μ, mount, and dry.
6. Deparaffinize, clear, and mount.

RESULTS

Golgi apparatus—black
yolk and fat—black (these may be bleached out with turpentine)

COMMENTS

If it is advantageous to have mitochondria stained on the same slide, follow deparaffinization (step 6) by hydrating to water (include cautious treatment with 0.125% potassium permanganate) and stain by the Altmann Method (page 370). Mitochondria will be crimson.

When using silver and osmic acid techniques, considerable experimentation may be necessary. During the impregnation keep solutions in a dark place and follow instructions for temperature carefully. When the solutions turn dark, renew them.

Nassonov-Kolatchew Method (Nassonov 1923, 1924)

This method does not use the graded osmic acid series.

FIXATION

24 hours in:

3% aqueous potassium dichromate	10.0 ml
1% aqueous chromic acid	10.0 ml
2% aqueous osmic acid	5.0 ml

PROCEDURE

1. Wash in running water: 24 hours.
2. Place in 2% aqueous osmic acid, 40°C: 8 hours, or 35°C: 3–5 days.
3. Wash in running water: overnight.
4. Dehydrate rapidly; clear; embed; and section, 2–4μ.
5. If sections are too dark, turpentine (preferably old and oxidized) will remove most of the excess color.
6. Proceed to deparaffinize, etc., according to desired process. Mounted sections may be counterstained with Altmann's stain, page 370.

RESULTS

Golgi apparatus—black
background—yellow
mitochondria after Altmann's—red

COMMENT

In interpreting the results, bear in mind that the mitochondria may become impregnated if the period in osmic acid is prolonged.

Saxena Method (1957)

FIXATION

3–6 hours in:

barium chloride	1.0 g
distilled water	85.0 ml

Add just before use:

formalin .... 15.0 ml

SOLUTIONS

Ramón y Cajal reducing solution

hydroquinone .... 1.5 g
formalin .... 15.0 ml
distilled water .... 100.0 ml
sodium sulfite, anhydrous .... 0.5 g

Gold chloride

gold chloride stock solution (page 532) .... 1.0 ml
distilled water .... 80–90.0 ml

PROCEDURE

1. Rinse blocks of tissue in distilled water.
2. Impregnate in 1.5% aqueous silver nitrate, room temperature: 1–2 days. (Use 1% for very small pieces or embryonic tissues, 2% for fatty tissues and spinal cord.)
3. Rinse, 2 changes, in distilled water.
4. Cut blocks into slices thinner than 2 mm. Reduce in developer: 5 hours.
5. Wash thoroughly in distilled water.
6. Dehydrate, infiltrate, and embed.
7. Section, 6–7μ; mount; and dry.
8. Deparaffinize and hydrate to water.
9. Tone in gold chloride: 2 hours.
10. Rinse in distilled water and fix in 5% aqueous sodium thiosulfate: 3 minutes.
11. Wash in running water: 5 minutes.
12. Counterstain, if desired, in hematoxylin, thionin, carmalum, etc.
13. Dehydrate, clear, and mount.

RESULTS

Golgi—black
cytoplasm—gray
mitochondria—medium to dark gray or black

COMMENTS

The silver preparations depend on fixation with salts of a heavy metal; barium in the Saxena method (page 366). Aoyama (1930) varies the method, using:

cadmium chloride	1.0 g
formalin	15.0 ml
distilled water	85.0 ml

The rest of the procedure is the same as that of Saxena.

Cold-blooded animal tissues require longer fixation and impregnation than do warm-blooded animal tissues (Aoyama 1930).

McDonald (1964) uses nitrocellulose embedding for silver-impregnated Golgi tissues, because the material is very friable and often difficult to handle when embedded with paraffin.

Direct Silver Method (Elftman 1952)

PROCEDURE

1. Immerse small blocks of fresh tissue in silver nitrate in formalin (2 g/100 ml 15% formalin): 2 hours.
2. Rinse briefly in distilled water.
3. Develop for 2 hours, in:

hydroquinone	2.0 g
formalin, 15% (15 ml/85 ml water)	100.0 ml

4. Return to 10% formalin to complete fixation: at least overnight.
5. Wash, dehydrate, and embed.
6. Section, 6–7μ; mount; and dry.
7. Deparaffinize, clear, and mount.

RESULT

Golgi—black

COMMENTS

Do not use buffered formalin, which may limit the solubility of the silver salts.

If the silver is too dense, Elftman suggests bleaching it with 0.7% iron alum. Check under the microscope and stop the reaction by washing thoroughly in running water.

Because the silver is readily oxidized, gold toning is usually preferable for a more permanent slide. This can follow deparaffinization; see other

Golgi methods. Also counterstaining may be included before dehydrating and clearing slides.

Elftman warns that all the black is not necessarily Golgi.

Sudan Black Method (Baker 1944)

FIXATION

Formalin-calcium (page 15): 3 days.

EMBEDDING and SECTIONING

Embed in gelatin (page 74). Harden the block in formalin-calcium-cadmium:

formalin, concentrated	10.0 ml
10 % aqueous calcium chloride	10.0 ml
10% aqueous cadmium chloride	10.0 ml
distilled water	70.0 ml

Wash in running water: 3–4 hours. Section on freezing microtome, 15μ.

PROCEDURE

1. Sections can be affixed to slides as for frozen sections, or they can be stained first and then mounted on slides.
2. Transfer through 50% and 70% alchohol: 1–2 minutes each.
3. Place in saturated solution of Sudan black in 70% alcohol: 7 minutes.
4. Remove excess stain in 3 changes of 50% alcohol: 1–2 minutes each.
5. Rinse in distilled water.
6. Counterstain with a light hematoxylin or a red nuclear stain.
7. Mount in glycerine jelly or Von Apathy gum syrup (page 142).

RESULT

Golgi—black. The vesicles, not the network, are stained. (See also Malhotra 1961.)

MITOCHONDRIA STAINING

Mitochondria (from Greek *mittos*, filament; and *kondria*, granule) are tiny complex cell organelles, which are bounded by a double membrane and include inner membranes (cristae), a ground substance (matrix), and occasionally dense granulations. Chemically the mitochondria matrix consists mainly of proteins and lipids, and the membranes are high in insoluble protein, phospholipid, and insoluble enzyme content. The osmiophilic state of the membranes is due to the lipid component. Mitochondria are able to concentrate a large number of substances such as proteins, lipids, metals, ferments, virus-like particles, and several chemical substances. They are known to contain numerous enzymes—oxidative enzymes, cytochrome oxidase, cytochrome C, succinic dehydrogenase, glutamic dehydrogenase, and adenosine triphosphatase, to name only a few. Mitochondria participate in forming cell organelles; probably of mitochondrial origin are the granules of granulocytes, platelets, neutrophilic myelocytes, zymogen bodies, yolk platelets, secreting granules in neurohypophysis, etc. The electron microscope is responsible for great contributions to the knowledge of the ultra-structure of mitochondria (Novikoff and Podber 1957; Rouiller 1960; *Symposium: The Structure and Biochemistry of Mitochondria*, 1953). (See also histochemistry, succinic dehydrogenase, page 444).

Altmann Method (Cowdry Modification, McClung 1939)

FIXATION

Regaud fixative (page 23): change every day for 4 days; store in refrigerator. Mordant in 3% potassium dichromate: 8 days, change every second day. Wash in running water overnight; dehydrate and embed.

SOLUTIONS

Altmann aniline fuchsin

Make a saturated solution of aniline in distilled water by shaking the two together. Filter. Add 10 g acid fuchsin (C.I. 42685) to 100 ml of filtrate. Let stand for 24 hours. Good for only 1 month.

Methyl green

methyl green (C.I. 42590)	1.0 g
distilled water	100.0 ml

PROCEDURE

1. Deparaffinize and hydrate slides to water.
2. Treat with 1% aqueous potassium permanganate: 30 seconds (see comments below).
3. Rinse briefly in distilled water and bleach in 5% aqueous oxalic acid: 30 seconds.
4. Rinse in several changes distilled water: 1–2 minutes total.

5. Dry slide around sections with filter paper, cover with Altmann fuchsin. Heat gently until fumes smelling of aniline come off. Cool. Allow stain to remain 6 minutes.
6. Dry off most of stain. Rinse in distilled water until sections barely hold stain. If a large amount is left in sections, it will form a troublesome precipitate with the methyl green. If too much is removed, mitochondria will be lightly colored.
7. Pipette methyl green on slide, watching colors by holding it over white paper: about 5 seconds.
8. Drain off stain; rinse briefly in 95% alcohol.
9. Dehydrate in absolute alcohol, clear, and mount.

RESULTS

mitochondria—bright red

COMMENTS

If the methyl green removes all fuchsin, mordanting by dichromate was incomplete. Omit steps 2 and 3, or treat sections for a few seconds with 2% potassium dichromate before staining.

If fuchsin is so intense that the methyl green removes it imperfectly, the mordanting was too great. Correct this in steps 2 and 4. If the methyl green washes out, omit rinsing in 95% alcohol; go directly to absolute.

A solution of aurantia,[1] saturated in 70% ethyl alcohol (30 seconds) may be used in place of methyl green. Go directly to absolute alcohol. (See Andrews and Johnson 1956, for blood smears.)

Iron Hematoxylin can also be used for mitochondria. Best fixative is Regaud, Mordant in 3% potassium dichromate: 7 days. Use the long method of staining: overnight in iron alum and overnight in hematoxylin. The method is not as specific as Altmann; other granules also stain.

Osmic Method (Newcomer 1940)

FIXATION

Zirkle solution: 48 hours.

potassium dichromate	1.25 g
ammonium dichromate	1.25 g
copper sulfate	1.0 g
distilled water	100.0 ml

---

[1] Michrome stains: Edward Gurr, London, England, or Grübler stains: Roboz Surgical Instrument Co., Washington, D.C.

PROCEDURE

1. Wash tissue blocks 8 hours to overnight.
2. Impregnate in 2% aqueous osmic acid: 4–6 days. Change solutions on alternate days.
3. Wash: 8 hours or overnight.
4. Dehydrate, clear, and embed. Use benzene, not xylene, for clearing.
5. Cut 5μ sections, mount on slides.
6. Deparaffinize and hydrate slides to water.
7. Bleach in 1% aqueous potassium permanganate: 5 minutes.
8. Rinse in distilled water.
9. Treat with 3% aqueous oxalic acid: 2–3 minutes.
10. Wash in running water: 15 minutes.
11. Dehydrate, clear, and mount.

RESULT

mitochondria—black

COMMENTS

Newcomer used his method on plant cells. A counterstain such as acid fuchsin can be added.

Short Acid–Fuchsin Method (Novelli 1962)

FIXATION

10% formalin or 1% osmic acid, room temperature: 24 hours. Formalin-fixed tissues are easier to process. Wash in running water: 4 hours.

EMBEDDING AND SECTIONING

Paraffin method, 1–4μ.

SOLUTION

acid fuchsin (C.I. 42685)	0.2	g
methyl blue (C.I. 42780)	0.1	g
N hydrochloric acid	100.0	ml

PROCEDURE

1. Deparaffinize and hydrate slides to water.
2. Stain: 5 minutes.
3. Rinse gently in distilled water.
4. Dehydrate quickly through 95% alcohol and 2 changes of absolute alcohol.
5. Clear and mount.

RESULTS

mitochondria—purple red with peripheral blue wall
chromatin and collagen—blue
nucleoli and erythrocytes—brilliant red

Acid-Hematein Method (Hori and Chang 1963)

SOLUTIONS

Acetone solution

acetone	50.0 ml
uranyl nitrate	0.01–0.02 g
chloral hydrate	0.01–0.02 g

Baker acid hematein. See Hori method, page 312.

Borax-ferricyanide solution

borax $(Na_2B_4O_7 \cdot 10H_2O)$	250.0 mg
potassium ferricyanide $(K_3Fe(CN)_6)$	250.0 mg
distilled water	100.0 ml

Keeps in refrigerator.

PROCEDURE

1. Prepare and section tissues by freeze-substitution method of Chang and Hori, page 423.
2. Transfer the sections into a screw-cap bottle of acetone mixture. Prechill the solution in crushed ice in a Dewar flask and leave buried in dry ice overnight. The presence of metal in the acetone intensifies the lipid reaction.
3. Rinse briefly in acetone at room temperature and mount on cover glasses.
4. Chromate in 5% aqueous potassium dichromate, 60°C: 5–6 hours.
5. Wash in distilled water: 5 minutes.
6. Stain in Baker acid hematein: 30 minutes.
7. Wash in distilled water.
8. Differentiate in borax-ferricyanide, room temperature: 18 hours.
9. Dehydrate, clear, and mount.

RESULTS

mitochondria—sharply stained blue-black
lipids in other cellular elements—may also stain

COMMENTS

Caulfield (1957) reports that tonicity has a direct effect on the appearance of the mitochondria, that low tonicity produced swollen mitochondria and high tonicity, shrank them. He recommends fixing 1 hour at 0°–4°C in the following.

Caulfield solution

stock buffer (see below)	5.0 ml
0.1 N hydrochloric acid	5.0 ml
distilled water	2.5 ml
2% osmic acid	12.5 ml

Adjust *p*H to 7.4, if necessary, by adding a few drops of 0.1 N HCl or stock buffer. Then add 0.045 g of sucrose per ml of solution.

Stock buffer

sodium veronal	14.714 g
sodium acetate (3H$_2$O)	9.714 g

Dilute to 500.0 ml.

See Chang (1956) for method with frozen dried tissues, Benés (1960) for method using amidoblack 10B, Avers (1963) for fixing and staining for the electron microscope, and page 444 of this book for demonstrating enzymes in mitochondria.

Takaya (1967) uses the Luxol fast blue method (page 238) but counterstains with 0.5% aqueous phloxin instead of PAS.

SUPRAVITAL STAINING

Certain dyes will penetrate living cells and stain specific organelles within the cells. Neutral red reacts on certain granulations, products of the cytoplasm, secretory granules, vacuoles of digestion, and others. These vacuoles or granules appear to be heterogeneous and contain a phospholipoprotein complex, acid phosphatase, lipase, and some alkaline phosphatase, and they apparently correspond to lysosomes (Ogawa and Okamoto 1961; Koenig 1963). Neutral red is commonly used in combination with Janus green B for the vital staining of Golgi and mitochondria in a single preparation. Neutral red, however, does not stain a reticular

Golgi apparatus, but does stain vesicles (lysosomes) in the region of the apparatus.

Janus green B does vitally stain the mitochondria and is dependent upon the enzymatic activities of the cell. The staining will take place under partial anaerobic conditions, but it will decolorize when all oxygen is removed, and it is reversibly inhibited by cyanide. Janus green B can be reduced in both mitochondrial and nonmitochondrial portions of the cell, but the localization of a cytochrome oxidase system in the mitochondria slows down the rate of reduction in that organelle, while reduction proceeds more rapidly in the other portions of the cell. It is erroneous to consider all isolated structures taking up Janus green B as mitochondria; secretion granules of the islets of Langerhans, and of enterochromaffin and others will stain as well.

Neutral Red–Janus Green Method

The dyes should be certified for vital staining so they are not toxic to living cells. The solutions used for staining are so weak that it is convenient to make them up as stock solutions of greater strength. They will keep indefinitely if stored in glass-stoppered bottles. If using a pipette with a rubber bulb, do not allow the dye to run up into the bulb. Make the diluted solutions in small quantities, and if they are mixed together, use immediately. The mixture is not stable. Slides must be clean. They should be cleaned in dichromate solution, washed well in running water and then in distilled water, and stored in 95% alcohol. Rub dry just before spreading with stain.

SOLUTIONS

Neutral red stock solution

neutral red (C.I. 50040) . . . .   0.5 g
neutral absolute ethyl alcohol . . . .   100.0 ml

Janus green B stock solution

Janus green B (C.I. 11050) . . . .   0.5 g
neutral absolute ethyl alcohol . . . .   100.0 ml

The alcohol must be neutral and should be distilled over calcium.

PROCEDURE

1. Dilute neutral red stock solution, 1 ml to 10 ml of neutral absolute ethyl alcohol.

2. Mix 0.4 ml of Janus green B stock with 3 ml of dilute neutral red solution.

3. Flood clean slides with mixed dyes and touch edge of slide to absorbent material to draw off excess solution. Allow to dry horizontally in warm air, but keep them protected from dust. Be sure to mark the stained surface. The stain should be distributed thinly and evenly. When dry, the slides can be stored in a dustfree box. Sometimes successful smears can be made by placing a drop of solution at the end of the slide and smearing it like a blood film.

   Another method is to make enough solution to fill a coplin jar or bottle of comparable size. Gently flame a clean slide and plunge it while hot into the stain. Withdraw the slide and drain it upright on absorbent paper. Wave it rapidly in the air to quick-dry it and to preserve even distribution on the slide. If the humidity is high, dry the smears in hot air by using a hair dryer or other drying equipment. The smear may not be as uniform as one following slower drying, but it is necessary to speed up the drying, because humidity allows the alcohol to take up moisture and ruin the dye film before the evaporation is complete.

4. Place a drop of live cells or organisms on a clean cover glass and carefully lower onto a stained slide. Seal with vaseline.

   Blood, bone marrow, and other tissue fluids or cells may be diluted with an equal amount of autogenous heparinized plasma and examined on a stained slide. After a few trials and errors, the proper size of drop will be realized. With blood, the red cells should spread in a single layer. If the preparations are too thick, the leukocytes will round up instead of flatten. If air bubbles are trapped under the cover glass, the cells do not spread properly. If the cover glass is dropped on the stain or if there is too little of the suspension, the cell membranes may rupture.

5. If a warm stage is not available, place slides of tissue cells in a 37°C incubator for 20 minutes. Then examine immediately. Cells begin to round up as the slide cools.

6. When the mitochondria or neutral red bodies are sufficiently stained, if it is desirable to make the slides semipermanent, clean off the vaseline and slip the cover glass toward the edge of the slide, so that the blood or other fluid is spread evenly on both the slide and the cover glass. Slip the cover glass off and rapidly dry both the slide and the cover glass. (Both carry cell preparations.) If dried too slowly, the cells will be distorted. The drying may be completed in a vacuum desiccator (2–4 hours) or by shaking in several changes

of anhydrous ether. The anhydrous ether is essential to remove fat from bone marrow. Clear in xylene and mount.

RESULTS

mitochondria—green
neutral red bodies—red

Schwind Method (1950)

Schwind replaces Janus green B with Pinacyanole, which does not interfere with the staining of vacuolar systems by neutral red, does not fade from the mitochondria as rapidly as Janus green B, and does not interfere with the cell motility.

SOLUTIONS

Neutral red stock solution

neutral red (C.I. 50040)	0.2 g
neutral absolute ethyl alcohol	50.0 ml

Keeps indefinitely.

Pinacyanole stock solution

Pinacyanole [2]	0.05 g
neutral absolute ethyl alcohol	50.0 ml

Keeps for 6 months.

Working solution

neutral red stock	30 drops
Pinacyanole stock	9 drops
neutral absolute ethyl alcohol	5.0 ml

Mix and make films.

RESULTS

mitochondria—deep blue
chromatin—purplish blue

---

[2] Eastman Kodak Company.

Jackson Fluorescent Method (1961)

SOLUTIONS

Acridine orange stock solution

acridine orange	0.1 g
95% ethyl alcohol	50.0 ml

Allow excess dye to settle on bottom of bottle.

Working solution

acridine orange stock	5 drops
95% ethyl alcohol	5.0 ml

Do not draw up any of excess dye from stock solution, use from top of solution.

PROCEDURE

Flood clean slides and drain off excess stain. Dry in vertical position. Store in dustfree box. Apply cover glass and cells as above. Examine with ultraviolet light (McClung 1939; Pienaar 1962; Schwind 1950; Scott 1928).

RESULTS

neutrophils—bright green nuclei, orange granules
eosinophils—same as neutrophils, but with larger granules
basophils—yellow granules
lymphocytes—green nucleus, usually one large red granule, occasionally more than one granule
monocytes—typically folded bright green nucleus, dense green cytoplasm
mature red cells—no fluorescence
reticulocytes—bright orange reticulum
platelets—pale green with pale orange granules
LE bodies—pale green

CHAPTER 23

Staining Microorganisms

In this chapter, to simplify the specificity of staining methods, the parasitic microorganisms will be broken down into the following groups: bacteria, spirochetes, fungi, rickettsia, and viruses (Burrows 1954).

BACTERIA STAINING

Bacteria are customarily studied by direct microscopic observation and differentiated by shape, grouping of cells, presence or absence of certain structures, and the reaction of their cells to differential stains. Bacteria may be stained with aniline dyes—in a single dye, in mixed dyes, in polychromed dyes, or by differential methods. One of the most universally used stains was developed by the histologist, Gram, while he was trying to differentiate the bacteria in tissue. His method separates bacteria into two groups: (1) those that retain crystal violet and are said to be Gram positive; (2) those that decolorize to be stained by a counterstain and are said to be Gram negative.

Some bacteria of high lipid content cannot be stained by the usual methods but require heat or long exposure to the stain. They are also difficult to decolorize. Because they resist acid alcohol, they have been given the name *acid fast*. The spiral forms will stain only faintly if at all and must be colored by silver methods.

Many bacteria form a capsule from the outer layer of the cell membrane, and the capsule appears like a halo around the organism, or over a chain of cells. This capsule will not stain in the customary stains; Hiss stain (page 390), however, is simple and usually effective in this situation. Some bacteria are able to form spores that can be extremely resistant to injurious conditions (heat, chemicals). Boiling will destroy some of these spores, but many are more resistant. Some bacteria have flagella—filamentous appendages for locomotion.

Bacteria can be classified according to shape: coccus, spherical; bacillus, rod shaped; spiral, a curved rod.

Gram Staining

When a Gram-staining procedure has been applied, a Gram-positive cell or organism retains a particular primary dye. This process includes mordanting with iodine to form (with the dye) a precipitate that is insoluble in water and is neither too soluble nor insoluble in alcohol, the differentiator. According to Bartholomew and Mittwer (1950, 1951), the mechanism of stainability can be explained by differences in the permeability of the cell membrane. The bacteria stain by linkage between acidic groups of the bacteria and alkaline groups of dye. Iodine forms a complex with the dye, and this complex is dissociated by alcohol. If alcohol passes easily through the membrane, decolorization is rapid and the reaction is Gram negative. If the membrane is hardly or not at all penetrated by alcohol, the reaction is Gram positive.

Pearse (1960) suggests that the initial acidic protein–crystal violet complex is broken by the iodine, which then combines with the basic groups at the ends of the triphenylmethane molecule to form a relatively insoluble crystal violet–iodine precipitate. This precipitate is not easily removed if the following conditions exist: (1) if more of the crystal violet–iodine complex has formed where originally there had been more crystal violet in combination with protein; (2) if there are physical barriers, consisting of lipid or lipoprotein membranes, that resist extraction of the crystal violet–iodine.

For more details about Gram staining, see Bartholomew and Finkelstein (1959) and Mittwer et al. (1950).

Gram-Weigert Method (Krajian and Gradwohl 1952)

FIXATION

10% formalin or Zenker formol.

SOLUTIONS

Eosin

eosin Y (C.I. 45380)	1.0 g
distilled water	100.0 ml

Sterling gentian violet

crystal violet (C.I. 42555) (gentian violet)	5.0 g
95% ethyl alcohol	10.0 ml
aniline oil	2.0 ml
distilled water	88.0 ml

Mix aniline oil with water and filter. Add the crystal violet dissolved in alcohol. Keeps for several weeks to months.

Gram iodine solution, see page 532.

PROCEDURE

1. Deparaffinize and hydrate slides to water. Remove $HgCl_2$.
2. Stain in eosin: 5 minutes.
3. Rinse in water.
4. Stain in Sterling gentian violet solution: 3 minutes for frozen sections; 10 minutes for paraffin sections.
5. Wash off with Gram iodine and then flood with more of same solution: 3 minutes.
6. Blot with filter paper.
7. Flood with equal parts of aniline oil and xylene; reflood until color ceases to rinse out of sections.
8. Clear in xylene and mount.

RESULTS

Gram-positive bacteria, and fungi—violet
Gram-negative organisms—not usually stained
fibrin—blue-black

Brown and Brenn Method (1931)

FIXATION

10% formalin preferred.

SOLUTIONS

Mayer hematoxylin, see page 150.

Crystal violet

crystal violet (C.I. 42555) (gentian violet) ...	1.0	g
distilled water	100.0	ml

Sodium bicarbonate

sodium bicarbonate	5.0	g
distilled water	100.0	ml

Gram iodine, see page 532.

Basic fuchsin

basic fuchsin (C.I. 42500)	0.25	g
distilled water	100.0	ml

Dilute 0.1 ml in 100 ml distilled water for use.

Acetone–picric acid

picric acid	0.1	g
acetone	100.0	ml

PROCEDURE

1. Deparaffinize and hydrate slides to water.
2. Stain in Mayer hematoxylin: 2–3 minutes.
3. Wash in running water: 5 minutes.
4. Mix 26 drops of crystal violet and 5 drops of sodium bicarbonate solution. Flood slide, agitate occasionally: 2 minutes.
5. Rinse in water.
6. Treat with Gram iodine: 1 minute.
7. Rinse in water and blot with moist filter paper.
8. Decolorize with 1 part ether and 3 parts acetone, flooding slide

until no more blue color comes off. Blot with filter paper moistened with ether-acetone.
9. Stain in basic-fuchsin working solution: 5 minutes.
10. Rinse quickly in water, blot with moist filter paper but do not dry.
11. Dip briefly in acetone and immediately differentiate in acetone–picric acid until sections turn yellowish pink. Overdifferentiation may take out Gram-negative stain.
12. Rinse quickly in acetone, transfer to acetone-xylene (equal parts); then to xylene, and mount.

RESULTS

Gram-positive organisms—deep violet or black
Gram-negative organisms—bright red
cell nuclei—reddish brown
cytoplasm—yellowish
red blood corpuscles—yellow or reddish yellow
cartilage—pink
basophilic granules—red

COMMENTS

It is easier to differentiate step 8 by using a coplin jar of ether-acetone (1:3) and dunking the slide.
In step 4 Taylor (1966) uses Hucker solution:

crystal violet (C.I. 42555)	2.0 g
95% ethanol	20.0 ml
ammonium oxalate	800.0 mg
distilled water	60.0 ml

Dissolve dye in alcohol and oxalate in water. Mix. Keeps 2–3 years.
In step 9 Taylor uses an alcoholic basic fuchsin:

basic fuchsin (C.I. 42510)	0.1 g
95% ethanol	100.0 ml

Dilute 5 ml in 60 ml distilled water for use.

Lillie's Quick Method (1928) (Bartholomew Modification 1962)

FIXATION

Any good general fixative.

SOLUTIONS

Crystal violet solution

crystal violet (C.I. 42555)	2.0 g
95% ethyl alcohol	20.0 ml
ammonium oxalate, 1% (1 g/100 ml water) . . .	80.0 ml

Filter. Keeps well.

Lugol solution, see page 532.

Safranin

safranin O (C.I. 50240)	0.5 g
distilled water	100.0 ml

PROCEDURE
1. Deparaffinize and hydrate slides to water.
2. Stain in crystal violet solution: 30 seconds.
3. Wash in running water: 5 seconds, dipping up and down.
4. Rinse off water with Lugol solution and treat with Lugol's: 1 minute.
5. Wash in running water: 5 seconds, dipping up and down.
6. Decolorize in *n*-propyl alcohol, 3 changes: 1 minute each. Agitate slide.
7. Wash in running water: 5 seconds, dipping up and down.
8. Counterstain in safranin: 1 minute.
9. Wash in running water: 5 seconds, dipping up and down.
10. Dehydrate and differentiate in acetone. Some of red differentiates out of cells, leaving nuclei red and cytoplasm pink. (Blot and air-dry smears.)
11. Clear and mount.

RESULTS
Gram-positive bacteria—blue-black
Gram-negative bacteria—red
cell nuclei—red
cytoplasm, fibrin, collagen—pink

COMMENTS
For tubercle and lepra bacilli, it may be necessary to stain for a longer period or heat for 90 seconds on a 50°–52°C hot plate.

Acid-Fast Staining

In acid-fast staining, the presence of phenol in the stain solution appears to be essential; it must, in some way, influence the dye or substance that imparts the acid-fast character. Lartique and Fite (1962) propose that phenol decreases the solubility of fuchsin in water and increases its solubility in the lipids of the bacillus. The acid-fastness can be enhanced by artificially coating the bacilli with oil. Actually, no specific chemical role can be assigned to phenol; it does not combine with the fuchsin, nor does it cause capsular disruption or protein denaturation.

Staining of lepra bacilli is improved if, when removing paraffin from sections, 15% of mineral oil is added to the xylene (5 minutes). Wash 30 seconds in detergent solution (Haemosol), wash in water, and proceed to stain.

Ziehl-Neelsen Method (Putt Modification 1951)

FIXATION

Any general fixative, but 10% formalin is best.

SOLUTIONS

New fuchsin

new fuchsin (C.I. 52520) (magenta III)	1.0 g
phenol (carbolic acid)	5.0 g
ethyl or methyl alcohol (absolute)	10.0 ml
distilled water to make	100.0 ml

Acetic alcohol

glacial acetic acid	5.0 ml
absolute ethyl alcohol	95.0 ml

Counterstain

methylene blue chloride (C.I. 52015)	0.5 g
absolute ethyl alcohol	100.0 ml

PROCEDURE

1. Deparaffinize and hydrate slides to water; remove $HgCl_2$.
2. Stain in new fuchsin, room temperature: 3 minutes.

3. Treat with lithium carbonate, saturated aqueous: 1 minute. Agitate gently, discard solution if it becomes blue.
4. Differentiate until pale pink in acetic alcohol.
5. Rinse in absolute alcohol, 2 changes: 1–2 minutes each.
6. Counterstain.
7. Decolorize in absolute alcohol, 2 changes: $\frac{1}{2}$ minute each.
8. Clear and mount.

RESULTS

acid-fast bacteria—red
red blood corpuscles—pink
mast cell granules—deep blue
other bacteria—blue
nuclei—blue

COMMENTS

Willigan et al. (1961) combine the new fuchsin method with H and E.

Kinyoun Carbol-Fuchsin Method (Marti and Johnson Modification, 1951)

FIXATION

Any general fixative, but 10% formalin is preferred.

SOLUTIONS

Carbol-fuchsin

basic fuchsin (C.I. 42500)	4.0 g
phenol (carbolic acid) crystals	8.0 g
95% ethyl alcohol	20.0 ml
distilled water	100.0 ml

Add 1 drop Turgitol 7 to every 30 ml of above.

Acid alcohol

nitric acid, concentrated	0.5 ml
95% ethyl alcohol	95.0 ml

Malachite green

malachite green oxalate (C.I. 42000)	1.0 g
distilled water	100.0 ml

PROCEDURE

1. Deparaffinize and hydrate slides to water.
2. Stain in carbol-fuchsin: 30 minutes, room temperature.
3. Wash in running water: 5 minutes.
4. Decolorize in acid alcohol: 3 minutes. Mixture should not be yellow.
5. Wash in running water: 5 minutes.
6. Rinse in 95% alcohol: 1 minute.
7. Wash in running water: 2 minutes.
8. Stain in malachite green: 30 seconds.
9. Wash in running water: few minutes.
10. Rinse in 95% alcohol: 10 dips.
11. Dehydrate in absolute alcohol, clear, and mount.

RESULTS

acid-fast bacteria—red
tissue elements—shades of green

COMMENTS

Instead of malachite green, methylene blue can be used (Lillie, 1944):
1% methylene blue in 0.5% aqueous acetic acid: 3 minutes.

Pottz et al. (1964) reduce procedure time by combining the decolorizer and counterstain. Also they add a wetting agent to the carbol-fuchsin stock solution.

Fite-Formaldehyde Method (Wade Modification, 1957)

FIXATION

Zenker preferred; removal of $HgCl_2$ not necessary, since it disappears during processing.

SOLUTIONS

Phenol new fuchsin

new fuchsin (C.I. 42520) (magenta III)	0.5 g
phenol (carbolic acid)	5.0 g
ethyl or methyl alcohol	10.0 ml
distilled water to make	100.0 ml

Van Gieson, modified

acid fuchsin (C.I. 42685)	0.01 g
picric acid	0.1 g
distilled water	100.0 ml

PROCEDURE

1. Deparaffinize in turpentine–paraffin oil (2 : 1) 2 changes: 5 minutes total.
2. Drain, wipe off excess fluid, blot to opacity, place in water.
3. Stain overnight in phenol-fuchsin, room temperature.
4. Rinse off excess stain in tap water.
5. Treat with formalin (full strength): 5 minutes. (The tissue will turn blue.)
6. Wash in running water: 3–5 minutes.
7. Treat with sulfuric acid, 5% (5 ml/95 ml water): 5 minutes.
8. Wash in running water: 5–10 minutes.
9. Treat with potassium permanganate, 1% (1 g/100 ml water): 3 minutes.
10. Wash in running water: 3 minutes.
11. Treat with oxalic acid, 2–5% (2–5 g/100 ml water) individually with agitation: not more than 30 seconds. Slides can remain in water while others are being treated.
12. Stain with modified Van Gieson: 3 minutes. *Do not wash.*
13. Rinse for few seconds in 95% alcohol.
14. Dehydrate, clear, and mount.

RESULTS

acid-fast bacteria—deep blue or blue-black
connective tissue—red
other tissue elements—yellowish

COMMENTS

Beamer and Firminger (1955) emphasize care in the use of formalin. Old formalin yields poor results; a redistilled form produces the most brilliant staining. The reagent grade in 16-oz. brown bottles, if kept tightly closed, gives a good stain.

Tilden and Tanaka (1945) outline a method for frozen sections (essentially the same as the Fite-formaldehyde method), after the sections have been mounted with a celloidin protective coat. They also emphasize the need for good formalin; if the sections fail to turn blue, try a different batch.

Fluorescent Method (Richards, et al. 1941; Richards 1941; Bogen 1941)

FIXATION

10% formalin for sections; smears by heat.

SOLUTIONS

Auramin stain

auramin O (C.I. 41000)	0.3 g
distilled water	97.0 ml
melted phenol (carbolic acid)	3.0 ml

Shake to dissolve dye or use gentle heat. Solution becomes cloudy on cooling, but is satisfactory. Shake before using.

Decolorizing solution

hydrochloric acid, concentrated	0.5 ml
70% ethyl alcohol	100.0 ml
sodium chloride	0.5 g

PROCEDURE

1. Deparaffinize and hydrate sections to water.
2. Stain in auramin, room temperature: 2–3 minutes.
3. Wash in running water: 2–3 minutes.
4. Decolorize: 1 minute.
5. Transfer to fresh decolorizer: 2–5 minutes.
6. Wash in running water: 2–3 minutes.
7. Dry smear and examine. Mount sections in fluorescent mountant (page 146) or Harleco Fluorescent Mountant for examining.

RESULT

bacilli—golden yellow

COMMENTS

Bogen counterstains with Loeffler alkaline methylene blue solution before mounting:

methylene blue (C.I. 52105)	3.0 g
absolute ethyl alcohol	30.0 ml
potassium hydroxide, 0.01% aqueous	100.0 ml

See also Braunstein and Adriano (1960); Moody et al. (1958); Yamaguchi and Braunstein (1965).

Carter and Leise (1958) use a single fluorescent antiglobulin for various bacteria.

Capsule Staining

Hiss Method (Burrows 1954)

FIXATION

Any good general fixative; 10% formalin is satisfactory.

SOLUTIONS

Stain solution

Either of the following may be used.

A. basic fuchsin (C.I. 42500)	0.15–0.3 g
distilled water	100.0 ml
B. crystal violet (C.I. 42555)	0.05–0.1 g
distilled water	100.0 ml

Copper sulfate solution

copper sulfate crystals	20.0 g
distilled water	100.0 ml

PROCEDURE

1. Deparaffinize and hydrate slides to water.
2. Flood with either staining solution and heat gently until the stain steams.
3. Wash off the stain with copper sulfate.
4. Blot, but do not dry.
5. Dehydrate in 99% isopropyl alchohol, 2 changes: 1 minute each.
6. Clear and mount.

RESULTS

capsules—light pink (basic fuchsin) or blue (crystal violet)
bacterial cells—dark purple surrounded by the capsule color

SPIROCHETE STAINING

These are spirally shaped organisms, which multiply by transverse fission. They cause relapsing fever, syphilis, and yaws.

Dieterle Method (Beamer and Firminger Modification 1955)

FIXATION

10% formalin.

SOLUTIONS

Dilute gum mastic

gum mastic, saturated in absolute alcohol	30 drops
95% ethyl alcohol	40.0 ml

Developing solution

distilled water	20.0 ml
hydroquinone	0.5 g
sodium sulfite	0.06 g

While stirring add:

formalin	4.0 ml
glycerine	5.0 ml

When thoroughly mixed add, drop by drop, with constant stirring:

gum mastic saturated in absolute alcohol	4.0 ml
absolute alcohol	4.0 ml
pyridine	2.0 ml

PROCEDURE

1. Deparaffinize and hydrate slides to water.
2. Remove any formalin pigment in 2% ammonium hydroxide (2 ml/98 ml water): few minutes.
3. Transfer to 80% alcohol: 2–3 minutes.
4. Wash in distilled water: 5 minutes.
5. Treat with uranium nitrate 2–3% (2–3 g/100 ml water), previously warmed to 60°C: 10 minutes.
6. Wash in distilled water: 1–2 minutes.

7. Transfer to 95% alcohol: 1–2 minutes.
8. Treat with dilute gum mastic: 5 minutes.
9. Wash in distilled water, 3–4 changes, until rinse is clear.
10. Impregnate with silver nitrate, 2% (2 g/100 ml water), previously warmed to 60°C: 30–40 minutes.
11. Warm developing solution to 60°C. Dip slide up and down in solution until sections turn light tan or pale brown.
12. Rinse in 95% alcohol, then in distilled water.
13. Treat with silver nitrate, 2% (2 g/100 ml water): 1–2 minutes.
14. Wash in distilled water: 1–2 minutes.
15. Dehydrate, clear, and mount.

RESULTS

spirochetes—black
background—yellow

COMMENTS

Use thin sections, 4–5μ.

The uranium nitrate prevents the impregnation of nerve fibers and reticulum.

Warthin–Starry Silver Method (Kerr 1938; Faulkner and Lillie 1945*a*; Bridges and Luna 1957)

Caution: All glassware must be cleaned with potassium dichromate-sulfuric acid. Avoid contamination; coat forceps with paraffin. Solutions must be fresh (no more than 1 week old) and made from triple-distilled water. Carry a known positive control slide with test slide through the process or, preferably, a control section on the same slide.

FIXATION

10% formalin.

SOLUTIONS

Acidified water

triple-distilled water .... 1000.0 ml
citric acid .... 10.0 g

The *p*H can range from 3.8 to 4.4.

2% silver nitrate

silver nitrate	2.0 g
acidified water	100.0 ml

1% silver nitrate

Dilute a portion of the 2% silver nitrate solution with equal volume of acidified water.

0.15% hydroquinone

hydroquinone	0.15 g
acidified water	100.0 ml

5% gelatin

sheet or granulated gelatin of high degree of purity	10.0 g
acidified water	200.0 ml

Developer

2% silver nitrate	1.5 ml
5% gelatin	3.75 ml
0.15% hydroquinone	2.0 ml

Warm solutions to 55°–60°C and mix in order given, with stirring. Use immediately.

PROCEDURE

1. Deparaffinize and hydrate slides to acidified water.
2. Impregnate in 1% silver nitrate, 55°–60°C: 30 minutes.
3. Place slides on glass rods, pour on warm developer (55°–60°C). When sections become golden brown or yellow, and the developer, brownish-black (3–5 minutes), pour off. The known positive can be checked under microscope for black organisms.
4. Rinse with warm (55°–60°C) tap water, then distilled water.
5. Dehydrate, clear, and mount.

RESULTS

spirochetes—black
background—yellow to light brown

melanin and hematogenous pigments—may darken

Underdevelopment will result in pale background, very slender and pale spirochetes.

Overdevelopment will result in dense background, heavily impregnated spirochetes, obstructed detail, sometimes precipitate.

COMMENTS

Faulkner and Lillie (1945*a*) use water buffered to *p*H 3.6–3.8 with Walpole *M*/5 sodium acetate–*M*/5 acetic acid buffer (page 540).

Levaditi Method for Block Staining (Mallory 1944)

FIXATION

10% formalin.

SOLUTIONS

Silver nitrate

silver nitrate	1.5–3.0 g
distilled water	100.0 ml

Reducing solution

pyrogallic acid	4.0 g
formalin	5.0 ml
distilled water	100.0 ml

PROCEDURE

1. Rinse blocks of tissue in tap water.
2. Transfer to 95% ethyl alcohol: 24 hours.
3. Place in distilled water until tissue sinks.
4. Impregnate with silver nitrate, 37°C, in dark: 3–5 days.
5. Wash in distilled water.
6. Reduce at room temperature in dark: 24–72 hours.
7. Wash in distilled water.
8. Dehydrate, clear in cedarwood oil, and infiltrate with paraffin.
9. Embed, section at 5μ, mount on slides, and dry.
10. Remove paraffin with xylene, 2 changes, and mount.

RESULTS

    spirochetes—black
background—brownish yellow

For a fluorescent method, see Kellogg and Deacon (1964).

FUNGI STAINING

Open or draining lesions are difficult to examine for fungi because of
heavy bacterial contamination, but dermatophytes are easily demonstrated. Scrapings from horny layers or nail plate or hair can be mounted
in 10%–20% hot sodium hydroxide. This dissolves or makes transparent the
tissue elements and then the preparation can be examined as a wet mount.
Fungi in tissue sections are readily stained.

Hotchkiss-McManus Method (McManus 1948)

FIXATION

    10% formalin or any general fixative.

SOLUTIONS

Periodic acid

periodic acid	1.0 g
distilled water	100.0 ml

Schiff reagent, see page 327.

Differentiator

potassium metabisulfite, 10% aqueous	5.0 ml
N HCl (page 531)	5.0 ml
distilled water	100.0 ml

Light green

light green SF, yellowish (C.I. 42095)	0.2 g
distilled water	100.0 ml
glacial acetic acid	0.2 ml

PROCEDURE

1. Deparaffinize and hydrate slides to water; remove $HgCl_2$ if present.
2. Oxidize in periodic acid: 5 minutes.
3. Wash in running water: 15 minutes.
4. Treat with Schiff reagent: 10–15 minutes.
5. Differentiate, 2 changes: total 5 minutes.
6. Wash in running water: 10 minutes.
7. Stain in light green: 3–5 minutes. If too dark, rinse in running water.
8. Dehydrate, clear, and mount.

RESULTS

fungi—red. Not specific, however—glycogen, mucin, amyloid, colloid, and others may show rose to purplish red
background—light green

COMMENTS

To remove glycogen, starch, mucin, or RNA, see page 329.

DePalma and Young (1963) use 0.1% basic fuchsin for 2 minutes with agitation instead of Schiff reagent. Differentiate only 1–2 minutes with agitation.

Gridley Method (1953)

FIXATION

Any good general fixative.

SOLUTIONS

Chromic acid

chromic acid	4.0 g
distilled water	100.0 ml

Schiff reagent, see page 327.

Sulfurous rinse

sodium metabisulfite, 10% (10 g/100 ml water)	6.0 ml
N hydrochloric acid	5.0 ml
distilled water	100.0 ml

Aldehyde-fuchsin, see page 190.

Metanil yellow

metanil yellow	0.25 g
distilled water	100.0 ml
glacial acetic acid	2 drops

PROCEDURE

1. Deparaffinize and hydrate slides to water; remove $HgCl_2$.
2. Oxidize in chromic acid: 1 hour.
3. Wash in running water: 5 minutes.
4. Place in Schiff reagent: 15 minutes.
5. Rinse in sulfurous acid, 3 changes: 1.5 minutes each.
6. Wash in running water: 15 minutes.
7. Stain in aldehyde-fuchsin: 15–30 minutes.
8. Rinse off excess stain in 95% alcohol.
9. Rinse in water.
10. Counterstain lightly in metanil yellow: 1 minute.
11. Rinse in water.
12. Dehydrate, clear, and mount.

RESULTS

mycelia—deep blue
conidia—deep rose to purple
background—yellow
elastic tissue, mucin—deep blue

Gomori Methenamine–Silver Nitrate Method (Grocott Adaptation, 1955; Mowry Modification, 1959)

FIXATION

10% formalin or any good general fixative.

SOLUTIONS

Methenamine–silver nitrate stock solution

silver nitrate, 5% aqueous	5.0 ml
methenamine, 3% aqueous	100.0 ml

Working solution

borax, 5% aqueous	2.0 ml
distilled water	25.0 ml
methenamine–silver nitrate stock solution	25.0 ml

Light green stock solution

light green SF, yellowish (C.I. 42095)	0.2–0.5 g
distilled water	100.0 ml
glacial acetic acid	0.2 ml

Working solution

light green stock solution	10.0 ml
distilled water	100.0 ml

PROCEDURE

1. Deparaffinize and hydrate slides to water; remove $HgCl_2$.
2. Oxidize in periodic acid, 0.5% (0.5 g/100 ml water): 10 minutes.
3. Wash in running water: 3 minutes.
4. Oxidize in chromic acid, 5% (5 g/100 ml water): 45 minutes.
5. Wash in running water: 2 minutes.
6. Treat with sodium bisulfite, 2% (2 g/100 ml water): 1 minute to remove chromic acid.
7. Wash in running water: 5 minutes.
8. Rinse in distilled water, 2–3 changes: 5 minutes total.
9. Place in methenamine–silver nitrate, 58°C: 30 minutes. Do not use metal forceps. Sections appear yellowish brown.
10. Wash thoroughly, several changes distilled water.
11. Tone in gold chloride (10 ml stock solution/90 ml water) until sections turn purplish gray, fungi are black.
12. Rinse in distilled water.
13. Fix in sodium thiosulfate, 5% (5 g/100 ml water): 3 minutes.
14. Wash in running water: 5 minutes.
15. Counterstain in light green: 30 seconds.
16. Dehydrate, clear, and mount.

RESULTS

fungi—black
background—light green

COMMENTS

The Mowry modification uses oxidation with both periodic and chromic acid (former methods use only the latter) with the result that the final staining is stronger and more consistent than that of the original Gomori method.

Metachromatic Method (Kelly et al. 1962)

FIXATION

10% formalin.

SOLUTIONS

Sulfation reagent

Slowly add cold concentrated sulfuric acid to an equal volume of iced diethyl ether.

Toluidine blue

toluidine blue O (C.I. 52040)	0.01 g
3% aqueous acetic acid	100.0 ml

PROCEDURE

1. Deparaffinize sections and transfer to absolute ethyl alcohol.
2. Dry sections in air: 5–10 minutes.
3. Sulfate: 5 minutes.
4. Wash in 3% aqueous acetic acid: 1 minute. Change solution after 10 slides.
5. Stain in toluidine blue: 5 minutes.
6. Wash in 3% aqueous acetic acid: 1 minute.
7. Dehydrate in absolute ethyl alcohol: 1 minute.
8. Clear and mount.

RESULTS

fungi—red
background—pale blue or colorless

COMMENTS

Some fungal forms are detected by this method only; sulfation reveals previously undetected mycelia of *Candida*. The 1,2 glycol groups (—CHOH—CHOH—) in fungi apparently are esterified by sulfuric acid to form ester sulfate groups ($-OSO_3$), which react metachromatically with toluidine blue.

Schneider (1963) simplifies a Gram stain for fungi.

Fluorescent Method (Pickett et al. 1960)

FIXATION

10% formalin, Zenker, alcohol, and other general fixatives.

SOLUTIONS

Acridine orange

acridine orange	0.1 g
distilled water	100.0 ml

Weigert hematoxylin, see page 151.

PROCEDURE

1. Deparaffinize and hydrate sections to water.
2. Stain in Weigert hematoxylin: 5 minutes.
3. Wash in running water: 5 minutes.
4. Stain in acridine orange: 2 minutes.
5. Rinse in tap or distilled water: 30 seconds.
6. Dehydrate in 95% alcohol: 1 minute.
7. Dehydrate in absolute alcohol, 2 changes: 2–3 minutes.
8. Clear in xylene, 2 changes: 2–3 minutes.
9. Mount in a nonfluorescing medium.

RESULTS

All fungi fluoresce except *Nocardia* and *Rhizopus;* colors of the fluorescing genera appears as follows:

Coccidioides, Rhinosporidum—red

Aspergillus—green

Actinomyces, Histoplasma—red to yellow

Candida, Blastomyces dermatitids, Monosporium—yellow-green

Blastomyces brasiliensis—yellow

COMMENTS

Old hematoxylin and eosin slides may be decolorized and restained as above.

Weigert hematoxylin staining is necessary as a quenching agent, because some fungi are difficult to see against a background which also fluoresces. With the hematoxylin, the fungi fluoresce brightly against a dark setting.

For a fluorescent method, see Clark and Hench (1962).

For fluorescent antibody techniques, see Procknow et al. (1962) and Batty and Walker (1963).

STAINING OF RICKETTSIAE AND INCLUSION BODIES

Rickettsiae are very small, Gram-negative coccobacillary type microorganisms associated with typhus and spotted fever and related diseases. They may appear as cocci or short bacilli, and may occur singly, in pairs, or in dense masses. Most of them are intracellular. Some species are found only in the cytoplasm; others prefer the nucleus. Rickettsiae stain best in a Giemsa-type stain or by the Ordway-Machiavello method.

Viruses are microorganisms too small to be visible under the microscope, and they are capable of passing through filters. Viruses are responsible for many diseases, including yellow fever, poxes, poliomyelitis, influenza, measles, mumps, shingles, rabies, colds, infectious hepatitis, infectious mononucleosis, trachoma, psittacosis, and foot-and-mouth disease. In tissue sections and smears, viruses are characterized by elementary bodies and inclusion bodies. Elementary bodies are infectious particles, and inclusion bodies are composed of numerous elementary bodies. Both types of bodies vary in size and appearance; some, such as rabies, psittacosis, and trachoma, are located in the cytoplasm of infected cells, and some, such as poliomyelitis, are intranuclear. Special staining methods can demonstrate them effectively.

Modified Pappenheim Stain (Castañeda 1939)

FIXATION

Any general fixative; Regaud recommended.

SOLUTIONS

Stock Jenner stain

Jenner stain	1.0 g
methyl alcohol, absolute	400.0 ml

Stock Giemsa stain

Giemsa stain	1.0 g
glycerine	66.0 ml

Mix and place in oven 2 hours, 60°C. Add:

methyl alcohol, absolute	66.0 ml

Working solution A

distilled water	100.0 ml
glacial acetic acid	1 drop
Jenner stock solution	20.0 ml

Working solution B

distilled water	100.0 ml
glacial acetic acid	1 drop
Giemsa stock solution	5.0 ml

PROCEDURE

1. Deparaffinize and hydrate slides to water.
2. Stain in solution A, 37°C: 15 minutes.
3. Transfer directly to solution B, 37°C: 30–60 minutes.
4. Dehydrate quickly, 2 changes absolute alcohol.
5. Clear and mount.

RESULT

rickettsiae—blue to purplish blue

Castañeda Method (Gradwohl 1956)

FIXATION

Any general fixative; Regaud recommended.

SOLUTIONS

Buffer solution

Solution A

dibasic sodium phosphate ($Na_2HPO_4 \cdot 12H_4O$)	23.86 g
distilled water	1000.0 ml

Solution B

monobasic sodium phosphate, anhydrous	
(NaH$_2$PO$_4$)	11.34 g
distilled water	1000.0 ml

Working solution

solution A	88.0 ml
solution B	12.0 ml
formalin	0.2 ml

Methylene blue solution

Dissolve methylene blue (C.I. 52015)	2.10 g
in 95% alcohol	30.00 ml
Dissolve potassium hydroxide	.01 g
in distilled water	100.00 ml

Mix the two solutions and let stand 24 hours.

Staining solution

Mix buffer working solution with 1.0 ml methylene blue solution.

PROCEDURE
1. Deparaffinize and hydrate slides as far as 50% alcohol.
2. Stain in methylene blue: 2–3 minutes.
3. Wash in running water: 30 seconds.
4. Counterstain in 1% aqueous safranin: 1–2 minutes.
5. Dip briefly in 95% alcohol.
6. Dehydrate in 2 changes absolute alcohol, clear, and mount.

RESULT
rickettsiae—light blue

COMMENTS
Burrows (1954) recommends this as one of the best methods for rickettsiae.
For a fluorescent method, see Anderson and Grieff (1964).

Ordway-Machiavello Method (Gradwohl 1956)

FIXATION
Regaud recommended.

SOLUTION

Poirrier blue (C.C.), 1% aqueous	10.0 ml
eosin bluish (C.I. 45400) 0.45% aqueous	15.0 ml

Mix just before use. Add slowly with constant shaking:

distilled water	25.0 ml

Use within 24 hours.

PROCEDURE
1. Deparaffinize and hydrate slides to water.
2. Stain: 6–8 minutes.
3. Decolorize in 95% ethyl alcohol until slides appear pale bluish pink.
4. Dehydrate in absolute alcohol: 1 minute.
5. Clear and mount.

RESULTS
rickettsiae and inclusion bodies—bright red
nuclei and cytoplasm—sky blue

Giemsa Method (American Public Health Association 1956)

FIXATION
Any general fixative; Regaud recommended.

SOLUTIONS

Giemsa stock solution, see page 402.

Buffer solutions

Solution A

dibasic sodium phosphate, anhydrous	
(Na_2HPO_4)	9.5 g
distilled water	1000.0 ml

Solution B

monobasic sodium phosphate	
($NaH_2PO_4 \cdot H_2O$)	9.2 g
distilled water	1000.0 ml

Working solution, *p*H 7.2

solution A	72.0 ml
solution B	28.0 ml
distilled water	900.0 ml

Giemsa working solution

Dilute 1 drop Giemsa stock solution with 5 ml of buffer working solution.

PROCEDURE

1. Deparaffinize and hydrate slides to water. (Smears can be carried directly to water: wash well in running water: 5–10 minutes.)
2. Leave slides in Giemsa working solution overnight, 37°C.
3. Rinse thoroughly in distilled water. Dry between blotters.
4. Dip rapidly in absolute alcohol. If overstained, use 95% alcohol to decolorize them, dip in absolute alcohol.
5. Clear and mount. (Smears, after treatment with absolute alcohol, can be washed in distilled water: 1–2 seconds, and blotted dry. Examine with oil or mount with a cover glass.)

RESULT

rickettsiae and inclusion bodies (psittacosis)—blue to purplish blue

Modified Gomori Method for Trachoma

SOLUTION

Dilute 1 drop of stock Giemsa with 2 ml of buffer working solution above.

PROCEDURE

1. Hydrate slides to water (these will be smears).
2. Stain in working solution, 37°C: 1 hour.
3. Rinse rapidly, 2 changes, 95% ethyl alcohol.
4. Dehydrate in absolute alcohol, clear, and mount.

RESULT

inclusion bodies—blue to purplish blue

Schleifstein Method for Negri Bodies (1937)

FIXATION

Zenker preferred.

SOLUTION

basic fuchsin (C.I. 42500)	1.8 g
methylene blue (C.I. 52015)	1.0 g
glycerine	100.0 ml
methyl alcohol	100.0 ml

For use add about 10 drops to 15–20 ml of dilute potassium hydroxide (1 g/40,000 ml water). Alkaline tap water may be used. Keeps indefinitely.

PROCEDURE

1. Deparaffinize and hydrate slides to water; remove $HgCl_2$.
2. Rinse in distilled water and place slides on warm electric hot plate.
3. Flood amply with stain and steam for 5 minutes. Do not allow stain to boil.
4. Cool and rinse in tap water.
5. Decolorize and differentiate each slide by agitating in 90% ethyl alcohol until the sections assume a pale violet color. This is important.
6. Dehydrate, clear, and mount.

RESULTS

Negri bodies—deep magenta red
granular inclusions—dark blue
nucleoli—bluish black
cytoplasm—blue-violet
erythrocytes—copper

COMMENTS

Schleifstein outlines a rapid method for fixing and embedding so the entire process can be handled in 8 hours, including fixing, embedding, and staining.

Massignani and Malferrari Method for Negri Bodies (1961)

FIXATION

10% formalin or saturated aqueous mercuric chloride–absolute alcohol (1 : 2).

EMBEDDING and SECTIONING

Paraffin method, sections 4μ.

SOLUTIONS

Harris hematoxylin, see page 150.

Dilute hydrochloric acid

hydrochloric acid, concentrated	1.0 ml
distilled water	200.0 ml

Dilute lithium carbonate

lithium carbonate, saturated aqueous	1.0 ml
distilled water	200.0 ml

Phosphotungstic acid–eosin stain

Grind together 1 g eosin Y (C.I. 45380) and 0.7 g phosphotungstic acid. Mix thoroughly into 10.0 ml of distilled water and then bring volume up to 200.0 ml with distilled water. Centrifuge at 1,500 rpm: 40 minutes. Pour off supernatant solution but do not throw it away. Dissolve the precipitate in 50 ml of absolute alcohol. When dissolved, add to the supernatant solution. The solution is ready to use. If the eosin and phosphotungstic acid are not ground together and then dissolved, it requires 24 hours for the dye-mordant combination to form.

PROCEDURE

1. Deparaffinize and hydrate sections to water; remove $HgCl_2$.
2. Stain in hematoxylin: 2 minutes.
3. Wash in running water: 5 minutes.
4. Dip 8 times in dilute hydrochloric acid.
5. Wash in running water: 5 minutes.
6. Blue in dilute lithium carbonate: 1 minute.
7. Wash in running water: 5 minutes.
8. Dehydrate to absolute alcohol.
9. Stain in phosphotungstic acid–eosin: 8 minutes.
10. Rinse in distilled water.
11. Dehydrate by quick dips in 50%, 70%, 80%, and 90% alcohol; then follow with 95% alcohol: 1 second.
12. Complete dehydration in absolute alcohol, 2–3 changes: 4 minutes each.
13. Clear and mount.

RESULT

Negri bodies—deep red

COMMENTS

Massignani and Refinetti (1958) adapted the Papanicolaou stain (page 456) for Negri bodies. Because further study led to the discovery that eosin combined with phosphotungstic acid is responsible for Negri body staining, the Massignani and Malferrari stain above was developed specifically for these bodies.

ANTIGENS AND ANTIBODIES

The subject of immunity is complex and has filled several large textbooks. It is mentioned briefly here to bring to mind certain principles used in this field, and to highlight one of the most important techniques developed for locating antigents and antibodies in tissues.

If foreign materials, living or nonliving—bacteria and viruses, for instance—invade the body, certain substances are formed in body fluids to combat these foreign materials and to "neutralize" them. The defense substances are proteins called *antibodies* that have the power to combine specifically with the invading foreign materials (*antigens*), which induced the formation of the antibodies. Eventual immunity to an infection can be induced by entrance of the antigen, either naturally or artificially (by "shots," for instance). For artificial immunity, serum, called antiserum (immune serum), from an artificially immunized animal can be injected into a nonimmune animal to induce immunity. Antiserum can be used in this way to combat an infection already present (mumps, measles, anthrax, tetanus), or it can be used to prevent infection (measles, poliomyelitis, tetanus, diphtheria).

If the fluid portion of blood plasma is allowed to clot (fibrinogen precipitates out), a relatively stable serum remains. This is made up of two protein fractions: albumin and globulin. The globulins consist of two alpha globulins, one beta globulin, and one gamma globulin. It is the gamma globulin that is associated with immunity. Antibodies have been classically identified with gamma globulin (to a lesser extent with the other globulins) and are considered modified serum globulin. It has become a common practice to use for immune serum the gamma-globulin fraction, which can be isolated from the other serum proteins and used in a concentrated form. The difference, therefore, between immune and normal serum globulin lies in the ability of the former to combine with an antigen.

If antigens can be labeled, sources of antibody production in tissues can be seen microscopically. Coons and his associates have coupled antibodies with fluorescein isocyanate and isothiocyanate to form fluorescent car-bamide-proteins. The *conjugates* thus formed from dye and antibodies retain the specific reaction of the antibody for the antigen that causes its formation. Labeled antibodies are poured over the tissues or cells and will leave in the tissues a labeled protein that can be observed under a fluorescent microscope.

Some of the solutions, antisera (immune serum), and conjugates have to be prepared in the laboratory; some can be purchased,[1] but it is recommended that all conjugates be purified by absorption against tissue powders. For the reasons and means of doing this, see Coons et al. (1955), Coons (1958), and Mellors (1959). The method is complicated and should not be undertaken without a thorough study of the source material.

With a few exceptions (some polysaccharides), chemical fixatives must be avoided in order to retain the specific activity of the antigen. Smears, touch preparations, tissue cultures, and cell suspensions can be used. If sections are preferred, freeze-dried or unfixed frozen sections cut in a cryo-stat (page 423) are mandatory. Tissues may be quick-frozen in petroleum ether previously chilled to $-65°C$ with a dry ice-alcohol mixture in a Dewar flask. When the tissues are completely frozen, blot and store in tightly stoppered test tube at $-20°C$ to $-70°C$ until used. Long storage is to be avoided because the tissues dehydrate (Tobie 1958). Coons et al. (1955) place small pieces of tissue against the wall of a small test tube and plunge it into alcohol cooled to $-70°C$ with solid carbon dioxide. Store at $-20°C$ until used. Saint-Marie (1962) details a paraffin-embed-ding method.

After the tissues have been sectioned by either method, sections must be fixed before the antigen-antibody solution is applied. Fixation depends on the antigen.

Coons Fluorescent Antibody Technique

FIXATION

Proteins: 95% ethyl alcohol, sometimes absolute methyl alcohol. Poly-saccharides: can be fixed in the block by picric acid–alcohol–formalin

---

[1] Arnel Products Co., N.Y. (Sylvania Chemicals); Baltimore Biological Laboratory, Baltimore, Maryland.

(Rossman fluid, page 23) and paraffin-embedded. Lipids: 10% formalin. Viruses: acetone.

DIRECT METHOD

The direct method consists of staining directly with fluorescein-labeled antibody against a specific antigen. Slides are dried after fixation, rinsed in buffered saline (0.8% sodium chloride with 0.01 M phosphate, pH 7.0) and the excess saline wiped off except over the sections. Labeled antibody is pipetted over the section; the slides, covered with a petri dish with moist cotton or filter paper attached to its undersurface, are allowed to stand at room temperature: 20–30 minutes. The slide is wiped dry except for the section and mounted in buffered glycerol (9 parts anhydrous glycerol, 1 part buffered saline), then covered with a cover glass. Examination is made under the fluorescent microscope.

INDIRECT METHOD

A tissue section containing antigen is covered with unlabeled specific antiserum to allow the antibody molecules to react with it and be fixed in situ (humid environment): 20 minutes. The slide is rinsed off in buffered saline: 10 minutes, then wiped dry except for the section, and fluorescein-labeled antiglobulin serum (prepared against the species that furnished the specific serum) added: 20 minutes. The slide is washed in buffered saline and mounted in buffered glycerol. Cells that reacted with the antiserum will be covered with antibody (globulin) and will have reacted with the labeled antiglobulin. This has been termed layering because the bottom layer is antibody, the middle layer the antigen, and the labeled antibody lies on top.

COMMENTS

Coons (1958) recommends control slides and uses a blocking technique. Unlabeled specific antiserum is added to a slide and allowed to react for 20 minutes. Then, if the section is washed with buffer solution and treated with a drop of labeled antibody, only a faint fluorescence, if any, should be exhibited.

Silverstein (1957) described a means of applying contrasting fluorescent labels for two antibodies, using rhodamine B with fluorescein.

The Coons methods have been applied for the detection of bacteria, viruses, rickettsiae, epidemic typhus, mumps, influenza, canine hepatitis, chicken pox, canine distemper, measles, psittacosis, and poliomyelitis.

This description is intended as only a brief résumé of Coons techniques in order to acquaint the technician with their potentialities. The following references contain extensive and pertinent discussions of the procedures: Coons (1956 and 1958), Goldman (1968), Mellors (1959) and Nairn (1962).

A tremendous bibliography is developing in this field; the following references will lead to many others: Buckley et al. (1955); Cohen et al. (1955); Coons et al. (1942, 1950, 1951, 1955); Coons and Kaplan (1950); Kaplan et al. (1950); Lacey and Davis (1959); Leduc et al. (1955); Marshall et al. (1958, 1961); Mellors et al. (1955); Noyes (1955); Noyes and Watson (1955); Tobie (1958); Watson (1952); White et al. (1955); Zwaan and Van Dam (1961).

Peroxidase-labelled Antibody Technique (Nakane and Pierce 1967)

An improvement over the fluorescent antibody technique has been developed by conjugating enzymes instead of the fluorescent compound to antibodies. An antibody (to a particular hormone) in rabbit serum attaches to that hormone when applied to sections on microscope slides. Then an anti-rabbit gamma globulin, to which a peroxidase is coupled, is applied to the sections and becomes attached to the antigen-antibody complex. By allowing the peroxidase to oxidate a substrate (3,3'-diaminobenzidine), an insoluble brown precipitate is left at the site of the reaction. The enzyme reaction products are visible both for light and electron microscopy.

This method overcomes some of the problems of Coons method. Darkfield ultraviolet is not required; the preparations are permanent; the peroxidase-antibody conjugates are stable and can be stored indefinitely in a frozen state for several months at 4°C.

PART **III**

HISTOCHEMISTRY AND MISCELLANEOUS SPECIAL PROCEDURES

CHAPTER 24

Histochemistry

By definition the field of histochemistry is concerned with the localization and identification of a chemical substance in a tissue. In a broad sense this might include staining, combining chemical and physical reactions, and using chemical methods to demonstrate basic and acidic properties of tissue. But, strictly speaking, histochemistry applies only to chemical methods that immobilize a chemical or enzyme at the site it occupies in living tissue. These methods can apply to inorganic substances, such as calcium, iron, barium, copper, zinc, lead, mercury, and others, as well as organic substances, such as saccharides, lipids, proteins, amino acids, nucleic acid, enterochromaffin substance, and some pigments. Some substances are soluble and react directly; others are insoluble and must be converted into soluble substances before a reaction takes place. Occult or masked materials are part of a complex organic molecule, which has to be destroyed by an unmasking agent before the chemical can react. Some chemicals may be fixed in place; others that are soluble or diffusible have to be frozen quickly and prepared by the freeze-drying method, without a liquid phase. Sometimes it is advisable to make control slides, thereby preventing confusion between a genuine reaction and a nonspecific one that gives a similar effect.

It is difficult and somewhat impractical to make a sharp distinction between staining and histochemical methods. Some will disagree with the arrangement of methods in this book, but a sequence according to similar tissue or cell types seems most adaptable to general laboratory application —the primary intent of the book. This section on histochemistry will therefore cover mainly procedures used to identify enzymatic activity.

The field is a tremendous and exciting one of increasingly extensive activity. It is impossible and impractical to include all histochemical methods and their variations in this type of manual. Because there are available many excellent books and journals on histochemistry that describe the newest developments, only methods most familiar (at least to the author) and practical are incorporated in this text. The procedures can be adapted to a minimum of glassware, equipment, chemicals, and time, and yet demonstrate some of the more cooperative of the enzymes. The student who enters research in any aspect of histochemistry or enzymology will have to consult the more specialized and the most recent literature as his interest continues to progress.

For greater details start with the following: Burstone (1962); Cassellman (1959); Danielli (1953); Davenport (1960); Glick (1949); Gomori (1952); Gurr (1958); Lillie (1965); Pearse (1968); Thompson (1966).

FIXATION

The use of fixed tissue for enzyme demonstration is debatable, but cutting frozen unfixed tissues may result in a greater loss of enzyme into the incubating medium and more cell damage than if fixed tissue is used. Many enzymes will tolerate some exposure to fixatives and should be so treated, if possible, for ease of processing and reduction of diffusion. Even succinic dehydrogenase can tolerate calcium-formalin fixation at 5°–10°C for 8–16 minutes, if the pieces are kept small. The addition of sucrose (10%–15%) to fixatives will often improve the preservation of some enzymes and should be tried.

Aldehydes

Calcium formalin (Baker 1944) is used cold, 4°C, and recommended for the preservation of many enzymes (acid phosphatase, esterase, and lipase, but not for alkaline phosphatase or aminopeptidase)

calcium chloride, anhydrous .... 1.0 g
distilled water .... 60.0 ml
formaldehyde, concentrated .... 10.0 ml

Adjust pH to 7.0–7.2 with 1 N NaOH. Add distilled water to make a total of 100.0 ml. Store in refrigerator. Check pH frequently, or just before use, and adjust if necessary.

Glutaraldehyde

Sabatini et al. (1964) discovered that 2–6% glutaraldehyde in a phosphate buffer can be used for some enzymes (alkaline phosphatase and partial preservation of acid phosphatase and esterase). Sea water can be used as diluent for marine tissues and tyrode solution for tissue culture cells.

50% solution of glutaraldehyde[1] .... 4–12 ml
0.1–0.2 M phosphate buffer .... 100.0 ml

The pH should be 7.4. If shrinkage occurs in the tissue, use the dilution at 2% (4 ml of glutaraldehyde per 100 ml of buffer, not a higher concentration). One percent of sucrose can be added to the glutaraldehyde solutions for better preservation of tissue constituents. Store in the refrigerator.

Cacodylate buffer, pH 7.2

1 M cacodylic acid (137.99 g/1000 ml) .... 54.6 ml
1 N NaOH (40 g/1000 ml) .... 50.0 ml
distilled water .... 895.4 ml

Glutaraldehyde-fixed tissue frequently is postfixed in 1% osmic acid buffered with 0.1 M phosphate buffer and glucose or veronal-acetate-sucrose after cacodylate-buffered glutaraldehyde.

Physiologic buffer

Solution A. 2.26% $NaH_2PO_4 \cdot H_2O$

Solution B. 2.52% NaOH

Solution C. 5.4% glucose

Solution D. Isotonic disodium phosphate

---

[1] Biological Grade, G-151, Fisher Scientific Co.

solution A	41.5 ml
solution B	8.5 ml

Adjust *p*H to 7.4–7.6 with solution B.

Phosphate-buffered fixative

solution D	45.0 ml
osmic acid	0.5 g
solution C	5.0 ml

Stable for several weeks at 4°C.

Wash tissues in buffer several hours or overnight. The maximal stability of glutaraldehyde is at concentrations between 2% and 10%. Low temperatures and storage in the dark also are advisable. See Anderson (1967) about the purification of glutaraldehyde.

Paraformaldehyde

0.2 M s-collidine buffer

Solution A

s-collidine (2,4,6 trimethylpyridine)[2]	2.64 ml
distilled water	47.36 ml

Solution B

1.0 *N* HCl	9.0 ml
distilled water	41.0 ml

Working solution

Equal parts of A and B, *p*H 7.4 to 7.45

Paraformaldehyde solution

paraformaldehyde	4.0 g
distilled water to make	66.0 ml

Add distilled water to the paraformaldehyde in a graduated cylinder until combined total is 66.0 ml. Warm in flask to 60°C. Depolymerize by adding 0.1 *N* NaOH, drop by drop; shake until solution clears.

---

[2] Eastman Organic Chemicals.

Working solution

paraformaldehyde solution	66.0 ml
0.2 *M* *s*-collidine buffer	33.0 ml
0.5 *M* $CaCl_2$ (5.55 g/100 ml)	1.0 ml

Adjust *p*H to 7.4 if necessary.

Acetone (absolute)

Acetone, 4°C or colder (refrigerated or kept in the freezer) is used for some enzymes, such as aminopeptidase and alkaline phosphatase.

Buffered acetone (Kaplow and Burstone 1963) is recommended for several enzymes, such as alkaline phosphatase, esterase, peroxidase, and dopa oxidase.

acetone	300.0 ml

Add with stirring to a mixture of

0.03 *M* sodium citrate (dihydrate)	32.0 ml
0.03 *M* citric acid (monohydrate)	168.0 ml

The *p*H should be 4.2. Store in freezer.

Preparation for Fixation and Storage of Tissues

For the best results, always fix very small pieces of tissue, no larger than 1–2 mm thick. They can be left in calcium formalin or glutaraldehyde in the refrigerator overnight. Then proceed to sectioning or store for several days or weeks in a gum sucrose solution or glycerine.

Gum Sucrose (Holt et al. 1960)

Store tissue at 0° to −2°C.

sucrose	30.2 g
distilled water	100.0 ml
gum acacia	1.0 g

Prepare only enough for immediate use. If prepared in excess, stock mixtures should be kept frozen to prevent growth of mold. This can be done in small vials, then only single vials are melted as needed.

The gum sucrose preserves the enzymes, facilitates cutting, and produces good morphological detail.

Glycerine (Turchini and Malet 1965)

Store tissue at −20°C.

glycerine	50.0 ml
distilled water	50.0 ml

Place fresh tissue immediately into the glycerine solution. It can be stored in this condition for at least nine months. For processing, remove the block of tissue and fix for routine histological staining or in calcium formalin. Nuclear and cytoplasmic alterations may occur, but enzymatic activities will be well preserved.

See specific enzymes for additional fixation and preservation recommendations.

DEHYDRATING AND EMBEDDING

Freeze-drying Method

If the apparatus is available and controllable, freeze-drying is considered the proper tool for the preparation of tissues for enzyme study. There are many types of freeze-drying apparatus on the market; many have been expensive but cheaper ones are now available. Temperatures sometimes are difficult to control and the drying time may be long. The principle of freeze-drying is that frozen tissue is kept under high vacuum until all water molecules are removed and condensed onto a cold surface or collected by a desiccant. Small pieces (approximately 1 mm) are frozen solid instantly with isopentane or Freon 12 (obtainable in a can) cooled by liquid nitrogen to a temperature of −150°C. The tissue must be frozen rapidly to prevent large crystal formation, which would disrupt the cells. The initial freezing is commonly called quenching; it stops all chemical reactions in the tissues. Immediately after freezing and while still frozen the tissue is dehydrated in a drying apparatus in vacuo at a temperature of −30° to −40°C, and the ice is sublimed into water vapor and removed. There is no liquid phase, and therefore no diffusion of enzymes.

When the material is dry, allow it to rise to room temperature, infiltrate with degassed paraffin or Carbowax, and embed.

Because of sensitivity to water, the sections cannot be floated on water, but must be applied directly to warm albumenized slides (Mendelow and Hamilton 1950). Embedded blocks must be kept stored in a desiccator.

Many types of freeze-dry apparatus are offered by the manufacturers.

Thieme (1965) gives directions for the construction of an inexpensive and convenient model that is easy to operate and dries tissue overnight. Stumpf and Roth's (1967) model is available as a kit.[3] Wijffels (1971) describes a model that is easy to construct and operate.

Freeze-Substitution Method

A simpler and cheaper method that produces excellent results is the freeze-substitution procedure. The tissue is rapidly frozen and then the ice formed within the tissue is slowly dissolved in a fluid solvent such as ethyl or methyl alcohol. Other polar solvents also have been tried (Patten and Brown 1958). When the tissue is free of ice and completely permeated by the cold solvent, it is brought to room temperature and embedded.

Advantages of this method are these: there are no disruptive steaming movements in the tissue—the ice simply dissolves—and the substitution solution can sometimes be chosen to contain a specific precipitant for a specific substance; and the method can be used for radioautography, for fluorescent preparations, for studying many enzymes, and for other cyto-histochemical preparations. There can, however, be cases in which the freeze-drying procedure is preferred—for instance, when an enzyme might be damaged by the denaturing action of the alcohol as it warms up.

Freeze-Substitution Technique (Feder and Sidman 1958; Hancox 1957; Patten and Brown 1958)

PROCEDURE

1. Cool a beaker of liquid propane-isopentane (3 : 1) or Freon 12 to $-170°$ to $-175°C$ with liquid nitrogen.
2. Place tissue specimen (small piece, 1.5–3 mm) on thin strip of aluminum foil and plunge into the cold propane-isopentane. Stir vigorously: 30 seconds.
3. Transfer frozen specimen to ice solvent, previously chilled to $-60°$ to $-70°C$ (see Comments below). After a few minutes the tissue can be shaken loose from the foil. Store in dry-ice chest for 1 or 2 weeks: has been stored for fifty-two days without harm (Patten and Brown).
4. Remove fluid and tissue to refrigerator and wash 12 hours or longer in 3 changes of absolute alcohol in refrigerator.

---

[3] Delmar Scientific Glass Products, Maywood, Ill., catalog no. 7126.

5. Transfer to chloroform: 6–12 hours in refrigerator, to remove any traces of water. (This step may be omitted in most cases.)
6. Transfer to cold xylene (or similar agent), $-20°C$; and remove to room temperature for 10 minutes.
7. Transfer to fresh xylene, room temperature: 10 minutes.
8. Transfer to xylene-tissuemat, $56°C$, vacuum: 15 minutes.
9. Infiltrate, Tissuemat no. 1, vacuum: 15 minutes.
10. Transfer to Tissuemat no. 2, vacuum: 15 minutes.
11. Transfer to Tissuemat no. 3, vacuum: 15 minutes.
12. Embed.

COMMENTS

Ice-solvents that can be used are absolute methyl, absolute ethyl alcohol, and *n*-butyl alcohol. Feder and Sidman include complete directions for obtaining, storing, and disposing of liquid nitrogen, isopentane, and propane; also details about equipment. *Warning:* There are potential explosion hazards.

See Moline and Glenner (1964) for a method that involves coating the tissue with powder to permit faster cooling of the tissue. It has the advantage of safety, simplicity, and speed over methods using hydrocarbon baths cooled in liquid nitrogen.

See Bullivant (1965) concerning the use of glycerine preservation prior to freeze-substitution.

Blank and Delamater (1951) use glycols if fixation is to be avoided, alcohol or acetone if fixation is permissible. This method is as follows:

1. Place slices (2–3 mm) of tissue in cold propylene glycol in deep freeze below $-20°C$: 1–2 hours.
2. Transfer to mixture of Carbowaxes, $55°C$: 2–3 hours.

Carbowax 4000 . . . .   9 parts
Carbowax 1500 . . . .   1 part

3. Prepare blocks (page 89).
4. Cut sections and float on slides according to Blank and McCarthy method (page 90).
5. Stain or use for autoradiographs. Water-soluble substances will not have leached out. For autoradiographs, before each cut, coat the block face with thin molten wax. Press wax-reinforced sections against emulsion for exposure. See also Feder and Sidman (1958); Freed (1955); Masek and Birns (1961); Meryman (1959, 1960); Woods and Pollister (1955).

Freeze-substitution of Sections (Patten and Brown 1958; Chang and Hori 1961)

PROCEDURE

1. Quench 1–2 mm slices in liquid nitrogen or isopentane prechilled by liquid nitrogen.
2. Section at −15°C in cryostat.
3. Sweep sections into screw-cap vial of acetone prechilled in dry ice.
4. Return vial to dry ice and complete dehydration.

Chang and Yokoyama Method (1970)

PROCEDURE

1. Cut small pieces of fresh tissue and section in cryostat. Transfer sections to cold cover glasses and immediately thaw by placing a finger on the glass under the section.
2. Immediately immerse in acetone at dry-ice temperature for freeze substitution: 2 hours to overnight.

This method permits the cutting of many tissue samples to be preserved at dry-ice temperature until they are processed or incubated. A number of enzymes can be preserved in this way.

CRYOSTAT SECTIONING

Carbowax or paraffin can be cut at room temperature, but many enzymes and other tissue constituents must be sectioned at low temperatures of −25° to −20°C or they are lost. Since embedding is impractical at these temperatures, tissue is frozen in isopentane or Freon 12 (−160°C) or in dry ice–ethyl alcohol mixtures (−70°C). The tissue must be maintained at least at dry-ice temperature (−20°C) and sectioned and mounted at this temperature. This is performed inside a cold-chamber cryostat[4] (−25° to −20°C). The object holder is precooled (object holders can be stored in the cold chamber) and the tissue is frozen on it with a few drops of water. Sectioning is facilitated by using O.C.T. compound (page 000) to support and protect the tissue. I usually use O.C.T.; it melts away against the slide and does not interfere with enzymatic reactions. *Caution:*

---

[4] Harris Refrigeration Co.; Fisher Scientific Co.; Lipshaw; Lab Tek; American Optical.

if the slides are to be used for autoradiography, remove the O.C.T. by washing slides in distilled water before dipping in emulsion.

The knife used in sectioning must be kept sharp and dry. Carefully wiping it with a finger tip before each section keeps it dry and clean and reduces compression in the section. If O.C.T. compound is used and the sections are to be mounted on slides, cut through the compound and through the tissue just until the knife edge reaches the O.C.T. along the opposite edge of the tissue. This means that the cut is complete through the tissue but not through the O.C.T. (Figure 25). Stop sectioning. (As the cut is being made, lightly hold the section against the knife with a small cold brush. This prevents curling. Flatten the section with the brush and finish cutting through the O.C.T. With the cold brush (sometimes two brushes) lift the section to a cold slide. Thaw the section by placing the undersurface of the glass against a finger or the cushiony part of the palm.

When thawed the sections are ready for processing. If the sections are to go directly into a substrate or fixative, lift them, curled or uncurled, with the cold brush and drop them into the solution. Usually a slight tap of the brush against the side of the container will shake them loose. Frozen tissues can be mounted on cover glasses in the same way as on slides and in some cases the section can be picked up by touching it with the cover glass. If the tissues have been suspended in gum sucrose, 6–10μ sections are easily cut and then dropped into the proper substrate for

FIGURE 25. *Flattening and correcting folds in section: cut through*
OCT compound (strippled area) is not yet complete.

incubation or mounted on cover glasses or slides. Gelatin embedding (Burkholder et al. 1961; Taylor 1965) can be tried for better tissue support; it does permit easier sectioning and handling.

Fitz-William et al. (1960) recommend that, immediately before each section is cut, the tissue block be coated with 20% polystyrene dissolved in methylene chloride (volatile at cryostat temperature). Wait until the highly reflecting surface that forms disappears and then cut section. Curling of sections is minimal and they can be picked up with fine forceps. The polystyrene solution should be stored in the cryostat. These authors also include complete directions on the care of the microtome used in the cryostat.

Consult the following references for more about cryostat sectioning: Blank and Delamater (1951); Feder and Sidman (1958); Freed (1955); Gersh (1932); Glick and Malstrom (1952); Hancox (1957); Hanson and Hermodsson (1960); Ibanez et al. (1960); Jennings (1951); Mellors (1959); Meryman (1959, 1960); Patten and Brown (1958); Woods and Pollister (1955); Zlotnik (1960).

SECTIONING WITHOUT A CRYOSTAT

If a cryostat is not part of the laboratory equipment, frozen sections of many of the enzymes can be made on the clinical freezing microtome, or by the method developed by Adamstone-Taylor.

Adamstone-Taylor (1948) Cold Knife Technique

PROCEDURE

1. Quickly freeze fresh tissue (2 mm blocks) on freezing block of microtome or in liquid nitrogen, isopentane or Freon 12 chilled with nitrogen, or petroleum ether containing solid carbon dioxide.
2. Store in dry ice ($-75°C$) or in deep freeze ($-25°C$).
3. Chill knife by fastening solid blocks of dry ice on each end. Use Scotch tape or thin sheets of metal foil.
4. When the knife is chilled, freeze the frozen block of tissue in place on freezing head.
5. Cut sections and hold them flat on the knife with a camel's hair brush. Do not allow them to thaw.
6. Remove section to a small metal scoop, which is kept cold with a few bits of dry ice. Transfer the section to a slide. If the slide is

cold, speed of transfer is not as essential as it is if the slide is at room temperature.

7. Press the section in place, and as it begins to soften but, before it completely thaws, immerse slide in fixative.

COMMENTS

Unthawed sections can be removed rapidly to a beaker of cold acetone surrounded by dry ice. With needles chilled in the dry ice, mount the sections in the acetone onto chilled cover glasses. Remove, allow to dry, then fix.

This method is difficult in warm and humid atmospheres; the sectioning should preferably be done in a cold room.

ACETONE FIXATION AND EMBEDDING

Some enzymes can be carried through rapid paraffin embedding with good results, but great care must be taken to keep the duration of time in solutions to a minimum and temperatures as low as possible. Even with extreme care, results can occasionally be disappointing.

Gomori Method (1952)

PROCEDURE

1. If possible, chill tissue for a short time before immersing in acetone. It may prevent some shrinkage artifacts. But fix within 10 minutes of death for best results. Cut very thin slices, 1–2 mm, no larger.
2. Fix in cold acetone, 2–3 changes: 12 hours each.
3. Transfer to ether–absolute alcohol (1:1): 2 hours in refrigerator.
4. Infiltrate with dilute celloidin or nitrocellulose (approximately 10%): overnight in refrigerator.
5. Drain on cleansing tissue. Place in benzene or chloroform: 30 minutes–1 hour. Stirring on a magnetic mixer aids penetration of fluid.
6. Infiltrate with paraffin 52°–56°C (52°C is best): 15–20 minutes in oven without vacuum; 30 minutes with vacuum; and 10–15 minutes without vacuum.
7. Embed. Store in refrigerator.
8. Section. Float on lukewarm water, briefly (10 minutes maximum).
9. Drain off water and dry.
10. If to be kept for some time, put in oven 5–10 minutes to melt paraffin, forming a protective coating against atmospheric influence,

Store in cold room or refrigerator. The safest way to store is in uncut paraffin block in refrigerator.

COMMENTS

Step 4 can be reduced to 2–3 hours in nitrocellulose, and the entire procedure following fixation can thus be completed in one day with good results.

Alternate Short Gomori Method

1. Fix in cold acetone overnight.
2. Transfer through 2 changes of cold petroleum ether: 1 hour each.
3. Remove petroleum ether and tissues to room temperature: 15 minutes.
4. Infiltrate in 45°C paraffin in vacuum: 30 minutes.

Novikoff et al. (1960) find acetone fixation is good for some cryostat preparations.

Frozen Sections Fixed in Acetone (Burstone 1962)

PROCEDURE

1. Cut unfixed frozen sections. Mount on slides.
2. Immerse in acetone: 1–5 minutes.
3. Transfer to 90–95% acetone-water: 1–3 minutes.
4. Transfer to 80–85% acetone-water: 30 seconds–1 minute.
5. Transfer to 70% acetone-water: 30 seconds.
6. Rinse in water: 10–20 seconds.
7. Incubate

COMMENTS

The quality of fixation is not the best but the method is simple and useful when cytological detail is not of great importance.

GENERAL SUGGESTIONS

Stock Solutions

The following solutions can be kept in storage and ready for immediate use: in the refrigerator—calcium-formalin and glutaraldehyde; in the freezer—buffered acetone, absolute acetone, absolute methyl alcohol, and gum sucrose. (See page 432 concerning methyl alcohol.)

Read labels on all substrates and diazonium salts. Some must be kept in the refrigerator or freezer. If in doubt, store such materials in the refrigerator for longer life.

Control Slides

Most enzyme reactions should be accompanied by a control slide in order to help detect false positives. An enzyme can be inactivated by any one of the following methods:

Place slide in distilled water and bring to boil: 10 minutes.
Immerse in 1 N HCl: 15 minutes; wash in tap water: 0.5 hour.
Incubate slide in substrate medium containing an inhibitor (NaCN or KCN).
Incubate untreated slide in a medium lacking substrate.

For additional information, see each specific enzyme method.

Preservation of Incubation Media

Klionsky and Marcoux (1960) prepared twelve different media, froze them in a mixture of dry ice and acetone, and stored them in a dry-ice chest. For at least 6 months, the solutions remained as good as freshly prepared solutions. The types of media tested were:

Azo dye and metal precipitate methods for acid and alkaline phosphatases.
Alpha-naphthyl and indoxyl acetate for esterase.
5-nucleotidase.
Nitro-BT for succinic dehydrogenase.
Postcoupling technique for beta-glucuronidase.
Triphosphopyridine nucleotide and diphosphopyridine nucleotide diaphorase.

Storage of Slides and Tissues

Storage for long periods of time is not generally recommended for enzymes; even at $-40°C$ they can be kept only about 6 hours. If, however, storage is unavoidable, quenching at $-160°C$ and storage at $-85°C$ will preserve many enzymes in the tissue. Some enzymes on slides can be kept for 3 months at $-25°C$, but $-85°C$ is even better and good for a year.

ALKALINE PHOSPHATASE

Sections containing active phosphatase are incubated in a mixture of calcium salt and phosphate ester, usually sodium glycerophosphate, so that calcium phosphate is precipitated at the sites. The enzyme liberates the phosphate ions, which are held on the spot by salts of metals whose phosphates are insoluble. (Magnesium ions are often required as an activator and are added in the form of magnesium sulfate or chloride. The mechanism of activation is unknown.) The insoluble phosphates are made visible by cobalt nitrate and ammonium sulfide, which produces cobalt phosphate and finally black cobalt sulfide.

Phosphatase is dissolved or destroyed by ordinary fixatives, but alcohol or acetone will preserve it. Celloidin or paraffin embedding can be used safely, but paraffin sections must be covered with a film of celloidin to protect against dissolution of enzyme. Freeze-substitution preserves it well. See Burstone (1960b) for fluorescent demonstration of alkaline phosphatase.

Gomori Method (Frankel and Peters Modification 1964)

FIXATION

Thin slices (1–2 mm) are fixed at once in chilled acetone, 3 changes: 24–48 hours. Other fixatives may be 80%, 95%, and absolute alcohol, or cold formalin (4°C) followed by freezing method.

INFILTRATION AND EMBEDDING

For paraffin embedding use chloroform or petroleum ether and paraffin at no higher than 56°C melting point, preferably 52°C. The short methods are recommended. Cut sections 6–10μ. Do not dry slides at higher than 37°C. Store all blocks at 6°C until needed.

INCUBATING SOLUTION

0.8% paranitrophenyl phosphate[5]	2.0 ml
2% sodium barbital	2.0 ml

---

[5] Sigma's 104 phosphatase substrate.

distilled water	1.0 ml
2% calcium chloride	4.0 ml
5% magnesium sulfate	0.2 ml

PROCEDURE

1. Deparaffinize in light petroleum or chloroform: 1 minute.
2. Hydrate with 3 or 4 dips in each of absolute acetone, 80% acetone–water, 50% acetone–water.
3. Rinse with several dips in water, or until surface of slide appears homogeneous.
4. Incubate in incubating solution, 37°C: 30–45 minutes.
5. Wash well in distilled water.
6. Treat with 2% cobalt nitrate: 3–5 minutes.
7. Rinse well in distilled water.
8. Treat with 2% yellow ammonium sulfide: 1–2 minutes.
9. Wash well in distilled water.
10. Counterstain with hematoxylin, safranin, Kernechtrot, or the like.
11. Dehydrate, clear, and mount.

RESULTS

sites—black, sharp and clear, nondiffuse
nucleus and nucleolus—show a minimum of gray deposit of cobalt sulfide

COMMENTS

Paranitrophenyl phosphate is split by alkaline phosphatase at a faster rate than other phosphates (phenyl phosphate, glycerol phosphate, and phenolphthalein phosphate) that have been used for this reaction. Since the paranitrophenyl phosphate is so rapidly hydrolyzed, the time of the reaction is reduced to 30 minutes and diffusion artifact formation is minimized. The sites correspond to those demonstrated by the azocoupling method below. The color in the coupling method tends to fade within a year, but both methods should be tried.

Freeze-substituted and cryostat-cut sections can be used.

The pH optimum is 8.5–10.0. The specific activator of alkaline phosphatase is Mg^{++}; other divalent cations, such as Mn^{++}, Zn^{++}, and Co^{++}, may activate in very low concentrations. Inhibitors include EDTA, zinc chloride, formaldehyde, oxidizing agents such as potassium permanganate and iodine, cysteine, and sodium arsenate.

False positives can be due to hemosiderin or melanin, and the sites should be checked for the presence of these pigments.

For additional reading about the Gomori method, see Ackerman (1958);

Gomori (1941a, 1946); Kabat and Furth (1941); Moffat (1958); and Pearse (1968).

Azo-Coupling Method

One of the very important naphthols is 3-hydroxy-2-naphthoic acid. It is made when sodium naphtholate is treated with carbon dioxide and is used to produce the so-called naphthol AS (naphthol anilid–säure) compounds. Naphthol AS phosphates are phosphate esters of these complex compounds and release highly insouble naphthols upon enzymatic hydrolysis. These naphthols couple immediately with diazonium salts to form insoluble azo dyes, thereby demonstrating microscopic localization of enzymes in properly fixed tissues. Several of these naphthol phosphate compounds are suitable for demonstrating alkaline phosphatase, but probably the most popular is AS-MX (others are AS-KB, AS-TR, AS-BS, AS-AN, AS-E, and AS-BI).

The azo-coupling method can be used on tissues prepared in the same manner as those for the modified Gomori technique. After deparaffinizing and hydrating, transfer the sections into the incubation solution below.

For the beginner in enzyme methodology, smears and touch preparations are simple to prepare and can be processed in conjunction with sections of the tissue to verify the presence or absence of an enzyme. For example, if tissue blocks are made of liver, also make and fix touch preparations from the same organ. Stain sections and touch preparations in the same incubating medium. If the sections are negative, but the touch preparations are positive, the enzyme is being lost somewhere in the processing for sectioning.

The occurrence of alkaline phosphatase in the neutrophils of the peripheral blood of most animals (not mice or chickens) makes these cells excellent and easily prepared subjects for the demonstration of this enzyme. Spin whole blood in a small tube (hematocrit or bacteriological culture tube, 11 mm in diameter) 2500–3000 rpm for 10 minutes. With a Pasteur pipette collect the buffy coat of white cells lying on top of the packed red cells and place a *small* drop on a slide. Immediately smear like a blood smear (page 245), or place another slide on top of the drop. It should be placed so the drop is close to one end of the slide, as shown in Figure 26. Wait until drop stops spreading; then quickly pull off the top slide. When preparing smears, do not allow the cells to be overcrowded; the cells must be well flattened to demonstrate precisely the alkaline phosphatase granules. Allow the smears to dry thoroughly;

a

b

c

FIGURE 26. *Type of smear useful for enzyme preparations:* (a) *place a slide on top of drop of white cells;* (b) *allow cells to spread between slides;* (c) *when spreading stops, pull top slide length of and off bottom slide.*

slight heat, such as that from a slide warmer, for a few seconds, causes no damage to the enzyme and helps flatten the cells.

When the smears are dry, place them in cold fixative ($0°$ to $-10°C$) consisting of 5.0 ml of concentrated formaldehyde and 45.0 ml of absolute methyl alcohol, and leave them overnight. The alcohol must be fresh and kept tightly closed. Alcohol from the final third of a gallon bottle may not preserve the enzyme adequately: I purchase only one-pound bottles, and store them in a refrigerator or freezer. Mixed fixative can be kept in a freezer, but not longer than one week.

Temperature is very important; if it is too high or too low, the fixation of the enzyme will be imperfect and either diffuse or reduced activity will result. Incubation of smears and touch preparations can be undertaken on the same day after 30 minutes of fixation. Such preparations can be kept in the cold fixative for 2 nights and then be processed with sections prepared from tissue blocks that have been fixed overnight, rapidly embedded, and sectioned. Do not leave smears and touch preparations in the fixative more than 2 days; to do so reduces the activity. Storing in gum-glucose does not preserve alkaline phosphatase activity.

When drawing blood for buffy coats, use heparin, not EDTA, as an anticoagulant. The EDTA is an inhibitor of alkaline phosphatase.

Although a phosphatase score will drop very slightly within the first couple of hours after blood withdrawal, good scores can be obtained within 6 hours after withdrawal. Keep the blood at room temperature or 37°C. Do not refrigerate it; low temperatures inhibit the activity. If a centrifuge is not handy, allow a buffy coat to form by sedimentation for 1–2 hours. Then collect the white cells.

Kaplow Method, Modified (1963)

SOLUTIONS

Propanediol stock solution

2 amino-2-methyl-1,3-propanediol	10.5 g
distilled water	500.0 ml

Buffer working solution, pH 9.7

propanediol solution	25.0 ml
0.1 N HCl	5.0 ml
distilled water to make	100.0 ml

Incubating solution

naphthol AS-MX phosphate[6]	5.0 mg

Dissolve in:

N-N dimethyl formamide (DMF)[7]	0.2–0.3 ml

---

[6] Nutritional Biochemicals Corporation, Cleveland, Ohio.
[7] Fisher Scientific Company.

Add:

buffer working solution	60.0 ml
fast red violet salt LB [8]	30.0–40.0 mg

Shake for 30 seconds and filter onto slides.

PROCEDURE

1. Deparaffinize sections in chloroform: 1 minute. Hydrate through acetone: few dips; through 70% acetone–water: few dips; through 50% acetone–water: few dips; then wash in distilled water: 30–60 seconds. Wash smears and touch preparation in distilled water: 20–30 seconds, and blot.
2. Incubate, room temperature: 20–25 minutes.
3. Wash in running water: 1 minute.
4. Stain in Mayer hematoxylin: 3 minutes.
5. Wash in running water: 1–2 minutes.
6. Blue in Scott solution: several dips.
7. Wash in running water: 1 minute.
8. Stain in 0.5% aqueous wool green S: $\frac{1}{2}$ minute.
9. Rinse in water (tap or distilled): 3–4 seconds.
10. Dehydrate in 2 changes of 95% ethyl alcohol: few seconds in each.
11. Dehydrate in 2 changes of absolute ethyl alcohol: 2–3 minutes in each.
12. Clear in 2 changes of xylene and mount.

RESULTS

sites—red
nuclei—blue
cytoplasmic staining—pale green
eosinophilic granules—brilliant green

COMMENTS

Fast green FCF or light green can be substituted for wool green, but the stain is paler and less interesting. Kaplow does not use the green, but addition of it provides background cytoplasmic staining and distinctive green color to the eosinophilic granules.

Kaplow blots and dries the slides after step 5, and mounts in glycerol jelly or Permount. Bluing in Scott solution and counterstaining does not alter the score of the phosphatase in the neutrophils.

Gelva (page 451) is satisfactory as a mounting medium. It may prevent some fading, but the resolution is not quite as precise as when a

---

[8] Verona Dyestuffs, Union City, N.J.

synthetic resin is used. The problem of fading is not critical; 8 months after slide preparation and mounting in synthetic resin, I obtained a score only a few points below the score obtained on the day of preparation.

For scoring the amount of alkaline phosphatase in the neutrophils, see Kaplow (1963).

Coupling can be done with other salts, such as fast blue RR (Sigma), which gives blue sites.

For references about alkaline phosphatase, see Burstone (1962); Gomori (1951); Manheimer and Seligman (1949); and Pearse (1968).

ACID PHOSPHATASE

This enzyme exhibits optimal activity at pH below 7.0, usually between 3.8 and 6.0. The best tissues for its demonstration are spleen, liver, kidney, and blood leukocytes. In many cells, acid phosphatase appears to be associated almost exclusively with the lysosomes, so much so that acid phosphatase is coming to be considered a "marker" for them. Lysosomes are rich in hydrolytic enzymes—ten or more of them, all showing an acid pH optimum. Esterase also can be demonstrated in lysosomes. The small granules of concentrated acid phosphatase found in the Golgi region are tentatively considered to be lysosomes.

Care must be taken in the preparation of tissues for acid phosphatase. The fact that the enzyme is readily made soluble and released from the tissue is a probable cause of failures in past methods. The simplest and most dependable fixation is to place small pieces (2 mm) in cold calcium-formalin (page 416), 4°–10°C, overnight. Large organs from large animals often require 1–2 minutes of perfusion with fixative before a small block of tissue is removed. The tissue should then be stored in gum sucrose, 4°C, for 24 hours.

Section on freezing microtome and drop sections into incubation medium. Mount sections on slides and allow to air dry: 2 hours. Proceed to substrate or store overnight in refrigerator. If sectioning cannot be done immediately after overnight fixation, transfer the blocks of tissue into gum sucrose (page 419) and refrigerate. Krus et al. (1961) say it can be preserved for 40 days. (Freeze-substitution preserves 75%–80% of the enzyme.)

Tissues can be acetone-fixed and paraffin-embedded (short method), but deparaffinizing the sections and mounting them on slides may inactivate the enzyme.

Cold 2% glutaraldehyde can be used for good preservation of vacuoles and other structures, but it should be accompanied by calcium-formalin fixed blocks. There are disadvantages to using glutaraldehyde. It penetrates slowly; even after overnight fixation the center of a block may remain unfixed, and diffusion of the enzyme into the outer fixed area may have taken place. (Never cut blocks thicker than 1–2 mm for enzyme studies.) Glutaraldehyde leaves a yellow background color, which may be considered undesirable. See Janigan (1965) for studies on aldehyde fixation for acid phosphatase.

As recommended for alkaline phosphatase, touch preparations and smears are useful adjuncts to sections. Two methods for fixing such preparations are suggested. Prepare duplicate slides; leave one dry and unfixed and fix the other for 20–30 seconds (no longer; do not store in fixative) in cold buffered acetone (page 419) or cold calcium-formalin. Drain against filter paper or cleansing tissue and allow to dry. Incubate both slides with tissue sections. Postfix the unfixed slide before counterstaining (see step 4, azo-coupling method).

Protozoans make fine subjects for lysosome studies and for other enzymes. Prepare some on subbed slides for better adherence but always make control slides that have no adhesive on them, since it is possible that "sub" material or albumen may inhibit an enzyme.

Gomori Method (1952)

SOLUTIONS

Sodium acetate buffer, pH 5.0

0.6% acetic acid	300.0 ml
sodium acetate (27 g/1000 ml water)	700.0 ml

Substrate

acetate buffer	25.0 ml
distilled water	100.0 ml
lead nitrate	0.12 g
3% sodium glycerophosphate (freshly prepared)	10.0 ml

Keep at 37°C for 24 hours. Filter. Add 2–5 ml of distilled water to prevent precipitation on evaporation. Keep in refrigerator. If the solution becomes turbid, discard. It is best when used within 3–4 days.

PROCEDURE

1. Deparaffinize sections as for alkaline phosphatase and hydrate to water. Smears and frozen sections can be placed directly into substrate.
2. Incubate in substrate, 37°C: 1–24 hours; average is 4 hours.
3. Rinse in distilled water: few seconds.
4. Wash in 1–2% aqueous acetic acid: 1 minute (see comments below).
5. Wash in distilled water: 1 minute.
6. Transfer to 1% aqueous ammonium sulfide: 2 minutes.
7. Wash in distilled water: several minutes.
8. Counterstain in hematoxylin, Kernechtrot, or the like.
9. Wash well in tap water to remove excess stain.
10. Mount in glycerine jelly; or dehydrate, clear, and mount in synthetic resin.

RESULT

sites—black

COMMENTS

Wachstein et al. (1962) use pH 6.0.

Gomori followed step 3 by a wash in dilute acetic acid to remove nonspecific bound lead, but Goldfischer et al. (1964) imply that the rinse can remove all the reaction product from an enzyme site of low activity. One slide should therefore be processed as a test slide with the acetic acid rinse omitted.

Gomori method is analogous to the alkaline phosphatase technique in that phosphate is released by enzymatic activity and is precipitated as lead phosphate, which is finally converted to a black lead sulfide.

Controls: since sodium fluoride inhibits acid phosphatase, add the fluoride to the incubating solution to a final concentration of 0.005 M. The sodium glycerophosphate can then be left out of the incubating solution as a control.

Thin blocks of tissue can be incubated, embedded in paraffin or some other medium, thin 2–3μ sections cut, mounted on slides, cleared, and cover slipped.

For unfixed frozen sections, see Bitensky (1963) for modification of Gomori method.

Azo-Coupling Method (Kaplow and Burstone 1964)

This also is analogous to the alkaline phosphatase method, but employs a buffer at pH 5.0.

SOLUTION

Substrate

naphthol AS-MS phosphate	5.0 mg

Dissolve in:

N-N dimethyl formamide (DMF)	0.2–0.3 ml

Add:

distilled water	25.0 ml
0.2 *M* acetate buffer (see Gomori method above)	25.0 ml
fast red violet salt LB	30.0–40.0 mg

Shake for 30 seconds and filter onto slides. One may substitute 60 ml of a 0.1 *M* citrate buffer, *p*H 5.2, for the water and acetate buffer.

PROCEDURE

1. Deparaffinize sections in petroleum ether or chloroform, and hydrate quickly through acetone solutions to water. Frozen sections, smears, and touch preparations go directly into substrate.
2. Incubate, 37°C: 2 hours.
3. Wash in running water: 1 minute.
4. Fix unfixed smears in 10% formalin: 30 seconds, and wash in running water: 2–3 minutes. This step may be omitted, but better nuclear staining is obtained after brief fixation.
5. Stain in Mayer hematoxylin: 3 minutes.
6. Wash in running water: 5 minutes.
7. Mount in glycerine jelly or polyvinyl alcohol (page 452).

RESULTS

sites—red
nuclei—blue

COMMENTS

Gelva preserves acid phosphatase, but bubbles develop in the mounting medium after a few weeks, even when the cover glass is sealed. PVP/VA does not preserve it adequately. See page 452 for plastic mountant.

Since the lysosomal membrane is less permeable to glycerophosphate than to naphthol AS phosphates, the reaction is slower in the glycerophosphate technique (Maggi and Riddle 1965).

Schajowicz and Cabrini (1959) say that, if decalcification is necessary,

2% formic acid and 20% sodium citrate (1:1), pH 5, is satisfactory up to 15 days. There is loss of enzyme if Versene is used.

For additional reading about acid phosphatase, see Burstone (1959*b*, 1962); De Duve (1959, 1963*a*, *b*); Gomori (1950*a*, 1956); Novikoff (1960, 1961*b*); Pearse (1968); Weissmann (1964).

AMINOPEPTIDASE (PROTEOLYTIC ENZYME)

The proteolytic enzymes hydrolyze proteins and amino acid compounds. The aminopeptidase (and exopeptidase) reaction depends on the hydrolysis of a peptide bond adjacent to a terminal alpha amino group, resulting in the liberation of a chromogenic moiety. The enzyme hydrolyzes *L*-leucinamide at a highly preferential rate, which makes leucyl naphthylamide a favored substrate. It can be activated by bound metal ions such as Mg^{++}, Mn^{++} and Co^{++}. It can be destroyed by a one-hour immersion in water at 70°C; it can be inhibited by heavy metal ions, Zn^{++}, Cu^{++}, Hg^{++}, Cd^{++}, Pb^{++}, and Fe^{++}, or by fixing in absolute methanol and 40% formaldehyde.

High activity will be found in the kidney, parathyroid, small intestine, and uterus; some activity in salivary glands, thyroid, liver, and thymus. Buffy coats make excellent preparations; fix in cold acetone: 1 hour. Fix tissue blocks in cold acetone ($-10°$ to $0°$C) overnight and embed by the short method (page 427). Xylene, 1–2 hours, can be substituted for petroleum ether. Freeze-dried cryostat sections or fresh tissue cut frozen can be used.

A large number of diazotates are stabilized as double zinc salts, and since aminopeptidase is sensitive to metallic ions like zinc, it is important to use the proper diazonium salt in the coupling reaction. Garnet GBC is available as a stabilized diazonium sulfate.

Burstone and Folk Method (1956)

SOLUTIONS

Substrate stock solution

L-leucyl-*b*-naphthylamide	1.0 g
distilled water	100.0 ml

Tris buffer, pH 7.19

0.2 M tris (hydroxymethyl) aminomethane	25.0 ml
0.1 N HCl	45.0 ml
dilute with distilled water to	100.0 ml

Working solution

stock substrate	1.0 ml
distilled water	40.0 ml
tris buffer	10.0 ml
garnet GBC [9] (diazotized o-aminoazo- toluene)	30.0 mg

Shake and filter.

For substrate, the alanyl compound may be substituted: 10 mg in 40 ml hot distilled water (90°C). Cool and add 10 ml of tris buffer and 30 mg garnet GBC.

PROCEDURE

1. Chloroform or petroleum ether, 2 changes: 4 minutes each.
2. Absolute acetone: 1 minute.
3. 95% acetone: 1 minute.
4. 85% acetone, several dips.
5. Distilled water, few dips.
6. Substrate: 15 minutes to 1 hour, room temperature, or 37°C.
7. Tap water: 5 minutes.
8. Hematoxylin: 3 minutes.
9. Tap water: 10 minutes.
10. Mount in glycerine jelly.

RESULT

sites—red

COMMENTS

If necessary to purify GBC salt (not necessary with Imperial Chemical Industries product):

1. Dissolve 10 g salt in 50 ml methyl alcohol. Let stand 30 minutes. Filter.
2. Repeat with residue.
3. Dissolve second residue in 500 ml ether. Let stand 2–3 hours.

---

[9] Imperial Chemical Industries, Manchester, England.

4. Filter. Wash residue 2–3 times with ether. Store dry powder in refrigerator.

Aminopeptidase preparations are only semipermanent. Bubbles develop in glycerine jelly mounts within a few weeks, and the reaction fades in PVP.

For further information about aminopeptidase, see Burstone (1962); Pearson et al. (1963), Zugibe (1970).

ESTERASES (NONSPECIFIC) AND LIPASES

These terms are used in reference to enzymes that hydrolyze esters of carboxylic acids. There is an overlapping between enzyme types. Some seem to occupy intermediate positions between lipases and esterases, and react like both forms (Burstone 1962; Chessick 1953; Gomori 1952). In general, esters of short-chained fatty acids are acted on by esterases, long-chained esters by lipases. A large number of presumably specific enzymes have been reported and demonstrated in lysosomes.

Esterases or lipases are not destroyed by acetone fixation, are resistant to heat, and can be embedded in paraffin if not used excessively (3 hours maximum, 58°C). Sections fixed overnight in calcium-formalin or 10% formalin and frozen sections can be used for esterase and lipase. Infiltration with gum sucrose, 12–24 hours or longer, facilitates sectioning. Freeze-drying gives good preservation.

Fix buffy coats in formalin vapor (1–2 ml of concentrated formaldehyde in a coplin jar) add dried smears, and cover tightly: 30 minutes. Store dry until use. Unfixed smears can accompany fixed smears through incubation; then, before counterstaining, postfix the unfixed smears in 10% formalin: 60 seconds; or methyl alcohol: 30 seconds.

Lipases are incubated with the long-chained fatty acid esters of sorbitan and mannitan, water-soluble commercial products by the name of Tween. On hydrolysis, the liberated fatty acids form insoluble calcium soaps deposited at the sites. These are converted into lead soaps and then into brown sulfide of lead.

Esters of naphthols also are used, and naphthol is liberated and demonstrated by azo-coupling.

Different esterases react to different inhibitors, but for nonspecific esterases, use E600 (diethyl-p-nitrophenyl phosphate), $10^6\ M$ to $10^{-3}\ M$, as an inhibitor, or leave the substrate out of the solution.

Esterase (Moloney et al. Modification 1960)

SOLUTIONS

Buffer, pH 7.4

Sodium barbiturate (1.03 g/50.0 ml water)	50.0 ml
0.1 N HCl	31.0 ml
distilled water to make	100.0 ml

Substrate

naphthol AS–D chloroacetate[10]	10.0 mg

Dissolve in:

acetone	0.5 ml

Add:

distilled water	25.0 ml
buffer	25.0 ml
fast blue RR [11] or fast garnet GBC salt	30.0 mg

Shake 30 seconds and filter onto slides.

PROCEDURE

1. Deparaffinize and quickly hydrate paraffin sections. Wash fixative out of fixed smears or frozen sections.
2. Incubate in substrate, room temperature: 30 minutes.
3. Rinse in running water: 1 minute. (2–3 changes of water for loose sections.)
4. Counterstain in Mayer hematoxylin: 3 minutes.
5. Wash in running water: 3 minutes. (2–3 changes of water for loose sections.) Do not blue; the hematoxylin will remain purplish and present fair contrast to the dark blue of fast blue RR.
6. Mount in glycerine jelly.

COMMENTS

Postfixing in 10% formalin improves cellular detail and nuclear staining of both fixed and unfixed smears.

Kernechtrot can be substituted for hematoxylin, but it fades in glycerol jelly. I prefer the fast blue RR salt; the fast garnet GBC produces an undesirable yellowish background. Glycerine jelly mounts keep well; PVP

---

[10] Verona Dyestuffs, Union City, N.J.   [11] Sigma.

and Gelva mounts show leaching and fading after standing for a few weeks.

Burstone (1962) uses a tris buffer at *p*H 7.1.

0.4 *M* tris (24.2 g/500 ml water)	500.0 ml
N HCl	189.0 ml
distilled water to make	1000.0 ml

See also Holt (1956) concerning the indoxyl methods.

Lipase (George and Iype 1960, and George and Ambadkar 1963)

SOLUTION

5% Tween 85 (see comment below)	2.0 ml
barbiturate buffer, *p*H 8.4 (page 541)	5.0 ml
10% calcium chloride	2.0 ml
distilled water	40.0 ml

Incubate mixture, 40°C for 8–10 hours before use, in order to precipitate free fatty acids. Filter. Add crystal of thymol.

PROCEDURE

1. Quickly freeze fresh tissue and cut 40μ sections. Drop into cold (4°C) 6% neutral formalin.
2. Wash in cold distilled water: 1 hour.
3. Mount on albumenized slides and dry.
4. Coat slides with a thin layer of 5%–10% aqueous gelatin. Preserve gelatin with thymol; phenol inactivates the enzyme. (See comment below.)
5. Place in freezer to solidify gelatin.
6. Fix in 6% cold formalin: 1 hour.
7. Wash in running water: 2 hours. Rinse in distilled water.
8. Incubate, 37°C: 8–12 hours.
9. Wash well in warm (40°C) distilled water.
10. Treat with 1% lead nitrate: 20 minutes.
11. Wash in warm distilled water.
12. Wash in running water: 10 minutes. Rinse in distilled water.
13. Treat with 1% aqueous yellow ammonium sulfide: 1–2 minutes.
14. Wash in 1% aqueous acetic acid to remove yellow color: few minutes.
15. Wash in distilled water and mount in glycerine jelly or polyvinyl alcohol (page 452).

RESULT

sites—brownish black

COMMENTS

Eapen (1960) proved that lipase is activated by gelatin; he coated sections to give better results.

If Tween 80 is used, it is acted on by nonspecific esterases as well as lipase.

Counterstaining with hematoxylin can be done.

Bokdawala and George (1964) refine the method by using alizarine red S stain for the calcium soap that is formed.

Additional pieces of tissue can be fixed in calcium-formalin and embedded in gelatin, and the frozen sections can be cut and stained for lipids.

To inactivate the lipase, use heat or Lugol iodine solution.

Sources of error: Brownish pigment or calcareous deposits, pigments visible in unincubated section. These may be hemosiderin (perform Prussian blue reaction) or calcium deposits (remove before incubation in citrate buffer of pH ± 4.5: 10–20 minutes).

SUCCINIC DEHYDROGENASE

Oxidation-reduction reactions, which take place in biological systems, are characterized by hydrogen atoms or electrons being transferred from molecule to molecule. Oxidation refers to the loss of an electron. It follows that the substance donating hydrogen or electrons is oxidized and the one accepting is reduced. These reactions are catalyzed by enzymes, and the living cell contains a series of electron donors and acceptors, which act in a chain. Molecular oxygen is the final electron acceptor in most reactions, and hydrogen combines with the oxygen to form water. Some enzymes are unable to use oxygen directly and the hydrogen atoms must be transferred by intermediate compounds. Dehydrogenases are the first compounds of the chain of enzymes, which transfer the hydrogen atoms from the substrate to the molecular oxygen. Succinic dehydrogenase acts as an intermediate carrier.

Among the techniques for the demonstration of dehydrogenases, the tetrazolium technique has become the preferred one. The tetrazolium salts, characterized by a heterocyclic ring structure containing one carbon and four nitrogen atoms, are colorless or pale and are readily reduced to form intensely colored water-insoluble formazans. The tetrazolium salts, however, do not accept the electrons from the dehydrogenases, but from subsequent links in the chain.

The tetrazolium salts are:

TPT—2,3,5-triphenyl tetrazolium chloride: simplest, but with limited sensitivity.

BT—2,2′,5,5′-tetraphenyl-3,3′ (3,3′-dimethoxy-4,4′-biphenylene) ditetrazolium chloride: better pigment qualities.

Nitro BT—2,2′-di-*p*-nitrophenyl-5,5′-diphenyl-3,3′-(3,3′-dimethoxy-4,4′-biphenylene) ditetrazolium chloride.

Most methods require anaerobic (absence of oxygen) conditions to prevent atmospheric oxygen from competing with the tetrazolium for the electrons of the dye and thus to decrease reduction. Sodium succinate is the substrate.

Mitochondria are a complete biochemical unit involved in biological oxidations and have the succinic oxidase (also cytochrome oxidase) system built within their structure. If mitochrondria are well preserved and the substrate is not oxidized by nonmitochondrial components, the staining could be interpreted as mitochondrial. But not all sites are necessarily mitochondria; formazan can form crystals on the surface of lipid droplets, lipofuscin granules, fat vacuoles, etc.

For further details concerning dehydrogenases and tetrazolium salts, see Burstone (1962).

Most sections are fresh frozen, are cut at 30μ (not less than 20μ), and are improved by cutting in a cryostat. They may be floated on a beaker of the incubating fluid inside the cryostat or mounted on slides and air dried in the cryostat. For anaerobic conditions, remove air from the solution with vacuum, or boil the substrate and then cool to 37°C. In either method, keep the solution in full, tightly closed bottles (small weighing bottles are excellent). Drop the sections in and close the bottle immediately.

Warning: The freezing and thawing of sections may disrupt the morphological structure of the tissue and may result in small formazan crystal artifacts, which can be confused with mitochondria. Cardiac muscle is the best tissue to use for an unfixed frozen preparation.

Novikoff (1961*a*) relates that localization is not possible unless the cell structure is better preserved than is that of cryostat-cut unfixed sections. Try fixing small (1 mm) slices of tissue in cold calcium-formalin, 0° to 5°C: 15 minutes. Quench in isopentane, −70°C: 1–2 minutes, and cut in cryostat at −25°C. Transfer to cover glasses and incubate or transfer directly into incubation solution.

Touch preparations and smears are always easy to prepare, and they make beautiful demonstrations of succinic dehydrogenase. Some cells may be broken and produce some diffusion, but those with unbroken membranes can always be recognized. Both fixed and unfixed slides can be incubated together. Fix the touch preparations and smears in cold calcium formalin: 15–20 seconds.

Nitro BT Method

SOLUTIONS

Tris buffer, pH 7.4

N HCl	17.0 ml
0.4 M tris (24.2 g/500 ml water)	83.0 ml

Substrate

Nitro BT [12]	5.0 mg

Dissolve in:

N-N dimethyl formamide (DMF)	0.25 ml

Add:

sodium succinate	500 mg
distilled water	10.0 ml

Add:

tris buffer	20.0 ml

Shake the solution after each addition. Filter onto slides.

PROCEDURE

1. Incubate fixed or unfixed smears, frozen or cryostat-cut sections, 37°C: 30–45 minutes.
2. Wash in running water: 1 minute. Following wash, unfixed preparations may be fixed in 10% formalin or calcium formalin, room temperature: 10–30 seconds.
3. Counterstain in Mayer hematoxylin: 2–3 minutes.
4. Wash in running water: 3 minutes. Do not blue the hematoxylin.
5. Mount in glycerine jelly or polyvinyl alcohol; or dehydrate, clear, and mount.

RESULTS

sites—blue
nuclei—purple-blue

---

[12] Dajac Laboratories, Philadelphia, Penn.

COMMENTS

Control slides: Replace succinate with distilled water and incubate as above.

Aronson and Pharmakis (1962) observed irregular staining in hearts and fatty degeneration. By adding 25.0 ml of 0.05% potassium cyanide to 75.0 ml of substrate, they improved the staining.

A 0.1 M to 0.06 M phosphate buffer, pH 7.5, may be substituted for the tris buffer.

For more information about succinic dehydrogenase, see: Burstone (1962); Farber and Bueling (1956); Farber and Louviere (1956); Morrison and Kronheim (1962); Nachlas et al. (1957); Novikoff (1961a); Novikoff et al. (1960); Pearson (1958); Rutenburg et al. (1950, 1953); and Zugibe (1970).

THE OXIDASES

The oxidases are enzymes that catalyze the transfer of electrons from a donor substrate to oxygen. Most oxidases contain iron or copper. Tyrosinase contains copper and catalyzes the aerobic oxidation of tyrosine and of some phenols. When tyrosine is acted upon by tyrosinase, dihydroxyphenylalanine (dopa) is formed. Dopa is converted to O-quinone, and is then condensed to form hallochrome, a red pigment, which is spontaneously transformed into a black pigment, melanin. The enzyme tyrosinase, therefore, is responsible for the conversion of tyrosine into melanin, but the so-called tyrosinase system is very complicated, and care should be taken in interpreting its reaction (Burstone, 1962; Pearse, 1960).

Laidlaw Dopa Oxidase Method (Glick 1949)

FIXATION

Only fresh tissue with no fixation (or 5% formalin for no longer than 3 hours, but results are inferior to those from fresh tissue).

SOLUTIONS

Dopa stock solution

DL b (3.4 dioxyphenylalanine) phenyl tyrosine[13]	0.3	g
distilled water	300.0	ml

Store in refrigerator. Discard when it turns red.

---
[13] Eastman Kodak Co. no. 4915.

Buffer solution A

disodium hydrogen phosphate	11.0 g
distilled water	1000.0 ml

Buffer solution B

potassium dihydrogen phosphate	9.0 g
distilled water	1000.0 ml

Buffered dopa solution

stock dopa solution	25.0 ml
buffer solution A	6.0 ml
buffer solution B	2.0 ml

Filter through fine filter paper. Should be *p*H 7.4.

Buffered control solution

distilled water	2.0 ml
buffer solution A	6.0 ml
buffer solution B	2.0 ml

Cresyl violet

cresyl violet acetate	0.5 g
distilled water	100.0 ml

PROCEDURE

1. Cut frozen sections, 10μ.
2. Wash in distilled water: few seconds.
3. Treat with buffered dopa solution, 30°–37°C. The solution becomes red in about 2 hours, and gradually turns sepia brown in 3–4 hours. Do not allow sections to remain in a sepia-colored solution; they may overstain. After first 30 minutes, change to a new solution.
4. Wash in distilled water.
5. Counterstain in cresyl violet.
6. Dehydrate, clear, and mount.

RESULTS

dopa oxidase—black
leukocytes, melanoblasts—gray to black
melanin—yellow brown

COMMENTS

Freeze-dried and embedded tissues give a good dopa reaction.

Tyrosinase is inhibited by compounds which form a stable complex with copper; potassium cyanide and hydrogen peroxide are good examples.

PEROXIDASE METHODS

Peroxidases are hemoproteins that catalyze the transfer of two electrons from substrates to hydrogen peroxide to form water and oxidized dyes. Benzidine was introduced in 1904 as a reagent for peroxidase and is used as an acceptor in the transfer of oxygen from hydrogen peroxide in the reaction catalyzed by this enzyme. This produces a blue product which in a short time fades into brown and diffuses. Treatment with nitroprusside stabilizes this blue reaction product by forming a more stable salt than is formed by the benzidine and peroxidase alone. Peroxidase activity occurs in the mitochondria of striated muscle and heart and the granules of myeloid and mast cells. It takes place also in the acinar cells of thyroid and salivary glands, the medulla of the kidney, the Kupffer cells of the liver, and the hair follicles.

Wachstein and Meisel (1964) use fresh frozen sections (cut 10–20μ), which they spread on lukewarm water and then put directly into the incubation mixture.

Smears, air-dried, can be fixed in formalin–95% alcohol (1:9) or acetone: 30 seconds.

Straus Method (1964)

FIXATION

Fix thin slices of tissue in 10% formalin plus 30% of sucrose, 0°–4°C: 18 hours, not less than 15 hours.

SOLUTIONS

Benzidine solution

50%–70% ethanol	50.0 ml
benzidine[14]	0.1–0.2 g
hydrogen peroxide	0.015 ml

---

[14] Dajac Laboratories.

Nitroprusside solution A

sodium nitroprusside	4.5 g
70% ethanol	50.0 ml

Nitroprusside solution B

30% aqueous sodium nitroprusside	15.0 ml
absolute methanol	35.0 ml
0.2 *M* acetate buffer, *p*H 5.0 (page 000)	2.5 ml

PROCEDURE

1. Cut frozen sections, 4–6μ, attach to slides and dry 1 hour.
2. Treat in ice-cold benzidine solution until peroxidase is distinctly blue. Stop reaction before blue pigment crystals form.
3. Transfer into nitroprusside solution A, ice-cold: few seconds.
4. Transfer into nitroprusside solution B, ice-cold in refrigerator: 1–2 hours.
5. Dehydrate through 3 changes 70% alcohol, 2 changes 95% alcohol, absolute alcohol, absolute alcohol–xylene, and 2 changes xylene: 2–3 minutes each. Mount.

RESULT

sites—gray-green to blue-black.

COMMENTS

For blood pictures the smears or sections can be counterstained in Giemsa: 10 minutes.

Giemsa stock solution (page 255)	4.0 ml
acetone	3.0 ml
0.1 *M* phosphate buffer, *p*H 6.5	2.0 ml
distilled water	31.0 ml

Then dehydrate through isopropanol. Do not use a buffered fixative or a calcium fixative; the use of salts appears to increase the formation of crystals in the blue reaction product. Washing the fixed sections for 24 hours in 30% sucrose also reduces this tendency. Storage in glucose at 0°–4°C for several weeks preserves the enzyme.

If catalase is added to the incubation mixture, no reaction occurs because the catalase destroys the peroxide. Boiling the tissue also destroys activity. If the slides are kept in the refrigerator, they are usable for several months. If kept at room temperature, they fade in a few days.

Wachstein and Meisel (1964) have found that frozen sections are satisfactory, but the red blood cells hemolyze in the aqueous medium and do not react. Calcium-formalin fixation, 4°C: 30 minutes, preserves a positive stain in the hemogloblin and improves activity in muscle. Longer fixation, however, suppresses the activity.

A peroxidase method can be used to demonstrate microbodies (cytoplasmic bodies Type II, peroxisomes) in hepatocytes of livers and renal proximal tubules of kidneys. See Essner (1970); Roels et al. (1970; with bibliography); and Straus (1967).

Cytochrome Oxidase

Cytochrome oxidase is an iron-porphyrin enzyme that catalyzes the reduction of oxygen to water and accounts for the reduction of most of the oxygen used by tissues. It is closely associated with cytochrome C, a hemoprotein capable of undergoing a reversible oxidation reduction. Cytochrome oxidase is demonstrated by the indophenol blue or "nadi" reaction. It was named nadi from the first two letters of *na*phthol and *di*amine (alpha-naphthol and N,N-dimethyl-p-phenylenediamine) the two chemicals that are used to produce the reaction (Burstone 1962).

For information about oxidases, see Burstone (1962), Lillie and Burtner (1953), and Wachstein and Meisel (1964).

MOUNTANTS

Gelva

Burstone (1957*a*) recommends the following procedure for some azo-dye methods:

1. Mount slides directly from 75–80% alcohol in 20% solution of polyvinyl acetate (Gelva 2.5)[15] in 80% alcohol. It is soluble also in acetic acid, *n*-butanol (90%), aniline, pyridine, and ethyl Cellosolve.
2. It preserves esterase, acid and alkaline phosphatase, PAS, hematoxylin, celestin blue, aldehyde fuchsin, eosin.
3. Cover glass is immovable after 1 hour. No bubble formation. Index of refraction is 1.3865, but it increases after evaporation of alcohol and water.

---

[15] Shawinigan Resins Corporation, Springfield, Mass.

Polyvinyl Alcohol (McCurdy and Burstone 1966)

Add 20 g AIRCO [16] to 100 ml cold or tepid water with stirring. When all particles are wet, heat the slurry to the boiling point for about 15 minutes or until all particles are dissolved. A pale yellow color forms. A thin film forms on the surface but can be removed when using the solution.

Polyvinyl alcohol is recommended for various enzyme systems—alkaline and acid phosphatase, esterase, succinic dehydrogenase, and cytochrome oxidase in particular.

McCurdy and Burstone also mention two acrylic resins.

PVP (vinyl pyrrolidone–vinyl acetate copolymers)

PVP/VA, E-735 (Plastone C),[17] is also recommended by Burstone (1962) for some of the enzymes. Mix the PVP with an equal volume of water, stirring well until the plastic is thoroughly mixed into the water. Mounting can be done from water or alcoholic solutions. In my experience, esterase fades in PCP, but acid and alkaline phosphatase are reasonably well preserved, but with some diffusion. If smears or mounted sections are taken from the water and dipped 1 or 2 times in 95% alcohol, mounting in PVP is expedited.

Other Mounting Media

Other mounting media are Nyssco-Plastic (Diacumakos et al. 1960) and prepolymerized butyl methacrylate (Izard 1964).

SUBSTRATE FILM METHODS

In Daoust's method, a film of gelatin containing a substrate is placed in contact with tissue sections. During the exposure, the tissue areas that contain the appropriate enzyme attack the substrate in the film. After exposure, either the unaltered substrate or the products of the reaction are visualized by staining. An "autograph" is made in the film and can

---

[16] Air Reduction Chemical and Carbide Co., N.Y.
[17] Antara Chemicals, General Aniline and Film Corporation, N.Y.

be compared with the corresponding tissue sections to reveal the sites of enzyme activity. For example, if checking for DNAse activity, add DNA to the gelatin. After exposure, the DNA remaining in the film is stained with toluidine blue. Unstained regions correspond to areas of tissue in which DNAse activity occurs (Daoust 1957, 1961; Daoust and Amano 1960; Mayner and Ackerman 1962, 1963).

In another method, a blackened photographic plate is used, and the gelatine film is digested by protease enzymes, thereby leaving an autograph (Adams and Tuqan 1961).

OSMIUM BLACK METHODS

Seligman and Hanker and their coworkers have developed substrates which use groups capable of reacting with osmic acid and can be used for demonstrating enzymes, such as esterase, phosphatases, dehydrogenase and other substances. These osmiophilic reagents can be used for stable preparations that are nonvolatile and electron-opaque for the electron microscope and have good pigment qualities for the light microscope. The end product is called "osmium black." A new principle of osmium bridging has been developed which uses thiocarbohydrizide (TCH) as a ligand to bond osmium black to other metals. The application of this principle has been described in numerous reports; start with Seligman et al. (1968) and their references.

CHAPTER 25

Special Procedures I

EXFOLIATIVE CYTOLOGY

Exfoliative cytology (study of cells, as opposed to histology, study of tissues) concerns the preparation and examination of desquamated cells. These are cells that have been shed or pulled off from a superficial epithelium, mucus membranes, renal tubules, or the like. For example, the horny layer of epidermis is shed normally, but in disease or inflammation the process may become exaggerated and form abnormal-size flakes or scales. These may be placed on slides, stained, and examined.

Methods in this field are especially useful in diagnostic pathology. By screening slides made from smears (vaginal, cervical, prostatic, etc.) or from body fluids (peritoneal, pleural, gastric, urine, spinal, etc.), rapid diagnosis of malignancies is possible.

The cells in question degenerate rapidly, and smears should be made and fixed immediately, in equal parts of 95% alcohol and ether, for a minimum of 15 minutes. The slides may remain in this fluid as long as a week before staining.

Add to body fluids an equal volume of 95% alcohol and centrifuge as soon as possible, 2000 rpm: 3 minutes. Remove supernatant fluid. (If smears cannot be made immediately, cover the sediment with absolute alcohol and place in refrigerator.) Drop some of the sediment on an albumen-coated slide and, using another slide, smear the sediment as evenly as possible over the albumenized area. Allow the smear to begin to dry around the edge but to remain still moist in the center. Fix in ether-alcohol: 1 hour.

After fixation, the slides should not be allowed to dry at any time before or during the staining procedure. In emergencies, they can be shipped dry, but some protection for the smear is advisable if possible. Ehrenrich and Kerpe (1959) add 5% of polyethylene glycol to the fixative. Following complete fixation, allow the slides to dry 5–10 minutes, and then prepare them for shipping. Papanicolaou (1957) protects smears for shipping by covering the fixed smears with Diaphane[1] solution: 2 parts of Diaphane to 3 parts of 95% alcohol. Allow to dry, 20–30 minutes, and prepare for shipping. The polyethylene glycol or the Diaphane form a protective coating over the smear.

Richardson (1960) uses the following solution for shipping purposes; it is economical, noncombustible, and nonvolatile. Biopsy pieces or slides can be shipped in it.

monoethylene glycol	344.6 ml
diethylene glycol	18.1 ml
borax	3.6 g
water	577.6 ml
glacial acetic acid	50.0 ml

Brandi et al. (1962) fix in 1% nitrocellulose in 95% ethyl alcohol–ether (1:1): 15 minutes. Drain until dry.

Spray-Cyte[2] can be sprayed on the slides from an aerosol can and allowed to dry.

Membrane filters are used for screening large volumes of fluids. The suspended cells are concentrated on the filter and processed there with relative ease. The Gelman membrane filters[3] were developed for this purpose and do not retain stain or clear more readily than do other cytological membranes. If possible, fluids should be filtered first and the cells washed and fixed after they are collected on the membrane. Directions and suggestions for the processing of various types of fluids can be obtained with the filters. For more about filter techniques, see Bernhardt et al. (1961); Hulton (1958); McCormick and Coleman (1962); Nedelkoff and Christopherson (1962); Thomison (1961).

---

[1] Will Corp., Rochester, N.Y.   [2] Clay Adams.
[3] Gelman Instrument Co., Chelsea, Mich.

Papanicolaou Method (1942, 1947, 1954, 1957)

This can be used as a routine method for paraffin sections. Excellent preparations of the following solutions can be purchased from the Ortho Pharmaceutical Corporation, Raritan, N.J.: Harris hematoxylin, EA 65, EA 50, and OF 6. Otherwise the stains can be prepared as follows:

SOLUTIONS

Harris hematoxylin, see page 150 (omit acetic acid).

Orange G 6 (OG 6):

orange G 6, 0.5% (0.5 g/100 ml 95% alcohol) . . .	100.0 ml
phosphotungstic acid	0.015 g

Eosin-azure 36 (EA 36 or EA 50, the commercial designation):

light green SF yelolwish (C.I. 42095), 0.5% (0.5 g/100 ml 95% alcohol)	45.0 ml
Bismarck brown (C.I. 21000), 0.5% (0.5 g/100 ml 95% alcohol)	10.0 ml
eosin Y (C.I. 45380), 0.5% (0.5 g/100 ml 95% alcohol)	45.0 ml
phosphotungstic acid	0.2 g
lithium carbonate, saturated aqueous	1 drop

EA 65 is the same, except for the light green content; 0.25% in 95% alcohol is used. This gives a lighter, more transparent stain, which is desirable in smears containing a lot of mucus. Differentiation between acidophilic and basophilic cells is better, however, with EA 36 (EA 50), and therefore it is preferable for vaginal, endocervical, and endometrial smears.

PROCEDURE

1. Out of fixative, hydrate slides to water. Deparaffinize sections before hydration. Rinse in distilled water. A few seconds in each solution is adequate, or dip slides up and down until their surface has a homogeneous appearance.
2. Stain in Harris hematoxylin, either of following methods:
    a. Papanicolaou dilutes Harris with an equal amount of distilled water: 8 minutes.
    b. Harris hematoxylin without acetic acid: 4 minutes.

3. Wash in tap water, running water if flowing only slightly; do not wash off or loosen parts of smear: 3–5 minutes.
4. Differentiate nuclei in 0.5% hydrochloric acid in 70% alcohol (0.5 ml/100 ml) until nuclei are sharp against a pale blue cytoplasm.
5. Wash in slightly running tap water: 5 minutes, or until nuclei are a clear blue.
6. Rinse in distilled water.
7. Transfer through 70%, 80%, and 95% alcohol: few seconds in each or until surface appears homogeneous.
8. Stain in OG 6: 1–2 minutes.
9. Rinse in 95% alcohol, 3 changes: few seconds each.
10. Stain in EA 36 (EA 50): 2–3 minutes.
11. Rinse in 95% alcohol, 3 changes: few seconds each.
12. Dehydrate in absolute alcohol, 2 changes: 1 minute each.
13. Clear and mount.

RESULTS

nuclei—blue
acidophilic cells—red to orange
basophilic cells—green or blue-green
cells or fragments of tissue penetrated by blood—orange or orange-green
blood vessels—orange-red

COMMENTS

Papanicolaou cautions against agitating slides excessively, or crowding them while staining. If parts should float off from a positive slide onto a negative one, a false positive can occur.

Johnson and Klein (1956) apply the method to paraffin sections.

Carson (1959) substitutes Technicon dehydrant[4] for ethyl alcohol, thereby eliminating the need for revenue-taxed ethyl alcohol, which is difficult for some laboratories to obtain.

MacLean (1960) describes processing Papanicolaou smears on the Technicon.

Fluorescent Method (von Bertalanffy and Bickis 1956; von Bertalanffy et al. 1958)

If proper equipment is available, the fluorescent procedure requires less time for preparation and less skill on the part of the examiner than the

---

[4] Technicon Chemical Co., Chauncey, N.Y.

Papanicolaou method. The polychrome staining is so brilliant that cells are seen in sharp contrast against a black background, rendering mass screening simple and quick. Under low power atypical cells show increased amounts of fluorescence; malignant cells are a brilliant, flaming, reddish orange (they seem to glow) in contrast to normal cells, which are usually greenish gray with whitish yellow nuclei. The method is based on the differentiation of RNA and DNA by acridine orange NO (AO). The RNA in the cytoplasm and nucleolus fluoresces red, and the DNA of the nucleus fluoresces green.

SOLUTIONS

Acridine orange stock solution

acridine orange (C.I. 46005)	0.1 g
distilled water	100.0 ml

Keep in dark bottle in refrigerator.

M/15 phosphate buffer, pH 6

$M/15$ sodium phosphate, dibasic ($Na_2H_2PO_4$) (11.876 g/1000 ml water)	7.0 ml
$M/15$ potassuim phosphate (KH_2PO_4) (9.078 g/1000 ml water)	43.0 ml

Keep in refrigerator.

Working solution

acridine orange stock solution	5.0 ml
buffer solution	45.0 ml

Calcium chloride, M/10:

calcium chloride	11.0 g
distilled water	1000.0 ml

PROCEDURE

1. Smears are fixed by Papanicolaou method, ether–95% alcohol: at least 15 minutes.
2. Hydrate to water: few seconds in each solution.
3. Rinse briefly in acetic acid, 1% aqueous (1 ml/99 ml water), followed by wash in distilled water: few minutes.

4. Stain in acridine orange: 3 minutes.
5. Destain in phosphate buffer: 1 minute.
6. Differentiate in calcium chloride: 30 seconds. Wash with buffer.
7. Repeat step 6. If smear is very thick, a longer differentiation may be necessary. Nuclei should be clearly defined.
8. Wash thoroughly with phosphate buffer. Calcium chloride must be removed.
9. Blot carefully and mount with a drop of buffer.

Frozen Sections, Method A

1. Wash out ether-alcohol, freeze, and cut.
2. Omit step 2 and proceed as usual. Use small, shallow dishes; mount sections on slide after staining.

Frozen Sections, Method B

1. Section immediately after biopsy or wrap tissue in saline-soaked gauze and wax paper surrounded by ice. Keep in refrigerator, 4°C, until processed. Cut, transfer to slides, and fix in alcohol or Carnoy's fluid.
2. Hydrate and stain.

RESULTS

DNA of chromatin of nucleus—green, from whitish to yellowish hues
connective tissue fibrils, cornified epithelia; structures—greenish
mucus—dull green
leukocytes—bright green
RNA of basophilic structures of cytoplasm, main portion of nucleoli
    —red and orange fluorescence
sites of strongly acidic groups—deep carmine
malignant cells under low power—brilliant flaming red-orange
proliferating malignant cells under high power—intense fluorescence
        cytoplasm—flaming red-orange containing reddish granules,
            patches, or fine granules
        nuclei—yellowish hues, yellow-orange to yellow-green
        nucleoli—brilliant orange-red
    degenerating and necrotic malignant cells—characterized by a gradual
        loss of cytoplasmic RNA and therefore of the flaming red-orange
            cytoplasm—faint orange or brick red, RNA often concentrated at
                periphery in a rim of dark brick red

nuclei—brilliant green, green-yellow or pale yellow-gray
nucleoli—often enlarged and multiple, orange-red

COMMENTS

Paraffin sections can also be stained by this method.

The buffer mount is a temporary mount, even when ringed. If Elvanol 51–50[5] is dissolved in the buffer until the solution is of thin, syrupy consistency, a clearer and more lasting mount results. Seal the cover slip, however, with one of the ringing cements (page 146). Many scientific supply houses are now advertising media for fluorescent mounting.

After the smear has been used for fluorescent examination, it can be stained by the Papanicolaou technique.

1. Remove cover glass in distilled water.
2. Transfer to 50% alcohol: approximately 1 minute total, 2 changes.
3. Proceed to Papanicolaou method.

Bacteria and parasites will fluoresce and can confuse the results; for example: bacteria—bright orange; *Trichomonas,* cytoplasm—bright orange, and nucleus—whitish yellow. These are easily recognized in vaginal smears. See Umiker and Pickle (1960) for lung cancer diagnosis.

See Taft and Lojananond (1962) for evaluation of the fluorescent method; see Sani et al. (1964) for an excellent article on the cytodiagnosis of cancer.

Dart and Turner Method (1959)

SOLUTIONS

McIlvaine buffer

sodium phosphate (Na_2HPO_4)	10.081 g
citric acid, monohydrate	13.554 g
distilled water	1000.0 ml

Keep in refrigerator. Good for 1 week.

Buffered acridine orange

Stock solution

acridine orange (C.I. 46005)	0.1 g
distilled water	100.0 ml

---

[5] PVA, polyvinyl alcohol, E. I. du Pont de Nemours and Co., Wilmington, Del.

Add 2 ml of Tween 80 to 1000 ml of stock.

Working solution

     acridine orange stock solution . . . .   1 part
     McIlvaine buffer . . . .   9 parts

PROCEDURE

1. Hydrate slides to water, 5 dips in each of 80%, 70%, and 50% alcohol.
2. Dip 4 times in acetic acid, 1% aqueous (1 ml/99 ml water).
3. Rinse in distilled water: 2 minutes.
4. Transfer to McIlvaine buffer: 3 minutes.
5. Stain in buffered acridine orange: 3 minutes.
6. Differentiate in McIlvaine buffer: 4 minutes.
7. Wipe excess material from ends of slides, gently blot with coarse blotting paper.
8. Apply cover glass with buffer. Let stand 2 minutes before examining.

COMMENT

Can be restained, as above, with Papanicolaou. After this, can be decolorized and restained with acridine orange.

SEX CHROMATIN

Barr and Bertram's (1949) demonstration of sex differences in cat neurological tissue created interest in the cells of all parts of the body, and sex chromatin has been identified in 50%–80% of the cells in thin (5μ) sections of various tissues from females. In males, a similar smaller chromocenter occurs in an average of 5% of the cells, but may not be homologous with the sex chromatin of the female (Hamerton 1961). The female chromatin is a dark-staining mass (Barr body) usually lying against the nuclear membrane.

A most interesting and easily observed form of this phenomenon is the drumstick of the neutrophil leukocyte. The drumstick is a round hyperchromatic appendage attached to one lobe of the nucleus by a stem of chromatin. The drumstick should not be confused with smaller clubs, sessile nodules, tags, or "racket" formations. Although the chromatin-rich sessile nodules and tags are more frequent in females, the small clubs and racket structures are perhaps more common to the male (Davidson and Smith 1954). Before a sample is considered chromatin-negative, 500 neutrophils should be examined. If sessile nodules are seen, it is advisable to search beyond the 500 cells for sex chromatin. There is wide variation in the number of cells displaying drumsticks among women: some may show them in as many as 6% of the neutrophils; others may have as few as 1 or 2 per 1000 neutrophils (Briggs 1958). In the case of a low count, the examination should be repeated several times (Caratzali et al. 1957; Kosenow 1962).

Since the sex chromatin is composed largely of DNA and histochemically resembles chromosomes, it can be stained by any stain specific for DNA or for chromatin. With a culture method, Riis (1957) unmasked differences in chromatin structure between male and female lymphocytes. Moldovanu (1961) and Seman (1961) demonstrated sex differences in lymphocytes and monocytes.

Greenblatt and Manautou Method (1957)

FIXATION

Peripheral blood smears in methyl alcohol: 1–2 minutes. See comment below for "buffy coat" suggestion. Oral mucosal and vaginal smears in absolute alcohol–ether (1:1): 2–24 hours.

SOLUTIONS

Pinacyanole

Pinacyanole [6]	0.5 g
70% ethyl or methyl alcohol	100.0 ml

Wright buffer

monobasic potassium phosphate	6.63 g
dibasic sodium phosphate	3.20 g
distilled water	1000.0 ml

PROCEDURE

1. Cover with Pinacyanole; blood smears: 30 seconds; oral mucosal and vaginal smears: 45 seconds.
2. Dilute with Wright buffer (equal to amount of stain); blood smears: 30 seconds; oral mucosal and vaginal smears: 45 seconds.
3. Wash in running water: 2–3 minutes.
4. Decolorize in 50%, 70%, 95%, and absolute alcohol: 30 seconds each.

---

[6] Eastman Kodak Co.

5. *a.* Air-dry blood smears and they are ready for use.

    b. Clear oral mucosal and vaginal smears in xylene, 2 changes; and mount.

RESULTS

Staining: chromatin—blue

Counting: Oral mucosal and vaginal smears—6 cells out of 100 counted must be positive in order for female sex chromatin to be considered female; if fewer than 4 per 100 are positive, the specimen is male.

Blood smears—at least 6 rich-staining "drumsticks" must be counted in 500 polymorph cells for a female count; no more than 2 drumsticks should appear in a male count. (See excellent pictures in Greenblatt and Manautou.)

COMMENT

For sex chromatin counts, the preparation of a buffy coat is recommended. It gives the examiner a fighting chance at a relatively easy cell count. Place the collected blood, plus an anticoagulant, in a glass tube and allow it to stand for several hours, or use mild centrifugation, 2500–3000 rpm. The cellular elements will settle out and leave a straw-colored solution on top. The leukocytes are not as dense as the erythrocytes and will settle out last, remaining as the so-called buffy coat between the red-cell layer and the plasma on top. Smear this buffy coat with its concentration of leukocytes on the slide and fix.

Klinger and Ludwig Method (1957)

FIXATION

95% alcohol for smears and thin embryonic membranes: 30 minutes to 4 hours. Tissue blocks in Davidson fixative (see below): 3–24 hours and transfer to 95% alcohol. If material was fixed in some other fixative, transfer immediately to 95% alcohol, but do not leave in first fixative for longer than 24 hours for good results.

SOLUTIONS

Davidson fixative

95% ethyl alcohol	30.0 ml
formalin	20.0 ml
glacial acetic acid	10.0 ml
distilled water	30.0 ml

Buffered thionin

Solution A

> thionin (C.I. 52000), saturated solution in 50% alcohol. Filter.

Solution B

sodium acetate	9.714 g
sodium barbiturate	14.714 g
distilled water, CO_2-free	500.0 ml

Solution C

hydrochloric acid, sp gr 1.19	8.50 ml
distilled water	991.5 ml

Working solution, pH 5.7

solution A	40.0 ml
solution B	28.0 ml
solution C	32.0 ml

PROCEDURE

1. Deparaffinize and hydrate slides to water.
2. Hydrolyze in 5 N HCl (page 531), 20°–25°C: 20 minutes.
3. Rinse thoroughly in distilled water, several changes; no acid should be carried over into thionin.
4. Stain in buffered thionin: 15–60 minutes.
5. Rinse in distilled water, and in 50% alcohol.
6. Rinse in 70% alcohol until clouds of stain cease to appear.
7. Dehydrate, clear, and mount.

RESULTS

sex chromatin—deep blue-violet
nuclear chromatin—lightly colored
cytoplasm—unstained
fibrin and related structures will show metachromasia

COMMENTS

The Feulgen technique may be used for sex chromatin body identification, but the Klinger and Ludwig method is simpler and quicker. The principle is the same in both methods: the sex chromatin body differentiates from nonspecific nuclear chromatin because of a higher content of DNA, which takes longer to extract. If the basophil shell of the nucleolus is not visible, no attempt at sexing should be made (Klinger and Ludwig 1957). In the Feulgen method, the sex chromatin stains more deeply than any other Feulgen-positive material present.

Skin biopsies are reliable if the sections are of good quality, but the simplest specimens are oral scrapings, easy to obtain and equally easy to stain. Scrape the mucosa of the cheek with a tongue depressor and smear scrapings on an albumen-coated slide. Fix immediately in Davidson's or alcohol. Papanicolaou fixative (equal parts of ether and 95% alcohol) also may be used.

For other methods, see Guard (1959) and Beckert & Garner (1966), who use Biebrich scarlet and fast green; and Lennox (1956), who uses gallocyanine. Romanovsky stains, like Wright, stain the drumstick intensely. Cresyl violet acetate (page 477) can be used to demonstrate the Barr body.

Makowski et al. (1956) identified sex chromatin in the cellular debris of amniotic fluid, spreading the centrifuged sediment on albumenized slides. Vernino and Laskin (1960) identified it in bone decalcified in EDTA. Barr et al. (1950) describe it in nerve cells. See also: Klinger (1958); Marberger et al. (1955); Marwah and Weinmann (1955); Moore and Barr (1955); Moore et al. (1953); Taylor (1963).

CHROMOSOMES

Chromosome Squashes

Giant chromosomes are found in some of the somatic tissues of two-winged flies and reach the largest size in the salivary glands of the larvae. In well-made squashes, the cross striations of the chromosomes stain in varying degrees of intensity. *Drosophila* larvae are the familiar organisms used for demonstrating these striations. But Barley (1964) found the larvae of black flies (*Simulium vittatum*) even easier to work with. Martin (1966) uses the larvae of *Chironomus* ("bloodworms") found in decaying vegetation along edges of ponds.

Acetocarmine or Aceto-orcein Staining

SOLUTIONS

Acetocarmine stock solution

Boil an excess (approximately 0.5 g/100 ml) of carmine in 45% acetic acid, aqueous: 2–4 minutes. Cool and filter.

Working solution

Dilute stock solution with 45% acetic acid, aqueous, 1 : 2. An iron-mordanted stain is often favored because of darker, bluish-tinged red. Belling (1926) adds a few drops of ferric hydrate in 50% acetic acid, but only a few drops: too much iron produces a precipitate in a short time. Moree (1944) determined quantitatively the amount of ferric chloride to add and includes tables of various normalities and volumes. 1% by volume of lactic acid can be added to the solution; this intensifies the staining of pachytene chromosomes (Yerganian 1963).

Aceto-orcein solution

Add 1–2 g orcein[7] to 45 ml of hot acetic acid. When cool, add 55 ml of distilled water. LaCour (1941) used 2% orcein in 70% acetic acid.

Alternate solution

Mix 85% lactic acid and acetic acid in equal amounts. Heat to boiling before adding stain (1–2 g/100 ml solution). Cool and filter. This stain does not precipitate (Yerganian 1963).

PROCEDURE

1. Choose large, sluggish, *Drosophila* larvae, which have crawled up the side of the jar. Hold the posterior end of the larva (in cold-blood saline) with forceps, and with a dissecting needle pull away the mouth parts. The salivary glands will be attached to the mouth parts.
2. Trim off the fat bodies and place the glands in Carnoy fixative or directly into a small amount of stain on an albumenized clean slide.
3. Place a cover glass on top of preparation and cover with bibulous paper. Apply pressure.
4. Seal edges with paraffin, vaseline, or dentists' sticky wax, and allow to stand one day.

RESULTS

acetocarmine—red
iron acetocarmine—deep bluish red
aceto-orcein—dark purple

---

[7] G. T. Gurr's synthetic orcein is expensive, but excellent.

COMMENTS

Sometimes it is advantageous to allow material to remain in stain 5–10 minutes. Then remove to a fresh batch of stain before squashing.

Fragments of animal tissue (ovaries, testes, biopsies) can be put into a tube with a large excess of stain for 2–7 days. Remove to fresh solution on slide and squash.

Barley (1964) stores in Carnoy's. Before making squashes, he treats the glands with acid alcohol (hydrochloric acid-absolute alcohol, 1 : 1) for 2–3 minutes, returns glands to Carnoy's for 5 minutes, and proceeds to stain. This method facilitates rupture of the cells.

Properly applying the pressure on the cover glass is very important for good squashes. Press the thumb in the middle of the cover glass (covered with bibulous paper). Gently roll the thumb toward an edge of the cover glass, so a small amount of stain oozes out and is absorbed by the paper. Continue to roll the thumb toward the edges of the cover glass until no fluid squeezes out. Always roll from the center toward the edges; the fluid must not roll back toward the center and disrupt the cells.

Moorman (1971) stains with a modification of iron hematoxylin.

Jona (1963) squashes under Scotch tape no. 665; Murín (1960) squashes under cellophane.

Lactopropionic Orcein Method (Dyer 1963)

SOLUTIONS

Stock solution

natural orcein	2.0 g
lactic acid	50.0 ml
propionic acid	50.0 ml

Working solution

stock solution	45.0 ml
distilled water	55.0 ml

PROCEDURE

1. Fix 5 minutes in the following:

absolute ethyl alcohol	10 parts
glacial acetic acid	2 parts
chloroform	2 parts
formalin	1 part

2. Macerate in 1 N HCl, 60°C: 5 minutes.

3. Tap out into stain: **2 minutes.**
4. Squash.

chromosomes—deep reddish purple

Permanent Mounts

If a slide is to be made into a permanent mount and there is a chance of losing the specimens, coat the slide with albumen before smearing.

Smith Method (1947)

1. Remove paraffin or vaseline seal with razor blade and xylene.
2. Soak in equal parts of acetic acid and 95% alcohol until cover glass comes off.
3. Place in equal parts of 95% ethyl alcohol and tertiary butyl alcohol: 2 minutes.
4. Transfer to tertiary butyl alcohol, 1–2 changes: 2 minutes each.
5. Briefly drain slide and cover glass against absorbent paper or blotter.
6. Add thin resin mountant to slide and return cover glass to same position as before.

Step 2 can be followed by 2 changes of 95% ethyl alcohol; mount in Euparal or Diaphane.

Nolte Method (1948)

1. Place slide in covered dish so edge of cover glass dips into 95% alcohol: 6–12 hours.
2. Immerse in 95% alcohol: 1–2 hours.
3. With sharp needle, gently pry cover glass free while immersed.
4. Drain off excess alcohol and add drop of Euparal or Diaphane and replace cover glass.

Lamination Method

See page 515.

Other Permanent Methods

Peary (1955) uses a frozen method of dehydration, also counterstains with fast green. See also McClintock (1929), Celarier (1956), and Bradley (1948*b*).

Delamater (1951) describes a freezing method.

Conger and Fairchild (1953) freeze smears while they are in stain on a block of dry ice: 30 seconds. Pry off cover glass with razor blade. Place before thawing in 95% or absolute alcohol, 2 changes: 5 minutes. Mount in Euparal or Diaphane.

Freeze slides in a mixture of absolute methyl or ethyl alcohol and dry ice. Pry off cover slip. If liquid nitrogen is available, freeze slides in it and pry off cover glass.

Hrushovitz and Harder (1962) make permanent mounts in Abopon. See Gardner and Punnett's (1964) method with Hoyer medium.

Elston and Sheehan (1967) spray the underside of the slide with Freon-aerosol mixture.

Restoration of Deteriorated Slides (Persidsky 1954)

If dye has precipitated as dark crystals because the preparation has dried:

1. Remove sealing material.
2. Place 1 drop of 2 N HCl at one edge of cover glass. Apply blotter to opposite edge to help draw the acid under: 3–5 minutes.
3. Replace HCl with acetocarmine by same method.
4. Heat gently; do not boil. Crystals are redissolved and specimen restained.
5. Reseal.

Pachytene Chromosomes

Pachytene chromosomes, the prophase stage of meiosis, are filaments somewhat irregular in outline, but they are clearly double, forming two parallel strands (chromatids) and are united at one point (the centromere). Squash preparations must be processed so that the chromosomes are well spread and all chromosomes can be identified. See Clendenin (1969), Gardner and Punnett (1964); Ohno (1965); Schultz and St. Lawrence (1949); Welshons et al. (1962); Williams et al. (1970); and Yerganian (1957).

Somatic Chromosomes: Culturing and Spreading Techniques

Progress in cytogenetics has advanced rapidly since 1956 with refinements in techniques for establishing the correct number of chromosomes in animals, including man (Tijo and Levan 1956; Ford and Hamerton 1956). Squash techniques have been applied for years, particularly to the chromosomes of plants and invertebrates, but tissue culture methods have now been simplified and spreading techniques have also been developed. Cultures now can be made from bits of tissue, bone marrow, or peripheral blood cells.

The bone marrow method is disadvantageous in that it causes some discomfort to the subject, if not its death. Cultures from the skin cause less discomfort, but the final preparations are the result of many divisions in vitro. The karyotype may possibly alter. Peripheral blood cultures are relatively simple and require only a short-term duration of culture, 2–3 days. The cells are likely to be in first mitosis, or perhaps at most in the second or third mitosis. Apparently, the lymphocytes are the cells that undergo mitosis and have to be separated out from the red cells. Between 5 and 10 ml of blood is ample for several cultures. Other methods have been developed for small samples: Brown and Fleming (1965), Edwards and Young (1961), Hungerford (1965), and Shelley (1963).

At metaphase, chromosomes are more concentrated and can be most easily recognized individually and be counted; in their natural state they lie closely packed and must be separated for identification. The mucoprotein plant extract, phytohemagglutinin, was used originally to agglutinate the erythrocytes as a means of separating the leukocytes from whole blood. Then it was found to be a specific initiator of mitotic activity as well; lymphocytes appeared to be altered to a state in which they were capable of division (Nowell 1960). This stimulation of mitoses is probably an antigenic effect (Goh 1965). Phytohemagglutinin now is used as a stimulant of mitosis as well as a precipitant of red cells. Two forms of phytohemagglutinin (PHA) can now be purchased, M and P. PHA-P is approximately 50 times more potent than PHA-M in its hemagglutinizing capacity and should be used with caution. The desiccated mucoprotein is rehydrated by adding 5 ml of bacto-hemagglutinin buffer or sterile triple-distilled water. Keep the PHA in the freezer and never let the stock reach room temperature. Discard rehydrated stock after one month. PHA-M is recommended for the beginner; if P is used, reduce the amount by at least 50%.

After a period of incubation, a spindle inhibitor is added to the culture to interfere with the formation of the mitotic spindle, arrest the dividing cells and allow them to accumulate at metaphase. Either colchicine or its analogue, Colcemid (CIBA), are used for this purpose. The concentration of colchicine or colcemid and the length of treat-

FIGURE 27. *Chromosome spread of human male.*

FIGURE 28. *Karyotype of human male chromosomes, identified by the London system.*

FIGURE 29. *Chromosome spread of human female.*

ment should be carefully controlled. (Minimum time: 90 minutes; maximum time: 14–18 hours if cells have a generation time of 12–24 hours.) The degree of chromatin condensation varies according to duration of treatment: the longer the action, the more contracted the chromosomes; the longer the chromosomes, the more affected they are by condensation action. Highly contracted chromosomes tend to have thin centromeres, which are located more medially than are those in less contracted chromosomes. Thus the degree of condensation can affect their relative lengths when considered in terms of percentage of the whole complement (Sasaki 1961).

The use of hypotonic solutions (or water) swells the cells and aids in the dispersal of the chromosomes. Squashing techniques have been improved but are surpassed by methods for spreading the chromosomes.

Identification and analysis for karyotypes are made from photographs of suitably spread metaphases (Figures 27, 29). The chromosomes are cut individually from the photographic print and arranged in decreasing size, measured, and paired. Human chromosomes are identified numerically and alphabetically according to a combination of the Denver

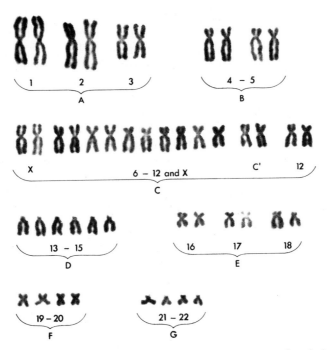

FIGURE 30. *Karyotype of human female chromosomes, identified by the London system.*

System ("Denver Report" 1960) and that of Patau (1960, 1961); this system of identification is called the London System ("London Conference" 1964) (Figures 28, 30).

Culture methods vary greatly but space does not permit inclusion of all of them. Only a few methods used successfully by myself and my colleagues at Los Alamos and the Oak Ridge Associated Universities are given here, although some of the other methods will be cited. Variations on techniques should be tried. Failures must be expected but should not discourage the beginner. When preparing cultures from abnormal blood, try two types of cell suspensions, one with and one without PHA. Experiment with cultures: vary the concentration and combination of the constituents of the medium; try different incubation temperatures; or sample the culture for metaphases at the end of 48, 72, and 96 hours, particularly if abnormal cells are present.

Leukocytes suspended in plasma can be kept 96 hours at 5°C, but do not leave them at room temperature (Hungerford and Nowell, 1962). Blood in a sterile container can be transported on ice for about 5 hours, or kept in the refrigerator at 5°C for 24 hours, without detriment to future culturing (Patau et al., 1961).

Since mammalian cells are sensitive to all conditions, a number of precautions must be taken when culturing them. Use only glassware or plasticware designed for tissue culture work. For washing use nontoxic cleaners sold specially for such containers. Rinse in distilled water. All containers must be sterilized and maintained in a sterile condition. Use stainless steel hypodermic needles; white, silicone, and most amber pure gum rubber stoppers are satisfactory. Keep a constant pH control; make certain the containers are gastight by using screw caps with rubber liners or tight rubber stoppers, or maintain cultures in a controlled CO_2 atmosphere.

Always flame (run through flame of a Bunsen burner) all instruments and openings of bottles used in culturing, before adding cells and after adding cells and before stoppering. When in doubt—flame. Wipe off stopper of heparin bottle with 70% alcohol before inserting syringe needle.

Culturing Peripheral Blood Leukocytes (Goh, 1965)

PREPARATION OF BLOOD

1. Using sterile techniques, draw 5–10 ml of venous blood into a syringe containing 0.1 ml of heparin (Lilly, USP, 100 units per ml)
2. Gently mix by inverting the syringe. With sterile technique, cap the needle or transfer the blood to a 15-ml centrifuge or other small tube, and stopper it with a sterile silicone stopper.

SEDIMENTATION

Must be done under sterile conditions.

1. Let syringe or tube stand vertically at room temperature: 1–2 hours, or until red cells are separated from leukocytes and plasma.
2. Inject from syringe, or aspirate from tube, approximately 1.0 ml of plasma and suspended leukocytes and place in sterile culture bottle (medicine bottles, milk dilution bottles, or others) containing this medium:

TC 199 [8]	7.0 ml
PHA-M	0.1 ml
autologous plasma	2.0 ml

---

[8] Microbiological Associates, Bethesda, Md. and Albany, Calif.; Baltimore Biological Laboratory, Baltimore, Md., Difco Laboratory, Detroit, Mich.; Hyland Laboratory, Los Angeles, Calif.

Add 100 units penicillin and 100 mg streptomycin for each ml of final concentration.

Gently shake bottle.

INCUBATION

1. Incubate in closed bottles, 37.5°C: 68–72 hours.
2. Add 0.1 ml of 10^{-6} M colchicine per 1 ml of culture: $1\frac{1}{2}$–2 hours (up to 4 hours is satisfactory). Leave at 37.5°C. (Sterile conditions no longer required.)
3. Centrifuge, 600 rpm: 6–10 minutes. Discard supernate.

HYPOTONIC TREATMENT

1. Add slowly 1.0 ml of 0.37% sodium citrate. Shake well after first few drops of citrate have been added and again after each addition of more citrate. Cells must be unclumped and resuspended. Let stand at room temperature: 20 minutes. At 10 minutes, bottle may be shaken again.
2. Centrifuge, 300 rpm: 6 minutes. Pipette off and discard supernate. Resuspend cells.

FIXATION

1. Slowly add, with shaking, 1 ml of freshly mixed methanol–acetic acid (3 : 1). Merck's glacial acetic acid is recommended. Ethyl alcohol may be substituted for methyl alcohol, but for best results with either, the alcohol must be 100 proof. Keep bottles tightly closed so contents cannot absorb moisture and keep only one-pound bottles as stock. Resuspend cells thoroughly after first few drops of fixative. Add remaining solution slowly, with shaking, and leave at room temperature: 10 minutes.
2. Centrifuge, 800 rpm: 5–10 minutes. Remove supernate and add fresh fixative.
3. Repeat step 2 at least once more. Several washings with fresh fixative improve the final spreads. Use about 0.5 ml of fixative for final wash.

SLIDE PREPARATION

1. Place clean slides in ice water. Remove a chilled slide, shake off excess moisture, and immediately add a drop or two of cell suspension. Tilt slide against blotter or cleansing tissue to draw off excess fluid and to help spread cells.
2. Dry rapidly over a flame or hot plate. Dry thoroughly before staining.

COMMENTS

Hungerford (1965) uses 0.075 *M* potassium chloride warmed to 37°C in place of sodium citrate and adds 16 USP units of heparin per ml of the hypotonic KCl to reduce the surface adhesion among suspended cells. The clumps then break up more readily.

Alternate methods are given by Ford and Woollam (1963) and Ford and Zeiss (1961). For squashing methods see page 466.

Staining Somatic Chromosomes

Aceto-orcein Method

1. Dissolve 2 g orcein in 45 ml of acetic acid. Boil with stirring. Cool to 50°C and add 55 ml distilled water. Cool to room temperature and filter. Refilter before use to eliminate precipitate.
2. Stain: 30 minutes.
3. Rinse off excess stain with tertiary butyl alcohol (isopropyl or ethyl alcohol may also be used): 3 washes.
4. Mount while still moist in Euparal, or clear in xylene and mount in synthetic resin.

Giemsa Method for Specific Banding (Drets and Shaw 1971)

SOLUTIONS

Alkaline solution, pH 12.0

sodium hydroxide	2.8 g
sodium chloride	6.2 g
distilled water	1000.0 ml

Saline-citrate solution, pH 7.0

sodium chloride	105.2 g
trisodium citrate	52.9 g
distilled water	1000.0 ml

Adjust *p*H to 7.0 with 0.1 *N* HCl (page 531) if necessary.

Buffered Giemsa

Giemsa stock solution	5.0 ml
absolute methanol	3.0 ml

0.1 *M* citric acid (page 544) ....   3.0 ml
distilled water ....   89.0 ml

Adjust *pH* to 6.6 with 0.2 *M* Na$_2$HPO$_4$ (page 544) if necessary.

PROCEDURE

1. Treat fixed and flamed slides in alkaline solution, room temperature: 30 seconds.
2. Rinse in saline-citrate solution, 3 changes: 5–10 minutes in each.
3. Incubate in saline-citrate solution, 65°C: 60–72 hours.
4. Treat with 3 changes of 70% ethanol and 3 changes of 95% ethanol: 3 minutes each.
5. Air dry.
6. Stain in buffered Giemsa: 5 minutes.
7. Rinse briefly in distilled water.
8. Air dry and mount.

COMMENTS

This method results in a highly specific banding pattern in characteristic regions on the arms of each chromosome. The specificity of the banding aids in differentiating the individual chromosomes and in de-
The fluorescent method (Caspersson et al. 1970) also effects discrete
tecting structural changes in them.
labeling and helps to identify members of the different chromosome groups, such as numbers 6–12 (Figures 28, 30). The reagent used in this method can be obtained from Polysciences, Inc., Warrington, Pa.

Cresyl Violet Acetate Method (Humason and Sanders 1963)

1. Place slides in 95% alcohol: 2 minutes or longer.
2. Rinse in distilled water.
3. Hydrolyze in 5 *N* hydrochloric acid: 5 minutes.
4. Rinse in tap water.
5. Wash in 95% alcohol, 2 changes: 15 minutes.
6. Rinse in tap water.
7. Stain in 1% aqueous cresyl violet acetate:[9] 2 hours. Filter the solution before each use.
8. Rinse, 4 or 5 dips in each of 3 changes of tap water. Excess stain must be washed off. Dry immediately on warm plate.
9. Dip in xylene and mount in synthetic resin.

---

[9] National Aniline, CC.

This method results in a deep blue stain especially good for photography.

For other culture methods, see Scherz and Louro (1963), Moorhead et al. (1960), and Humason and Sanders (1963).

Stains for Squashes and Spreads

Aceto–basic fuchsin: Tanaka (1961)
Aceto–iron hematoxylin: Lowry (1963), Wittman (1962, 1963, 1965)
Acridine orange: Schiffer and Vaharu (1962)
Carbol-fuchsin: Carr and Walker (1961)
Feulgen: Rafalko (1946) and Sachs (1953)
Gomori's hematoxylin: Melander and Wingstrand (1953)
Inverted Feulgen: Koulischer and Mulnard (1962)
Lactic-acetic orcein: Welshons et al. (1962)
Lacto-propionic orcein: Dyer (1963)
Iron hematoxylin: Griffin and McQuarrie (1942) and Chen (1944*a*)
Iron hematoxylin with acetocarmine: Austin (1959)
Pinacyanole: Klinger and Hammond (1971)
Silver-hexamethylentetramine: de Martino et al. (1965)
Sudan black B: Cohen (1949)
Sudan black with acetocarmine: Bradley (1957)

Methods of Culturing for Animals

Birds: Krishan (1962)
Dogs: Ford (1965)
Domestic fowl: Newcomber and Donnelly (1963)
Drosophila: Horikawa and Kuroda (1959)
Macaques: Egozcue and Egozcue (1966), Sanders and Humason (1964)
Opossums: Shaver (1962)
Snakes and other cold bloods: Beçak et al. (1962, 1964)
Spider monkeys: Bender and Eide (1962)
Insects: Crozier (1968)
Pigs: Srivastava and Lasley (1968)
Mouse embryos: Wroblewska (1969)

Hunsaker Method for Lizards[10]

SOLUTIONS

Medium

TC 199	100.0 ml
chicken serum	25.0 ml
PHA-M	2.5 ml

Antibiotic solution

penicillin G	350.000 units
streptomycin	350.0 mg
Hank solution (sterile)	100.0 ml

Incubate at 30°C, swirling bottles twice daily.

PROCEDURE

Draw 1.0 ml blood from postcaval or mesenteric vein or heart puncture and sediment it 30 minutes. Plant 0.1 ml of plasma and leukocytes in 5.0 ml of medium and 0.1 ml of antibiotics.

Techniques for culturing biopsies of tissues such as skin will not be included here. The methods are not as simple as those for peripheral blood and require longer periods of culturing. See Berton and Phillips (1961); Ford and Woollam (1963); Harnden (1959); Harnden and Brunton (1965); Hsu and Kellogg (1960); Puck et al. (1958); Swanson and McKee (1964); Tÿo and Puck (1958).

For an in vivo method of cell culture see Berman and Kaplan (1959); Berman and Newby (1963); and Gengozian et al. (1964). Tiny chambers are made of millipore filters and lucite rings. Various kinds of cell suspensions can be placed in these chambers, which are then planted in the peritoneal cavities of mice. The cells will grow in vivo, with the mice serving as a tissue culture medium.

New methods are continually appearing in the literature.

Bone Marrow Preparations

Bone marrow provides the simplest of all methods for preparing chromosome spreads. Only a small sample is needed, and sterile conditions

---

[10] Personal communication.
[11] Courtesy of Mrs. Ruby I. Thompson Zuelzer, Childrens Hospital of Michigan.

can be forgotten. The following method [11] has been modified at Oak Ridge Associated Universities[12] to give excellent well-spread and well-defined chromosomes.

Oak Ridge Modification of Zuelger Method

PROCEDURE

1. Collect 1–5 drops of fresh marrow; place in 10 ml of warm TC 199 (37°C) containing 10 units of heparin and 0.002% colchimide/1 ml.
2. Incubate at 37°C: 1 hour.
3. Centrifuge, 600 rpm: 6–10 minutes. Discard supernate.
4. Add 2–3 ml 0.075% potassium chloride with shaking to break up the cell-button. Make certain the cells are well dispersed in solution. Let stand 7 minutes.
5. Centrifuge and discard supernate.
6. Add fresh methanol–acetic acid (3 : 1) with continuous shaking. The cell button must be broken and the cells dispersed in the fixative.
7. Repeat step 6 at least 2–3 times. In the last fixing solution, stir the cells into suspension (the solution should appear cloudy) and scatter several large drops on a clean slide. Immediately turn the slide upside down and run it through the flame of a Bunsen burner. When the solution ignites, turn the slide right side up and set it aside to cool.
8. Stain with cresyl violet, orcein, buffered Giesma, or any desired stain.

COMMENTS

Do not use "bloody" marrow; a high content of red cells reduces the chances of a successful preparation. In step 1, break up any large clumps of cells; very small groups of cells and, ideally, individual cells make the best spreads.

Chromosome Analysis

Human chromosomes are classified in 7 groups on the basis of total length and arm ratio. Certain indices are relied on for analysis: (1) the length of each chromosome relative to total length of a normal X-containing haploid set (the sum of the lengths of 22 autosomes and the X chromosome expressed per 1000); (2) arm ratio of the chromosomes expressed

---

[12] Dr. Littlefield and staff, Cytogenetics, Medical Division.

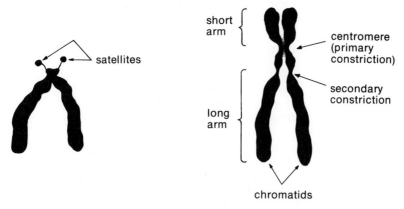

short arm

satellites

centromere (primary constriction)

secondary constriction

long arm

chromatids

FIGURE 31. *Chromosomes and their parts.*

as the length of the longer arm relative to the shorter one; (3) centromeric index (used sometimes), expressed as the ratio of the length of the shorter arm to the length of the whole chromosome; (4) the presence of additional distinguishing features such as satellites and secondary constrictions (Figure 31).

All is not as simple and reliable as it may sound. Considerable variation does exist even between the homologs in the same nucleus. One may be more contracted than the other, and chromosomes lying near the periphery of a spread tend to be larger than those in the center. Thus, the value of measurements is questionable. Each arm should be measured along its axis, but the choice of exact termination of the arm can be arbitrary. In the best of photographs, the ends of the arms are always fuzzy. If one person, however, makes all measurements on a set of chromosomes, some constancy in results can be achieved. Measurements may have to be relied on when other distinguishing features are absent. Labeling with tritiated thymidine will identify some chromosomes. In many chromosome analyses, two or more methods may be required to corroborate the final results.

The members of each of the 7 groups are easily distinguished from those of the other 6, but it is not always easy to distinguish those within a group from one another. Often it becomes necessary to speak of a chromosome as a C chromosome, for example, rather than to designate it by number. They should be arranged according to length, decreasing in size. This form of alphabetical-numerical arrangement is being adopted for other animals.

Exaggerating the constrictions has been tried by Kaback et al. (1964); Patau (1965); Saksela and Moorhead (1962); Sasaki and Makeno (1963); and Upadhya (1963).

For the measurements of normal human chromosomes, see Burdette (1962).

Preparation of Karyotypes

A karyotype is the systematized arrangement of the chromosomes of a single cell prepared by a camera lucida drawing or a photograph (see Figures 28, 30). An idiogram is a diagrammatic representation of a karyotype and may be based on the measurements of the chromosomes of several or many cells.

Contrast process panchromatic film can be used for photographing the spreads. Enlargement prints can be made on Velour black A1, A2, A3, or A4 paper, depending on the contrast required for the film. See Christenson (1965) for photographic suggestions.

Two enlargements are developed and printed, one for determining arm measurements and one for cut-ups for the karyotype. The chromosomes can be cut from the enlarged photograph and be arranged and pasted on Bristol board or the like. Photographic mounting tissue (Kodak) can be lightly attached to the back of the photograph with a mounting iron. Then cut out the chromosomes and apply them to the Bristol board with a mounting iron. Cut-out chromosomes can be lightly mounted by one edge with double-coated Scotch tape on Bristol board (sprayed with plastic) or on a heavy sheet of plastic. Then if necessary, they can be pulled loose with fine forceps and rearranged.

If precise arm measurements are required, they can be made according to Puck.[13] Apply a strip of Scotch linerless no. 665 double-coated tape over each arm of each chromosome, so it follows the central axis of the arm. Using surgical silk[14] follow the axis of each arm, including all curves from proximal to distal end. Cut the silk with a sharp razor, pick up the silk with fine forceps, and stretch it on a section of white Scotch tape no. 666. Measure it to tenths of a millimeter with a millimeter ruler. From these measurements, total lengths, arm ratios, centromeric index, percentage total, and complement lengths can be determined. Select metaphases that show a minimum of contraction; slender chromosomes with well defined centromeres are easier to measure and have more constant arm ratios than short fat chromosomes.

---

[13] Personal communication.   [14] 3–0, Deknatel and Son.

FIGURE 32. *Measuring a chromosome arm with calipers.*

Measuring and calculating by the above method is slow and time consuming. La Bauve et al. (1965) have developed a digitized comparator to obtain the initial linear measurements on punchcards. Final matching, however, still requires some visual evaluation as in the above method.

If chromosomes are well spread with reasonably straight arms and with a minimum of curls and wiggles, and precise measurements are unnecessary, calipers with fine pointed tips can check the arm lengths (Figure 32) or total chromosome length of one chromosome against that of another.

Computer analysis of chromosome spreads is also possible (Neurath 1968).

CHAPTER 26

Special Procedures II

PREPARATION OF INVERTEBRATES FOR WHOLE MOUNTS AND SECTIONS

It is impossible to cover specifically all the members of this huge group of organisms, but certain generalizations can be made that will be applicable to the common forms and perhaps be adaptable to the less common ones. Fixatives can be applied directly to many of the invertebrates, but not to others that, because of their propensity to contract or ball up, pull in their tentacles or other appendages, and thereby make whole mounts or sections practically worthless. Careful anesthetizing or narcotizing of these invertebrates must precede killing and fixation. As soon as the narcotization is complete and before death, if possible, fixation can be successful.

Anesthetizing and Narcotizing Agents

Magnesium chloride or magnesium sulfate

Either is widely and successfully used on sea anemones, corals, annelids, tunicates, and nudibranchs, to name a few. Crystalline magnesium sulfate can be tied in a bag suspended above and just touching the water surface, or a 33% aqueous solution siphoned slowly in, controlled by a screw clamp. When the organisms are anesthetized (no reaction to the touch of a needle), siphon off the water until the animals are barely covered and carefully add fixative. Disturb the animals as little as possible. When partially hardened, transfer to fresh fixative.

Cocaine

This can be used for cilitates, rotifers, bryozoans, hydras, some worms, and nudibranchs. A 1% aqueous solution is added to the water in proportions of about 1.0 ml to 100.0 ml of the water containing the animals. Eucaine hydrochloride can be used in the same manner. Check for contraction and fix.

Menthol

Sprinkle on the water surface and leave overnight. Good for difficult-to-narcotize sessile marine animals, coelenterates, some bryozoans, hydroids, also flukes. It is more efficient when combined with chloral hydrate in proportions of 45.0 g menthol and 55.0 g chloral hydrate. Grind together in a mortar with a little water. Drop on surface of the water. Fix animals when they no longer contract. Large marine forms probably will require overnight treatment. Chloral hydrate can be used alone, sprinkled on water surface for annelids, molluscs, tunicates, bryozoans, and turbellarians.

Chloretone

A 0.33 to 1% solution can be used, but it is slow acting.

Chloroform

This can be dropped on the water surface for many aquatic forms and is used in special bottles for insects and arachnids (see below).

Ether and alcohol

These can be used by dropping on the water, or alcohol can be added gradually to the water by a tube controlled with a screw clamp until the proportion of alcohol to water is approximately 10%. It is particularly good for fresh water forms and earthworms. Ether can be used like chloroform for insects.

Asphyxiation

Boil water to remove the air and seal it in a jar. Particularly good for gastropods (snails); place them in the boiled water after it has cooled.

Cold

Partially freeze organisms in salt and ice mixture or in freezing compartment of refrigerator, or in icewater until they are relaxed. Good for tapeworms. Transfer to lukewarm water and then fix.

Hanley solution (Gray 1954)

water	90.0 ml
ethyl Cellosolve	10.0 ml
eucaine hydrochloride	0.3 g

Add 1 drop per 10 ml of water in which animals are living. Good for rotifers and bryozoans.

Propylene phenoxetol[1] (Owen 1955; Owen and Steedman 1956, 1958; Rosewater 1963)

Introduce a large globule of the compound into the water containing the animals in an amount equally 1% of water volume, or shake vigorously 5 ml of the compound with 15–20 ml of sea water. Add to water containing animals. Good for molluscs.

Tips on Special Handling

Porifera

Small forms can be dropped directly into osmic–mercuric chloride (water, 250.0 ml; osmic acid, 2.5 g; mercuric chloride, 9.0 g). Large forms fix better in alcoholic sublimate (Gilson, Carnoy).

Calcareous sponges can be decalcified in 70%–80% alcohol plus 3% of hydrochloric or nitric acid. Siliceous sponges can be desilicified in 80% alcohol plus 5% hydrofluoric acid added gradually. Perform the latter in a glass dish coated inside with paraffin. After a few hours, transfer to 80%

---

[1] Goldschmidt Chemical Corp., N.Y., N.Y.

alcohol. Do not breathe hydrofluoric acid fumes. Small spicules will section easily without desilification.

Coelenterates

Hydrae

Place in a small amount of water (few drops) in a shallow dish. When animals are extended, rapidly pipette warm mercuric chloride–acetic acid (95.0 ml saturated aqueous mercuric chloride, 5.0 ml acetic acid) at them. Work the fixative from base toward oral region, thereby preventing tentacles from contracting or curling.

Sea anemones

Anesthetize overnight with menthol; or 30% magnesium chloride (50 to 100 ml) can be added gradually over a period of one hour: leave in the solution until there is no more contraction of tentacles. Siphon off the solution until anemones are just barely covered. Add fixative (Susa, Bouin, mercuric-chloride combinations, 10% formalin–sea water) slowly down the side of the container. Also pipette some directly into the throats of the anemones. When the anemones are partially hardened, transfer into fresh undiluted fixative.

Jellyfish

Place in sea water; while stirring, add fixative (10% formalin) down side of container until proportions are approximately 10 ml or more to 100.0 ml of sea water. Stir for 3 or 4 minutes. After 2–3 hours, change to fresh formalin.

Medusae

These must be anesthetized; magnesium solution or cocaine is best. Allow them to expand in a small dish of water; treat with magnesium solution or add 3–4 ml of 1% cocaine per 100.0 ml of water. When they no longer contract, add fixative (10% formalin) and stir.

Corals with extended polyps

Narcotize with magnesium sulfate and fix in hot saturated mercuric chloride plus 5% acetic acid.

Coelenterates

These are sometimes frozen.

Platyhelminths

Planaria

Starve for a few days before killing. Place in a small amount of water on a glass plate. When the worm is extended, add a drop of 2% aqueous nitric acid directly on it. Follow by pipetting fixative (Gilson or saturated mercuric chloride in saline) on it and then transfer into a dish of fixative.

Trematodes

Anesthetize large trematodes with menthol: 30 minutes. Then flatten by the following method: saturate filter paper with Gilson and lay on a plate of glass. Lay worms on paper and quickly cover with another sheet of saturated paper and a second plate of glass. Add a weight, but not such a heavy one that the worms are crushed: 8–12 hours. Remove glass and paper and transfer worms to fresh fixative. A single specimen can be flattened between slides, tied together, and dropped into fixative. After 2–3 hours, remove the string, but do not disturb the slides. Leave overnight. Remove slides and transfer animal to fresh fixative.

Small trematodes can be shaken in a small amount of 0.5%–1.0% salt solution: 3 minutes. Add saturated mercuric chloride plus 2% acetic acid. Shake for several minutes. Change to fresh fixative: 6–12 hours. If they should be flattened, they can be laid on a slide and covered with another slide or, if they are very delicate, with a cover glass. Pipette fixative carefully along the side of the cover and then lower carefully into fixative.

Cestodes

Place a large tapeworm in icewater until relaxed (overnight). Then flatten between glass plates for microscope-slide whole mounts (as for trematodes—see above). The worm can be wound spirally around slides. Then place a slide against both sides, press, and tie together (Demke 1952). For museum mounts, wind the worm around a tall bottle or 100-ml graduated cylinder and pour fixative over it.

Small cestodes can be stretched from an applicator stick supported over a tall container and fixative poured down the length until they hang straight. Immerse them completely; then they can be flattened between microscope slides.

Nemertines

Drop into saturated mercuric chloride–acetic acid (95:5).

Rotiferae

Narcotize by adding 3–5 drops of the following solution for every ml of culture:

2% benzamine hydrochloride	3 parts
distilled water	6 parts
Cellosolve	1 part

When the animals are narcotized so they do not contract when touched, but the cilia are still beating, add 10% of formalin to the culture.

Nemathelminths

Nematodes (hookworms)

Shake 3 minutes in physiological saline. Pour off and drop worms into hot glycerine alcohol (70%–80% alcohol plus 10% of glycerine). A sublimate fixative can be used. (This procedure may be used also for ascarids.)

Very small nematodes can be relaxed and killed in a depression slide or watch glass by gentle heat. An incubator regulated to 50°–52°C is most reliable. If using a flame or hot plate, be careful not to boil the worms. Transfer to fixative.

Formalin-acetic-alcohol is an excellent fixative for small nematodes:

95% ethyl alcohol	12–20.0 ml
glacial acetic acid	1.0 ml
formalin	6.0 ml
distilled water	40.0 ml

If a few drops of saturated aqueous picric acid are added, the worms take up a little color.

Slide whole mounts are easy to prepare. Place the worms in the alcohol-glycerine mixture in an incubator (35°–37°C) or on top of an oven to permit evaporation of the alcohol. When the solution is almost pure glycerine, mount in glycerine jelly (page 130). Helminth ova can also be mounted in this way. For other methods, see whole mounts, page 496.

Friend (1963) makes a microstrainer for handling small ova. Electron microscope grids are attached to the ends of 3.5-mm (internal diameter) glass tubing with an insoluble resin. This method can also be used for processing small larvae.

Microfilaria

Blood infected with microfilaria is smeared and dried as for any blood smear. Then fix in any mercuric chloride fixative and stain with Delafield hematoxylin. If it is preferable to have the slides dehemoglobinized, after fixation treat them with 2% formalin plus 1% of acetic acid: 5 minutes. Wash and stain. Alternate method: smear and dry the blood; dehemoglobinize in 5% acetic acid and air dry; fix in methyl alcohol and stain with Giemsa (page 256).

Bryozoa

Anesthetize with menthol. Fix salt water forms in chromo-acetic solution (7–10 ml 10% chromic acid; 10 ml 10% acetic acid; water to make 100 ml). Fix fresh water forms in 10% formalin.

Brachiopoda

The shells of brachiopods are so fragile the animals must be treated to remain open without forcing the shells. Leave them just covered with sea water and gradually add 95% alcohol until the alcohol concentration is approximately 5%. Let stand for 30–60 minutes. When the shell is easily opened, prop it open with a piece of glass rod and transfer to fixative.

Mollusca

Snails

Place in boiled water or add propylene phenoxetol to the water until snails are limp. Fix.

Mussels

Sections of undecalcified shell can be made by grinding. Decalcification can be undertaken with 3%–4% nitric acid. If the soft parts are to be fixed, wedge the valves apart with a small length of glass rod and place entire animal in fixative. Dissect out after it is fixed. Gills can be removed and fixed flat in a mercuric chloride fixative.

Nudibranchs

Add 1% formalin, a few drops at a time every 15 minutes, to the sea water containing them.

Freshwater molluscs

Gradually warming the water will make them extend their feet. Fix when they no longer respond to the prick of a needle. Bouin, Zenker, or Gilson are satisfactory.

Annelida

If sections of worms (aquatic or terrestrial) are desired, the intestines must be freed of grit and other tough particles.

Cocke (1938) feeds earthworms on cornmeal and agar (1:1) and some chopped lettuce for 3 days, changing the food every day. Becker and Roudabush (1935) recommend a container with the bottom covered with agar. Wash off the agar twice a day for 3–4 days. Moistened blotting paper can be used. When the animals are free from grit, place them in a flat dish with just enough water to cover them. Slowly siphon in 50% alcohol until the strength of the solution is about 10% alcohol. Chloroform can also be used for narcotizing. Fix in Bouin's or mercuric chloride saturated in 80% alcohol plus 5% acetic acid. The worms may be dipped up and down in the fixative and then supported by wire through a posterior segment, hanging down in the fixative, or placed in short lengths of glass tubing in fixative to keep them straight. This is necessary if perfect sagittal sections are to be cut. Embedding is probably most successful by the butyl alcohol method. After removal of mercuric chloride in iodized 80% alcohol, transfer to *n-* or tertiary butyl alcohol: 24 hours (change once). Transfer to butyl alcohol saturated with paraffin (in 50°–60°C oven): 24 hours; pure paraffin: 24 hours; and embed.

Sea worms can be kept in a container of clean sea water, changed every day for 2 or 3 days, then anesthetized with chloroform and fixed, using fast-penetrating fixative (Bouin or a mercuric chloride fixative).

Arthropoda

Insects and arachnids

Ether, chloroform, or potassium cyanide are used for killing. Simplest method: Place a wad of cotton in the bottom of a wide-mouthed jar and cover the cotton with a piece of wire screen. Dampen the cotton with ether or chloroform or lay a few lumps of potassium cyanide on it before adding the screen wire. Keep tightly closed with a cork or screw cap. A piece of rubber tubing soaked in chloroform until it swells and placed under the screen wire will hold chloroform for several days. If the appendages should be spread when fixed, as soon as the insect is dead place it on a glass slide with another slide on top of it. Run fixative in between slides.

For whole organisms, rapidly penetrating fixatives should be used: picro-sulfuric, sublimate fixatives, mixtures containing nitric acid (Carnoy), alcoholic and ordinary Bouin, or Sinha (1953) fixative.

For whole mounts, clearing of exoskeleton is sometimes difficult.

Body contents have to be made transparent or have to be removed. Lactophenol (page 144) mounting will serve the purpose in the first case, but heavily pigmented arthropods probably will have to be bleached in hydrogen peroxide for 12 hours or longer. Fleas, ticks, and the like make better demonstrations if not engorged; in any case they should be treated with 10% potassium hydroxide, 8–12 hours, to swell and dissolve the soft tissues, thus clearing out the body contents. Wash well to remove the potassium hydroxide. Acid corrosives, such as that which follows, are preferred by some because they do not soften the integument as much as alkaline corrosives.

glacial acetic acid	1.0 ml
chloral hydrate	1.0 g
water	1.0 ml

Do not use a fixative containing alcohol or formalin if a corrosive is necessary. The organism will not clear.

Avoid the use of potassium hydroxide for very small or delicate insects. Transfer them into equal parts of chloral hydrate and phenol for about 2 weeks. If they are not cleared at the end of this period, place insects and solution in a 40°C oven for 2 days. This should complete the clearing. Transfer to absolute alcohol and mount in Berlese mountant or Turtox mountant (page 143).

For methods of mounting whole mounts, see page 496.

Because of chitin, sectioning insects can be difficult. Avoid the high-ethyl alcohols. Dioxane methods (page 39), butyl alcohol (above), or double embedding (page 92) must be used and provide excellent results. Soaking the tissue blocks overnight or for several days in water (page 61) simplifies sectioning.

A modified Carnoy is recommended for fixation (Bilstad [2]):

absolute isopropyl alcohol	6 parts
chloroform	3 parts
formic acid	1 part

Fixation, dehydration, and infiltration with Paraplast are all performed under reduced pressure (DeGiusti and Ezman 1955).

For mounting heavily chitinized sections, if a protective cover of a nitrocellulose is undesirable, Bilstad suggests Land mounting medium (page 538). Subbed slides (page 538) may be most efficient for mounts.

See Beckel (1959) for a sectioning method for heavily sclerotized in-

---

[2] Personal communication.

sects. Farnsworth (1963) uses Scotch tape to transport insect eggs through processing.

Echinodermata

Narcotize echinoderms by placing them in fresh water or sprinkle menthol on the water surface. Magnesium chloride or magnesium sulfate can be gradually added to the water.

Inject fixative into the tip of the rays of starfish. This will extend the feet. Then drop the animal in fixative; mercuric chloride–acetic is good.

See Moore (1962) for collecting and preserving starfish, ova, and larvae.

Staining Invertebrates

Sections can be stained in any manner, depending on the fixatives and the desired results.

For beautiful, transparent, whole mounts, the carmine stains (page 498) are the usual choice for most of the invertebrates: obelias, hydras and hydroids, daphnids, Bryozoans, medusae, flukes, tapeworms, small annelids, tunicates, and ammocoetes (larvae), to name a few. But Kornhauser hematein (page 500) is an excellent substitute, particularly for flukes and tapeworms.

If the cuticle and muscle layers of flukes and tapeworms tend to remain opaque and obscure the details of internal anatomy, either of the two methods can be tried to correct the condition. Dehydrate and clear in an oil, such as cedarwood oil or terpinol. Place in a flat dish under a dissecting microscope and carefully scrape away some of the tissue from both surfaces. With care, this can free the animal of some of the dense tissue material and not harm the internal structures. An alternate method is to finish staining and then wash in water. Transfer to a potassium permanganate solution made from a few drops of 0.5% solution added to water until only a pink color develops. When the worm begins to show a greenish brown sheen, remove it immediately to distilled water: 5 minutes. Transfer the worm to 2%–3% aqueous oxalic acid until it is bleached and the sheen is lost. Wash thoroughly in running water at least one hour. Dehydrate, clear, and mount.

References for perparing invertebrates are Becker and Roudabush (1935); Galigher and Kozloff (1964); Gatenby (1950); Gray (1952); Guyer (1953); Hegner, Cort, and Root (1927); Mahoney (1968); and Pantin (1946).

PREPARATION OF CHICK EMBRYOS

The following procedure is a standard and simple one for removing, fixing, and staining chick embryos of any size, from primitive streak to age 96 hours.

With the handle of the scissors, break the shell at the air space end, turn scissors, and cut the shell around the long axis, being careful to keep the tip of the scissors pressed against the inside of the shell while cutting. This will prevent penetration of the yolk or the embryo. With egg submerged in physiological saline at 37°C in a finger bowl, remove top half of the shell. (With rapidity developed from experience, warm water serves just as well as saline, with no ultimate harm to the chick if it is to be fixed immediately.) The chick will be found floating on the upper surface of the yolk. It can be a waste of time to remove the lower half of the shell, but many workers prefer to do so. This is a matter of individual preference. With small scissors, cut quickly around the outside of the vascular area. Except for large embryos, do not cut through the vascular area. Keep one edge gripped by forceps, and slip a syracuse watch glass under the chick. Withdraw the watch glass and embryo with a little of the saline. Bring as little yolk as possible with the chick.

The vitelline membrane most likely will come with the embryo; occasionally it will float free. If not, wave the embryo gently back and forth in the saline until the membrane begins to float loose. Release grip on embryo and remove the membrane. Make certain that the embryo is wrong side up, that is, the side that was against the yolk is now uppermost. Straighten out the chick, taking care that there are no folds. Sometimes washing with saline in a pipette across the top helps to clean and flatten it. Pull off remaining saline and yolk, carefully, not just from one edge, but slowly and with gentle pressure on the pipette bulb around the vascular area. Do not get so close that the embryo is drawn up into the pipette.

Have a circle of filter paper cut with the inside hole approximately the size of the vascular area if it is to be retained, or a little larger than the embryo if it only is to be used. Being careful not to disturb the embryo, drop the circle of paper around it. Press gently around the paper to make it adhere to the blastoderm. With a pipette apply fixative (Gilson or Bouin), carefully dropping it first directly on the embryo; ease it on (do not squirt it on) and then work outward toward and over the paper circle, finally adding enough fixative to completely immerse embryo and paper. Leave overnight. The embryo should remain adhering to the paper, and it can be transported in this fashion from solution to solution. Follow fixation by proper washing, depending upon choice of fixative. Stain the

FIGURE 33. *Development of a chick embryo.* [*Data from Patten* (1952).]

embryo in a carmine, hematoxylin, or hematein stain (see whole mounts, page 500), and dehydrate. Remove the chick from the circle, clear, and mount. The cover glass may have to be supported (page 497) to protect the delicate embryo from destructive pressure. After washing, the embryos can be dehydrated, cleared, and embedded in paraffin.

Incubation time and determination of age can be puzzling. A perfectly aged chick sometimes is difficult to obtain, because the age is determined not only by the number of hours in, and the temperature of, the incubator, but the length of time the egg was in the hen and the temperature at which the egg was stored. The temperature of the incubator should be 37.5°C, and about 4 hours are usually required to warm up the egg once it is placed in the incubator. Thus, a 36-hour chick would require about 40 hours in the incubator for proper development. The accompanying chart (Figure 33) indicates how counting the number of pairs of *somites* in the chick will determine its age; thus 10 pairs indicates an age of 30 hours. This, rather than the number of hours in the incubator, is the best criterion for development.

Any bird or reptilian embryo can be fixed as outlined for the chick. Amphibian larvae can be dropped directly into fixative, after any surrounding yolk or jelly has been dissected off.

Mammalian embryos should be dissected from the uteri and fixed in Bouin. If the uterine site also is to be preserved, cut through the uterus on either side of the embryo and fix. A little fixative can be injected into the uterine cavity to quicken the fixing action.

Ascarid uteri are fixed in alcoholic Bouin; sometimes Carnoy will give desired results.

WHOLE MOUNTS

Whole mounts are slide mounts of entire specimens small enough to be studied in toto and mounted in resins in toluene or water, in gums or glycerine as aqueous mounts, or as dry mounts. Some examples from the innumerable methods will follow, with the hope that a technician can adapt them to almost any whole-mount venture.

General Preparation

FIXATION

95% alcohol, followed by bleaching if necessary. From 70% alcohol into bleaching fluid (3–4 drops of Clorox in 10 ml of 70% alcohol, or water plus H_2O_2 (1 : 1) with a trace of ammonia): 24 hours.

PROCEDURE

1. Wash with 70% alcohol, 3 changes.
2. Dehydrate in 95% and absolute alcohol.
3. Clearing is often difficult. Try mixture of xylene and beechwood creosote (or aniline or phenol) (prevents whole mounts from becoming brittle), 2 changes. If opaque patches persist, return them to absolute alcohol.
4. Impregnating with mountant can be difficult. Specimen may collapse and turn black or opaque when taken directly from alcohol or clearer to mountant. Impregnate gradually by adding each day a drop or two of mountant to the clearer containing the specimen. Carefully stir the mountant into the solution.

    Galigher (1934) uses a cone of filter paper containing the mountant, and allows it to enter the solution gradually and continuously.

    When the clearer has obviously thickened, allow concentration to continue by evaporation.
5. Mounting sometimes requires considerable experience (and patience) in judging the correct method and amount of mountant. Large specimens should have supports under the cover glass to prevent tilting of the cover glass; delicate specimens should have supports under the cover glass for protection against smashing. Various materials may be used to support the cover glass: bits of cover glass or slide, thin glass rods, circles, squares, or strip stamped or cut out of heavy aluminum foil, bits of glass wool.

    Place a drop of mountant on slide and place specimen in it. Lower cover glass carefully; keep it flat (not tilted); have specimen in center of mount. If several specimens are to be mounted under one cover glass, chance of their slipping out of place is lessened by allowing the mountant to air dry for a few hours. Place a bit more mountant over them and add cover glass. Warming the cover glass aids in bubble prevention.

COMMENTS

Demke (1952) embeds helminths in celloidin to support them. Dehydrate through absolute alcohol, through alcohol-ether, and into thin celloidin. Pour celloidin and specimens into flat dish (petri dish); allow solvent to evaporate slowly. Cut out squares of celloidin containing specimens, dehydrate, clear, and mount. No other support of cover glass required.

Rubin (1951) uses PVA mounting medium, page 143. See other mounting media in that section.

Glycerine Jelly Mounts

Many materials, including frozen sections, can be mounted directly from water into glycerine jelly. If there is danger of an object's collapsing, transfer it from 70% alcohol or water into a mixture of 10%–15% glycerine in alcohol or water. Leave the dish uncovered until most of the alcohol or water has evaporated. Mount in glycerine jelly. Sections are mounted without cover glass support. Thicker specimens probably will require a supported cover glass. With a turntable, spin a ring of gold size on the slide. Allow it to dry. If it is not high enough, add more layers to the height desired.

Melt glycerine jelly in water bath or in oven. Add a drop or two of jelly onto specimen inside ring. Use just enough to fill ring. (Warm the slide briefly, and the glycerine jelly will spread readily under the cover glass.) Warm cover glass and ease horizontally into place. Carefully wipe off any glycerine jelly that works out from cover glass. After a few tries, it becomes relatively easy to estimate the correct amount of jelly to make a clean mount. Seal cover glass, supported or unsupported, with gold size or other cover glass sealer (page 146). See page 146 for a method using 2 cover glasses.

Stained Whole Mounts: Hydras, Embryos, and Flukes

Many fixatives are suitable: Carnoy or Gilson for worms, Bouin or Zenker for embryos, formol acetic or saturated $HgCl_2$ acetic (95/5) for flukes. Follow by proper washing and extraction of any undesirable pigments or crystals. But before fixing consult the previous section on invertebrates or an authoritative source concerning the problem at hand.

Choice of stain and method also depends upon the type of material. A few of wide usage are included here.

Grenacher Borax Carmine (Galigher 1934)

SOLUTION

carmine (C.I. 75470)	3.0 g
borax	4.0 g
distilled water	100.0 ml

Boil approximately 30 minutes or until carmine is dissolved. Mixture may be allowed to stand until this occurs. Add:

70% ethyl alcohol	100.0 ml

Allow to stand 1–2 days. Filter.

PROCEDURE

1. Transfer from 50% alcohol to borax carmine: 3–4 hours or overnight.
2. Add concentrated HCl, slowly, drop at a time, stirring, until carmine has precipitated and is brick red. Let stand 6–8 hours or overnight.
3. Add equal volume of 3% HCl in 70% alcohol and thoroughly mix. Let stand 2–3 minutes until specimens have settled. Draw off precipitated carmine with pipette. Add more acid alcohol, mix, allow to settle, and draw off fluid. Repeat until most of carmine is removed.
4. Add fresh acid alcohol and allow tissue to destain, checking at intervals under microscope: may require 2 hours or more. If destaining is exceedingly slow, increase percentage of acid in alcohol.
5. When destained, replace acid alcohol with 80% alcohol, several changes, to remove acid: over a period of 1 hour.
6. Dehydrate in 90% alcohol: 30 minutes or more, depending on size of specimen; absolute alcohol: 30 minutes or more.
7. Clear. For delicate objects clearing probably should be done gradually through the following concentration of absolute alcohol and creosote (or aniline or carbol) xylene, and then into pure creosote-xylene.

absolute alcohol	*creosote (aniline or carbol) xylene*
80 parts	20 parts
60	40
50	50
40	60
20	80
0	100

8. Mount. For exceedingly delicate objects and roundworms, heed the warning above concerning slow impregnation with the mounting medium. Judgment concerning its use rests with experience, but slowness is invariably necessary for roundworms.

Mayer Carmalum (Cowdry 1952)

SOLUTIONS

Stock solution

carminic acid (C.P. carmine) (C.I. 75470) ...	1.0	g
ammonium alum	10.0	g
distilled water	200.0	ml

When dissolved, filter and add:

formalin .... 1.0 ml

Working solution

carmalum stock .... 5.0 ml
glacial acetic acid .... 0.4 ml
distilled water .... 100.0 ml

PROCEDURE

Stain for 48 hours, no destaining. Dehydrate as above. Carmalum is good and requires no destaining.

Hematein (Kornhauser 1930)

SOLUTION

Stock solution

hematein .... 0.5 g
95% ethyl alcohol .... 10 ml
potassium aluminum sulfate, 5% aqueous ...... 500 ml

Grind hematein with alcohol and add to the aqueous sulfate.

Working solution

Dilute above 1:10 with distilled water.

PROCEDURE
1. Stain overnight.
2. Transfer to 70% alcohol.
3. Destain in acid alcohol, 5% HCl in 70% alcohol.
4. Blue in alkaline alcohol, ammonia, or sodium bicarbonate in 70% alcohol.
5. Dehydrate and mount as above.

Hematein is good for flatworms. Alum cochineal used to be popular but has been largely replaced by carmine stains. Alum hematoxylin may be used for small organisms if they are not too dense. The Celestin blue method (Demke 1952) and trichrome stain (Chubb 1963) give good results.

Protozoa

An easy method for handling protozoa is the following: it gives excellent results and rarely fails.

Chen (1944*a*) Cover-glass Method

FIXATION

Hot Schaudinn, 50°C: 5–15 minutes. Other fixatives that can be used are Bouin, Champy, Flemming, or Worcester. See Merton (1932) for fixing stentors, spirostomids, and vorticellas.

Organisms such as paramecia should be concentrated by centrifuging. Quickly pour off some of the solution and add fixative. Amoebae will settle on the bottom of a clean culture dish. Decant off most of the culture media, leaving the bottom barely covered. Quickly pour hot Schaudinn or Bouin (50°–60°C) over the organisms. After a few minutes add an equal amount of 85% alcohol. Carefully loosen any amoebae clinging to the bottom and collect the solution in a centrifuge tube. For both paramecia and amoebae, allow to settle or centrifuge down at low speed. Pour off fixative and wash several times with 70% alcohol, centrifuging after each wash. If a mercuric chloride fixative was used, one of the washes must contain some iodine.

PROCEDURE

1. Follow the 70% washing with 80%–85% alcohol: 10 minutes.
2. Smear cover glass with albumen fixative. Place it albumen-side up on slides with a bit of edge projecting beyond the slide. The projecting edge can be easily grasped with forceps when it becomes necessary to move the cover glass. With a pipette pick up a few drops of alcohol containing organisms. Drop them in the center of the cover glass, where they will spread over its surface. The alcohol will begin to evaporate and bring the specimens in contact with the albumen. Avoid complete drying; the edges may dry, but the center will remain slightly moist. Carefully add a couple of drops of 95% alcohol onto the specimens, and then transfer the cover glass to a petri dish of 95% alcohol. Slides lying in the bottom of the dish will help to handle the cover glass, as above.
3. Carefully remove the cover glass from the 95% alcohol, dip it gently in absolute alcohol, and flood it with 1% celloidin or nitrocellulose. Drain off excess celloidin against filter paper. Wave it back and forth a few seconds until it begins to turn dull and place it in 70%–80% alcohol. It can remain in this solution until ready for staining.

4. Any stain can be applied—hematoxylin, carmine, Feulgen—depending upon study to be undertaken.
5. The cover glasses are finally dehydrated, cleared, and mounted with the specimens albumen-side down on a drop of mounting resin.

COMMENTS

Subbed slides (page 538) are excellent. Cells adhere well and rarely rupture. Apply a drop of cells, drain off excess fluid, and allow to dry.

Smyth (1944) uses a quicker method and directly on slides. After fixation he carries the organisms through graded alcohols into absolute alcohol. This is dropped on a film of albumen on a slide. Place the slide in absolute alcohol and continue from there to the stain.

Agrell (1958) places suspensions of minute embryos on albumen-coated slides. Allowed to almost, but not quite, dry, they become flattened and in close contact with albumen. Dip into absolute alcohol, and then into fixative; place them horizontally in 95% alcohol vapor: 1 minute. Then fix. This coagulates the embryos and attaches them.

Paramecia can be fixed to preserve its normal shape without contraction by first adding copper sulfate or acetate to the cell suspension.

Merton Method (Kirby 1947)

1. Add a drop of paramecia to an albumenized or subbed slide: 30 seconds.
2. Add an equal-sized drop of 1% copper sulfate: 7–8 minutes or 3% copper acetate: 45 seconds.
3. Draw off part of fluid and suspend slide over 2% osmic acid: 45 seconds.
4. Add saturated aqueous mercuric chloride to the drop of organisms: 10 minutes.
5. Transfer to 70% alcohol plus small amount of iodine: 10 minutes.
6. Wash with distilled water.
7. Stain.

Prescott and Carrier Method (1964)

Amoebae, well flattened for tritium or other special studies, can be prepared by this method.

1. Place a small drop of some amoebae on a subbed slide (page 538).
2. Add a small drop of fixative (70% alcohol or acetic alcohol) to a cover glass and place over the amoebas.

3. Immediately freeze in liquid nitrogen: 15 seconds. Flip off cover glass, rinse slide in 95% alcohol, and air dry. If liquid nitrogen is not available, fix in 50% aqueous acetic acid, and freeze in dry ice. (Alcohol fixative will not freeze in dry ice.)

Staining Protozoa

There are many special and routine stains adapted to protozoa. Trichrome and hematoxylin stains are always good. To stain fibrillar elements with iron-hematoxylin, use a fixative containing chromium, warm the stain to 50°C, and destain with 10% commercial H_2O_2 (Kidder 1933). Rothenbacher and Hitchcock (1962) use a Giemsa method for flagella. Loeffler's stain also can be used (Kirby 1947). In ciliates, the surface pattern, cilia, basal granules, and connecting filaments are clearly shown by the Gelei osmium–toluidine blue method (Kirby 1947 and Pitelka 1945). For the silver line system, the classic techniques of Klein (1926), Gelei (1932, 1935), and Chatton and Lwoff (1930, 1935, and 1936) are outstanding. Also see Corliss (1953) for a silver method and Bodian (1937) for an activated protein silver method. Protozoans can be tagged for radioautographs and are excellent subjects for vital staining and enzyme techniques. Schiff et al. (1967) use a safranin–fast-green stain to differentiate the micro- and macronuclei of protozoa. Nigrosin demonstrates various organelles.

Borror (1968) Nigrosin Stain

SOLUTIONS

Solution A

saturated aqueous $HgCl_2$	10.0 ml
glacial acetic acid	2.0 ml
formalin	2.0 ml
tertiary butanol	10.0 ml

Solution B

formalin	20.0 ml
nigrosin (C.I. 50420), water soluble	4.0 g
distilled water	100.0 ml

Working solution

solution A	12 parts
solution B	1 part

PROCEDURE

1. Place a drop of concentrated organisms on slide.
2. Add from a height of 2–3 cm a drop of working solution. Leave 3–4 seconds.
3. Gently wash culture and stain to one end of slide with additional drops of working solution. Stained fixed organisms will remain attached to slide. Leave 15 seconds.
4. Dehydrate, clear, and mount.

RESULTS

ciliary organelles—black against gray background

COMMENTS

A drop of 10% aqueous nigrosin can be mixed with a culture of living organisms and the staining of the various organs observed under the microscope. After allowing the slide to dry, the dye will adhere to some of the structures and the slide can be mounted with resin and cover glass. (Repak and Levine 1967)

The food vacuoles of parameciums are demonstrated by adding a small amount of carmine to the culture. After about 5 minutes the food vacuoles are filled with carmine. Mix a couple of drops of the culture with an equal amount of saturated aqueous solution of nigrosin. Air dry and cover. Do not use too much fluid; it slows down the drying. (Wilhelm and Smoot 1966)

Sectioning Protozoa

Stone and Cameron (1964) **Agar Method** (Kimball and Perdue Modification 1962)

1. Pour a small amount of melted agar (page 532) into a short length of glass tubing sealed at one end. Chill until agar is hardened.
2. Pipette a concentration of organisms on top of agar and add fixative.
3. When cells are fixed and settled on the agar, draw off the fixative.
4. Add more agar mixed with a little eosin or other counterstain and chill to solidify agar.

5. With a pipette, force water down side of agar and under it, thereby forcing the block loose and out of the tubing.
6. Dehydrate, clear, and embed block of agar like a piece of tissue.
7. When the tissue is being sectioned, the cells will be found at the junction of colored agar with the uncolored layer.

Alternate method

Form a block of agar and hollow out a small cavity in it. Pipette a drop of concentrated fixed protozoa into the cavity and allow the organisms to settle. Remove as much fluid as possible and seal the cavity with warm agar. Embed in paraffin. Eosin can be added to the organisms to make them more visible in the block. Eosin in the added agar also helps to locate them.

Dry Mounts of Foraminifera and Radiolaria

An opaque type of slide mounting is used for dry objects that are to be examined by reflected light. Glue the shells on a black background, place a cover glass and supporting ring around them, and cement the cover glass and ring in place. Gray (1964) and Kirby (1947) suggest a simpler and perhaps more practical method. Cut two pieces of cardboard to slide size. Punch a ⅝-inch hole in the center of one piece. Paint a black ⅞-inch square in the center of the other piece or paste a ⅞-inch square of black paper on it. Stick the two pieces of cardboard together with dry mounting tissue (used in photography) and a flatiron. The cavity thus formed can be covered with a cover glass glued on with household cement, or with a slide laid over it, and the parts can be held together with a specially designed aluminum holder.[3] Gray recommends gum tragacanth as the best adhesive for sticking the shells to the background. Place little dabs of the gum on the black background, and with a moistened fine brush (red sable) pick up a good specimen and place it in the center of a drop of gum. Breathe on it; moisture is necessary to make it adhere. When all dabs of gum have a shell applied, breathe again several times on the preparation. Then let the slide dry. Turn it over and tap it several times. If some of the shells are not adhering, it may be necessary to add a drop of water to the gum to make the shells stick. But make certain the slide is dry before covering it. Special opaque slides can be purchased

---

[3] Ward's Natural Science Establishment #C5054.

for making dry mounts (Ward's), or regular depression slides can be used if the depression is painted black.

ANIMAL PARASITES

Animal organisms parasitic in or on man include protozoans, platyhelminths, nemathelminths, and arthropods. Clinical parasitology is extensive and therefore only a few methods are incorporated here. Since the Romanovsky-type stains are used on blood parasites (malaria, trypanosomes, filaria, etc.) the preparation of smears is included in the chapter on hematologic elements (p. 245). Tissue sections can be stained in a similar manner. Parasitic roundworms (pinworms, *Trichuris*, *Ascaris*, hookworms), flatworms (lung, intestinal, bile duct, and blood flukes, tapeworms), and arthropods (ticks, lice, mites) are discussed in the sections on invertebrates and whole mounts in this chapter. Methods given there can be adapted for most parasitic helminths and arthropods. Sections of tissue parasitized by protozoa or helminths are effectively stained by hematoxylin methods, also by periodic acid–Schiff; protozoa and worms are strongly PAS positive, because of stored glycogen in both forms, and PAS-positive cuticle in the helminths. The scolices hooks in hydatid disease (Echinococcus), however, do not show with PAS and are better demonstrated against a hematoxylin background. Ova and larvae can be handled according to directions given on page 145 or 496. Intestinal protozoa (amoebae, flagellates, ciliates, and coccidia) require the following special methods, both for smears and tissue sections.

Only permanent slide mounts are described. Consult clinical laboratory manuals for temporary and rapid examination methods for immediate diagnosis. Gradwohl (1956) is comprehensive.

Intestinal Protozoa: Smear Techniques

Preparing Concentrate Smears (Arensburger and Markell 1960)

1. Add 1 ml of feces to 10–15 times its volume of tap water. Mix well and strain through 2 layers of wet gauze in a funnel. Collect in a small centrifuge tube. Add 1–2 ml of ether. Using a cork or thumb for a stopper, cautiously shake the tube. Fill with water to 1 cm from top.

2. Centrifuge 45 seconds, 2500 rpm. Break up any plug at top and decant supernatant fluid.
3. Add 2–3 ml of normal saline and shake to resuspend the sediment. Fill tube with normal saline to within 1 cm of top and centrifuge.
4. Decant supernatant fluid. Take a small quantity of the original fecal specimen on the end of an applicator stick and mix well with sediment at bottom of tube.
5. With an applicator stick, transfer as much as possible of the material to a clean slide. Smear as for a conventionally made slide and fix immediately in Schaudinn fluid.
6. Stain as preferred, or with one of the following stains:

Goldman (1949) Smears

FIXATION

Schaudinn (page 24), 40°C: 5–15 minutes.

SOLUTIONS

Stock solution A

hematoxylin, 10% in 95% alcohol (10 g/100 ml 95% alcohol)	1.0 ml
95% ethyl alcohol	99.0 ml

Stock solution B

ferric ammonium sulfate	4.0 g
glacial acetic acid	1.0 ml
sulfuric acid, concentrated	0.12 ml
distilled water	100.0 ml

Working solution

Equal parts of A and B. It turns purple, but within a few hours changes to dark brown. It should then be filtered, after which it is ready for use. If it turns greenish black, it is unsuitable for staining; the hematoxylin was too ripe and became overoxidized.

Hansen iron trioxyhematein (page 152) may be used in the same way.

PROCEDURE

1. Rinse slides in 70% alcohol and treat with iodized alcohol.
2. Wash in several changes of 50% alcohol until brown color is removed.

3. Stain in hematoxylin progressively: 3–5 minutes.
4. Wash in running water: 15–30 minutes.
5. Dehydrate, clear, and mount.

RESULT

protozoans—black nuclear stain

Kessel (1925) and Chen (1944a) Smears (Modified)

FIXATION

Schaudinn (page 24), 40°C: 10–15 minutes or more.

SOLUTIONS

Iron alum

ferric ammonium sulfate	4.0 g
distilled water	100.0 ml

Hematoxylin stock solution

hematoxylin	1.0 g
absolute ethyl alcohol	10.0 ml

Allow to ripen several months or hasten process (page 153).

Hematoxylin working solution

hematoxylin stock solution	0.5 ml
distilled water	99.5 ml

Add 3 drops of saturated aqueous lithium carbonate. If the hematoxylin working solution looks rusty or muddy brown, it is unsatisfactory and will not stain efficiently.

PROCEDURE

1. Transfer slides into 70% alcohol (from fixative): 2–3 minutes.
2. Treat with Lugol solution: 2–3 minutes.
3. Wash in running water: 3 minutes.
4. Decolorize with 5% sodium thiosulfate: 2 minutes.
5. Wash in running water: 3–5 minutes.
6. Mordant in iron alum, 40°C: 15 minutes.
7. Wash in running water: 5 minutes.

8. Stain in hematoxylin, 40°C: 15 minutes.
9. Wash in running water: 5 minutes.
10. Destain in iron alum 2% (dilute stock 4% with distilled water, 1 : 1) until nuclei and chromatoidal bodies are sharp against colorless cytoplasm. Check under high dry objective.
11. Wash thoroughly in running water: 15–30 minutes.
12. Dehydrate, clear, and mount.

RESULT

nuclei, chromatoidal bodies—sharp blue-black

COMMENTS

Saturated aqueous picric acid may be used for destaining. Follow it with a rinse in dilute ammonia (2–3 drops/100 ml water) and thorough washing in water.

Diamond (1945) added 1 drop of Turgitol 7[4] to the diluted hematoxylin solution just before use (1 drop per 30–40 ml solution). He substituted the Turgitol treatment for heat; it reduces surface tension and increases cell penetration. Staining time was reduced to 5 minutes.

This method may be used for amoebae in tissue as well as in smears.

Lawless Rapid Method (1953)

SOLUTIONS

Schaudinn-PVA fixative (Burrows 1967)

Dissolve 4.5 g $HgCl_2$ in 31 ml of 95% ethanol in stoppered flask. Shake at intervals. Add 5 ml acetic acid and mix. Set aside until needed. Add 5 g PVA powder[5] to 1.5 ml glycerine and mix thoroughly until all particles of PVA appear to be coated with glycerine. Transfer to 125 ml flask, add 62.5 ml distilled water. Stopper and leave at room temperature: 3 hours to overnight. Occasionally swirl the mixture. Place flask in 70°–75°C water bath: 10 minutes, swirling frequently. When PVA is mostly dissolved add fixative mixture. Continue to mix in bath: 2–3 minutes, or until PVA is dissolved and solution clears. Cool at room temperature. If solution gels during storage, warm in 56°C water bath. Some protozoans fix better in warm PVA fixative.

---

[4] Carbide and Carbon Chemical Corporation.
[5] Delkote, Wilmington, Del.: specify the grade as pretested for use in Brooke and Goldman fixative.

Stain

chromotrope 2R (C.I. 16570)	0.6 g
light green SF, yellowish (C.I. 42095)	0.15 g
fast green FCF (C.I. 42053)	0.15 g
phosphotungstic acid	0.7 g
glacial acetic acid	1.0 ml
distilled water	100.0 ml

Add acetic acid to dyes and phosphotungstic acid; let stand 15–30 minutes. Add water.

PROCEDURE

A small portion of stool is fixed in PVA fixative (1 part stool to 3 parts fixative): 15 minutes to 1 hour or more. In this form it can be shipped in a vial. When ready to make slides, decant off excess PVA solution. Replace cap of vial and shake the emulsion. Remove cap and cover vial opening with gauze. Place 3 or 4 drops of strained material on cleansing tissue or blotter. Allow absorption of PVA for about 5 minutes. Scrape up moist residue and spread with an applicator stick on slide or cover glass. Dry thoroughly and drop into iodine-alcohol (page 13). If smears wash off, too much PVA was carried over; leave material for longer time on cleansing tissue. (In the dry condition, smears can be stored for several months.)

Smears may be fixed directly on slides. Take 1 drop of fecal material to 3 drops of PVA fixative. Smear over a large area of the slide and dry in oven overnight. Immerse in iodine-alcohol.

1. Leave in iodine-alcohol: 1 minute.
2. Decolorize in 70% alcohol, 2 changes: 1–2 minutes each.
3. Stain: 5–10 minutes.
4. Differentiate in acidified 90% alcohol (1 drop acetic acid/10 ml alcohol): 10–20 seconds or until stain no longer runs from smear.
5. Dehydrate in absolute alcohol; rinse twice, dipping up and down.
6. Dehydrate in second change of absolute alcohol: 1 minute.
7. Clear and mount.

Warning: Press the cover glass in place very gently; the smear is somewhat brittle and is easily loosened. Too much pressure on the cover glass may cause parts of the smear to float around on the slide.

RESULTS

background—predominantly green
cysts—bluish green cytoplasm, purplish red nuclei; green is more intense than background stain

engulfed red cells—vary: either green, red, or black
chromatoidal bodies—intense green, but may be reddish
karyosomes—ruby-red
helminth eggs and larvae—usually red

COMMENTS

If cysts do not stain, or do not stain predominantly red, fixation is incomplete. Warming the fixative may help, although cold fixation yields more critical staining. A newly prepared stain is predominantly red in color and reaction; older stains show more violets and greens. The stain is more transparent than hematoxylin stain, but it fades after a few years.

Hajian (1961) fixes in Bouin for better staining of karyosomes.

Tomlinson and Grocott (1944) describe a phloxine–toluidine blue stain for malaria, leishmania, microfilaria, and intestinal protozoa in tissue.

Luthringer and Glenner (1961) modify a methenamine silver nitrate method for endamoebae in tissue.

Silva (1961) combines PAS with iron hematoxylin and light green. Meriwether (1934) uses carmine.

CHAPTER **27**

Special Procedures III

SPECIAL MOUNTS

Sections Mounted on Film (Pickett and Sommer 1960; Sommer and Pickett 1961)

Pickett and Sommer use film in place of glass slides for supporting tissue sections. The flexibility of film and the low cost of material and time can be advantageous. Many sections mounted on a long strip of film are stained in one process by using a developing reel and then are sprayed with plastic as a unit instead of having to be individually coverslipped. A length of film can be stored in a small space, and single sections or groups of sections can be clipped or stapled to records or reports and filed or shipped in envelopes. Film with section also can be mounted in film holders. If desired, one section of a series can be mounted on film and a successive section mounted on glass. Sections already mounted on film can be cut out of it and mounted on glass slides in the conventional manner. Section details are clear under low and medium power. For oil-immersion or high-power examination, a cover glass should be mounted on the section with a nondrying immersion oil (cedarwood oil dissolves plastic). Only low-power projection can be used with this film. Pickett and Sommer include the design of a holder to keep the film flat while it is being scanned under the microscope.

REQUIRED MATERIALS

35-mm film, Cronar, P-40B leader film[1]
Krylon Plastic Spray, Crystal Clear Spray Coating no. 1302 [2]
Nikor 35-mm developing reel, stainless steel

PROCEDURE

1. Cut strips of film of desired length; clean by dipping in acid alcohol; dry with soft cloth (not gauze).
2. Coat film with albumen fixative.
3. Place tissue sections on water bath, maneuver film under ribbon and pick up sections. Dry in 56°C oven. Film can be cut in 3 x 1-inch strips; lay on glass slide; smear with albumen. Place section on film; pipette water under it, and dry on slide warmer as usual for slides.
4. After drying, load lengths of film in reel and stain as usual. Short pieces can be handled individually. To guard against loose sections, protect with coating of 0.5% celloidin or nitrocellulose. Proceed as follows:
   a. Xylene: 1–2 minutes.
   b. Absolute alcohol: 1–2 minutes.
   c. 0.5% celloidin: 5 minutes.
   d. Air dry: 3–4 minutes.
   e. 80% alcohol: 5 minutes.
   f. Wash and stain.
   g. Dehydrate through 95% alcohol.
   h. Mixture of $\frac{1}{3}$ chloroform and $\frac{2}{3}$ absolute alcohol, 2 changes.
   i. Xylene.
5. Place face up on a blotter; do not allow to dry. The film must lie flat; buckling results in uneven plastic coat. Use weights along edges. Apply spray quickly from end to end, 2–3 times until surface appears smooth. Hold can about 6 inches above film. Allow to harden: 5–10 minutes. Repeat 2–3 times. Allow to dry.
6. Label with India ink if desired.

COMMENTS

Mylar polyester film (Du Pont), 0.25 mm thick, also has been recommended (Johnston 1960).

---

[1] E. I. du Pont de Nemours and Co., Wilmington, Del.
[2] Krylon, Norristown, Pa.

Burstone and Flemming (1959) recommend film for smears, touch preparations, and sprayed-on suspensions. Adherence is excellent, and the preparations can be floated on incubation media, staining solutions, etc. The film can be mounted in glycerine jelly or water-soluble plastics.

Berton and Phillips (1961) use film for tissue cultures. The live cells attach and adhere firmly to film.

Shanklin and Laite (1963) immerse the reel of film (after clearing) in Krylon crystal clear 150 base plastic material: 10–15 minutes. Drain: 5–10 minutes. Dip again: 1–2 minutes. Several more short dips may be added, 5–10 minutes between dips. Dry on blotting paper but strip the film from the reel before film dries completely. Stand film on edge to dry.

Cellulose Tape Mounting (Palmgren 1954)

This method has been devised for difficult sections, such as tissues containing a large amount of yolk, chitinous material, or hard tissue that breaks away from soft parts and falls out of sections. Cellulose tape is pressed firmly on the section surface of the paraffin block or the tissue frozen on the freezing microtome. The sticky surface of the tape attaches to all parts of the section and prevents its wrinkling or shattering during sectioning; 1-μ to 100-μ cuts will perform equally well. Using this method, Palmgren has cut whole adult mice for radiography on the freezing microtome without loss of parts. Since the sticky material is soluble in xylene, the paraffin cannot be dissolved while on the tape, and it is preferable to transfer the sections to a glass slide or plate.

PROCEDURE

1. Mount section and tape with section-side up on glass. Hold it in place with small strips of tape, but do not cover the section.
2. Spread a thin layer of 0.5% nitrocellulose over the section, but not over the tape. Let dry.
3. Immerse in xylene in a petri dish, and leave until the nitrocellulose film and section can be loosened easily from the tape. Leave in xylene a little longer to remove all sticky material.
4. Smear clean slide liberally with albumen fixative or use subbed slides (page 538). Float section on it and press in place. Blot dry.
5. Transfer immediately to absolute alcohol–ether (1 : 1) and leave until nitrocellulose is dissolved. Section should be adhering to slide.
6. Transfer to 95% alcohol, hydrate to water, and stain.

COMMENTS

Frozen sections must be either dried (for autoradiographs) or carried from water up to absolute alcohol before the nitrocellulose is applied for histological staining.

Serial sections can be made by adding sections in sequence to continuous tape, rolled at each end of tissue block.

Laminated Thermoplastic Mounts (LaCroix and Preiss 1960)

LaCroix and Preiss describe a practical and simple plastic mounting method for chromosome smears, whole mounts, and cross and sagittal sections. The cellular detail is good, and the plastic mounting makes these preparations practical for class study and demonstration. There is no danger of breakage, and if the plastic gets marred, the scratches can be polished away. See Tonna and Love (1961) for a high-pressure method.

AUTORADIOGRAPHY

"Autoradiography" will be retained here in preference to the alternate form "radioautography" (Tauxe et al. 1954). "Auto-" is added to the noun radiography to describe such techniques as gamma radiography and X-radiography, in which a radioactive substance is placed in direct contact with a photographic plate. Autoradiography can be broken down further into micro-, macro-, or even histo-autoradiography.

Radioactive isotopes of stable compounds can be incorporated in some tissue compounds, and autoradiographs (ARGs) will reveal the location of the compound that was metabolically active during the time of the application of the labeled material. Autoradiographs have become one of the most useful methods for studying biochemical reactions in cells. Many of the most dependable tracer isotopes are those of elements normally found in the body, and the selection of isotope will depend on the study to be undertaken: for example, strontium or calcium for bone, iodine for thyroid, thymidine for DNA.

After proper preparation, the specimen containing the radioactive material is placed in contact with a photographic emulsion. Ionizing radiations are emitted during decay of the isotope and change the emulsion to produce blackening during its development. The number of developed grains will depend on the emulsion and the kind of ionizing particles emitted from the specimen. Alpha (α) particle tracks are large and dense, are not easily deflected, and move in a straight path only a few micra in length. Beta (β) particles are small and produce longer tracks, but they are easily deflected and produce grains far apart and more scattered. Gamma (γ) rays are useless with long random paths, but many isotopes that give off γ rays also produce low-energy β particles, and these will give the picture on film. High-energy β particles (^{32}P) may travel a millimeter or so before producing a grain, but low-energy β particles (^{131}I and ^{14}C) have short paths and give very satisfactory resolution on photographic emulsion.

Many varieties of material can be used—tissue sections, chromosome squashes and spreads, microorganisms, whole cells, smears, cultures, sections of whole animals, etc. The specimen and film are placed in contact for a certain length of exposure time. The radioactive atoms are decaying, and the emitted radiation strikes the photographic emulsion.

Monochrome film emulsions consist of mixtures of gelatin and crystalline silver halides, primarily bromides. The halide crystals are imperfect and provide electron traps to produce a concentration of electrons. The positively charged silver atoms are attracted by the negatively charged electrons and form a latent image. Development reduces the silver halide to metallic silver, and the unreduced grains of silver are dissolved away in the hypo solution. Also, the emulsion is hardened by the hypo. Ordinary photographic emulsions cannot be used, because they contain only about 30% silver halide and have large grains. Special emulsions have been developed to intercept the maximum number of rays. The nuclear emulsions contain up to 90% and 95% silver halide and have small grains. American Kodak nuclear emulsions are NTA, NTB, NTB2, and NTB3, arranged according to ascending sensitivity. The last two are the most commonly used, particularly for tritium and ^{14}C. Emulsions of Ilford Ltd., England, are also popular for use, because of their sensitivity and grain size.

As mentioned above, the grains are not uniform; some carry a silver speck, and others are relatively insensitive. Some grains develop in an unexposed emulsion, and some become accidentally developable by darkroom light, cosmic rays, or other sources of ionizing radiation. When evaluating autoradiographs, such "background" values have to be considered and checked. Take care to avoid unnecessary background accumulation.

The following are good texts on autoradiography:

Baserga, Renato, and Malamud, Daniel. *Autoradiography. Techniques and Application.* New York: Harper and Row Publishers 1969.

Gude, William D. *Autoradiographic Techniques*. Biological Techniques Series, Englewood Cliffs, N.J.: Prentice-Hall 1968.

Rogers, Andrew W. *Techniques of Autoradiography*. New York: Elsevier Publishing Company 1967.

Roth, Lloyd J., and Stump, Walter E. *Autoradiography of Diffusible Substances*. New York: Academic Press 1969.

PROCEDURES FOR AUTORADIOGRAPHS

Fixation and Slide Preparation

Knowing the half-life of the isotope that will be used for autoradiographs helps to determine the approximate time of exposure required for ARGs and the length of time that the tissue can be stored before processing. The half-life is the time required for disintegration of one-half of the atoms present. For example, ^{32}P has a half-life of 14.2 days and must be processed immediately. Within 42 days only 12% of the original radioactivity will remain. ^{3}H has a half-line of more than 12 years and loses only 0.5% of its activity each month. ^{14}C has a half-life of 5568 years, so you have your lifetime plus for processing. ^{125}I has 60 days, ^{131}I only 8.07 days. For the latter, fix in ethanol and process in a hurry.

The aldehydes are usually satisfactory for fixation, but avoid solutions containing mercuric chloride, which will produce artifacts on the emulsion. If a fixative such as Zenker has been used, treat the sections with iodine and thiosulfate followed by 0.2 M cysteine for 10 minutes. This will reduce desensitization of the emulsion. Carnoy, methanol, or acetic-alcohol (1 : 3) can be used for smears and tissue cultures. Allow smears to dry and flatten the cells before fixation. In all cases the fixative must be washed out of the tissue; some fixatives cause air bubbles. This applies as well to smears; wash them in several changes of 70% alcohol and distilled water.

Solubilities of isotopes must be checked carefully; some leach out in water or paraffin solvents and must be prepared by freeze-drying methods (Holt and Warren 1953). In some cases Carbowax can be used, but it will dissolve water-soluble isotopes (Holt et al. 1949, 1952; Holt and Warren 1950, 1953). Mounting must be handled with care to prevent leaching (Gallimore et al. 1954).

Witten and Holmstrom (1953) freeze the tissue at the microtome immediately after excision. The knife is kept cold with dry ice, keeping the section frozen after it is cut. Then the section is carried still frozen to the photographic emulsion. As the section thaws it produces a bit of moisture and thereby adheres to the emulsion.

Bone must *not* be decalcified for this process. In some of the best techniques for bone, methacrylate or epoxy embedding is used before sectioning (Arnold and Jee 1954*a,b;* Norris and Jenkins 1960; and Woodruff and Norris 1955).

If the isotope leaches out slowly and quantitative results are not required, paraffin embedding can be undertaken, but keep the periods in fixative and alcohol solutions to a minimum. Butyl alcohol extracts less of some isotopes (^{125}I and ^{131}I) than does isopropyl or ethyl alcohol. Use xylene, not toluene, for clearing before paraffin infiltration. A brownish discoloration of the sections indicates improper infiltration with paraffin. Vacuum infiltration will correct this, but if vacuum is not available, when the sections are mounted on slides later, leave them on a hot plate or run them through a Bunsen burner just long enough to melt the paraffin (page 65). Check for leaching by counting the reagents after use in an appropriate counter. If a significant count is given off, avoid the use of that reagent. When the tissue is embedded in paraffin, leaching is no longer a problem.

Slides used for mounting must be clean; dust can cause air bubbles and distortion of the emulsion. To clean, soak the slides in cleaning solution (page 555). Wash them overnight in running tap water; one hour in distilled water, changing the solution several times; and in 2 changes of 80% alcohol. Dry. They also can be cleaned by boiling in Alconox. Treat with sub solution (page 538) and dry. Store the subbed slides in dustfree boxes and use within a month. In the darkroom, frosted ends on slides help to determine the side holding the sections. Add identifying code with Labink.[2] Other marks may wash off in solutions.

Mount the sections, drain off the water, dry for 10–15 minutes on a slide warmer, and store in dustfree boxes until application of the emulsion. Subbed slides do not require as long a drying period as do albumenized slides, and the sections and emulsion adhere better to the subbed surface than to albumen. (*Suggestion:* Mount subbed slides from a water bath or place a puddle of water on slide before adding sections. Do not try to place sections on dry subbed slides and then add water.)

Sometimes staining is used before emulsion application, but many stains are affected by photographic processing and some stains may remove the tracer. If staining is done first, a protective coating (celloidin, Formvar, or nylon) usually has to be applied to prevent decolorization during de-

[2] Black no. 6249, Arthur H. Thomas.

velopment. This layer will reduce the amount of radiation that reaches the emulsion from weak emitters, e.g., tritium and ^{14}C. Feulgen stain is commonly used before emulsion is applied, but most staining is done after development.

Emulsion Application

The two methods of application are the "stripping film" and the "liquid emulsion" or "dipping" methods. Stripping film is a little more difficult to master than dipping, but it produces a more uniform emulsion. The dipping technique is given in detail here, since it is simpler and less likely to go wrong.

Stripping Film Method

Stripping film emulsion is removed from its glass plate. First, cut the edges of the emulsion with a razor blade, peel off the emulsion, and float it on distilled water with 1% Dupanol (wetting agent) added. Then slip the slide with the section uppermost under the emulsion and lift it out, with the emulsion. To hold the emulsion in place without slipping, fold it under three sides of the slide, dry it in place, and make the exposure (Bogoroch 1951; Pelc 1956; Simmel 1957).

Liquid Emulsion Method

1. Store liquid emulsion at 5°C; 4 months is maximum time. Remove to room temperature, but do not open until in darkroom. The emulsion is a gel at room temperature and has to be melted in a water bath at 42°–45°C. This must be done in the darkroom under Wratten series no. 1 safelight, red filter, 15-watt bulb. Work no closer than 3 feet from light.
2. When emulsion is melted, pour the amount to be used into the dipping vessel. Never reuse emulsion. For a few slides a coplin jar is convenient; if it is filled to the top of the slots, the depth will be sufficient for most of the slides. Leave at 42°–45°C for 30–45 minutes to attain proper fluidity and to allow bubbles to escape.
3. Test each new bottle of emulsion. Dip a blank slide slowly into the emulsion and withdraw it with a slow uniform motion (4–5 seconds total). Check it for lumpiness, uneven coating, and bubbles. Drain against tissue or gauze. Wipe the back dry and air dry in a vertical position in a light, tight box: 1–2 hours. Develop and fix (see below)

FIGURE 34. *Slide holder for autoradiograph processing.*

and check for background. If the background is objectionable, see Caro (1964) and Brenner (1962) for ways of reducing it.

4. Squashes, spreads, etc. can be dipped as is, but paraffin must be removed from sections. Use 2 changes of xylene, remove the xylene with 2 changes of absolute alcohol, and hydrate the sections to water. If Carbowax has been used, remove it in water. The slides with mounted tissue are dipped as described above in step 3. Two slides can be dipped back to back. Coleman (1965) dips 5 at a time by using a molded plastic holder[3] and completes all processing in the holder (Figure 34). To reduce cooling of the emulsion, arrange the slides back to back in an empty coplin jar in the water bath to warm them before dipping.

5. Drain and separate slides if they have been dipped back to back. Slides in plastic holder can be stood on end, holder uppermost, to drain. Prescott (1964) arranges them in a neoprene-coated test tube rack, where they fit diagonally across the square openings. Place in a light, tight box for drying overnight. Allow them to dry slowly. Do not use hot air; fast drying increases the background (Sawicki and Pawinska 1965), and 40–50% humidity is desirable. An 80% relative humidity incubator for slow drying is recommended by Kopriwa and Leblond (1962).

For supporting slides while drying, I use lengths (12–15") of wood or plastic (1½–2" square) lying on large blotters (Figure 35).

---

[3] Disposable Supply Co., Long Beach, Calif.

FIGURE 35. *Slides drying while resting against a plastic or wood block or slide box.*

The slides lean in a vertical position, backside against the wood, and drain on the blotter. For a small number of slides, the side of a closed wooden or plastic slide box is a satisfactory back rest. A cool darkroom with water running in the sink usually provides an adequate drying atmosphere, but must be completely dark.

6. The slides must be dry before they are boxed for exposure, and they must be kept dry, because a wet emulsion is practically insensitive. When dry, transfer them to bakelite slide boxes (25-slide capacity) no more than 8–9 to a box, allowing 2–3 slots of spacing between slides. Place 15–25 g of Drierite in a tissue or gauze bag in the bottom of the box. Do not allow Drierite to get on the emulsion. Sikov (1965) recommends an encapsulated desiccant, "Humi-caps,[4] that can be rejuvenated in a 60–70°C oven or discarded after use. Seal the four sides of the box with two lengths of black photographic tape and wrap in aluminum foil. Store in refrigerator, 40°C during exposure. Keep the boxes standing on end with the slides in horizontal position, emulsion side up.

    Slides in a molded plastic holder (Figure 34) can be wrapped in several layers of heavy foil for exposure.

7. At intervals of 1, 2, 3, or more weeks, test slides can be removed from the exposure box and be developed. If the slides are cold, allow them to reach darkroom temperature before development. All solutions should be of uniform temperature, 17–18°C. At temperature above 20°C, reticulation occurs in the emulsion.

    a. Develop in D11, D19, or Dektol: 2 minutes.

    b. Rinse in water or treat in a stop-bath of 1% acetic acid: 10 seconds.

---

[4] Dri-Aire, East Norwalk, Conn.

c. Transfer to Kodak Rapid Fixer: 5 minutes. (Some stains react better with the tissue if a hardener is not included in the fixer.)

d. Wash in running water: 10–15 minutes.

e. Rinse in distilled water.

f. Dry, store in dustfree box, or proceed to stain.

General references for autoradiography are Boyd (1955); Caro (1964); Controls for Radiation, Inc.,[5] instruction manual; Fitzgerald et al. (1953); Joftes (1959); Joftes and Warren (1955); Pelc (1958); Perry (1964); Prescott (1964); Sacks (1965).

For dry mounting of water-soluble materials, see Branton and Jacobson (1961); Fitzgerald (1961); Gallimore et al. (1954); Miller et al. (1964); Sterling and Chichester (1956); Stumpf and Roth (1964).

For double emulsion techniques, see Baserga (1961); Baserga and Nemeroff (1962); Pickworth et al. (1963).

Staining

For most experimental work, I find that 5–10 minutes staining with Mayer hematoxylin or Kernechtrot provides satisfactory backgrounds for ARGs. The red of Kernechtrot is a good contrast color for the black of the developed silver grains. Sams and Davies (1967) add Alcian blue and tartrazine to their staining sequence for additional colors. Gude stain (below) is excellent. Keep all solutions cool; never use warm ones.

After mounting ARGs in synthetic resin, dark opaque areas occasionally develop. The resin, instead of penetrating the emulsion, is drawing the clearing agent out of the tissue and leaving it opaque. If this happens, return the slides to the clearing agent (xylene or toluene) long enough to remove all the mounting resin and the tissues will look clear again. Then transfer the slides through the following solutions:

xylene/resin (2 : 1) for 1–2 hours
xylene/resin (1 : 1) for 1–2 hours
xylene/resin (1 : 2) for 1–2 hours

Mount.

The slides can be left overnight in any of the solutions.

---

[5] 130 Alewife Brook Parkway, Cambridge, Mass.

Giemsa Stain (Gude et al. Modification 1955)

PROCEDURE

1. Air dry slides of smears, sections, chromosomes, etc.
2. Stain 1–2 hours in freshly mixed:

Giemsa stock solution	2.5 ml
absolute methyl alcohol	3.0 ml
distilled water	100.0 ml
$M/10$ citric acid	11.0 ml
$M/5$ disodium phosphate	6.0 ml

Use only once.

3. Rinse off stain in distilled water.
4. Air dry and cover with synthetic medium and cover glass.

RESULT

The light blue staining of nuclear material furnishes a good contrast for the black-silver grains of the chromatin tag.

COMMENTS

If this stain is used on tissue sections, follow step 3 with 3 changes isopropanol, 2–3 minutes each; xylene-isopropanol, 2–3 minutes; xylene, 2 changes, 2–3 minutes each (xylene-resin mixes if necessary); and mount.

Excess emulsion can be removed from the back of slides with a razor blade.

Other stains may be used: Darrow red–light green (Wolberg 1965); methylene blue–basic fuchsin (Bélanger 1961); nuclear fast red–indigo carmine (Mortreuil-Langlois 1962); peroxidase (Popp et al. 1962); toluidine blue–methyl green–pyronin (Stone and Cameron 1964); metanil yellow–hematoxylin (Simmel et al. 1951). Celestin blue and gallocyanin have been recommended, but Deuchar (1962) reports a dense over-all graining of film with celestin blue, and Stenram (1962) found that gallocyanin causes a loss of silver grains. See Thurston and Joftes (1963) for a review of staining.

The Use of an Isotope

The isotope in popular use for small cytological structures is tritium (^{3}H). It emits only β radiation, and in water the maximum range of its emitted electrons is 3μ. If an emulsion lies over a point-like source containing tritium, only an area averaging 3μ in diameter will be exposed. Tritium, therefore, has become the isotope of choice for tagging chromosomes.

Mitotic chromosomes cannot be labeled, since DNA is not replicated during mitosis. But mitosis is preceded by the synthesis of DNA, and if radioactive DNA precursor is administered during synthesis the nucleus becomes radioactive and retains the tag during mitosis. It also passes the tag on to its daughter cells. Thymidine (deoxyriboside of thymine, one of the bases of DNA) can be effectively incorporated into DNA, and when it is labeled with tritium it enters the synthetic chain to label the DNA. To obtain efficiently labeled metaphases, therefore, the radioactive thymidine must be given to the cells several hours before they divide, that is, during the DNA-synthetic or S-period. This period varies but is species-specific. Different types of cultures may influence the length of periods—the S-period and the lag period that follows. The amount of thymidine influences the rate of uptake, and incubation temperature is critical. It is necessary to explore all these influences when first using thymidine on cultured cells.

Chromosome Cultures

PROCEDURE

1. Add tritiated thymidine[6] to a final concentration of 1 μc/ml (specific activity 1.9 curies/mmole): 5–6 hours before termination of the culture. Maintain culture at 37°C.
2. Harvest cells at 3, 4, 5, and 6 hours (also 7 and 8 if desired) after introduction of thymidine, but add colchicine or colcemid 2 hours before each harvesting.
3. The thymidine is removed with the medium. Treat with hypotonic solution: 10 minutes.
4. Fix, squash, or spread as usual (page 475). Stain some preparations if desired.
5. Dried spreads or squashes can be stored at 4°C for about 4 weeks. At room temperature background develops.
6. Coating or stripping techniques may be done with NTB2 (Kodak) or Ilford K5 emulsions.
7. Expose 2 days.
8. Develop: 2 minutes. Rinse, fix, and wash.
9. Dry and stain in Gude's Giemsa.

---

[6] Volk Radiochemical Co., Chicago, Los Angeles, Washington, D.C., and New York; Nuclear Chicago Corp., Des Plaines, Ill.; New England Nuclear Corp., Boston; Schwarz BioResearch Inc., Orangeburg, N.Y.

COMMENTS

Caution: Handle the tritium thymidine with care. Since it does not emit γ radiation, contamination cannot be easily detected with scintillation counters. Work with the labeled thymidine only on protected areas, covered with "hot mats" [7] or a large sheet of heavy aluminum foil or plastic under a large blotter.

For details about the use of tritiated thymidine, see Frøland (1965), Leblond et al. (1959), Schmid (1965), and Sisken (1964).

---

[7] 19 × 23 paper towels with neoprene backing (300 to package); B. H. Jordan Co. Inc., 10 S. Middle Neck Road, Great Neck, N.Y., or protective absorbent paper (polyethylene backing), rolls or sheets, New England Nuclear, Boston, Mass.

PART IV

SOLUTION PREPARATION AND GENERAL LABORATORY AIDS

CHAPTER 28

Solution Preparation

ABBREVIATIONS AND TERMS

Laboratory solutions, defined in terms of the methods of measurement used in their preparation, can be grouped as molecular solutions, normal solutions, and percentage solutions. In discussing these the following abbreviations are used:

ml	= milliliter	mg	= milligram
cc	= cubic centimeter	M	= molecular solution
g	= gram	N	= normal solution

Molecular Solutions

A molecular solution (M) contains the molecular weight in grams of the substance made up to 1 liter with distilled water (Cowdry 1952). For example: M oxalic acid $(COOH)_2 \cdot 2H_2O$ is 126 (molecular weight) grams in 1 liter of water.

Molecular weight expressed in grams is called the gram-molecular weight or mole. A millimole is 1/1000 of a mole.

Normal Solutions

A normal solution (N) contains 1 gram-molecular weight of dissolved substance divided by the hydrogen equivalent of the substance (that is, 1 gram equivalent) per liter of solution. N oxalic acid is half the concentration of the M solution above.

Percentage Solutions

In the preparation of percentage solutions, it should be made clear whether the percentage is determined by weight or volume. The percentage should be written out in grams and milliliters, or it should be expressed as follows:

w/v = weight in grams in a 100-ml volume of solution
v/v = volume in milliliters in a 100-ml volume of solution

Although it is erroneous, a long-established practice of technicians is to dilute liquids as though the reagent solution was 100% concentration. In most procedures the correct dilution is indicated in parentheses following the percentage desired. That is, a 1% solution of acetic acid is 1 ml of glacial acetic acid in 99 ml of distilled water.

The table on the facing page gives dilutions for 1 N solutions which are sufficiently accurate for histological preparations. To make a normal solution add enough distilled water to the number of ml shown in the right-hand column to make a combined total of 1 liter.

STOCK SOLUTIONS

Solutions normally found on histology laboratory shelves are referred to as stock solutions. Included here are solutions for general use, physiological solutions, environmental solutions, a few adhesive and buffer solutions, and some stain solubilities.

*Dilutions for 1 N Solutions**

Reagent	Molecular weight	Percentage assay	Specific gravity	g per liter	ml per liter
acetic acid	60.05	99.7–*100*†	1.0498	1050	57.2
(CH₃COOH)		99.0	1.0524	1042	57.6
		98.0	1.0549	1034	58.0
ammonium	17.03	26	0.904	235	72.4
hydroxide		28†	0.8980	251.4	67.7
(NH₄OH)		30	0.8920	267.6	63.63
formic acid	46.03	96	1.217	1180	39.3
(HCOOH)		98†	1.2183	1194	38.4
		99	1.2202	1208	38.0
		100	1.2212	1221	37.6
hydrochloric acid	36.46	36	1.1789	424.4	85.9
(HCl)		37–*38*†	1.188–1.192	451.6	80.4
		40	1.1980	479.2	76.0
nitric acid	63.02	69	1.4091	972.3	64.8
(HNO₃)		70†	1.4134	989.4	63.6
		71	1.4176	1006	62.6
		72	1.4218	1024	61.5
sulfuric acid	98.075	95	1.8337	1742	28.1
(H₂SO₄)		96†	1.8355	1762	27.8
		97	1.8364	1781	27.5

* Calculated from Norbert Adolph Lange, *Handbook of Chemistry* (Sandusky, Ohio: Handbook Publishers, Inc., 1956). *Italicized* figures were used in calculations.

† Indicates most common form of the reagent. In addition, a calculation has been made for the nearest lower and higher percentages.

General-purpose Solutions

Acid Alcohol

70% ethyl alcohol .... 100.0 ml

hydrochloric acid, concentrated .... 1.0 ml

Alkaline Alcohol

70% ethyl alcohol .... 100.0 ml

ammonia, concentrated .... 1.0 ml

Or use saturated aqueous sodium or lithium bicarbonate.

Carbol-Xylol

phenol (carbolic acid) melted . . . .   1 part
xylene . . . .   3 parts

During use, keep covered to reduce evaporation of the xylene.

For creosote-xylol use beechwood creosote in place of phenol. For aniline-xylol use aniline.

Gold Chloride Stock Solution

gold chloride (15 grains) . . . .   1.0 g
distilled water . . . .   100.0 ml

Lugol Solution

There are various formulas to which this name has been applied; the following are frequently used.

The strongest concentration

iodine crystals . . . .   1.0 g
potassium iodide . . . .   2.0 g
distilled water . . . .   12.0 ml

Weigert variation

iodine crystals . . . .   1.0 g
potassium iodide . . . .   2.0 g
distilled water . . . .   100.0 ml

Gram variation

iodine crystals . . . .   1.0 g
potassium iodide . . . .   2.0 g
distilled water . . . .   300.0 ml

For all solutions, first dissolve the potassium iodide, then the iodine will go into solution readily.

Chatton Agar for Blocking Small Organisms (Gray 1954)

agar . . . .   1.3 g

Add agar to:

      boiling distilled water ....   100.0 ml

Stir until dissolved, then add:

      formalin, concentrated ....   2.5 ml

Store in refrigerator.

Scott Solution

      sodium bicarbonate ....   2.0 g
      magnesium sulfate ....   20.0 g
      distilled water ....   1000.0 ml

Add a pinch of thymol to retard molds.

Sodium Thiosulfate Solution (hypo), 5%

      sodium thiosulfate ($Na_2S_2O_3$) ....   5.0 g
      distilled water ....   100.0 ml

Physiological Solutions (Balanced Salt Solutions, BSS)

Several physiological solutions should be available to satisfy the needs of different types of research. These solutions are useful for land vertebrates and can be imperative at times. A normal or *isotonic* solution is one which contains the right amount of salts to maintain tissues in a normal condition. For example, red blood cells in isotonic fluid will remain unaltered in form and will not lose their hemoglobin, because the osmotic pressure and salt content of both the solution and the blood fluid are the same. If the solution is *hypotonic,* the blood cells will swell because the osmotic pressure and salt content are less than that of the blood fluid. If the solution is *hypertonic,* the blood cells will shrink, because the osmotic pressure and salt content are greater than that of the blood fluid. For example, distilled water is hypotonic without the addition of the correct quantity of salt.

Physiological Saline

      mammals, 0.9% (9 g NaCl/1000 ml water)
      birds, 0.75% (7.5 g NaCl/1000 ml water)
      salamanders, 0.8% (8 g NaCl/1000 ml water)

frogs, 0.64% (6.4 g NaCl/1000 ml water)
insects, 0.6–0.8% (6–8 g NaCl/1000 ml water)
invertebrates, 0.75% (7.5 g NaCl/1000 ml water)

Earle Solution

sodium chloride	0.68 g
calcium chloride	0.02 g
magnesium sulfate	0.01 g
potassium chloride	0.04 g
sodium bicarbonate	0.014 g
sodium phosphate, monobasic	0.22 g
glucose	0.1 g
distilled water	100.0 ml

Hank Solution

	(1)	(2)
sodium chloride	0.8 g	0.8 g
calcium chloride	0.02 g	0.014 g
magnesium sulfate	0.02 g	0.02 g
potassium chloride	0.04 g	0.04 g
potassium phosphate, monobasic	0.01 g	0.006 g
sodium bicarbonate	0.127 g	0.035 g
sodium phosphate, dibasic	0.01 g	0.006 g
glucose	0.2 g	0.1 g
distilled water	100.0 ml	100.0 ml

If formula 1 gives too strong a solution, try formula 2.

Locke Solution

	(1)	(2)
sodium chloride	0.9 g	0.95 g
potassium chloride	0.042 g	.02 g
sodium bicarbonate	0.03 g	.02 g
calcium chloride	0.024 g	.02 g
glucose	0.1 g	0.1 g
distilled water	100.0 ml	100.0 ml

Which of the two formulas is used is a matter of personal preference. In both, decrease sodium chloride content to 0.65 g for cold-bloods; to 0.85 g for birds.

Ringer Solution

sodium chloride	0.9 g
potassium chloride	0.042 g
calcium chloride	0.025 g
distilled water	100.0 ml

Best prepared fresh. For cold-bloods, use 0.65 g of sodium chloride; for birds, 0.85 g.

Environmental Solutions

Just as physiological solutions are useful for land vertebrates, environmental solutions are convenient for some invertebrates. In the case of marine invertebrates, such solutions can become necessary for successful preparation and sometimes are recommended as the basic fluid for fixatives. Among the physical properties of sea water, the salt content must be considered; it varies and is complex, but an arbitrary definition of salt content (*salinity*) has been calculated. Salinity (S 0/00) equals the weight in grams (in vacuo) of solids in 1 kilogram of sea water. The major constituent, which is easily determined, is a silver-precipitating halide. *Chlorinity* (Cl 0/00) is defined as the weight in grams (in vacuo) of chlorides in 1 kilogram of sea water. Standard sea water is about 34.3243 0/00 salinity and 19.4 0/00 chlorinity (Barnes 1954).

Artificial Sea Water (Hale 1958)

Chlorinity, 19.0/00; *salinity*, 34.33 0/00. The weights in the right-hand column are for the anhydrous form of the salt; those in parentheses include crystallization water as indicated on the reagent bottle.

$NaCl$		23.991 g
KCl		0.742 g
$CaCl_2$		1.135 g
($CaCl_2 \cdot 6H_2O$	2.240 g)	
$MgCl_2$		5.102 g
($MgCl_2 \cdot 6H_2O$	10.893 g)	
Na_2SO_4		4.012 g
($Na_2SO_4 \cdot 10H_2O$	9.1 g)	
$NaHCO_3$		0.197 g
$NaBr$		0.085 g
($NaBr \cdot 2H_2O$	0.115 g)	
$SrCl_2$		0.011 g
($SrCl_2 \cdot 6H_2O$	0.018 g)	
H_3BO_3		0.027 g

Dissolve in enough distilled water to make 1 liter.

This solution is to be used only for technical purposes, not for aquaria, for which a proper pH condition is critical and must be maintained (Kelley 1965).

Synthetic Spring Water (Kirby 1947)

	(1)	(2)
Na_2SiO_3	15.0 mg	100.0 mg
NaCl	12.0 mg	12.0 mg
Na_2SO_4	6.0 mg	6.0 mg
$CaCl_2$	6.5 mg	6.5 mg
$MgCl_2$	3.5 mg	3.5 mg
$FeCl_3$	4.0 mg	4.0 mg
distilled water	1000.0 ml	1000.0 ml

Adjust with HCl to pH 6.8 to 7.0.

Use formula 1 for first transfer from nature; for succeeding transfers use 2.

Adhesive Solutions

Not all of the adhesive or buffer solutions will be found or required in every laboratory; however, it is practical to keep formulas of the more widely used solutions at hand, so that the information is available if it is needed.

Mayer Albumen Fixative

Beat white of an egg until well broken up, but not stiff, with egg beater and pour into tall cylinder. Let stand until the air brings suspended material to the top (overnight). Pour off liquid from bottom and to it add an equal volume of glycerine. A bit of sodium salicylate, thymol, merthiolate, or formalin (1:100) will prevent growth of molds. Many technicians filter their solution through glass wool, but I have found this of no advantage.

Faulkner and Lillie (1945*b*) substitute dried egg white. A 5% solution of dried egg white in 0.5% NaCl is shaken at intervals for one day. Do not allow it to froth. Filter in a small Büchner funnel with vacuum. Add an equal amount of glycerine and 0.5 ml of 1.10,000 merthiolate to each 100 ml of solution.

Masson Gelatin Fixative

gelatin	50.0 g
distilled water	25.0 ml

This is recommended by many for alkaline silver techniques, when sections tend to float off during or after impregnation. Float sections on solution on slide, and place on warm plate. When sections have spread, drain excess gelatin and blot dry with filter paper. Place in formalin vapor, 40–50°C: overnight.

Haupt Gelatin Fixative (1930)

gelatin	1.0 g
distilled water	100.0 ml

Dissolve at 30°C (not above) in water bath or oven. Add:

phenol (carbolic acid) crystals	2.0 g
glycerol	15.0 ml

Stir well and filter.

Use 2% formalin when mounting sections. This hardens the gelatin; water is not adequate. Some may find the formalin fumes irritating to the eyes and nostrils.

Haupt suggests that, if sections tend to loosen, a small uncovered dish of concentrated formalin be placed in an oven with the slides while they are being dried. The formalin tends to make the gelatin insoluble and helps to hold the sections in place.

Weaver Gelatin Fixative (1955)

Solution A

gelatin	1.0 g
calcium propionate[1]	1.0 g
Roccal (1% benzalkonium chloride)[2]	1.0 ml
distilled water	100.0 ml

Solution B

chrome alum [$Cr_2K_2(SO_4)_4 \cdot 24H_2O$]	1.0 g
distilled water	90.0 ml
formalin	10.0 ml

---

[1] Mycoba, E. I. du Pont de Nemours and Co.
[2] Rohm and Haas, Philadelpia, Penn.

Mix 1 part of A with 9 parts of B. Flood slide with adhesive mixture, add paraffin ribbons, and allow to stretch as usual. Drain off excess adhesive and blot. Wipe edges close to paraffin. Deposits of adhesive should be removed because they pick up stain. Good for sections that are difficult to affix.

Blood Serum Fixative (Priman 1954)

fresh human blood serum (only from negative Wassermann's)	15.0 ml
distilled water	10.0 ml
formalin, 5%	6.0 ml

Filter through filter paper and use same as albumen fixative. Does not stain and sections do not loosen.

"Subbed" Slides (Boyd 1955)

Dissolve 1 g of gelatin in 1 liter of hot distilled water. Cool and add 0.1 g of chromium potassium sulfate. Store in refrigerator. Dip slides several times in the solution. Drain and dry in a vertical position. Store in dustfree box.

Land Adhesive (Gray 1954)

Solution A

gum arabic	0.5 g
distilled water	50.0 ml

Solution B

potassium dichromate	0.5 g
distilled water	100.0 ml

Smear a small amount of A on a slide. Flood slide with solution B, add sections, and warm to flatten. Drain off fluid against cleansing tissue and dry. Bilstad [3] dries slides in bright sunlight for several hours; she found it more effective than artificial light.

---

[3] Personal communication.

Buffer Solutions

The following are a few of the chemicals most commonly used for buffer solutions. For others consult a chemical handbook.

Molecular Weights of Buffer Ingredients

acetic acid, CH_3COOH	60.05
borax, $Na_2B_4O_7 \cdot 10H_2O$ (sodium tetraborate)	381.43
boric acid, $B(OH)_3$	61.84
citric acid (anhydrous), $C_3H_4(OH)(COOH)_3$	192.12
citric acid crystals, $C_3H_4(OH)(COOH)_3 \cdot H_2O$	210.14
formic acid, $HCOOH$	46.03
hydrochloric acid, HCl	36.465
maleic acid, $HOOCCH\text{-}CHOOH$	116.07
potassium acid phosphate, KH_2PO_4	136.09
sodium acetate, CH_3COONa	82.04
sodium acetate, crystals, $CH_3COONa \cdot 3H_2O$	136.09
sodium barbital, $C_8H_{11}O_3N_2Na$	206.18
sodium bicarbonate, $NaHCO_3$	84.02
sodium carbonate, $NaCO_3$	106.00
sodium citrate, crystals, $C_3H_4OH(COONa)_3 \begin{cases} \cdot 5H_2O \\ 5\frac{1}{2}H_2O \end{cases}$	348.17 357.18
sodium citrate, granular, $C_3H_4OH(COONa)_3 \cdot 2H_2O$	294.12
sodium hydroxide, $NaOH$	40.005
sodium phosphate, monobasic, $NaH_2PO_4 \cdot H_2O$	138.01
sodium phosphate, dibasic, $Na_2HPO_4 \begin{cases} 7H_2O \\ \text{anhydrous} \end{cases}$	268.14 141.98
sulfuric acid, H_2SO_4	98.082
tris (hydroxymethyl) aminomethane $C_4H_{11}NO_3$	121.14

0.2 *M* Acetate Buffer (Gomori)

STOCK SOLUTIONS

Acetic acid

12.0 ml made up to 1000 ml with distilled water.

Sodium acetate

27.2 g made up to 1000 ml with distilled water.

Add a few crystals of *camphor* to both solutions. For desired *p*H mix correct amounts as indicated below:

pH	acetic acid (ml)	sodium acetate (ml)
3.8	87.0	13.0
4.0	80.0	20.0
4.2	73.0	27.0
4.4	62.0	38.0
4.6	51.0	49.0
4.8	40.0	60.0
5.0	30.0	70.0
5.2	21.0	79.0
5.4	14.5	85.5
5.6	11.0	89.0

Acetate–Acetic Acid Buffer (Walpole)

STOCK SOLUTIONS

M/5 acetic acid

12.0 ml (99% assay) made up to 1000 ml with distilled water.

M/5 sodium acetate

27.2 g made up to 1000 ml with distilled water.

For desired *p*H mix correct amounts as indicated below.

pH	M/5 acetic acid (ml)	M/5 sodium acetate (ml)
2.70	200.0	0.0
2.80	199.0	1.0
2.91	198.0	2.0
3.08	196.0	4.0
3.14	195.0	5.0
3.20	194.0	6.0
3.31	192.0	8.0
3.41	190.0	10.0
3.59	185.0	15.0
3.72	180.0	20.0
3.90	170.0	30.0

pH	M/5 acetic acid (ml)	M/5 sodium acetate (ml)
4.04	160.0	40.0
4.16	150.0	50.0
4.27	140.0	60.0
4.36	130.0	70.0
4.45	120.0	80.0
4.53	110.0	90.0
4.62	100.0	100.0
4.71	90.0	110.0
4.80	80.0	120.0
4.90	70.0	130.0
4.99	60.0	140.0
5.11	50.0	150.0
5.22	40.0	160.0
5.38	30.0	170.0
5.57	20.0	180.0
5.89	10.0	190.0
6.21	5.0	195.0
6.50	0.0	200.0

0.05 *M* Barbital Buffer (Gomori)

STOCK SOLUTION

Sodium barbital

1.03 g in 50 ml distilled water.

To 50 ml sodium barbital add 0.1 *N* HCl according to table below to obtain desired pH; dilute to total of 100 ml.

pH	0.1 N HCl (ml)
8.7	5.0
8.5	7.5
8.3	11.0
8.1	15.0
7.9	19.0
7.65	26.0
7.45	31.0
7.3	36.0
7.15	41.0
6.9	43.5

Boric Acid–Borax Buffer (Holmes)

STOCK SOLUTIONS

M/5 boric acid

12.368 g made up to 1000 ml with distilled water.

M/20 borax

19.071 g made up to 1000 ml with distilled water.

For desired *p*H mix correct amounts as indicated below.

pH	M/5 boric acid (ml)	M/20 borax (ml)
7.4	90.0	10.0
7.6	85.0	15.0
7.8	80.0	20.0
8.0	70.0	30.0
8.2	65.0	35.0
8.4	55.0	45.0
8.7	40.0	60.0
9.0	20.0	80.0

0.2 *M* Phosphate Buffer (Gomori)

STOCK SOLUTIONS

Monobasic sodium phosphate

27.6 g made up to 1000 ml with distilled water.

Dibasic sodium phosphate

53.6 g made up to 1000 ml with distilled water.

For desired *p*H mix correct amounts as indicated below.

pH	monobasic sodium phosphate (ml)	dibasic sodium phosphate (ml)
5.9	90.0	10.0
6.1	85.0	15.0
6.3	77.0	23.0
6.5	68.0	32.0
6.7	57.0	43.0
6.9	45.0	55.0
7.1	33.0	67.0
7.3	23.0	77.0
7.4	19.0	81.0
7.5	16.0	84.0
7.7	10.0	90.0

Phosphate Buffer (Sörensen)

STOCK SOLUTIONS

M/15 dibasic sodium phosphate

9.465 g made up to 1000 ml with distilled water.

M/15 potassium acid phosphate

9.07 g made up to 1000 ml with distilled water.

For desired pH mix correct amounts as indicated below: some techniques will require dilution with up to 1000 ml of water.

pH	M/15 dibasic sodium phosphate (ml)	M/15 potassium acid phosphate (ml)
5.29	2.5	97.5
5.59	5.0	95.0
5.91	10.0	90.0
6.24	20.0	80.0
6.47	30.0	70.0
6.64	40.0	60.0
6.81	50.0	50.0
6.98	60.0	40.0
7.17	70.0	30.0
7.38	80.0	20.0
7.73	90.0	10.0
8.04	95.0	5.0

Standard Buffer (McIlvaine)

STOCK SOLUTIONS

0.1 M citric acid (anhydrous)

19.212 g made up to 1000 ml with distilled water.

0.2 M disodium phosphate (anhydrous)

28.396 g ($\cdot 7H_2O$, 53.628 g) made up to 1000 ml with distilled water.

For desired *p*H mix correct amounts as indicated below.

pH	citric acid (ml)	disodium phosphate (ml)
2.2	19.6	0.4
2.4	17.76	1.24
2.6	17.82	2.18
2.8	16.83	3.17
3.0	15.89	4.11
3.2	15.06	4.94
3.4	14.3	5.7
3.6	13.56	6.44
3.8	12.9	7.1
4.0	12.29	7.71
4.2	11.72	8.28
4.4	11.18	8.82
4.6	10.65	9.35
4.8	10.14	9.86
5.0	9.7	10.3
5.2	9.28	10.72
5.4	8.85	11.15
5.6	8.4	11.6
5.8	7.91	12.09
6.0	7.37	12.63
6.2	6.78	13.22
6.4	6.15	13.85
6.6	5.45	14.55
6.8	4.55	15.45
7.0	3.53	16.47
7.2	2.61	17.39
7.4	1.83	18.17
7.6	1.27	18.73
7.8	0.85	19.15
8.0	0.55	19.45

0.2 *M* Tris Buffer (Hale)

STOCK SOLUTIONS

0.2 M tris (hydroxymethyl) aminomethane

24.228 g made up to 1000 ml with distilled water.

0.1 N HCl (38% assay)

8.08 ml made up to 1000 ml with distilled water.

To 25 ml 0.2 *M* tris add 0.1 *N* HCl as indicated in the following table below and dilute to 100 ml.

pH	0.1 N HCl (ml)
7.19	45.0
7.36	42.5
7.54	40.0
7.66	37.5
7.77	35.0
7.87	32.5
7.96	30.0
8.05	27.5
8.14	25.0
8.23	22.5
8.32	20.0
8.41	17.5
8.51	15.0
8.62	12.5
8.74	10.0
8.92	7.5
9.1	5.0

Tris Maleate Buffer (Gomori)

STOCK SOLUTION

maleic acid	29.0 g
tris (hydroxymethyl) aminomethane	30.3 g
distilled water	500.0 ml

Add 2 g charcoal, shake, let stand 10 minutes, and filter.

To 40 ml of stock solution add *N NaOH* (4%) as indicated below and dilute to 100 ml.

pH	sodium hydroxide (ml)
5.8	9.0
6.0	10.5
6.2	13.0
6.4	15.0
6.6	16.5
6.8	18.0
7.0	19.0
7.2	20.0
7.6	22.5
7.8	24.2
8.0	26.0
8.2	29.0

STAIN SOLUBILITIES

Stain solubilities are useful only occasionally, but they are definitely needed when a saturated solution must be made. Solubilities of different batches of dyes can vary, but the table on the next two pages can be consulted when preparing saturated solutions. Certain dyes whose solubilities were not available are not included, such as Pinacyanole (b), Darrow red (b), Sirius supra blue (a), fast garnet GB (b) and the azures (b). Gallocyanin is considered an amphoteric dye. For complete details concerning these dyes, see Conn, *Biological Stains* (1969).

Stain Solubilities

Stain	*In water (percent)*		*In absolute ethyl alcohol (percent)*		*Basic or Acidic*	*C.I. number*
	*26°C**	*15°C†*	*26°C**	*15°C†*		
acid alizarine blue	—	1.0	—	0.75	a	63015
acid fuchsin	—	45.0	—	3.0	a	42685
acridine orange	—	5.0	—	0.75	b	46005
acridine yellow	—	0.5	—	0.75	b	46025
acriflavine	—	15.0	—	1.0	b	46000
alcian blue 8GX	—	9.5	—	6.0	b	74240
alizarine red S	7.69	6.5	0.15	0.15	a	58005
amido black 10B	—	3.0	—	3.15	a	20470
aniline blue WS	—	50.0	—	0.0	a	—
auramin O	0.74	1.0	4.49	4.0	b	41000
aurantia	0.0	0.1	0.33	0.55	a	10360
azocarmine G	—	1.0	—	0.1	a	50085
basic fuchsin	0.26–0.39	1.0	5.93–8.16	8.0	b	42500
Biebrich scarlet	—	5.0	0.5	0.25	a	26905
Bismarck brown Y	1.36	1.5	1.08	3.0	b	21000
brilliant cresyl blue	—	3.0	—	2.0	b	51010
celestin blue B	—	2.0	—	1.5	b	51050
chlorantine fast red	—	1.0	—	0.45	a	28160
chlorazol black E	—	6.0	—	0.1	a	30235
chromotrope 2R	19.3	19.0	0.17	0.15	a	16570
chrysoidin Y	0.86	5.5	2.21	4.75	b	11270
congo red	—	5.0	0.19	0.75	a	22120
coriphosphine O	—	4.75	—	0.6	b	46020
cresyl violet (cresylecht violet)	0.38	9.5	0.25	6.0	b	—
crystal violet (methyl violet 10B)	1.68	9.0	13.87	8.75	b	42555
eosin B	39.11	10.0	0.75	3.0	a	45400
eosin (S) alc. sol. (ethyl eosin)	0.03	0.0	1.13	1.0	a	45386
eosin Y, WS	44.20	44.0	2.18	2.0	a	45380
erythrosin B	11.10	10.0	1.87	5.0	a	45430
erythrosin Y	—	8.5	—	4.5	a	45425
fast green FCF	16.04	4.0	0.35	9.0	a	42053
fluorescein (uranin)	50.02	50.0	7.19	7.0	a	45350
gallocyanin (bisulfite)	—	3.0	—	0.5	b	51030
gallocyanin (hypochlorite)	—	0.5	—	1.25	b	51030
hematein	—	1.5	—	7.5	b	75290
hematoxylin	1.75	10.0	60.0	10.0	b	75290
janus green B	5.18	5.0	1.12	1.0	b	11050
light green SF yellowish	20.35	20.0	0.82	4.0	a	42095
Luxol fast blue MBS	—	0.0	—	3.0	a	—

* From Conn (1969). † From Gurr (1960). *(continued)*

Stain Solubilities (cont.)

Stain	In water (percent)		In absolute ethyl alcohol (percent)		Basic or Acidic	C.I. number
	26°C*	15°C†	26°C*	15°C†		
malachite green	—	10.0	—	8.5	b	42000
martius yellow	4.57	1.0	0.16	0.0	a	10315
metanil yellow	5.36	5.0	1.45	1.5	a	13065
methyl blue	—	50.0	—	0.0	a	42780
methyl violet 2B	2.93	—	15.21	—	b	42535
methyl violet 6B	—	4.7	—	9.5	b	42535
methylene blue	3.55	9.5	1.48	6.0	b	52015
naphthol yellow S	—	12.5	—	0.65	a	10316
neutral red	5.64	4.0	2.45	1.8	b	50040
new methylene blue	13.32	—	1.65	—	b	52030
new fuchsin	1.13	1.0	3.2	8.0	b	42520
nigrosin, alc. sol.	—	0.0	—	2.5	b	50415
nigrosin WS	—	10.0	—	0.0	a	50420
Nile blue sulfate	—	6.0	—	5.0	b	51180
oil red O	—	0.0	—	0.5	a	26125
orange G	10.86	8.0	0.22	0.22	a	16230
orcein	—	2.0	—	4.2	b	—
phloxine B	—	10.5	—	5.0	a	45410
picric acid	1.18	1.2	8.96	9.0	a	10305
ponceau 2R	—	5.0	—	0.1	a	16150
ponceau S	—	1.35	—	1.2	a	27195
ponceau de xylidine	—	5.0	—	0.1	a	16150
Poirrier blue C4B (cotton blue)	s	—	s	—	a	42780
primilin	—	0.25	—	0.03	a	49000
pyronin B	—	10.0	—	0.5	b	45010
pyronin Y	8.96	9.0	0.6	0.5	b	45005
quinoline yellow	—	0.5	—	0.0	a	47005
rhodamin B	0.78	2.0	1.47	1.75	b	45170
rhodamin 6G	—	1.5	—	6.5	b	45160
safranin O	5.45	4.5	3.41	3.5	b	50240
Sudan black B	—	0.0	—	0.25	a	26150
Sudan III	0.0	0.0	0.15	0.25	a	26100
Sudan IV	0.0	0.0	0.09	0.5	a	26105
thioflavine S	—	1.0	—	0.4	a	49010
thioflavine T	—	2.0	—	1.0	b	49005
thionin	0.25	1.0	0.25	1.0	b	52000
titian yellow	—	1.0	—	0.02	a	19540
toluidine blue O	3.82	3.25	0.57	1.75	b	52040
trypan blue	—	1.0	—	0.02	a	23850
Victoria blue B	—	4.3	—	8.25	b	44045
Victoria blue 4R	3.23	3.0	20.49	20.0	b	42563
water blue I	—	50.0	—	0.0	a	42755
wool green S	—	4.0	—	—	a	44090

CHAPTER **29**

General Laboratory Aids

LABELING AND CLEANING SLIDES

All slide labels should contain complete information: the name or number of the tissue, the section thickness, the stain and date, and the fixative used.

Before ordinary slide labels are glued on, processed slides must be cleaned to prevent loosening of the labels. Dip the slides (when mountant is thoroughly hardened) in water to which has been added a small amount of ammonia and household cleanser (or glass cleaner). Wipe dry and paste on label.

Labels applied by pressure are easier to use and do not require slide cleaning. They are called Time Microscopic Slide Labels[1] and are made in a standard thickness, $\frac{7}{8}$ by $\frac{7}{8}$ inch, and a "tissue-high" thickness, $\frac{3}{8}$ by $\frac{7}{8}$ inch. The latter is used as an end label to protect the cover glass from sticking to an object laid on it.

Mounting can be untidy, and some of the resin may ooze over the

---

[1] Professional Tape Co., Inc., Riverside, Ill.

cover glass. If this happens, scrape off excess resin with a razor blade when the slide is completely dry, taking care not to chip or break the cover. Do not clean away the resin too close to the glass; in fact, leave a small band of it to protect the cover from chipping or catching against an object.

RESTAINING FADED SLIDES

Slides that have been standing for a long time or that have been exposed to bright light frequently fade. If they must be recovered, McCormick offers the following method of bleaching them for restaining.

McCormick Method (1959a)

1. Remove cover glass by soaking slides in xylene until cover glass slips off. Do not force it off; the sections might get torn.
2. Soak longer in xylene to make certain that all of the resin is removed.
3. Hydrate to water.
4. Treat with 0.5% potassium permanganate (0.5 g/100 ml water): 5 minutes.
5. Wash in running water: 5 minutes.
6. Bleach with 0.5% oxalic acid (0.5 g/100 ml water) until colorless: 1–2 minutes. If old stain still remains, repeat steps 4, 5, and 6.
7. Wash thoroughly in running water: 5 or more minutes.
8. Restain, using more dilute stains or staining for a shorter time. Potassium permanganate and oxalic acid make tissues especially sensitive to hematoxylin and aniline nuclear stains.

RECOVERING BROKEN SLIDES

Usually it is practical to remove the bits of broken cover glass by soaking the slide in xylene overnight or longer, until the cover glass pieces slide off unaided. Then mount the broken pieces of slide on another slide (as thin a slide as obtainable) with mounting medium, just as if mounting a cover glass. Do not allow the tissue on the slide to dry out during this process. Cover with a new cover glass and mountant. Allow to dry on warm plate overnight or several days. With a razor blade carefully clean away any mountant exuding from between the slides.

See Geil (1961) for salvaging histologic sections from broken slides.

RESTORING BASOPHILIC PROPERTIES

Luna Method (1968)

When poor staining is due to overlong storage in fixative, overexposure to decalcifying solutions, or dried or burned tissue, Luna suggests overnight soaking in one of the following: (1) 5% aqueous sodium bicarbonate, (2) 5% aqueous ammonium sulfide, or (3) 5% aqueous periodic acid. Follow by thorough washing for 10–15 minutes.

TWO DIFFERENT STAINS ON ONE SLIDE

Feder and Sidman Method (1957)

This method is good for a quick checking of a tissue against two stains, side by side.

1. Hydrate sections to water.
2. Blot carefully, and while they are still moist, coat alternate sections with silicone grease. A soft brush may be used, or more efficiently use a 1-ml syringe and no. 15 needle that has been clipped off and flattened.
3. Apply first staining procedure.
4. Dehydrate and leave in xylene 15–30 minutes to remove coating. (After sections have been in xylene 5 minutes, blotting with filter paper helps to remove grease.)
5. Rehydrate; apply coating to stained sections.
6. Apply second staining procedure.
7. Repeat step 4.
8. Rinse briefly in absolute alcohol and clear in fresh xylene. Mount.

RECLAIMING AND STORING SPECIMENS

Reclaiming Biopsy Material

Graves Method (1943)

If biopsy material has dried, do not try to process it before first softening and rehydrating it. Place it in physiological saline solution for 1 hour, then fix, dehydrate, clear, and embed as usual.

See Haviland (1963) for recovery of rust-stained formalin-preserved gross tissues.

Reclaiming Dried Gross Specimens[2]

Tissues that have been stored in alcohol or formalin often become completely or partially desiccated and shrunken. If a major catastrophe arises and tissue must be reclaimed, partial recovery can be made with fair returns for microscopic identification. Van Cleave and Ross (1947) restored desiccated helminths and invertebrates to normal size by soaking them in a 0.5% aqueous solution of trisodium phosphate. Trial runs of this nature were made on dried formalin-fixed specimens for 4 hours to 30 days, depending on the hardness and size of tissue. Occasional changing of the fluid seemed to help if the pieces were exceptionally dry; also, warming in a $\pm 40°C$ oven was helpful. Increasing the concentration of the phosphate did not speed recovery. Finally the tissues were washed for 30 minutes to 2–3 hours, again depending on size.

If the specimen was fixed in a mercuric chloride fixative, the trisodium phosphate was found to be ineffective until the tissue had been pretreated as follows: (1) soaked in water with enough Lugol's solution added to color the solution a deep brown: 1–2 days (if the solution became colorless, it was renewed); (2) washed in running water: 2 hours; (3) treated with 5% sodium thiosulfate: $\frac{1}{2}$–1 hour; and (4) washed: 2 hours.

Only fair results were obtained by the above methods. Considerable shrinkage remained in the cells, and the nuclei stained only lightly. The latter condition improved somewhat after mordanting in mercuric chloride or potassium dichromate.

Since recovery was based on the detergent action of trisodium phosphate, the inevitable question arose: Why not try one of the modern household detergents? A 1% solution of Trend in water was added to dried tissues and kept overnight in a paraffin oven at approximately 58°C. This was followed by 6–8 hours washing and then processing overnight in the Technicon for embedding the next morning. The results were creditable—good enough for tissue identification and some pathological reading. The tissues in this run averaged 10–25 cm in size and were fixed in either formalin or a mercuric chloride fixative. A longer stay in detergent might be required for larger pieces. No pretense is made that results

[2] Special acknowledgment is made to Robert Ingersoll Howes, Jr., for hours of effort in developing a recovery process for the Los Alamos Medical Center in 1959.

are exceptional: considerable shrinkage remains in the cells, and staining is not brilliant, but it is better than after trisodium phosphate. Therefore, if it is essential to check back on a tissue concerning its identity, a malignancy, or some other "matter of life or death," this method of recovery may save the day.

Sandison Method (1955)

Sandison treated mummified material with the following solution and prepared histological sections with good results.

95% ethyl alcohol	30 volumes
1% formalin	50 volumes
5% aqueous sodium carbonate	20 volumes

Leave tissues in solution overnight or until they become soft. Replace one-third of the fluid with 95% alcohol. Repeat until the tissues become firm. Dehydrate, clear, infiltrate, and embed.

Lhotka and Ferreira (1950) deformalizing technique will work on some tissues.

Preserving Gross Specimens in a Pliable Condition

Palkowski Method (1960)

It is economical to preserve some gross specimens, pathological ones in particular, for future teaching or demonstration purposes. Fix as soon as possible in Kaiserling solution: 3–4 hours.

Kaiserling solution

chlorate hydrate	300.0 g
potassium sulfate	6.0 g
potassium nitrate	114.0 g
sodium sulfate	54.0 g
sodium chloride	66.0 g
sodium bicarbonate	60.0 g
formalin	300.0 ml
distilled water	10,000.0 ml

Drain off excess Kaiserling and place specimens in four times their volume of 80% alcohol: 18–24 hours. Deep freeze at about −29°C, sealed in airtight polyethylene bags not much larger than the specimen.

Because of contained air, lungs will float on the fixative; submerge them under cotton soaked with fixative.

The specimens will keep indefinitely this way, retain their color, and become pliable after thawing. After use, they can be returned to deep-freeze with little deterioration.

Color Preservation in Gross Specimens

McCormick Fixation (1961)

formaldehyde, concentrated	50.0 ml
Sörenson's buffer, pH 6.4	950.0 ml
Prague powder[3]	8.5 g
ascorbic acid	1.7 g

Storing Gross Specimens

Sealing tissues for storage has been a serious problem. Bottled storage has risks; any seal can spring a leak, and the resulting evaporation culminates in desiccated tissue. Storing in plastic bags is a more reliable system than the use of bottles. A heavy quality of polyethylene plastic is recommended and is widely available. Lighter grades can split or unseal, whereas the heavy grade, once well sealed, almost supports itself without collapsing. The plastic can be cut to any size and sealed on all four edges, if necessary, with the Pack-Rite Poly Jaw Sealer.[4] A small amount of formalin included in the bag will keep the tissue moist as long as the bag remains sealed. A quantity of small containers can be sealed together in a large bag, affording additional protection against drying. Plastic tags with data can be enclosed or attached to the outside. Storing See also Lieb (1959), Gordon (1953), and Broadway and Koelle in bags also saves storage space, preservative, and containers. (1960).

---

[3] Griffith Laboratories, Chicago, Ill.
[4] Pack-Rite Machines, Milwaukee, Wisc. (The CRC "Plastiseal" Bar Sealer is smaller and not as sturdy as the Pack-Rite machine.)

REMOVING LABORATORY STAINS FROM HANDS AND GLASSWARE

The most frequently used laboratory stains, with suggested treatment for removal, are listed here.

basic fuchsin—difficult to remove; try strong acetic acid in 95% alcohol, or dilute HCl.

carmine—strong ammonia water or weak HCl; if stain resists, use them alternately.

chromic acid—dilute sulfurous acid in water, or concentrated sodium thiosulfate and a few drops of sulfuric acid added.

fast green and similar acid stains—ammonia water.

hematoxylin—weak acid or lemon juice.

hemoglobin—fresh stains: lukewarm to cool water (never hot). Older stains: soften with borax solution, dilute ammonia, or tincture of green soap; finally, treat with 2% aqueous oxalic acid.

iodine—sodium thiosulfate.

iron alum stains on glassware—strong NaOH (sticks) in water, followed by strong HCl.

methylene blue—acid alcohol or tincture of green soap.

most dyes—tincture of green soap.

osmic acid on glassware—3% H_2O_2 (Carr and Bacsich 1958).

PAS—to remove from clothing, soak in weak solution of potassium permanganate, approximately 0.5%; rinse off and bleach in 1% oxalic acid. Warning: some fabric dyes also bleach.

picric acid—lithium iodide or carbonate, aqueous.

potassium permanganate—dilute sulfurous acid, HCl, oxalic acid, or hyposulfite.

safranin and gentian violet—difficult to remove; try acid alcohol.

silver—Lugol or tincture of iodine, followed by sodium thiosulfate.

Verhoeff—4% aqueous citric acid (Hull and Wegner 1952).

Cleaning Solution for Glassware

Strong

potassium dichromate	20.0 g
water	200.0 ml

Dissolve dichromate in water; when cool add very slowly:

sulfuric acid, concentrated	20.0 ml

FIGURE 36. *A cabinet of drawers designed for use in microtechnique classes at the University of California at Los Angeles. The cabinets are not a part of the working table, but are arranged along a wall of the classroom. The drawers are designed to hold coplin jars, mountant jars, and similar items. The student can carry the drawer to the work table, turn it around with its backside facing forward, remove the back panel, and do the staining right in the drawer. This relieves the work table of staining equipment, and the student can carry all the staining jars at once in one "basket." If locking the drawers is unnecessary, the front panel may be made removable; this eliminates having to turn the drawer around. Drawers may be left out of the upper compartment, or upper two compartments, to make convenient storage nooks. These cabinets are custom made. Standard cabinets can be adapted to this purpose, but their drawers do not have a removable panel. A good cabinet (Unit no. P3) can be obtained from the Kewanee Manufacturing Company, Adrian, Michigan. The interior dimensions of its drawers are 20 15/16 inches long, 5 1/8 inches high, and 12 1/2 inches wide.*

Weak

potassium dichromate, 2% aqueous	9 parts
sulfuric acid	1 part

This is strong enough for most purposes.

TEACHING FILMS

Motion pictures can be of considerable value to instructors of laboratory methods. See especially: "Pathology Film Reviews" in *Laboratory Investigation* (vol. 9:5, part II [1960]). This article lists motion pictures for education, research, and recruitment, reviewing the content of the films, type of audience, where distributed or loaned. Samples are: "Autopsy Technique," "Compound Microscope," "Cytotechnologist's Role," and "Career: Medical Technologist." Volume 12:11 (1963) lists films on technique and pathology; volume 13:10 (1964) lists Armed Forces Institute of Pathology films, including films on tissue preparation.

For information concerning additional films write to The Director, Armed Forces Institute of Pathology, Audiovisual Communication Center, Washington, D.C., 20305.

SUPPLIERS OF EQUIPMENT, GLASSWARE, AND CHEMICALS

These companies will forward a catalog upon request. Specialized items have been noted in text footnotes.

Aldrich Chemical Co., Fairfield, N.J.
Arthur H. Thomas Co., Philadelphia, Penn.
Carolina Biological Supply Co., Burlington, N.C.
Cole-Parmer Instrument Co., Chicago, Ill.
Curtin Scientific Co., Atlanta, Boston, Dallas, Houston, Los Angeles, Minneapolis, New Orleans, Pittsburgh, San Francisco, Jacksonville, St. Louis, Washington, D.C.
Eastman Kodak Co., Rochester, N.Y.
Fisher Scientific Co., Boston, Chicago, Fort Worth, Houston, New York, Philadelphia, Pittsburgh, St. Louis, Atlanta, Washington, D.C.
General Biological Supply Co., Chicago, Ill.
Hartman-Leddon Co., Philadelphia, Penn.
Kessler Chemical Co., Philadelphia, Penn.
K and K Laboratories, Inc., Plainview, N.Y.
La Pine Scientific Co., Chicago, Irvington-on-Hudson, Berkeley
Lederle Laboratories, New York
Lipshaw Manufacturing Co., Detroit, Mich.
Matheson Scientific Co., Cincinnati, Ohio

Preiser Scientific, Inc., Charleston, W.Va., Cincinnati, Louisville, Kentucky

Sargent, Welch Scientific Co., Birmingham, Ala.

Scientific Products, Evanston, Ill.

Sigma Chemical Co., St. Louis, Missouri

Wards Natural Science Establishment, Rochester, N.Y.

Will Corporation, Rochester, Atlanta, New York, Baltimore, Buffalo, Charleston, W.Va.

References

Abadia-Fenoll, F.
  1968 "Staining pericellular boutons of glutaraldehyde-fixed nervous tissue, a chromat 」n-silvering sequence for frozen sections." *Stain Technology* 43:191–195.
Abrams, Harvey A., and Elias, Hans.
  1962 "Acridine orange fluorescence in paraffin sections." *Stain Technology* 37:50–52.
Ackerman, Adolph G.
  1958 "A combined alkaline phosphatase-PAS staining method." *Stain Technology* 33:269–271.
Adams, C. W. M.
  1956 "A stricter interpretation of the ferric ferricyanide reaction with particular reference to the demonstration of protein-bound sulphydryl and disulphide groups." *Journal of Histochemistry and Cytochemistry* 4:32–35.
  1957 "A p-dimethylaminobenzaldehydenitrite method for the histochemical demonstration of tryptophane and related compounds." *Journal of Clinical Pathology* 10:56–62.
Adams, C. W. M., and Sloper, J. C.
  1955 "Technique for demonstrating neurosecretory material in the human hypothalamus." *Lancet* 268:651–652.
  1956 "The hypothalamic elaboration of posterior pituitary principles in man, the rat and dog; histochemical evidence derived from a performic acid-alcian blue reaction for cystine." *Journal of Endocrinology* 13:221–228.
Adams, C. W. M., and Tuqan, N. A.
  1961 "The histochemical demonstration of protease by a gelatin-silver film substrate." *Journal of Histochemistry and Cytochemistry* 9:469–472.
Adamstone, F. B., and Taylor, A. B.
  1948 "The rapid preparation of frozen tissue sections." *Stain Technology* 23:109.
Adey, W. R.; Rudolph, Alice F.; Hine, I. F.; and Harritt, Nancy J.
  1958 "Glees staining of monkey hypothalamus: a critical appraisal of normal and experimental material." *Journal of Anatomy* 92:219–235.
Agrell, Ivar P. S.
  1958 "Whole mounts of small embryos attached directly to glass slides." *Stain Technology* 33:265–267.
Albrecht, Mildred.
  1954 "Mounting frozen sections with gelatin. *Stain Technology* 29:89–90.

1956 "A simple mechanical aid in staining frozen sections in quantity." *Stain Technology* 31:231–233.

Aldridge, William G., and Watson, Michael L.
1963 "Perchloric acid extraction as a histochemical technique." *Journal of Histochemistry and Cytochemistry* 11:773–781.

Alfert, Max, and Geschwind, I. I.
1953 "A selective staining method for basic proteins of cell nuclei." *Proceedings of National Academy of Sciences* 39:991–999.

Alpert, Morton; Jacobowitz, David; and Marks, Bernard H.
1960 "A simple method for the demonstration of lipofuscin pigment." *Journal of Histochemistry and Cytochemistry* 8:153–158.

Altman, Philip L., and Dittman, Dorothy S., eds.
1964 *Biology Data Book.* Washington, D.C.: Federation of American Societies for Experimental Biology.

Amano, M.
1962 "Improved techniques for the enzymatic extraction of nucleic acids from tissue sections." *Journal of Histochemistry and Cytochemistry* 10:204–212.

American Public Health Association.
1956 *Diagnostic Procedures for Virus and Rickettsial Diseases.* New York: The Association.

Ammerman, Frank.
1950 "A chrome-alum preparation for delicate and difficult fixations." *Stain Technology* 25:197–199.

Anderson, D. L.
1961 "Selective staining of ribonucleic acid." *Journal of Histochemistry and Cytochemistry* 9:619–620.

Anderson, Doris, and Grieff, Donald.
1964 "Direct fluorochroming of Rickettsiae." *Journal of Histochemistry and Cytochemistry* 12:194–196.

Anderson, F. D.
1959 "Dichromate-chlorate perfusion prior to staining degeneration in brain and spinal cord." *Stain Technology* 34:65–67.

Anderson, Paul J.
1967 "Purification and quantitation of glutaraldehyde and its effect on several enzyme activities in skeletal muscle." *Journal of Histochemistry and Cytochemistry* 15:652–661.

Andrews, Warren, and Johnson, Henry.
1956 "Staining mitochondria in fixed blood smears." *Stain Technology* 31:21–23.

Anthony, Adam, and Clater, Merlin.
1959 "Alcohol-xylene versus Dioxane in the shrinkage of tissue." *Stain Technology* 34:9–13.

Aoyama, Fumic
1930 "Eine Modifikation der Cajakschen Methode zur Darstellung des Golgischen Binnenapparates." *Zeitschrift für Wissenschaftliche Mikroskopie und Mikroskopische Technik* 46:489–491.

Aparicio, S. R., and Marsden, P.
1969 "Application of standard micro-anatomical staining methods to epoxy resin-embedded sections." *Journal of Clinical Pathology* 22:589–592.

Arensburger, Konstantin E., and Markell, Edward K.
1960 "A simple combination direct smear and fecal concentrate for permanent stained preparations." *American Journal of Clinical Pathology* 34:50–51.

Armstrong, J. A.
1956 "Histochemical differentiation of nucleic acids by means of induced fluorescence." *Experimental Cell Research* 11:640–643.

Armstrong, J. A.; Niven, Janet; and Anderson, S. S.
    1957 "Fluorescence microscopy in the study of nucleic acids." *Nature* 80: 1335–1336.
Arnold, James S., and Jee, Webster
    1954*a* "Preparing bone sections for radioautography." *Stain Technology* 29:49–54.
    1954*b* "Embedding and sectioning undercalcified bone and its application to radioautography." *Stain Technology* 29:225–239.
Arnold, Zach M.
    1952 "A rapid method for concentrating small organisms for sectioning." *Stain Technology* 27:199–200.
Aronson, Willard, and Pharmakis, Thomas.
    1962 "Enchancement of neotetrazolium staining for succinic dehydrogenase activity with cyanide." *Stain Technology* 37:321.
Arzac, J. P.
    1950 "A simple histochemical reaction for aldehydes." *Stain Technology* 25:187–194.
Ashley, C. A., and Feder, N.
    1966 "Glycol methacrylate in Histopathology." *Archives of Pathology* 81:391–397.
Atkinson, William B.
    1952 "Studies on the preparation and recoloration of fuchsin sulfurous acid." *Stain Technology* 27:153–160.
Austin, A. P.
    1959 "Iron alum aceto-carmine staining for chromosomes and other anatomical features of Rhodophyceae." *Stain Technology* 34:69–75.
Avers, Charlotte J.
    1963 "An evaluation of various fixings and staining procedures for mitochondria in plant root tissues." *Stain Technology* 38:29–35.

Baker, J. R.
    1932 "A new method for mitochondria." *Nature* 30:134.
    1944 "The structure and chemical composition of the Golgi element." *Quarterly Journal of Microscopical Science* 85:1–72.
    1945 *Cytological Technique.* London: Methuen.
    1946 "The histochemical recognition of lipine." *Quarterly Journal of Microscopical Science* 87:441–470.
    1947 "The histochemical recognition of certain guanidine derivatives." *Quarterly Journal of Microscopical Science* 88:115–121.
    1958 *Principles of Biological Technique.* London: Methuen; New York: John Wiley.
Bancroft, J. D.
    1963 "Methyl green as a differentiator and counterstain in the methyl violet technique for demonstration of amyloid in fresh cryostat sections." *Stain Technology* 38:336–337.
Bangle, Raymond.
    1954 "Gomori's paraldehyde fuchsin stain." *Journal of Histochemistry and Cytochemistry* 2:291–299.
Banny, Theresa M., and Clark, George.
    1949 "A simplified method for the double embedding of tissue." *Stain Technology* 24:223–224.
    1950 "New domestic cresyl echt violet." *Stain Technology* 25:195–196.
Barer, R.
    1956 "Phase contrast and interference microscopy in cytology." *Physical Techniques in Biological Research*, Vol. III. Ed. by Gerald Oster and Arthur W. Pollister. New York: Academic Press.

1959 "Interference and Polarizing Microscopy." *Analytical Cytology,* Ed. by Robert C. Mellors. New York: Blakiston.

Barka, T., and Anderson, P.J.
1963 *Histochemistry.* New York: Harper and Row.

Barley, David A.
1964 "Salivary gland chromosomes of black flies." *Turtox News* 44:298–300.

Barlow, R. M.
1957 "A tribasic stain for the granules of mast cells." *Journal of Pathology and Bacteriology* 73:272–274.

Barnard, E. A.
1961 "Acylation and diazonium coupling in protein cytochemistry with special reference to the benzoylation-tetrazonium method." *General Cytochemical Methods,* II. Ed. by J.F. Danielli. New York: Academic Press.

Barnard, John W.; Roberts, J.O.; and Brown, Joseph G.
1949 "A simple macroscopic staining and mounting procedure for wet sections from cadaver brains." *Anatomical Record* 105:11–17.

Barnes, H.
1954 "Some tables for the ionic composition of sea water." *Journal of Experimental Biology* 31:582–588.

Barr, M. L., and Bertram, E. G.
1949 "A morphological distinction between neurones of the male and female, and the behavior of the nucleolar satellite during accelerated nucleoprotein synthesis." *Nature* 163:676.

Barr, M. L.; Bertram, Fraser; and Lindsay, H. A.
1950 "The morphology of the nerve cell nucleus, according to sex." *Anatomical Record* 107:283–292.

Barrett, A. M.
1944 "On the removal of formaldehyde-produced precipitate from sections." *Journal of Pathology and Bacteriology* 56:135–136.

Barnett, R. J., and Seligman, A. M.
1952 "Demonstration of protein-bound sulfhydryl and disulfide groups by two new histochemical methods." *Journal of National Cancer Institute* 13:215–216.

Bartholomew, James W.
1962 "Variables influencing results, and the precise definition of steps in Gram staining as a means of standardizing the results obtained. *Stain Technology* 37:139–155.

Bartholomew, James W., and Mittwer, Tod.
1950 "The mechanism of the Gram reaction. 1. The specificity of the primary dye." *Stain Technology* 25:103–110.
1951 "The mechanism of the Gram reaction. 3. Solubilities of dye-iodine precipitates, and further studies of primary dye substitutes. *Stain Technology* 26:231–240.

Bartholomew, James W.; Mittwer, Tod; and Finklestein, Harold.
1959 "The phenomenon of Gram-positivity; its definition and some negative evidence on the causative role of sulfhydryl groups." *Stain Technology* 34:147–154.

Barton, A. A.
1959 "The examination of ultrathin sections with a light microscope after electron microscopy." *Stain Technology* 34:348–349.
1960 "Carbon coated grids for electron microscopy." *Stain Technology* 35:287–289.

Baserga, Renato.
1961 "Two-emulsion autoradiography for the simultaneous demonstration of precursors of deoxyribonucleic and ribonuclei acids." *Journal of Histochemistry and Cytochemistry* 9:586.

Baserga, Renato, and Nemeroff, Ken.
    1962 "Two-emulsion radioautography." *Journal of Histochemistry and Cytochemistry* 10:628–635.
Batty, I., and Walker, P. D.
    1963 "Differentiation of *Clostridium septicum* and *Clostridium chauvoei* by the use of fluorescent labelled antibodies." *Journal of Pathology and Bacteriology* 85:517–520.
Bauer, H.
    1933 "Mikroskopisch-chemischer Nachweis von Glykogen un einigen anderen Polysacchariden." *Zeitschrift für Mikroskopisch-Anatomische Forschung* 33:143–160.
Bayley, John.
    1955 "'Cracking' and 'chatter' in paraffin sections." *Journal of Pathology and Bacteriology* 70:548–549.
Beamer, Parker R., and Firminger, Harlan I.
    1955 "Improved methods for demonstrating acid-fast and spirochaetal organisms in histological sections." *Laboratory Investigation* 4:9–17.
Beçak, W.; Beçak, M. L., and Nazareth, H. R. S.
    1962 "Karyotypic studies of two species of South American snakes." *Cytogenetics* 1:305–313.
    1964 "Chromosomes of cold blood animals from whole blood short-term cultures." *Mammalian Chromosomes Newsletter* 14:55–56.
Beck, Conrad.
    1938 *The Microscope. Theory and Practice.* London: R. J. Beck.
Beckel, W. E.
    1959 "Sectioning large heavily sclerotized whole insects." *Nature* 184:1584–1585.
Becker, E. R., and Roudabush, R. L.
    1935 *Brief Directions in Histological Technique.* Ames, Iowa: Collegiate Press.
Beckert, William H., and Garner, James G.
    1966 "Staining sex chromatin; Biebrich scarlet and Fast green FCF as a mixture versus their use in sequence." *Stain Technology* 41:141–148.
Beech, R. H., and Davenport, H. A.
    1933 "The Bielschowsky staining technic. A study of the factors influencing its specificity for nerve fibers." *Stain Technology* 8:11–30.
Beek, R. M.
    1955 "Improvements in the squash technic for plant chromosomes." *Aliso* 3:131–133.
Behnke, O., and Rostgaard, J.
    1963 "A device for mounting glass knives in the ordinary rotary microtome for sectioning plastic-embedded material." *Stain Technology* 38:299–300.
Bélanger, Leonard F.
    1950 "A method for routine detection of radio-phosphates and other radioactive compounds in tissues. The inverted autograph." *Anatomical Record* 107:149–156.
    1952 "Improvements in the melted emulsion technique of autoradiography." *Nature* 170:165.
    1961 "Staining processed radioautographs." *Stain Technology* 36:313–317.
Bélanger, Leonard F., and Bois, Pierre.
    1964 "A histochemical and histophysiological survey of the effects of different fixatives on the thyroid gland of rats." *Anatomical Record* 148:573–580.
Bélanger, Leonard F., and Hartnett, Anna.
    1960 "Persistent toluidine blue metachromasia." *Journal of Histochemistry and Cytochemistry* 8:75.
Belling, John.
    1926 "The iron-aceto-carmine method of fixing and staining chromosomes." *Biological Bulletin, Woods Hole* 50:160–162.

1930 *The Use of the Microscope.* New York: McGraw-Hill.

Bencosme, Sergie; Stone, Robert S.; Latta, Harrison; and Madden, Sidney C.
  1959 "A rapid method for localization of tissue structures or lesions for electron microscopy." *Journal of Biophysical and Biochemical Cytology* 5:508–509.

Bender, M. A., and Eide, P. E.
  1962 "Separation of monkey (*Ateles*) leukocytes for culture." *Mammalian Chromosomes Newsletter* 8.

Benés, Karel.
  1960 "The chemical specificity of staining mitochondria in frozen dried and frozen substituted tissue with buffered Amidoblack 10B." *Acta Histochemica* 10:255–264.

Benge, W. P. J.
  1960 "Staining autoradiographs at low temperatures." *Stain Technology* 35:106–108.

Bennett, H. Stanley.
  1950 "The microscopical investigation of biological materials with polarized light." *Handbook of Microscopical Technique,* 3d ed. Ed. by Ruth McClung Jones. New York: Hoeber.
  1951 "The demonstration of thiol groups in certain tissues by means of a new colored sulfhydryl reagent." *Anatomical Record* 110:231–247.

Bennett, H. Stanley, and Watts, Ruth M.
  1958 "The cytochemical demonstration and measurement of sulfhydryl groups by azo-aryl mercaptide coupling with special reference to mercury orange." *General Cytochemical Methods,* I. Ed. by J. F. Danielli. New York: Academic Press.

Bensley, R. R., and Bensley, S. H.
  1938 *Handbook of Histological and Cytological Technique.* Chicago: University of Chicago Press.

Bensley, Sylvia H.
  1952 "Pinacyanol erythrosinate as a stain for mast cells." *Stain Technology* 27:269–273.
  1959 "The scope and limitations of histochemistry." *American Journal of Medical Technology* 25:15–32.

Bergeron, John A., and Singer, Marcus.
  1958 "Metachromasy: an experimental and theoretical re-evaluation." *Journal of Biophysical and Biochemical Cytology* 4:433–457.

Berkowitz, Lillian R.; Fiorello, Olga; Kruger, Lawrence; and Maxwell, David S.
  1968 "Selective staining of nervous tissue for light microscopy following preparation for electron microscopy." *Journal of Histochemistry and Cytochemistry* 16:808–814.

Berman, Irwin, and Kaplan, Henry S.
  1959 "The cultivation of mouse bone marrow in vivo." *Blood* 14:1040–1046.

Berman, Irwin, and Newby, Earlene J.
  1963 "Autoradiography and staining of hermatopoietic cells grown on Millipore membranes in vivo." *Stain Technology* 38:62–65.

Bernhardt, H.; Gourley, R. D.; Young, J. M.; Shepherd, M. C.; and Killian, J. J.
  1961 "A modified membrane-filter technic for detection of cancer cells in body fluids." *American Journal of Clinical Pathology* 36:462–464.

Berthrong, Morgan, and Barhite, Melvin.
  1964 "A technic for preparation of tissue sections of needle-aspirated material from bone marrow." *American Journal of Clinical Pathology* 42:207–211.

Berton, William M., and Phillips, W. R.
  1961 "A film leader technic for the study of cells in culture." *Laboratory Investigation* 10:373–380.

Bertram, E. G.
  1958 "A non-sticking coverglass weight." *Stain Technology* 33:143–144.

Berube, G. R.; Powers, Margaret M.; Kerkay, Julius; and Clark, George.
    1966 "The gallocyanin-chrome alum stain: influence of methods of preparation
    on its activity and separation of active staining compound." *Stain Technology* 41:73–81.
Betts, Anthony.
    1961 "The substitution of acridine orange in the periodic acid-Schiff stain."
    American Journal of Clinical Pathology 36:240–243.
Bharadwaj, T. P., and Love, Robert.
    1959 "Staining mitochondria with hematoxylin after formalin-sublimate fixation;
    a rapid method." *Stain Technology* 34:331–334.
Bielschowsky, M.
    1902 "Die Silber impragnation der Axencylinder." *Neurologisches Centralblatt*
    21:579–584.
Birge, Wesley J., and Tibbits, Donald F.
    1961 "The use of sodium chloride containing fixatives in minimizing cellular
    distortion in histological and cytochemical preparations." *Journal of Histochemistry and Cytochemistry* 9:409–414.
Bitensky, Lucille.
    1963 "Modification to the Gomori phosphatase technique for controlled temperature frozen sections." *Quarterly Journal of Microscopical Science* 104:193–196.
Black, M. M., and Ansley, H. R.
    1964 "Histone staining with ammoniacal silver." *Science* 43:693.
Black, M. M.; Ansley, H. R.; and Mandl, R.
    1964 "On cell specificity of histones." *Archives of Pathology* 78:350–368.
Blandau, R. J.
    1938 "A method of eliminating the electrification of paraffin ribbons." *Stain Technology* 13:139–141.
Blank, Harvey, and McCarthy, Philip L.
    1950 "A general method for preparing histologic sections with a water-soluble
    wax." *Journal of Laboratory and Clinical Medicine* 36:776.
Blank, Harvey; McCarthy, Philip L.; and Delamater, Edward D.
    1951 "A non-vacuum freezing-dehydrating technic for histology, autoradiography,
    and microbial cytology." *Stain Technology* 26:193–197.
Block, David P., and Godman, Gabriel C.
    1955 "A microphotometric study of the synthesis of desoxyribonucleic acid and
    nuclear histone." *Journal of Biophysical and Biochemical Cytology* 1:17–28.
Block, M.; Smaller, Victoria; and Brown, Jessie.
    1953 "An adaptation of the Maximow technique for preparation of sections of
    hematopoietic tissue." *Journal of Laboratory and Clinical Medicine* 42:
    145–151.
Bodian, David.
    1936 "A new method for staining nerve fibers and nerve endings in mounted
    paraffin sections." *Anatomical Record* 65:89–97.
    1937 "The staining of paraffin sections of nervous tissues with activated Protargol.
    The role of fixatives." *Anatomical Record* 69:153–162.
Bogen, Emil.
    1941 "Detection of tubercle bacilli by fluorescence microscopy." *American Review of Tuberculosis* 44:267–271.
Bogoroch, Rita.
    1951 "Detection of radio-elements in histological slides by coating with stripping
    emulsion—the strip-coating technic." *Stain Technology* 26:43–50.
Bohorfoush, Joseph G.
    1963 "A thiosulfate diluent for Wright's stain." *Stain Technology* 38:292–293.
Bokdawala, F. D., and George, J. C.
    1964 "Histochemical demonstration of muscle lipase." *Journal of Histochemistry
    and Cytochemistry* 12:768–771.

Borror, Arthur C.
   1968 "Nigrosin-HgCl₂-formalin; a stain-fixative for ciliates (Protozoa, Cilio-phora)." *Stain Technology* 43:293–295.
Borysko, Emil.
   1956 "Recent developments in methacrylate embedding. 1. A study of the polymerization damage phenomenon by phase contrast microscopy." *Journal of Biophysical and Biochemical Cytology* 2, Part 2, Supplement: 3–14.
Borysko, Emil, and Sapranauskas, P.
   1954 "A technique for comparison phase-contrast and electron microscope studies of cells grown in tissue culture with evaluation of time-lapse cinemicro-graphs." *Johns Hopkins Hospital Bulletin* 95:68–80.
Bouchard, Frances M.
   1963 "Acceleration of Giemsa staining by heat." *Stain Technology* 38:288.
Boyd, G. A.
   1955 *Autoradiography in Biology and Medicine.* New York: Academic Press.
Boyd, I. A.
   1962 "Uniform staining of nerve endings in skeletal muscle with gold chloride." *Stain Technology* 37:225–230.
Bradley, Muriel V.
   1948a "An aceto-carmine squash technic for mature embryo sacs." *Stain Technology* 23:29–40.
   1948b "A method for making aceto-carmine squashes permanent without re-moval of the cover slip." *Stain Technology* 23:41–44.
   1957 "Sudan black B and aceto-carmine as a combination stain." *Stain Technology* 32:85–86.
Brandi, A. James; Malferrari, Rosella; and Massignani, Adrian.
   1962 "Collodion fixation and protection of cytological smears for transportation and staining by Papanicolaou's method." *Stain Technology* 37:183–185.
Branton, Daniel, and Jacobson, Louis J.
   1962 "Dry, high resolution autoradiography." *Stain Technology* 37:239–242.
Braunstein, H., and Adriano, S. M.
   1961 "Fluorescent stain for tubercle bacilli in histologic sections." *American Journal of Clinical Pathology* 36:37–40.
Brecher, George.
   1949 "New methylene blue as a reticulocyte stain." *American Journal of Clinical Pathology* 19:895–896.
Brenner, R. M.
   1962 "Controlled oxidation of background grains in radioautographs." *Journal of Histochemistry and Cytochemistry* 10:678.
Bridges, C. B.
   1937 "The vapour method of changing reagents and of dehydration." *Stain Technology* 12:51–52.
Bridges, Charles H., and Luna, Lee.
   1957 "Kerr's improved Warthin-Starry technic. Study of permissible variations." *Laboratory Investigation* 6:357–367.
Briggs, Donald K.
   1958 "Recent studies in chromatin sex determination." *Transactions New York Academy of Sciences* Ser. II, 20:500–504.
Briggs, Norman L.
   1958 "A double embedding technic for fibrous connective tissue." *Stain Technology* 33:299.
Broadway, Charles B., and Koelle, Donald G.
   1960 "Use of polyethylene plastic bags for storage of pathologic tissue." *Technical Bulletin of the Registry of Medical Technologists* 30:17–21.
Brooke, M. M., and Donaldson, A. W.
   1950 "Use of a surface active agent to prevent transfer of malarial parasites

between blood films during mass staining procedures." *Journal of Parasitology* 36:84.

Brown, Charles D., and Fleming, Loraine.
1965 "A modified micromethod for the culturing of lymphocytes from peripheral blood for the assessment of chromosome morphology." *Mammalian Chromosomes Newsletter* 15:110–113.

Brown, J. H., and Brenn, L.
1931 "A method for differential staining of gram positive and gram negative bacteria in tissue sections." *Johns Hopkins Hospital Bulletin* 48:69–73.

Brown, James O.
1948 "A simplified and convenient method for the double embedding of tissues." *Stain Technology* 23:83–89.

Bruemmer, Nancy C.; Carver, Michael J.; and Thomas, Lloyd E.
1957 "A tryptophan histochemical method." *Journal of Histochemistry and Cytochemistry* 5:140–144.

Buck, R. C., and Jarvic, C. E.
1959 "A simple automatic ultramicrotome." *Stain Technology* 34:109–111.

Buckley, Sonja M.; Whitney, Elinor; and Rapp, Fred.
1955 "Identification by fluorescent antibody of development forms of psittacosis virus in tissue culture." *Proceedings of the Society of Experimental Biology and Medicine* 90:226–230.

Bullivant, S.
1965 "Freeze substitution and supporting techniques." *Laboratory Investigation* 14:1178–1195.

Bullivant, S., and Hotchin, J.
1960 "Chromyl chloride, a new stain for electron microscopy." *Experimental Cell Research* 21:211–214.

Bulmer, D.
1962 "Observations on histological methods involving the use of phosphotungstic and phosphomolybdic acids, with particular reference to staining with phosphotungstic acid hematoxylin." *Quarterly Journal of Microscopical Science* 103:311–324.

Burdette, W. J.
1962 *Methodology in Human Genetics.* San Francisco: Holden-Day.

Burkholder, Peter M.; Littell, Andrew H.; and Klein, Paul G.
1961 "Sectioning at room temperature of unfixed tissues, frozen in a gelatin matrix, for immunohistologic procedures." *Stain Technology* 36:89–91.

Burns, J., and Neame, P. B.
1966 "Staining of blood cells with periodic acid/salicylol hydrazide (PA-SH). A fluorescent method for demonstrating glycogen." *Blood* 28:674–682.

Burrows, Robert B.
1967 "Improved preparation of polyvinyl alcohol-HgCl₂ fixative used for fecal smears." *Stain Technology* 42:93.

Burrows, William.
1954 *Textbook of Microbiology.* Philadelphia: W. B. Saunders C.

Burstone, M. S.
1957*a* "Polyvinyl acetate as a mounting medium for azo-dye procedures." *Journal of Histochemistry and Cytochemistry* 5:196.
1957*b* "Polyvinyl pyrrolidone as a mounting medium for stains for fat and for azo-dye procedures." *American Journal of Clinical Pathology* 28:429–430.
1958 "The relationship between fixation and techniques for the histochemical localization of hydrolytic enzymes." *Journal of Histochemistry and Cytochemistry* 6:322–339.
1959*a* "New histochemical techniques for the demonstration of tissue oxidase (cytochrome oxidase)." *Journal of Histochemistry and Cytochemistry* 7:112–121.

1959b "Acid phosphatase activity of calcifying bone and dentin matrices." *Journal of Histochemistry and Cytochemistry* 7:147–148.
1960a "Histochemical demonstration of cytochrome oxidase with new amine reagents." *Journal of Histochemistry and Cytochemistry* 8:63–70.
1960b "Postcoupling, noncoupling, and fluorescence techniques for the demonstration of alkaline phosphatase." *Journal of the National Cancer Institute* 24:1199–1218.
1962 *Enzyme Histochemistry and its Application in the Study of Neoplasms.* New York: Academic Press.

Burstone, M. S., and Flemming, T. J.
1959 "A new technique for the histochemical study of smears." *Journal of Histochemistry and Cytochemistry* 7:203.

Burstone, M. S., and Folk, J. E.
1956 "Histochemical demonstration of aminopeptidase." *Journal of Histochemistry and Cytochemistry* 4:217–226.

Burton, George J.
1958 "Preparation of thick malarial blood smears by the tapping method." *Mosquito News* 18:228–229.

Bush, Vannevar, and Hewitt, Richard E.
1952 "Frozen sectioning: A new and rapid method." *American Journal of Pathology* 28:863–867.

Cameron, D. A.
1956 "A note on breaking glass knives." *Journal of Histochemistry and Cytochemistry* 2:57–59.

Cameron, M. L., and Steele, J. E.
1959 "Simplified aldehyde-fuchsin staining of neurosecretory cells." *Stain Technology* 34:265–266.

Caratzali, Alexandre; Phleps, Annchesi; and Turpin, Raymond.
1957 "Variations quantitatives des corpuscules sexuels des granulocytes neutrophiles au sours du cycle menstruel." *Bulletin de l'Académie de Medicine* 141:496–501.

Cares, A.
1945 "A note on stored formaldehyde and its easy reconditioning." *Journal of Technical Methods* 25:67–70.

Carey, Eben J.
1941 "Experimental pleomorphism of motor nerve plates as a mode of functional protoplasmic movement." *Anatomical Record* 81:393–413.

Carleton, H. M., and Leach, E. H.
1947 *Histological Technique.* New York: Oxford University Press.

Carlo, Ravetto.
1964 "Alcian blue-alcian yellow: a new method for the identification of different acidic groups." *Journal of Histochemistry and Cytochemistry* 12:44–45.

Carmichael, G. G.
1963 "A tetrazolium salt reduction method for demonstrating lipo-proteins in tissue sections." *Journal of Histochemistry and Cytochemistry* 11:738–740.

Caro, Lucian G.
1964 "High resolution autoradiography." *Methods in Cell Physiology,* I. Ed. by David M. Prescott. New York: Academic Press.

Carr, D. H. and Walker, J. E.
1961 "Carbol fuchsin as a stain for human chromosomes." *Stain Technology* 36:233–236.

Carr, L. A. and Bacsich, P.
1958 "Removal of osmic acid stains." *Nature* 182:1108.

Carr, Lawrence B.; Rambo, Oscar N.; and Feichtmeier, Thomas V.
1961 "A method of demonstrating calcium in tissue sections using chloranilic acid." *Journal of Histochemistry and Cytochemistry* 4:415–417.

Carson, Ruth H.
1959 "Staining of cytologic smears without ethyl alcohol." *American Journal of Clinical Pathology* 32:370.

Carter, C. H., and Leise, J. M.
1958 "Specific staining of various bacteria with a single fluorescent antiglobulin." *Journal of Bacteriology* 76:152–154.

Carver, M. Joseph; Brown, F. Christine; and Thomas, Lloyd E.
1953 "An arginine histochemical method using Sakaguchi's new reagent." *Stain Technology* 28:89–91.

Case, Norman M.
1953 "The use of a cation exchange in decalcification." *Stain Technology* 28:155–158.

Cason, Jane E.
1950 "A rapid one-step Mallory-Heidenhain stain for connective tissue." *Stain Technology* 25:225–226.

Caspersson, T.; Zech, L.; and Johansson, C.
1970 "Differential binding of alkylating fluorochromes in human chromosomes." *Experimental Cell Research* 60:315–319.

Casselman, W. B. Bruce.
1959 *Histochemical Technique.* London: Methuen; New York: Wiley.

Castañeda, M. Ruiz.
1939 "Experimental pneumonia produced by typhus Rickettsia." *American Journal of Pathology* 15:467–475.

Caulfield, James B.
1957 "Effects of varying the vehicle for osmic acid in tissue fixation." *Journal of Biophysical and Biochemical Cytology* 3:827–829.

Cavanagh, J. B.; Passingham, R. J.; and Vogt, J. A.
1964 "Staining of sensory and motor nerves in muscles with Sudan black B." *Journal of Pathology and Bacteriology* 88:89–92.

Celarier, Robert P.
1956 "Tertiary butyl alcohol dehydration of chromosome smears." *Stain Technology* 31:155–157.

Chang, J. P.
1956 "Staining mitochondria in frozen dried tissues." *Experimental Cell Research* 11:643–646.

Chang, J. P. and Hori, Samuel H.
1961 "The section-freeze-substitution technique. I. Method." *Journal of Histochemistry and Cytochemistry* 9:292–300.
1961 "The section-freeze-substitution technique. II. Application to localization of enzymes and other chemicals." *Annales d' Histochemie* 6:419–432.

Chang, J. P.; Hori, Samuel H.; and Yokoyama, Masao.
1970 "A modified section freeze-substitution technique." *Journal of Histochemistry and Cytochemistry* 18:683–684.

Chatton, E., and Lwoff, A.
1930 "Imprégnation, par diffusion argentique, de l'infraciliature des ciliés marins et d'eau douce, après fixation cytologique et sans dessiccation." *Comptes Rendus Hebdomadaires des Séances et Mémoires de la Société de Biologie* 104:834–836.
1935 "Les ciliés apostomes-première partie. Aperçu historique et général étude monographique des genres et des espèces." *Archives de Zoologie Expérimentale et Générale* 77:1–453.

1936 "Technique pour l'étude des protozaires, spécialement de leurs structures superficielles (cenétome et argyrome)." *Bulletin de la Société Française de Microscopie* 5:25–39.

Chen, Tze-Tuan.

1942 "A staining rack for handling cover-glass preparations." *Stain Technology* 17:129–130.

1944a "Staining nuclei and chromosomes in protozoa." *Stain Technology* 19:83–90.

1944b "The nuclei in avian parasites. I. The structure of nuclei in *Plasmodium elongatum* with some considerations on technique." *American Journal of Hygiene* 40:26–34.

Chessick, Richard D.

1953 "Histochemical study of the distribution of esterases." *Journal of Histochemistry and Cytochemistry* 1:471–485.

Chesterman, W., and Leach, E. H.

1958 "A bleaching method for melanin and two staining methods." *Quarterly Journal Microscopical Science* 99:65–66.

Chèvremont, M., and Fréderic, J.

1943 "Une nouvelle méthode histochimique de mise en évidence des substances à fonction sulfhydrile. Application à l'épiderme, au poil et à la levure." *Archives de Biologie* 54:589–605.

Chiffelle, Thomas L., and Putt, Frederick A.

1951 "Propylene and ethylene glycol as solvents for Sudan IV and Sudan black B." *Stain Technology* 26:51–56.

Chipps, A. D., and Duff, G. L.

1942 "Glycogen infiltration of the liver cell nuclei." *American Journal of Pathology* 18:645–660.

Christenson, Leroy P.

1965 *Human Chromosome Methodology.* Ed. by Jorge J. Yunis. New York: Academic Press.

Christophers, A. J.

1956 "The differential leukocyte count: observations on the error due to method of spreading." *Medical Journal of Australia* 1:533–536.

Chubb, James C.

1963 "Acetic acid as a diluent and dehydrant in the preparation of whole, stained helminths. "*Stain Technology* 37:179–182.

Chung, C. F., and Chen, Christian, M. C.

1970 "Restoring exhausted Schiff's reagent." *Stain Technology* 45:91–92.

Churg, Jacob, and Prado, Artie.

1956 "A rapid Mallory trichrome stain (chromotrope-aniline blue)." *Archives of Pathology* 62:505–506.

Clark, R. F., and Hench, M. E.

1962 "Practical application of acridine orange stain in the demonstration of fungi." *American Journal of Clinical Pathology* 37:237–238.

Clark, George.

1947 "A simplified method for embedding cellular contents of body fluids in paraffin." *American Journal of Clinical Pathology* 17:256.

1945 "A simplified Nissl stain with thionin." *Stain Technology* 20:23–24.

Clayden, E. C.

1952 "A discussion on the preparation of bone sections by the paraffin wax method with special reference to the control of decalcification." *Journal of Medical Laboratory Technologists* 10:103.

Clendenin, Thomas M.

1969 "Intraperitoneal colchicine and hypotonic KCL for enhancement of abundance and quality of meiotic chromosome spreads from hamster testes." *Stain Technology* 44:63–69.

Cocke, E. C.
  1938 "A method for fixing and staining earthworms." *Science* 87:443–444.
Cohen, Isadore.
  1949 "Sudan black B—a new stain for chromosome smear preparation." *Stain Technology* 24:117–184.
Cohen, Sophia M.; Gordon, Irving; Rapp, Fred; Macaulay, John C.; and Buckley, Sonja M.
  1955 "Fluorescent antibody and complement-fixation tests of agents isolated in tissue culture from measles patients." *Proceedings of the Society for Experimental Biology and Medicine* 90:118–122.
Cole, E. C.
  1933 "Ferric chloride as a mordant for phosphate ripened hematoxylin." *Mimeo from author.*
  1943 "Studies on hematoxylin stains." *Stain Technology* 18:125–142.
Cole, Wilbur V.
  1946 "A gold chloride method for motor-end plates." *Stain Technology* 21:23–24.
Cole, Wilbur V., and Mielcarek, J. E.
  1962 "Fluorochroming nuclei of gold chloride-stained motor endings." *Stain Technology* 37:35–39.
Coleman, Edward J.
  1965 "A simplified autoradiographic dipping procedure—slides handled in groups of five." *Stain Technology* 40:240–241.
Collins, Doris N.; Katz, Sidney S.; and Harris, Albert H.
  1961 "Preserving sputum for examination in the cytology laboratory." *American Journal of Clinical Pathology* 36:92–93.
Collins, E. M.
  1969 "Improved paraffin sectioning with etched microtome knife edges." *Stain Technology* 44:33–37.
Conger, Alan D.
  1960 "Dentist's Sticky wax: a cover sealing compound for temporary slides." *Stain Technology* 35:225.
Conger, Alan D., and Fairchild, Lucile M.
  1953 "A quick-freeze method for making smear slides permanent." *Stain Technology* 28:281–283.
Conklin, James S.
  1963 "Staining reactions of mucopolysaccharides after formalin containing fixatives." *Stain Technology* 38:56–59.
Conn, H. J.
  1939 "Cautions in the use of Dioxane." *Stain Technology* 14:152
  1946 "The development of histological staining." *Ciba Symposia* 7.
  1948 *History of Staining.* Geneva, New York: Biotech Publications.
  1969 *Biological Stains.* 8th ed. Rev. by R. D. Lillie. Baltimore: Williams and Wilkins.
Conn, H. J., and Emmel, Victor M.
  1960 *Staining Procedures.* Baltimore: Williams and Wilkins.
Controls for Radiation, Inc.
  n.d. *Con-Rad/Joftes Fluid Emulsion Radioautography Instruction Manual.* Cambridge, Massachusetts.
Conrad, M. E., Jr., and Crosby, W. H.
  1961 "Bone marrow biopsy: modification of the Vim-Silverman needle." *Journal of Laboratory and Clinical Medicine* 57:642–645.
Coons, Albert H.
  1956 "Histochemistry with labeled antibody." *International Review of Cytology.* Ed. by G. H. Bourne and J. F. Danielli, New York: Academic Press, V:1–23.
  1958 "Fluorescent antibody methods." *General Cytochemical Methods.* Ed. by J. F. Danielli. New York: Academic Press, I:399–422.

Coons, Albert H.; Creech, H. J.; Jones, Norman; and Berliner, Ernest.
  1942 "The demonstration of pneumococcal antigen in tissues by the use of fluo-
    rescent antibody." *Journal of Immunology* 45:159–170.
Coons, Albert H., and Kaplan, Melvin H.
  1950 "Localization of antigen in tissue cells. II. Improvements in a method for
    the detection of antigen by means of fluorescent antibody." *Journal of Ex-
    perimental Medicine* 91:1–13.
Coons, Albert H.; Leduc, Elizabeth H.; and Connolly, Jeanne M.
  1955 "Studies on antibody production. I. A method for the histochemical demon-
    stration of specific antibody and its application to a study of the hyperim-
    mune rabbit." *Journal of Experimental Medicine* 102:49–60.
Coons, Albert H.; Leduc, Elizabeth H.; and Kaplan, Melvin H.
  1951 "Localization of antigen in tissue cells. VI. The fate of injected foreign
    proteins in the mouse." *Journal of Experimental Medicine* 93:173–188.
Coons, Albert H.; Snyder, John C.; Cheever, F. Sargent; and Murry, Edward S.
  1950 "Localization of antigen in tissue cells. IV. Antigens of rickettsiae and
    mumps virus." *Journal of Experimental Medicine* 91:31–37.
Cordova, Margarito R., and Polanco, E. R.
  1962 "The use of plain agar in attaching paraffin sections to slides." *Technical
    Bulletin of the Registry of Medical Technologists* 32:129.
Corliss, John O.
  1953 "Silver impregnation of ciliated protozoa by the Chatton-Lwoff technic."
    Stain Technology 28:97–100.
Cosslett, V. E.
  1947 *The Electron Microscope.* New York: Interscience Publishers.
Cosslett, V. E., and Nixon, W. C.
  1960 *X-ray Microscopy.* Cambridge, England: Cambridge University Press.
Cowdry, E. V.
  1952 *Laboratory Technique in Biology and Medicine.* 3d ed. Baltimore: Wil-
    liams and Wilkins Co.
Crabb, Edward D.
  1949 "A rapid rosin-celloidin-paraffin method for embedding tissues." *Stain Tech-
    nology* 24:87–91.
Craig, Elson L.; Frajola, Walter J.; and Greider, Marie H.
  1962 "An embedding technique for electron microscopy using Epon 812." *Jour-
    nal of Cell Biology* 12:190–194.
Crandall, Frank B.
  1961 "An improved diamond knife holder for ultra-microtomy." *Stain Technol-
    ogy* 36:34–36.
Crary, D. D.
  1962 "Modified benzyl alcohol clearing on alizarin-stained specimens without loss
    of flexibility." *Stain Technology* 37:124–125.
Crozier, R. H.
  1968 "An acetic acid dissociation air-drying technique for insect chromosomes,
    with aceto-lactic orcein staining." *Stain Technology* 43:171–173.
Culling, C. E. A.
  1957 *Handbook of Histopathological Technique.* London: Butterworth.
Culling, Charles, and Vassar, Philip.
  1961 "Desoxyribose nucleic acid. A fluorescent histochemical technique." *Ar-
    chives of Pathology* 71:88/76–92/80.
Cumley, R. W.; Crow, J. F.; and Griffin, A. B.
  1939 "Clearing specimens for demonstration of bone." *Stain Technology* 14:7–11.
Cunningham, G. J.; Bitensky, L.; Chayen, J.; and Silcox, A. A.
  1961 "The preservation of cytological and histochemical detail by a controlled
    temperature freezing and sectioning technique." *Annales d'Histochimie*
    6:433–436.

Dalton, A. J.
1955 "A chrome-osmium fixative for electron microscopy." *Anatomical Record* 121:281.
Danielli, J. F.
1953 *Cytochemistry.* New York: Wiley.
Daoust, R.
1957 "Localization of deoxyribonuclease in tissue sections. A new approach to the histochemistry of enzymes." *Experimental Cell Research* 12:203–211.
1961 "Localization of Deoxyribonuclease activity by the substrate film method." *General Cytochemical Methods.* Vol. II. Ed. by J. F. Danielli. New York: Academic Press.
1964 "In vitro binding of nucleic acids to tissue sections after removal of tissue nucleic acids." *Journal of Histochemistry and Cytochemistry* 12:640–645.
1968 "The localization of enzyme activities by substrate film methods—Evaluation and perspectives." *Journal of Histochemistry and Cytochemistry* 16: 540–545.
Daoust, R., and Amano, Hamko.
1960 "The localization of ribonuclease activity in tissue secretions." *Journal of Histochemistry and Cytochemistry* 8:131–134.
Darrow, Mary A.
1952 "Synthetic orcein as an elastic tissue stain." *Stain Technology* 27:329–332.
Dart, Leroy H., and Turner, Thomas R.
1959 "Fluorescence microscopy in exfoliative cytology." *Laboratory Investigation* 8, Part II:1513–1522.
Davenport, H. A.
1933 "Block staining of nervous tissue with silver. III. Pericellular end-bulbs or boutons." *Stain Technology* 8:143–147
1948 "Protargol: old and new." *Stain Technology* 23:219–220.
1960 *Histological and Histochemical Technics.* Philadelphia: W. B. Saunders.
Davenport, H. A., and Combs, C. M.
1954 "Golgi's dichromate silver method. 3. Chromating fluids." *Stain Technology* 29:165–173.
Davenport, H. A.; Windle, W. F., and Buch, R. H.
1934 "Block staining of nervous tissue. IV. Embryos." *Stain Technology* 9:5–10.
Davidson, William M., and Smith, D. Robertson.
1954 "A morphological sex difference in the polymorphonuclear neutrophil leucocytes." *British Medical Journal,* 2 July: 6–7.
Davies, Howard G.
1958 "The determination of mass and concentration by microscopic interferometry." *General Cytochemical Methods,* Vol. I. Ed by J. F. Danielli. New York, Academic Press.
Davis, Helen, and Harmon, Pinkney J.
1949 "A suggestion for prevention of loose sections in the Bodian Protargol method." *Stain Technology* 24:249.
Debruyn, P. P. H.; Farr, R. S.; Banks, Hilda; and Morthland, F. W.
1953 "In vivo and in vitro affinity of diaminoacridine for nucleoproteins." *Experimental Cell Research* 4:174–180.
Debruyn, P. P. H.; Robertson, C.; and Farr, Richard S.
1950 "In vivo affinity of diamino-acridines for nuclei." *Anatomical Record* 108: 279–307.
Deck, J. David, and Desouza, G.
1959 "A disrupting factor in silver staining techniques." *Stain Technology* 34: 287.
Decosse, J. J., and Aiello, N.
1966 "Feulgen hydrolysis: effect of acid and temperature." *Journal of Histochemistry and Cytochemistry* 14:601–604.

DeDuve, C.
   1959 "Lysosomes, a new group of cytoplasmic particles." *Subcellular Particles.* Ed. by T. Hayashi. New York: Ronald Press.
   1963 "The lysosome." *Scientific American* 208 (May):64.
   1963 "General properties of lysosomes: the lysosome concept." *Lysosomes.* Ed. by A. V. S. deReuck and M. P. Cameron. Boston: Little, Brown.
De Giusti, Dominic, and Ezman, Leon.
   1955 "Two methods for serial sectioning of arthropods and insects." *Transactions of American Microscopical Society* 74:197–201.
De Harven, Etienne.
   1958 "A new technique for carbon films." *Journal of Biophysical and Biochemical Cytology* 4:133–134.
Deitch, Arline Douglas.
   1955 "Microspectrophotometric study of the binding of the anionic dye, naphthol yellow S, by tissue sections and by purified protein." *Laboratory Investigation* 4:324–351.
   1961 "An improved Sakaguchi reaction for microspectrophotometric use." *Journal of Histochemistry and Cytochemistry* 9:477–483.
   1964 "A method for the cytophotometric estimation of nucleic acids using methylene blue." *Journal of Histochemistry and Cytochemistry* 12:451–461.
Deitch, Arline Douglas; Wagner, Dieter; and Richart, Ralph M.
   1968 "Conditions influencing the intensity of the Feulgen reaction." *Journal of Histochemistry and Cytochemistry* 16:371–379.
Delameter, Edward D.
   1948 "Basic fuchsin as a nuclear stain." *Stain Technology* 23:161–176.
   1951 "A staining and dehydrating procedure for handling of micro-organisms." *Stain Technology* 26:199–204.
De La Torre, Luis, and Salisbury, G. W.
   1962 "Fading of Feulgen-stained bovine spermatozoa." *Journal of Histochemistry and Cytochemistry* 10:39–41.
Delez, Arthur L., and Davis, Olive Stull.
   1950 "The use of oxalic acid in staining with phloxine and hematoxylin." *Stain Technology* 25:111–112.
Del Vecchio, P. R.; Dewitt, S. H.; Borelli, J. I.; Ward, J. B.; Wood, T. A., Jr.; and Malmgren, R. A.
   1959 "Application of millipore fixation technique to cytologic material." *Journal of the National Cancer Institute* 22:427–432.
de Martino, C.; Capanna, E.; Civitelli, M. V.; and Procicchiani, G.
   1965 "A silver staining reaction for chromosomes and nuclei." *Histochemie* 5:78–85.
Demke, Donald D.
   1952 "Staining and mounting helminths." *Stain Technology* 27:135–139.
Dempsey, Edward W., and Lansing, Albert I.
   1954 "Elastic Tissue." *International Review of Cytology.* Ed. by G. H. Bourne and J. B. Danielli. New York: Academic Press.
Dempster, Wilfred Taylor.
   1944a "Principles of microscopic illumination and the problem of glare." *Journal of the Optical Society of America* 34:695–710.
   1944b "Visual factors in microscopy." *Journal of the Optical Society of America* 34:711–717.
   1944c "Properties of paraffin relating to microtechnique." *Michigan Academy of Science, Arts, and Letters* 29:251–264.
Denver Report
   1960 "A proposed standard system of nomenclature of human mitotic chromosomes." *Lancet* 1:1063–1065.

DePalma, P. A., and Young, G. G.
    1963 "Rapid staining of *Candida albicans* in tissue by periodic acid oxidation, basic fuchsin, and light green." *Stain Technology* 38:257–259.
DeSouza, E. J., and Kothare, S. N.
    1959 "A method for the cytochemical demonstration of succinic dehydrogenase in human leukocytes." *Journal of Histochemistry and Cytochemistry* 7: 77–79.
Deuchar, E. M.
    1962 "Staining sections before autoradiographic exposure: excessive background graining caused by celesting blue." *Stain Technology* 37:324.
Dewitt, S. H.; Del Vecchio, P. R.; Borelli, J. I.; and Hilberg, A. W.
    1957 "A method for preparing wound washings and bloody fluids for cytologic evaluation." *Journal of the National Cancer Institute* 19:115–122.
Diacumakos, E. G.; Day, Emerson; and Kopac, M. J.
    1960 "A new plastic mounting medium for cytologic and histologic preparations." *Laboratory Investigation* 9:499–502.
Diamond, L. S.
    1945 "A new rapid stain technic for intestinal protozoa, using Tergitol-hematoxylin." *American Journal of Clinical Pathology* 15:68–69.
Diegenbach, P. C.
    1970 "Commercial fabric softeners for facilitating sectioning of refractory paraffin-embedded tissues," *Stain Technology* 45:303–304.
Dowding, Grace L.
    1959 "Plastic embedding of undecalcified bone." *American Journal of Clinical Pathology* 32:245–249.
Drets, Maximo E., and Shaw, Margery W.
    1971 "Specific banding patterns of human chromosomes." *Proceedings of the National Academy of Sciences* 68:2073–2077.
Duddy, James A., and Curran, Charles S.
    1962 "Mechanical counting of serial sections cut by a sliding microtome." *Stain Technology* 37:113–114.
Dunn, R. C.
    1946 "A hemoglobin stain for histologic use based on the cyanol-hemoglobin reaction." *Archives of Pathology* 41:676–677.
DuPont de Nemours and Company
    n.d.   *"Elvanol" polyvinyl alcohol for adhesives and binders.* Dupont Vinyl Products Bulletin V 2–254.
Dyer, A. F.
    1963 "The use of lacto-propionic orcein in rapid squash methods for chromosome preparations." *Stain Technology* 38:85–90.
Dziabis, Marvin Dean.
    1958 "Luxol fast blue MBS, a stain for gross brain sections." *Stain Technology* 33:96–97.

Eapen, J.
    1960 "The effect of alcohol-acetic formalin, Zenker's fluid, and gelatin on the activity of lipase." *Stain Technology* 35:227–228.
Earle, W. R.
    1939 "Iron hematoxylin stain containing high concentration of ferrous iron." *Science* 89:323–324.
Eayrs, J. T.
    1950 "An apparatus for fixation and supravital staining of tissues by perfusion method." *Stain Technology* 25:137–142.
Edwards, J. H., and Young, R. B.
    1961 "Chromosome analysis from small volumes of blood." *Lancet* 2:48–49.

Egozcue, J., and Vilarasau de Egozcue, M.
    1966 "Simplified culture and chromosome preparations of primate leukocytes." *Stain Technology* 41:173–178.
Ehrenrich, Theodore, and Kerpe, Stase.
    1959 "A new rapid method of obtaining dry fixed cytological smears." *Journal of American Medical Association* 170:94–95.
Einarson, L.
    1951 "On the theory of gallocyanin-chromalum staining and its application for quantitative estimation of basophilia. A selective staining of exquisite progressivity." *Acta Pathologica et Microbiologica Scandinavica* 28:82–102.
Electron Miscroscopy Society of America
    1953 "Proceedings." *Journal of Applied Physiology* 24.
Elftman, Herbert.
    1952 "A direct silver method for the Golgi apparatus." *Stain Technology* 27:47–52.
    1954 "Controlled chromation." *Journal Histochemistry and Cytochemistry* 2:1–8.
    1956 "Response of the anterior pituitary to dischromate oxidation." *Journal of Histochemistry and Cytochemistry* 4:410.
    1957a "A chrome-alum fixative for the pituitary." *Stain Technology* 32:25–28.
    1957b "Phospholipid fixation by dichromate-sublimate." *Stain Technology* 32:29–31.
    1958 "Effects of fixation in lipoid histochemistry." *Journal of Histochemistry and Cytochemistry* 6:317–321.
    1959a "Combined aldehyde-fuchsin and PAS staining of the pituitary." *Stain Technology* 34:77–80.
    1959b "Aldehyde-fuchsin for pituitary cytochemistry," *Journal of Histochemistry and Cytochemistry* 7:98–100.
    1959c "A Schiff reagent of calibrated sensitivity," *Journal of Histochemistry and Cytochemistry* 7:93–97.
    1960 "Hematoxylin as a pituitary stain." *Stain Technology* 35:97–101.
    1963 "Combined Schiff procedures." *Stain Technology* 38:127–130.
Elftman, Herbert, and Elftman, Alice G.
    1945 "Histological methods for the demonstration of gold in tissues." *Stain Technology* 20:59–62.
Elias, Julius M.
    1969 "Effects of temperature, poststaining rinses and ethanol-butanol dehydrating mixtures on methyl green-pyronin staining." *Stain Technology* 44:201–204.
Ellinger, P.
    1940 "Fluorescence microscopy in biology." *Biological Reviews of the Cambridge Philosophical Society* 15:323–350.
Elston, Robert N., and Sheehan, John F.
    1967 "Freon-Aerosol freezing of squashes or smears for removal of cover glass." *Stain Technology* 42:317.
Emig, William H.
    1959 *Microtechnique; Test and Laboratory Exercises.* Colorado Springs: The author.
Emmel, Victor M., and Cowdry, E. V.
    1964 *Laboratory Technique in Biology and Medicine.* Baltimore: Williams and Wilkins.
Endicott, K. M.
    1945 "Plasma or serum as a diluting fluid for thin smears of bone marrow." *Stain Technology* 20:25–26.
Enerbäck, Lennart.
    1969 "Detection of histamine in mast cells by O-phthalaldehyde reaction after liquid fixation." *Journal of Histochemistry and Cytochemistry* 17:757–759.

Engström, Arne.
   1956 "Historadiography." *Physical Techniques in Biological Research,* Vol. III.
        Ed. by Gerald Oster and Arthur W. Pollister. New York: Academic Press.
   1959 "X-ray microscopy." *Analytical Cytology.* Ed. by Robert C. Mellors. New
        York: Blakiston.
Enlow, Donald H.
   1954 "A plastic seal method for mounting sections of ground bones." *Stain Tech-
        nology* 29:21-22.
   1961 "Decalcification and staining of ground thin sections of bone." *Stain Tech-
        nology* 36:250-251.
Epple, August.
   1967 "A staining sequence for A, B, and D cells of pancreatic islets." *Stain Tech-
        nology* 42:53-61.
Ericsson, Jan L. E.
   1965 "Transport and digestion of hemoglobin in the proximal tubule. I. Light
        microscopy and cytochemistry of acid phosphatase." *Laboratory Investiga-
        tion* 14:1-15.
Ericsson, Jan L. E., and Biberfeld, P.
   1967 "Studies on aldehyde fixation. Fixation rates and their relation to fine struc-
        ture and some histochemical reactions in liver." *Laboratory Investigation*
        17:281-298.
Essner, Edward.
   1970 "Observations on hepatic and renal peroximes (microbodies) in the de-
        veloping chick." *Journal of Histochemistry and Cytochemistry* 18:80-92.
Evans, T. C.
   1947 "Radioautographs in which tissue is mounted directly on photographic
        plate." *Proceedings of the Society for Experimental Biology and Medicine*
        64:313-315.
Ewen, A. B.
   1962 "An improved aldehyde-fuchsin staining technique for neurosecretory prod-
        ucts in insects." *Transactions of the American Microscopical Society* 81:
        94-96.

Fahimi, H. Daruish.
   1967 "Perfusion and immersion fixation of rat liver with glutaraldehyde." *Labora-
        tory Investigation* 16:737-750.
Fairley, A.; Linton, E. C.; and Forn-Moore, A. H.
   1936 "Toxicity to animals of some oxidation products of 1:4 Dioxan." *Journal of
        Hygiene* 36:341-347.
Falck, Bengt, and Hillarp, Nils-ake.
   1959 "A note on the chromaffin reaction." *Journal of Histochemistry and Cyto-
        chemistry* 7:149.
Farber, Emmanuel, and Bueling, Ernest.
   1956 "Histochemical localization of specific oxidative enzymes. V. The dissocia-
        tion of succinic dehydrogenase from carriers by lipase and the specific histo-
        chemical localization of the dehydrogenase with phenazine methosulfate
        and tetrazolium salts." *Journal of Histochemistry and Cytochemistry* 4:357-
        362.
Farber, Emmanuel, and Louvriere, Connie D.
   1956 "Histochemical localization of specific oxidative enzymes. IV. Soluble oxida-
        tion-reduction dyes as aids in histochemical localization of oxidative en-
        zymes with tetrazolium salts." *Journal of Histochemistry and Cytochemistry*
        4:347-356.
Farnsworth, Marjorie.
   1956 "Rapid embedding of minute objects in paraffin." *Stain Technology* 31:295-
        296.

1963 "Handling insect eggs during fixation." *Stain Technology* 38:300–301.

Farquhar, Marilyn G.
1956 "Preparation of ultrathin tissue sections for electron microscopy." *Laboratory Investigation* 5:317–337.

Farquhar, Marilyn G., and Rinehart, J. F.
1954 "Electron microscopic studies on the anterior pituitary glands of castrate rats." *Endocrinology* 54:516–541.

Faulkner, R. R., and Lillie, R. D.
1945a "A buffer modification of the Warthin-Starry silver method for spirochetes in single paraffin sections." *Stain Technology* 20:81–82.
1945b "Dried egg white for Mayer's albumin fixative." Stain Technology 20:99–100.

Favorsky, B. A.
1930 "Eine Modifikation des silber Impregnations-verfahrens Rámon y Cajal für das periphere Nervensystem." *Anatomischer Anzeiger* 70:376–378.

Feder, Ned.
1962 "Polyvinyl alcohol as an embedding medium for lipid and enzyme histochemistry." *Journal of Histochemistry and Cytochemistry* 10:341–347.

Feder, Ned, and Sidman, Richard.
1957 "A method for applying different stains to alternate serial sections on a single microscope slide." *Stain Technology* 32:271–273.
1958 "Methods and principles of fixation by freeze-substitution." *Journal of Biophysical and Biochemical Cytology* 4:593–600.

Fenton, J. C. B., and Innes, James.
1945 "A staining method for malaria parasites in thick blood films." *Transactions of the Royal Society of Tropical Medicine and Hygiene* 39:87–90.

Ferguson-Smith, Malcolm A.
1961 "Chromosomes and Human Disease." *Progress in Medical Genetics.* Ed. by Arthur G. Steinberg. New York: Grune and Stratton.
1962 "The identification of human chromosomes." *Proceedings of the Royal Society of Medicine* 55:471–475.

Férnandez-Moran, H.
1953 "A diamond knife for ultrathin sectioning." *Experimental Cell Research* 5:255.
1956 "Application of a diamond knife for ultrathin sectioning to the study of the fine structure of biological tissues and metals." *Journal of Biophysical and Biochemical Cytology, Supplement* 2:29–31.
1960 "Low temperature preparation techniques for electron microscopy of biological specimens based on rapid freezing with liquid helium II. Freezing and drying of biological materials. Part III." *Annals of New York Academy of Sciences* 85:689–713.

Ferreira, Alberto Vaz, and Combs, C. Murphy.
1951 "Deterioration of nitrocellulose solutions caused by light." *Stain Technology* 26:81–84.

Feulgen, R.
1914 "Über die Kohlenwassenstoffgruppe der echten Nukleinsaüre." *Zeitschrift für physiologische chemie* 92:154–158.

Feulgen, R., and Rossenbeck, H.
1924 "Mikroskopisch-Chemischer Nachweis einer Nukleinsaüre von Typus Thymusnucleinsaüre und die darauf beruhende elektive Färbung von Zellkernen in mikroskopischen Präparaten." *Zeitschrift für physiologische chemie* 135:203–248.

Field, J. W.
1941 "Further notes on a method of staining malarial parasites in thick blood films." *Transactions of the Royal Society of Tropical Medicine and Hygiene* 35:35–42.

Firminger, Harlan I.
1950 "Carbowax embedding for obtaining thin tissue sections and study of intra-cellular lipids." *Stain Technology* 25:121–123.

Fischer, H. W.
1957 "A technic for radiography of lymph nodes and vessels." *Laboratory Investigation* 6:522–527.

Fisher, E. R., and Haskell, A. E.
1954 "Combined Gomori methods for demonstration of pancreatic alpha and beta cells." *American Journal of Clinical Pathology* 24:1433–1434.

Fitzgerald, Patrick J.
1961 "Dry-mounting autoradiographic technic for intracellular localization of water-soluble compounds in tissue sections." *Laboratory Investigation* 10: 846–856.

Fitzgerald, Patrick J., and Pohlmann, M.
1966 "Use of the silver-hydroquinone sequence for the display of reticular fibers." *Stain Technology* 41:267–272.

Fitzgerald, Patrick J.; Simmel, Eva; Winstein, Jerry; and Martin, Cynthia.
1953 "Radioautography: theory, technic and applications." *Laboratory Investigation* 2:181–222.

Fitz-William, William G.; Jones, Georgeanna Seegar; and Goldberg, Benjamin.
1960 "Cryostat techniques: methods for improving conservation and sectioning of tissues." *Stain Technology* 35:195–204.

Flax, Martin, and Caulfield, James.
1962 "Use of methacrylate embedding in light microscopy." *Archives of Pathology* 74:387–395.

Flax, Martin, and Pollister, Arthur W.
1949 "Staining of nucleic acids by Azure A." *Anatomical Record* 105:536–537.

Foley, James O.
1943 "A Protargol method for staining nerve fibers in frozen or celloidin sections." *Stain Technology* 18:27–33.

Foot, Nathan Chandler.
1929 "Comments on the impregnation of neuroglia with ammoniacal silver salts." *American Journal of Pathology* 51:223–238.

Ford, C. E.
1962 "Methodology of chromosomal analysis in man." *National Cancer Institute Monograph* 7:105–113.

Ford, C. E., and Hamerton, J. E.
1956 "The chromosomes of man." *Nature* 178:1020–1023.

Ford, E. H. R., and Woollam, D. H. M.
1963 "A colchicine, hypotonic citrate, air drying sequence for foetal mammalian chromosomes." *Stain Technology* 38:271–274.

Ford, Lee.
1965 "Leukocyte culture and chromosome preparations from adult dog blood." *Stain Technology* 40:317–320.

Fox, M., and Zeiss, J. M.
1961 "Chromosome preparation from fresh and cultural tissues using a modification of the drying technique." *Nature* 192:1213–1214.

Frandsen, John C.
1964 "Serial sections: simplified handling of paraffin ribbons on a floating-out bath. *Stain Technology* 39:279–282.

Frankel, Howard H., and Peters, Robert L.
1964 "A modified calcium-cobalt method for the demonstration of alkaline phosphatase." *American Journal of Clinical Pathology* 42:324–327.

Freed, Jerome J.
1955 "Freeze-drying technics in cytology and cytochemistry." *Laboratory Investigation* 4:106–121.

Freeman, Barbara L.; Moyer, Elizabeth K.; and Lassek, Arthur M.
  1955 "The pH of fixing fluids during fixation of tissues. *Anatomical Record* 121: 593–600.
Freeman, James A.
  1964 *Cellular Fine Structure.* New York: McGraw-Hill.
Friedland, Lester M.
  1951 "A note on frozen section technic." *American Journal of Clinical Pathology* 21:797.
Friend, William G.
  1963 "A microstrainer for handling small ova." *Stain Technology* 38:205–206.
Frigerio, Norman A., and Shaw, Michael J.
  1969 "A simple method for determination of glutaraldehyde." *Journal of Histochemistry and Cytochemistry* 17:176.
Frøland, Anders.
  1965 "Photographic recording and dye staining of chromosomes for autoradiography and morphology." *Stain Technology* 40:41–43.
Frost, H. M.
  1959 "Staining of fresh, undecalcified thin bone sections." *Stain Technology* 34: 135–146.
Fullmer, Harold M., and Lillie, R. D.
  1956 "A selective stain for elastic tissue(orcinol-new fuchsin)." *Stain Technology* 31:27–29.
  1958 "The peracetic acid-aldehyde fuchsin stain." *Journal of Histochemistry and Cytochemistry* 6:391.

Gabor, D.
  1948 *The Electron Microscope.* New York: Chemical Publishing Co.
Gage, Simon Henry.
  1943 *The Microscope.* Ithaca, New York: Comstock.
Gairdner, B. M.
  1969 "Deteriorated paraformaldehyde: an insidious cause of failure in aldehyde-fuchsin staining. "*Stain Technology* 44:52–53.
Galigher, A. E.
  1934 *The Essentials of Practical Microtechnique.* Privately published.
Galigher, A. E., and Kozloff, Eugene N.
  1964 *Essentials of Practical Microtechnique.* Philadelphia: Lea and Febiger.
Gallimore, John C.; Bauer, E. C.; and Boyd, George A.
  1954 "A non-leaching technic for autoradiography." *Stain Technology* 29:95–98.
Galtsoff, Paul S.
  1956 "Simple method of making frozen sections." *Stain Technology* 31:231.
Gardner, D. L.
  1958 "Preparation of bone marrow sections." *Stain Technology* 33:295–297.
Gardner, Harold H., and Punnett, Hope H.
  1964 "An improved squash technique for human male meiotic chromosomes: softening and concentration of cells; mounting in Hoyer's medium." *Stain Technology* 39:245–248.
Gatenby, J. Brontë, and Beams, H. W.
  1950 *The Microtomists's Vade-Mecum.* London: J. and A. Churchill.
Gavin, Mary Ann, and Lloyd, Bolivar J., Jr.
  1959 "Knives of high silica content glass for thin sectioning." *Journal of Biophysical and Biochemical Cytology* 5:507.
Gavin, Thelma.
  1938 "Spirochetal stain on paraffin sections." *American Journal of Clinical Pathology, Technical Supplement* 2:144–145.

Gay, H., and Anderson, T. F.
 1954 "Serial sections for electron microscopy." *Science* 120:1071–1073.
Geil, Robert G.
 1961 "A simple technic for salvage of histologic sections from broken micro-slides." *Technical Bulletin of the Registry of Medical Technologists* 31: 195–196.
Gelei, J. V.
 1932 "Eine neue Goldmethode zur Ciliatenforschung und eine neue Ciliate: *Colpidium pannonicum.*" *Archiv für Protistenkunde* 77:219–230.
 1935 "Eine neue Abänderung der Klein'schen trockenen Silbermethode und das Silberliniensystem von *Glaucoma scintillans.*" *Archiv für Protistenkunde* 84: 446–455.
Gengozian, N.; Batson, J. S.; and Eide, P.
 1964 "Hematologic and cytogenetic evidence for hematopoietic chimerism in the marmoset, *Tamarinus nigricollis.*" *Cytogenetics* 3:384–393.
George, J. C.; and Ambadkar, P. M.
 1963 "Histochemical demonstration of lipids and lipase activity in rat testis." *Journal of Histochemistry and Cytochemistry* 11:420–425.
George, J. C., and Iype, P. Thomas.
 1960 "Improved histochemical demonstration of lipase activity." *Stain Technology* 35:151–152.
Geren, B. B., and McCullock, D.
 1951 "Development and use of the Minot Rotary microtome for thin sectioning." *Experimental Cell Research* 2:97–102.
German, W. M.
 1939 "Hortega's silver impregnation methods. Technique and applications." *American Journal of Clinical Pathology Technical Supplement* 3:13–19.
Gersh, Isidore.
 1932 "The Altman technique for fixation by drying while freezing." *Anatomical Record* 53:309–337.
 1956 "The preparation of frozen-dried tissue for electron microscopy." *Journal of Biophysical and Biochemical Cytology, Supplement* 2:37–43.
 1959 "Fixation and staining." *The Cell.* Vol. I. Ed. by Jean Brachet and Alfred E. Mirsky. New York: Academic Press.
Gettner, I. R., and Ornstein, L.
 1956 *Physical Techniques in Biological Research.* New York: Academic Press.
Geyer, Günther.
 1962 "Histochemical-methylation with methanol and thionylchloride." *Acta Histochemica* 14:284–296.
Giolli, Roland A.
 1965 "A note on the chemical mechanism of the Nauta-Gygax technique." *Journal of Histochemistry and Cytochemistry* 13:206–210.
Giovacchini, Rubert P.
 1958 "Affixing Carbowax sections to slides for routine staining." *Stain Technology* 33:274–278.
Gladden, Margaret H.
 1970 "A modified pyridine-silver stain for teased preparation of motor and sensory nerve endings in skeletal muscle." *Stain Technology* 45:161–164.
Glauert, Audrey M.
 1965 "The fixation and embedding of biological specimens." *Techniques for Electron Microscopy.* Ed. by Desmond H. Kay. Philadelphia: F. A. Davis.
Glauert, Audrey M., and Glauert, R. H.
 1958 "Araldite as an embedding medium for electron microscopy." *Journal of Biophysical and Biochemical Cytology* 4:191–194.
Glegg, R. E.; Clermont, Y.; and Leblond, C. P.
 1952 "The use of lead tetra-acetate, benzidine, O-dianisidine, and a 'Film Test'

in investigating the periodic-acid-Schiff technic." *Stain Technology* 27: 277–305.

Glenner, George G.

1957 "Simultaneous demonstration of bilirubin, hemosiderin and lipofuscin pigments in tissue sections." *American Journal of Clinical Pathology* 27:1–5.

1963 "A re-evaluation of the ninhydrin-Schiff reaction." *Journal of Histochemistry and Cytochemistry* 11:285–286.

Glenner, George G., and Lillie, R. D.

1957a "A rhodocyan technic for staining anterior pituitary." *Stain Technology* 32:187–190.

1957b "The histochemical demonstration of indole derivatives by the post-coupled p-dimethylaminobenzylidene reaction." *Journal of Histochemistry and Cytochemistry* 5:279–296.

1959 "Observations on the daizonium coupling reaction for the histochemical demonstration of tyrosine: metal chelation and formazon variants." *Journal of Histochemistry and Cytochemistry* 7:416–421.

Glick, David.

1949 *Techniques of Histo- and Cytochemistry.* New York: Interscience Publishers.

1962 *Quantitative Chemical Techniques of Histo- and Cytochemistry.* New York: Interscience Publishers. 2 vols.

Glick, David, and Malstrom, B. G.

1952 "Studies in histochemistry. XXIII. A simple and efficient freezing-drying apparatus for the preparation of embedded tissue." *Experimental Cell Research* 3:125–235.

Glynn, J. H.

1935 "The application of the Gram stain to paraffin sections." *Archives of Pathology* 20:896–899.

Goh, Kong-OO.

1965 "Human cytogenetics." *DM, Disease-a-Month,* April.

Goland, Philip P.; Jason, Robert S.; and Berry, Kathryn P.

1954 "Combined Carbowax-paraffin technic for microsectioning fixed tissues." *Stain Technology* 29:5–8.

Goldberg, Arthur F.

1964 "Acid phosphatase activity in Auer bodies." *Blood* 24:305–308.

Goldfischer, Sidney; Essner, Edward; and Novikoff, Alex B.

1964 "The localization of phosphatase activities at the level of ultrastructure." *Journal of Histochemistry and Cytochemistry* 12:72–95.

Goldman, Morris.

1949 "A single solution ironhematoxylin stain for intestinal protozoa." *Stain Technology* 24:57–60.

1968 *Fluorescent Antibody Methods.* New York: Academic Press.

Goldstein, D. J.

1962 "Ionic and non-ionic bonds in staining, with special reference to the action of urea and sodium chloride on the staining of elastic fibers and glycogen." *Quarterly Journal of Microscopical Science* 103:477–492.

Gomori, George.

1936 "Microtechnical demonstration of iron." *American Journal of Pathology* 12:655–663.

1941a "The distribution of phosphatase in normal organs and tissues." *Journal of Cellular and Comparative Physiology* 17:71–84.

1941b "Observations with differential stains on human islets of Langerhans." *American Journal of Pathology* 17:395–406.

1946 "The study of enzymes in tissue sections." *American Journal of Clinical Pathology* 16:347–352.

1948 "Chemical character of enterochromaffin cells." *Archives of Pathology* 45: 48–55.
1950a "An improved histochemical technic for acid phosphatase." *Stain Technology* 25:81–85.
1950b "A rapid one-step trichrome stain." *American Journal of Clinical Pathology* 20:662–664.
1950c "Aldehyde-fuchsin; a new stain for elastic tissue." *American Journal of Clinical Pathology* 20:665–666.
1950d "Sources of error in enzymatic histochemistry." *Journal of Laboratory and Clinical Medicine* 35:802–809.
1951 "Alkaline phosphatase of cell nuclei." *Journal of Laboratory and Clinical Medicine* 37:526–531.
1952 *Microscopic Histochemistry*. Chicago: University of Chicago Press.
1953 "Human esterases." *Journal of Laboratory and Clinical Medicine* 42:445–453.
1954a "The histochemistry of mucopolysaccharides." *British Journal of Experimental Pathology* 35:377–380.
1954b "Histochemistry of the enterochromaffin substance." *Journal of Histochemistry and Cytochemistry* 2:50–53.
1955 "Histochemistry of human esterases." *Journal of Histochemistry and Cytochemistry* 3:479–484.
1956 "Histochemical methods for acid phosphatase." *Journal of Histochemistry and Cytochemistry* 4:453–461.
Gonzalez, Romeo.
1959a "The removal of mercury after fixation in sublimate-containing mixtures." *Stain Technology* 34:111–112.
1959b "Differentiation of mastocyte granules by tartrazine counterstaining after PAS procedure." *Stain Technology* 34:173–174.
Gordon, Harold.
1953 "Method for storing wet histologic accessions and disposing of autopsy material." *Laboratory Investigation* 2:152–153.
Gough, J., and Fulton, J. D.
1929 "A new fixative for mitochondria." *Journal of Pathology and Bacteriology* 32:765–769.
Gower, W. Carl.
1939 "A modified stain and procedure for trematods." *Stain Technology* 14: 31–32.
Gradwohl, R. B. H.
1956 *Clinical Laboratory Methods and Diagnosis*. Vol. II. St. Louis: C. V. Mosby.
Graupner, Heinz, and Weissberger, Arnold.
1931 "Über der Verwendung des Dioxanes beim Einbetten mikroskopischer Objekte. Mitteilungen zur mikroskopischen Technik I." *Zoologischer Anzeiger* 96:204–206.
1933 "Die Verwendung von Lösungen in Dioxan als Fiexierungsmittel für Gefrierschnitte." *Zoologischer Anzeiger* 102:39–44.
Graves, K. D.
1943 "Restoration of dried biopsy tissue." *American Journal of Clinical Pathology, Technical Section* 7:111.
Gray, Peter.
1954 *The Microtomists's Formulary and Guide*. New York: Blakiston.
1964a "Making dry mounts of Foraminifera and Radiolaria." *Ward's Bulletin* 6.
1964b *Handbook of Basic Microtechnique*. 3d ed. New York: Blakiston.
Green, James A.
1956 "Luxol Fast Blue MBS: a stain for phase contrast microscopy." *Stain Technology* 31:219–221.

Greenblatt, Robert A.
   1956 "Method for preparation of blood smear for chromosomal sex determination." *Endocrinology and Metabolism* (Eighth Annual Postgraduate Assembly of the Endocrine Society) Houston: Texas Medical Center.
Greenblat, Robert A., and Manautou, Jorge Martinez.
   1957 "A simplified staining technique for the study of chromosomal sex in oral mucosal and peripheral blood smears." *American Journal of Obstetrics and Gynecology* 74:629–634.
Greenstein, J. S.
   1957 "A rapid phloxine-methylene blue oversight stain for formalin-fixed material." *Stain Technology* 32:75–77.
   1961 "A simplified five-dye stain for sections and smears." *Stain Technology* 36:87–88.
Gregg, V. R., and Puckett, W. O.
   1943 "A corrosive sublimate fixing solution for yolk-laden amphibian eggs." *Stain Technology* 18:179–180.
Gridley, Mary Francis.
   1951 "A modification of the silver impregnation method of staining reticular fibers." *American Journal of Clinical Pathology* 21:897–899.
   1953 "A stain for fungi in tissue sections." *American Journal of Clinical Pathology* 23:303–307.
   1957 *Manual of Histologic and Special Staining Technics*. Washington, D.C.: Armed Forces Institute of Pathology.
Griffin, A. B.
   1960 "Mammalian pachytene chromosome mapping and somatic chromosome identification." *Journal of Cellular and Comparative Physiology* 56:113–121.
Griffin, Lawrence E., and McQuarrie, Agnes M.
   1942 "Iron hematoxylin staining of salivary gland chromosomes in Drosophila." *Stain Technology* 17:41–42.
Grimley, Philip M.
   1964 "A tribasic stain for thin sections of plastic-embedded, osmic acid fixed tissues." *Stain Technology* 39:229–233.
Grimley, Philip M.; Albrecht, Joseph M.; and Michelitch, Herman J.
   1965 "Preparation of large Epoxy sections for light microscopy as an adjunct to fine-structure studies." *Stain Technology* 40:357–366.
Groat, Richard A.
   1949 "Initial and persisting staining power of solutions of iron hematoxylin lake." *Stain Technology* 24:157–163.
Grocott, R. G.
   1955 "Stain for fungi in tissue sections and smears, using Gomori's methenamine-silver nitrate method." *American Journal of Clinical Pathology* 25:975–979.
Grunbaum, Benjamin; Geary, John R., Jr.; and Glick, David.
   1956 "Studies in Histochemistry: the design and use of improved apparatus for the preparation and freezing-drying of fresh-frozen sections of tissue." *Journal of Histochemistry and Cytochemistry* 4:555–560.
Guard, Hormez, R.
   1959 "A new technic for differential staining of the sex chromatin and the determination of its incidence in exfoliated vaginal epithelial cells." *American Journal of Clinical Pathology* 32:145–151.
Gude, W. D.
   1968 *Autoradiographic Techniques*. Englewood Cliffs: Prentice-Hall.
Gude, W. D., and Odell, T. T.
   1955 "Vinisil as a diluent in making bone marrow smears." *Stain Technology* 30:27–28.

Gude, W. D.; Upton, Arthur C.; and Odell, T. T.
    1955 "Giemsa staining of autoradiograms prepared with stripping film." *Stain Technology* 30:161–162.
Guillery, R. W.; Shirra, B.; and Webster, K. E.
    1961 "Differential impregnation of degenerating nerve fibers in paraffin-embedded material." *Stain Technology* 36:9–13.
Gurr, Edward.
    1956 *A Practical Manual of Medical and Biological Staining Techniques.* New York: Interscience.
    1958 *Methods of Analytical Histology and Histochemistry.* London: Leonard Hill.
    1960 *Encyclopaedia of Microscopic Stains.* London: Edward Gurr; Baltimore: William and Wilkins.
Gurr, G. T.
    1953 *Biological Staining Methods.* London: G. T. Gurr.
Guyer, M. F.
    1953 *Animal Micrology.* Chicago: University of Chicago Press.

Hack, M. H.
    1952 "A new histochemical technique for lipids applied to plasmal." *Anatomical Record* 112:275–301.
Hajian, Ahmad.
    1961 "Note on trichrome stain." *Technical Bulletin of the Registry of Medical Technologists* 31:92.
Hale, Arthur J.
    1952 "The effect of temperature and of relative humidity on sectioning of tissues embedded in polyethylene glycol wax." *Stain Technology* 27:189–192.
    1955 "The effect of formalin on the periodic acid Schiff staining of certain types of mucus." *Journal of Histochemistry and Cytochemistry* 3:421–429.
    1957 "The histochemistry of polysaccharides." *International Review of Cytology,* VI. Ed. by G. H. Bourne and J. F. Danielli. New York: Academic Press.
Hale, C. W.
    1946 "Histochemical demonstration of acid polysaccharides in animal tissue." *Nature* 157:802.
Hale, Dean M.; Cromartie, William J.; and Dobson, Richard L.
    1960 "Luxol fast blue as a selective stain for dermal collagen." *Journal of Investigative Dermatology* 35:293–294.
Hale, L. J.
    1958 *Biological Laboratory Data.* London: Methuen; New York: John Wiley.
Hall, C. E.
    1953 *Introduction to Electron Microscopy.* New York: McGraw-Hill.
Halmi, Nicholas S.
    1950 "Two types of basophils in the anterior pituitary of the rat and their respective cytophysiological significance." *Endocrinology* 47:289–299.
Ham, Arthur Worth.
    1957 *Histology.* 3d ed. Philadelphia: J. B. Lippincott.
Hamerton, John L.
    1961 "Sex chromatin and human chromosomes. *International Review of Cytology,* XII. Ed by G. H. Bourne and J. F. Danielli. New York: Academic Press.
Hamlyn, J. H.
    1957 "Application of the Nauta-Gygax technic for degenerating axons to mounted sections." *Stain Technology* 32:123–126.
Hance, Robert T., and Green, Floyd J.
    1959 "Rapid ripening of hematoxylin solutions." *Stain Technology* 34:237–238.
    1961 "Behavior of rapidly oxidized hematoxylin." *Stain Technology* 36:253.

Hancox, N. M.
  1957 "Experiments on the fundamental effects of freeze substitution." *Experimental Cell Research* 13:263-275.
Hanson, A. A., and Oldemeyer, D. L.
  1951 "Staining root tip smears with aceto-carmine." *Stain Technology* 26:241-242.
Hanson, V., and Hermodsson, L. H.
  1960 "Freeze-drying of tissues for light and electron microscopy." *Journal of Ultrastructure Research* 4:332-348.
Harada, Kiyoshi.
  1957 "Selective staining of mast cell granules with chrysoidin stain." *Stain Technology* 32:183-186.
Hardonk, M. J., and vanDuijan, P.
  1964a "The mechanism of the Schiff reaction as studied with histochemical model systems. *Journal of Histochemistry and Cytochemistry* 12:748-751.
  1964b "A quantitative study of the Feulgen reaction with the aid of histochemical model systems." *Journal of Histochemistry and Cytochemistry* 12:752-757.
  1964c "Studies on the Feulgen reaction with histochemical model systems." *Journal of Histochemistry and Cytochemistry* 12:758-767.
Harmon, John W.
  1950 "The selective staining of mitochondria." *Stain Technology* 25:69-72.
Harnden, D. G.
  1959 "A human skin culture technique used for cytological examinations." *British Journal of Experimental Pathology* 41:31-37.
Harnden, D. G., and Brunton, Sheila.
  1965 "The skin culture technique." *Human Chromosome Methodology.* Ed. by Jorge J. Yunis. New York: Academic Press.
Harris, G. W., and Donovan, B. T.
  1966 *The Pituitary Gland.* Berkeley: University of California Press. 3 vols.
Hartman-Leddon Company.
  1952 *Informative Bulletin 306-2.* Philadelphia.
Hartroft, W. S.
  1951 "Fluorchromy as an aid in the resolution of the specific granules of the islets of Langerhans." *Nature* 168:1000.
Hartz, P. H.
  1945 "Frozen sections from Bouin-fixed material in histopathology." *Stain Technology* 20:113-114.
  1947 "Simultaneous histologic fixation and gross demonstration of calcification." *American Journal of Clinical Pathology* 17:750.
Haupt, A. W.
  1930 "A gelatin fixative for paraffin sections." *Stain Technology* 5:97.
Hause, Welland A.
  1959 "Saw for preparation of blocks of bone." *Technical Bulletin of the Registry of Medical Technicians* 29:101.
Haust, M. Daria.
  1958 "Tetrahydrofuran (THF) for dehydration and infiltration." *Laboratory Investigation* 7:58-67.
  1959 "Tetrahydrofuran (THF) for routine dehydration clearing and infiltration." *Technical Bulletin of the Registry of Medical Technicians* 29:33-37.
Haviland, Thomas N.
  1963 "Restoration of rust-stained formalin-preserved gross specimens." *American Journal of Clinical Pathology* 39:364.
Heady, Judith, and Rogers, T. Edwin.
  1962 "Turtle blood cell morphology." *Proceedings of the Iowa Academy of Sciences* 69:587-590.

Hegner, Robert W.; Cort, William W.; and Root, Francis M.
    1927 *Outlines of Medical Zoology.* New York: Macmillan.
Hendrickson, Anita; Kunz, Sandra; and Kelly, Douglas E.
    1968 "NaOH-HIO₄ treatment of osmium-collidine fixed Epoxy sections to facilitate staining after autoradiography." *Stain Technology* 43:175–176.
Herlant, Marc.
    1960 "Étude critique de deux techniques nouvelles destinées a mettre en évidence les différentes catégories cellulaires présentes dans la glande pituitaire." *Bulletin de Microscopie Appliquée* 10:37–44.
Hetherington, Duncan C.
    1936 "Pinacyanol as a supravital stain for blood." *Stain Technology* 11:153–154.
Hicks, J. D., and Matthaei, E.
    1955 "Fluorescence in Histology." *Journal of Pathology and Bacteriology* 70:1–12.
    1958 "A selective fluorescence stain for mucin." *Journal of Pathology and Bacteriology* 75:473–476.
Hillary, B. B.
    1938 "Permanent preparations from rapid cytological technics." *Stain Technology* 13:161–167.
Hilleman, Howard H., and Lee, C. H.
    1953 "Organic chelating agents for decalcification of bones and teeth." *Stain Technology* 28:285–286.
Himes, Marion, and Moriber, Louis.
    1956 "A triple stain for desoxyribonucleic acid, polysaccharides and proteins." *Stain Technology* 31:67–70.
Histochemical Society
    1953 "Proceedings." *Journal of Histochemistry and Cytochemistry* 1:387–388.
Hodgman, Charles D.
    1957 *Handbook of Chemistry and Physics.* Cleveland: Chemical Rubber Publishing Co.
Hoefert, Lynn L.
    1968 "Polychromatic stains for thin sections of *Beta* embedded in Epoxy resin." *Stain Technology* 43:145–151.
Holczinger, L., and Bálint, Zsuzsa.
    1961 "The staining properties of 'masked' lipids." Acta Histochemica 11:284–288.
Holde, P., and Isler, H.
    1958 "The effect of phosphomolybdic acid on the stainability of connective tissue by various dyes." *Journal of Histochemistry and Cytochemistry* 6:265–270.
Hollander, David H.
    1963 "An oil-soluble anti-oxidant in resinous mounting media to inhibit fading of Romanowsky stains." *Stain Technology* 38:288–289.
Hollister, Gloria.
    1934 "Clearing and dyeing fish for bone study." *Zoologica* 12:89–101.
Holmes, W. C.
    1929 "The mechanism of staining. The case for the physical theories." *Stain Technology* 4:75–80.
Holt, Margaret; Cowing, R. F.; and Warren, S.
    1949 "Preparation of radioautographs of tissues without loss of water soluble P³²." *Science* 110:328–329.
Holt, Margaret; Sommers, S. C.; and Warren, S.
    1952 "Preparation of tissue sections for quantitative histochemical studies." *Anatomical Record* 112:177–186.
Holt, Margaret, and Warren, Shields.
    1950 "A radioautograph method for detailed localization of radioactive isotopes in tissues without isotope loss." *Proceedings of the Society for Experimental Biology and Medicine* 73:545.

1953 "Freeze-drying tissues for autoradiography." *Laboratory Investigation* 2: 1–14.

Holt, S. J.
1956 "The value of fundamental studies of staining reactions in enzyme histochemistry, with reference to indoxyl methods for esterases." *Journal of Histochemistry and Cytochemistry* 4:541–554.

Holt, S. J.; Hobbiger, Eluned E.; and Pawan, G. L. S.
1960 "Preservation of integrity of rat tissues for cytochemical staining purposes." *Journal of Biophysical and Biochemical Cytology* 7:383–386.

Hood, R. C. W. S., and Neill, W. M.
1948 "A modification of alizarine red S technic for demonstrating bone formation." *Stain Technology* 23:209–218.

Hori, Samuel H.
1963 "A simplified acid hematein test for phospholipids." *Stain Technology* 38: 221–225.

Hori, Samuel H., and Chang, Jeffrey P.
1963 "Demonstration of lipids in mitochondria and other cellular elements." *Journal of Histochemistry and Cytochemistry* 11:115–116.

Horikawa, Masakatsu, and Kuroda, Yukiaki.
1959 "In vitro cultivation of blood cells of *Drosophila melanogaster* in a synthetic medium." *Nature* 184:2017–2018.

Hörmann, H.; Grassman, W.; and Fries, G.
1958 "Über den Mechanismus der Schiffschen Reaktion." *Justus Liebigs Annalen der Chemie* 616:125–147.

Horn, Robert G., and Spicer, S. S.
1964a "Sulfated mucopolysaccharide and basic protein in certain granules of rabbit leukocytes." *Laboratory Investigation* 13:1–15.
1964b "Sulfated mucopolysaccharides in azurophile granules of immature granulocytes of the rabbit." *Journal of Histochemistry and Cytochemistry* 12:33.

Horobin, R. W., and Kevill-Davies, I. M.
1971 "Basic fuchsin in acid alcohol: a simplified alternative to Schiff reagent." *Stain Technology* 46:53–58.

Hotchkiss, R. D.
1948 "A microchemical reaction resulting in the staining of polysaccharide structures in fixed tissues." *Archives of Biochemistry* 16:131–141.

Houck, C. E., and Dempsey, E. W.
1954 "Cytological staining procedures applicable to methacrylate embedded tissues." *Stain Technology* 29:207–212.

Hrushovetz, S. B., and Harder, C. Elizabeth.
1962 "Permanent mounting of unstained and aceto-orcein stained cells in the water soluble medium, Abopon." *Stain Technology* 37:307–311.

Hsu, T. C., and Kellogg, Douglas.
1960 "Primary cultivation and continuous propagation in vitro of tissues from small biopsy specimens." *Journal of the National Cancer Institute* 25:221–235.

Huber, William M., and Caplin, Samuel M.
1947 "Simple plastic mount for preservation of fungi and small arthropods." *Archives of Dermatology and Syphilology* 56:763–765.

Hukill, Peter, and Putt, Frederick A.
1962 "A specific stain for iron using 4,7-diphenyl-1,10-phenanthroline (Bathophenanthroline)." *Journal of Histochemistry and Cytochemistry* 10:490–494.

Hull, Susan, and Wegner, Sally.
1952 "Removal of stains." *Stain Technology* 27:224.

Hulton, William E.
1958 "A survey of the application of the 'molecular' membrane filter to the study of cerebrospinal fluid cytology." *American Journal of Clinical Pathology* 30:407–410.

Humason, Gretchen L., and Lushbaugh, C. C.
1960 "Selective demonstration of elastin, reticulum, and collagen by silver, or-cein, and aniline blue." *Stain Technology* 35:209–214.
1961 "A quick pinacyanole stain for frozen sections." *Stain Technology* 36:257–258.
1969 "Sirius supra blue FGL-CF; superior to aniline blue in the combined elastin, reticulum and collagen stain." *Stain Technology* 44:105–106.
Humason, Gretchen L., and Sanders, Phyllis, C.
1963 "Culture and slide preparation of leukocytes from peripheral blood." *Stain Technology* 38:338–340.
Humphrey, A. A.
1935 "Di-nitrosoresorcinol—a new specific for iron in tissue." *Archives of Pathology* 20:256–258.
1936 "A new rapid method for frozen section diagnosis." *Journal of Laboratory and Clinical Medicine* 22:198–199.
Hungerford, David A.
1965 "Leukocytes cultures from small inocula of whole blood and the preparation of metaphast chromosomes by treatment with hypotonic KCl." *Stain Technology* 40:333–338.
Hungerford, David A., and Nowell, Peter C.
1962 "Chromosome studies in human leukemia." *Journal of the National Cancer Institute* 29:545–565.
Hutchinson, H. E.
1953 "The significance of stainable iron in sternal marrow sections." *Blood* 8:236–248.
Hutner, S. H.
1934 "Destaining agents for iron alum hematoxylin." *Stain Technology* 9:57–59.
Hutton, W. E.
1953 "Ninhydrin staining of tissue sections." *Stain Technology* 28:173–175.

Ibanez, Michael L.; Russell, William O.; Chang, Jeffrey P.; Speece, Arthur J.
1960 "Cold chamber frozen sections for operating room diagnosis routine surgical stains." *Laboratory Investigation* 9:275–278.
Idelman, Simon.
1964 "Modification de la technique de Luft en vue de la conservation des lipides en microscopie électronique." *Journal de Microscopie* 3:715–718.
Ingram, R. L.; Otken, L. B.; and Jumper, J. R.
1961 "Staining of malarial parasites by the fluorescent antibody technic." *Proceedings of the Society for Experimental Biology and Medicine* 106:52–54.
Irugalbandara, Z. E.
1960 "Simplified differentiation of Nissl granules stained by toluidine blue in paraffin sections." *Stain Technology* 35:47–48.
Izard, Jacques.
1964 "The retention of toluidine blue metachromasia by mounting in pre-polymerized butyl methacrylate." *Journal of Histochemistry and Cytochemistry* 12:486–487.

Jackson, John F.
1961 "Supravital blood studies, using acridine orange fluorescence." *Blood* 17:643–649.
Jacobson, Stanley.
1963 "Handling sections in bulk, with special reference to the Nauta technic." *Stain Technology* 38:262–263.
Jagatic, Juraj, and Weiskopf, Robert.
1966 "A fluorescent method for staining mast cells." *Archives of Pathology* 82:430–433.

Janigan, David T.
   1965 "The effects of aldehyde fixation on acid phosphatase activity in tissue block." *Journal of Histochemistry and Cytochemistry* 13:476–483.
Jennings, Robert B.
   1951 "A simple apparatus for dehydration of frozen tissues." *Archives of Pathology* 52:195–197.
Jimenez, Fidelio A., and Schneider, Ellen.
   1962 "A Schain frozen section clarifier." *Stain Technology* 37:121–123.
Joftes, David L.
   1959 "Liquid emulsion autoradiography with tritium." *Laboratory Investigation* 8:131–148.
Joftes, David L., and Warren, S.
   1955 "Simplified liquid emulsion radioautography." *Journal of the Biological Photographic Association* 23:145.
Johanson, Donald.
   1940 *Plant Microtechnique.* New York: McGraw-Hill.
Johnson, Percy L., and Klein, Morris N.
   1956 "Application of the Papanicolaou stain to paraffin sections." *Stain Technology* 31:223–225.
Johnston, Muriel E.
   1960 "Film as a histological tissue support." *Journal of Histochemistry and Cytochemistry* 8:139.
Jona, Roberto.
   1963 "Squashing under Scotch Tape 665 for autoradiographic and permanent histologic preparations." *Stain Technology* 38:91–95.
Jones, Rose M.; Thomas, Wilbur A.; and O'Neal, Robert M.
   1959 "Embedding of tissues in Carbowax." *Technical Bulletin of the Registry of Medical Technologists* 29:49–52.

Kaback, M. M.; Saksela, E.; and Mellman, W. J.
   1964 "The effect of 5-bromide-oxyuridine on human chromosomes." *Experimental Cell Research* 34:182–186.
Kabat, Elvin A., and Furth, Jacob.
   1941 "A histochemical study of the distribution of alkaline phosphatase in various normal and neoplastic tissues." *American Journal of Pathology* 17:303–318.
Kaniwar, K. C.
   1960 "Note on the specificity of mercuric bromophenol blue for the cytochemical detection of proteins." *Experientia* 16:355.
Kaplan, Melvin H.; Coons, Albert H.; and Deane, Helen Wendler.
   1950 "Localization of antigen in tissue cells. III. Cellular distribution of pneumococcal polysaccharides, types II and III, in the mouse." *Journal of Experimental Medicine* 91:15–29.
Kaplow, Leonard S.
   1963 "Cytochemistry of leukocyte alkaline phosphatase." *American Journal of Clinical Pathology* 39:439–449.
Kaplow, Leonard S., and Burstone, M. S.
   1963 "Acid-buffered acetone as a fixative for enzyme cytochemistry." *Nature* 200:690–691.
   1964 "Cytochemical demonstration of acid phosphatase in hematopoietic cells in health and in various hematological disorders using azo dye techniques." *Journal of Histochemistry and Cytochemistry* 12:805–811.
Kassel, Robert, and Melnitsky, Ida.
   1951 "Embedding and staining ameboid forms." *Stain Technology* 26:167–171.

Kasten, Frederick H.
1960 "The chemistry of Schiff's reagent." *International Review of Cytology,* X. Ed. by G. H. Bourne and J. F. Danielli. New York: Academic Press.
1962 "Some comments on a recent criticism of the ninhydrin-Schiff reaction." *Journal of Histochemistry and Cytochemistry* 10:769–770.
1965 "Loss of DNA and protein, and changes in DNA during a 30-hour cold perchloric acid extraction of cultured cells." *Stain Technology* 40:127–135.

Kasten, Frederick H., and Burton, Vivian.
1959 "A modified Schiff's solution." *Stain Technology* 34:289.

Kasten, Frederick H.; Burton, Vivian; and Glover, P.
1959 "Fluorescent Schiff-type reagents for cytochemical detection of polyaldehyde moieties in sections and smears." *Nature* 184:1797–1798.

Kasten, Frederick H.; Burton, Vivian; and Lofland, Sue.
1962 "Schiff-type reagents in cytochemistry. Detection of primary amine dye impurities in pyronin B and pyronin Y (G)." *Stain Technology* 37:277–291.

Kasten, Frederick H., and Sandritter, W.
1962 "Crystal violet contamination of methyl green and purification of methyl green—a historical note." *Stain Technology* 37:253–255.

Katline, Vicki Castrogiovanni.
1962 "Retention of nuclear staining by phosphomolybdic-phosphotungistic mordanting." *Stain Technology* 37:193–195.

Kay, Desmond.
1965 *Techniques for Electron Microscopy.* Philadelphia: F. A. Davis.

Keeble, S. A., and Jay, R. F.
1962 "Fluorescent staining for the differentiation of intracellular ribonucleic acid and deoxyribonucleic acid." *Nature* 193:695–696.

Keller, G. J.
1945 "A reliable Nissl stain." *Journal of Technical Methods* 25:77–78.

Kelley, William E.
1965 "New aquaria and synthetic sea salts provide a notable advance in marine biology." *Ward's Bulletin* 5.

Kellogg, D. S., and Deacon, W. E.
1964 "A new rapid immunofluorescent staining technic for identification of *Treponema pallidum* and *Neisseria gonorrhoeae.*" *Proceedings of the Society for Experimental Biology and Medicine* 115:963–965.

Kelly, John W.; Morgan, Paul N.; and Saini, Nirmal.
1962 "Detection of tissue fungi by sulfation and meachromatic staining." *Archives of Pathology* 73:70–73.

Kent, Sidney P.
1961 "A study of mucins in tissue sections using the fluorescent antibody technique. I. The preparation and specificity of bovine submaxillary gland mucin antibody." *Journal of Histochemistry and Cytochemistry* 9:491–497.

Kerbaugh, Mildred A.
1960 "Identification of beta hemolytic streptococci, group A, by the fluorescent method." *Official Journal of American Medical Technologists* 22:13–16.

Kerenyi, N.
1959 "Congo red as a simple stain for the beta cells of the hypophysis." *Stain Technology* 34:343–346.

Kerenyi, N., and Taylor, W. A.
1961 "Niagra blue 4B as a fast simple stain for the adenohypophysis." *Stain Technology* 36:169–172.

Kerr, Donald A.
1938 "Improved Warthin-Starry method of staining spirochetes in tissue sections." *American Journal of Clinical Pathology, Technical Supplement* 2:63–67.

Kessel, J. F.
  1925 "The distinguishing characteristics of the intestinal protozoa of man." *China Medical Journal*, Feb.: 1–57.
Kidder, George W.
  1933 "Studies on *Conchophthirius mytili* De Morgan. I. Morphology and division." *Archiv für Protistenkunde* 79:1–24.
Kimball, R. F., and Perdue, Stella W.
  1962 "Quantitative cytochemical studies on *Paramecium. Experimental Cell Research* 27:405–415.
Kiossoglou, Kosmos A.; Wolman, Irving J.; and Garrison, Mortimer, Jr.
  1963 "Fetal hemoglobin-containing erythrocytes. I. Counts of cells stained by the acid elution method compared with alkali denaturation methods." *Blood* 21:553–560.
Kirby, Harold.
  1947 *Methods in the Study of Protozoa.* University of California Syllabus Series. Berkeley: University of California Press.
Klein, B. M.
  1926 "Über eine neue Eugentumlichkeit der Pellicula von Chilodon uncinatus Ehrbg." *Zoologischer Anzeiger* 67:160–162.
Klinger, H. B.
  1958 "The fine structure of the sex chromatin body." *Experimental Cell Research* 14:207–211.
Klinger, H. B., and Hammond, Daniel O.
  1971 "Rapid chromosome and sex-chromatin staining with pinacyanol." *Stain Technology* 46:43–47.
Klinger, H. B., and Ludwig, Kurt S.
  1957 "A universal stain for the sex chromatin body." *Stain Technology* 32:235–244.
Klionsky, B., and Marcoux, L.
  1960 "Frozen storage of incubation media for enzyme histochemistry." *Journal of Histochemistry and Cytochemistry* 8:329.
Kloeck, John M., and Sweaney, Henry C.
  1943 "Binocular fluorescent microscopy." *American Journal of Clinical Pathology, Technical Section* 7:96–98.
Klüver, H., and Barrera, E.
  1953 "A method for the combined staining cells and fibers in the nervous system." *Journal of Neuropathology and Experimental Neurology* 12:400–403.
Koenig, Harold.
  1963 "Intravital staining of lysosomes by basic dyes and metallic ions." *Journal of Histochemistry and Cytochemistry* 11:120–121.
Koenig, Harold; Groat, Richard A.; and Windle, William F.
  1945 "A physiological approach to perfusion-fixation of tissues with formalin." *Stain Technology* 20:13–22.
Koneff, Alexei A.
  1936 "An iron hematoxylin-aniline blue staining method for routine laboratory use." *Anatomical Record* 66:173.
  1938 "Adaptation of the Mallory-Azan staining method to the anterior pituitary of the rat." *Stain Technology* 13:49–52.
Koneff, Alexei A.. and Lyons, W. R.
  1937 "Rapid embedding with hot low-viscosity nitrocellulose." *Stain Technology* 12:57–59.
Kopriwa, Beatrix Markus, and Leblond, C. P.
  1962 "Improvements in the coating technique of radioautography." *Journal of Histochemistry and Cytochemistry* 10:269–284.
Kornhauser, S. I.
  1930 "Hematein. Its advantages for general laboratory usage." *Stain Technology* 5:13–15.

1943 "A quadruple stain for strong color contrasts." *Stain Technology* 18:95–97.
1945 "A revised method for the 'Quad' stain." *Stain Technology* 20:33–35.
1952 "Orcein and elastic fibers." *Stain Technology* 27:131–134.
Korson, Roy.
1964 "A silver stain for deoxyribonucleic acid." *Journal of Histochemistry and Cytochemistry* 12:875–879.
Kosenow, W.
1961 "Nuclear appendages in leucocytes and the sex pattern of chromosomes." *The Physiology and Pathology of Leukocytes.* Ed. by Dorothea Zucker-Franklin and Herbert Braunsteiner. New York: Grune and Stratton.
Koulischer, L., and Mulnard, J.
1962 "Staining of Chromosomes." *Lancet* 1:917.
Kraijian, Aram A., and Gradwohl, R. B. H.
1952 *Histopathological Technic.* St Louis: C. V. Mosby.
Krichesky, Boris.
1931 "A modification of Mallory's triple stain." *Stain Technology* 6:97–98.
Krishan, Awtar.
1962 "Avian microchromosomes: as shown by prefixation treatment with colchicine, squashing, and hematoxylin staining." *Stain Technology* 37:335–337.
Kristensen, Harold K.
1948 "An improved method of decalcification." *Stain Technology* 23:151–154.
Kropp, B.
1954 "Grinding thin sections of plastic embedded bone." *Stain Technology* 29:77–80.
Krus, Stefan; Andrade, Zilton A.; and Barka, Tibor.
1961 "Histochemical demonstration of specific phosphatases of the liver preserved in hypertonic sucrose solution." *Journal of Histochemistry and Cytochemistry* 9:487–490.
Kruszynski, J.
1954 "Selective demonstration of the Golgi structure and of mitochondria." *Stain Technology* 29:151–155.
Krutsay, M.
1960 "A versatile Resorcin-fuschin formula: I: Combined with formaldehyde. II: Used after periodic acid. III: Used after HCL hydrolysis." *Stain Technology* 35:283–284.
1962a "Permanganate-resorcin-fuchsin: a selective stain for elastic tissue." *Stain Technology* 37:250–251.
1962b "The preparation of iron-hematoxylin from alum hematoxylin." *Stain Technology* 37:249.
Kubie, Lawrence S., and Davidson, David.
1928 "The ammoniacal silver solution as used in neuropathology." *Archives of Neurology and Psychiatry* 19:888–903.
Kuhn, Geraldine D., and Lutz, Earnest L.
1958 "A modified polyester embedding medium for sectioning." *Stain Technology* 33:1–7.
Kurnick, N. B.
1952 "Histological staining with methyl green-pyronin." *Stain Technology* 27:233–242.
1955a "Pyronin Y in the methyl green-pyronin histological stain." *Stain Technology* 30:213–230.
1955b "Histochemistry of nucleic acids." *International Review of Cytology, IV.* Ed. by G. H. Bourne and J. F. Danielli. New York: Academic Press.
Kutlík, Von Igor E.
1958 "Über die Argentaffinität des geformten Bilirubins in den Geweben." *Acta Histochemica* 5:213–224.

1970 "Nachweis von Eisen in den Geweben mittels Chlorathämatoxylinfärbung." *Acta Histochemica* 37:259–267.

Kuyper, Ch. M. A.
    1957 "Identification of mucopolysaccharides by means of fluorescent basic dyes." *Experimental Cell Research* 13:198–200.

Kwan, S. K.
    1970 "Sticky wax infiltration in the preparation of sawed undecalcified bone sections." *Stain Technology* 45:177–181.

LaBauve, P. M.; LaBauve, R. J.; and Petersen, D. F.
    1965 "A digitized comparator for karyotype analysis." *Journal of Heredity* 56: 47–52.

Lacey, Paule E., and Davies, J.
    1959 "Demonstration of insulin in mammalian pancreas by the fluorescent antibody method." *Stain Technology* 34:85–89.

LaCour, L.
    1935 "Technic for studying chromosome structure." *Stain Technology* 10:57–60.
    1941 "Acetic-orcein: a new stain-fixative for chromosomes." *Stain Technology* 16:169–174.

LaCroix, J. Donald, and Preiss, Sister Rose Frederick.
    1960 "Lamination of thermoplastic sheets as a means of mounting histological material." *Stain Technology* 35:331–337.

Lagunoff, David; Phillips, Michael; and Benditt, Earl P.
    1961 "The histochemical demonstration of histamine in mast cells." *Journal of Histochemistry and Cytochemistry* 9:534–541.

Lamkie, N. Joan, and Burstone, M. S.
    1962 "Vinylpyrrolidonevinyl acetate copolymers as mounting media for azo and other dyes." *Stain Technology* 37:109–110.

Lampros, Steve J.
    1962 "A rapid method for preparing paraffin sections." *American Journal of Clinical Pathology* 37:77.

Landing, Benjamin H.
    1954 "Histological study of the anterior pituitary." *Laboratory Investigation* 3:348–368.

Landing, Benjamin H., and Hall, Hazel E.
    1955 "Differentiation of human anterior pituitary cells by combined metal-mordant and mucoprotein stains." *Laboratory Investigation* 4:275–278.
    1956 "Selective demonstration of histidine." *Stain Technology* 31:197–200.

Lane, Bernard P., and Europa, Dominic L.
    1965 "Differential staining of ultrathin sections of Epon-embedded tissues for light microscopy." *Journal of Histochemistry and Cytochemistry* 13:579–582.

Lange, Norbert Adolph.
    1956 *Handbook of Chemistry*. Sandusky, Ohio: Handbook Publishers.

Lartique, Donald J., and Fite, George L.
    1962 "The chemistry of the acid-fast reaction." *Journal of Histochemistry and Cytochemistry* 10:611–618.

Lascano, Eduardo F.
    1959 "A new silver method for the Golgi apparatus." *Archives of Pathology* 68: 499–500.

Lasky, A., and Greco, J.
    1948 "Argentaffin cells of the human appendix." *Archives of Pathology* 46:83–84.

Lawless, D. K.
    1953 "A rapid permanent mount stain technic for the diagnosis of the intestinal protozoa." *American Journal of Tropical Medicine and Hygiene* 2:1137.

Lazarus, Sidney S.
    1958 "A combined periodic acid-Schiff trichrome stain." *Archives of Pathology*
        66:767–772.
Leach, E. H.
    1946 "Curtis' substitute for van Gieson stain." *Stain Technology* 21:107–109.
Leach, W. B.
    1960 "A method for the histological examination of bone marrow granules."
        Canadian Medical Association Journal 83:717–719.
Leblond, C. P.; Messier, B.; and Kopriwa, Beatrix.
    1959 "Thymidine-H³ as a tool for the investigation of the renewal of cell popu-
        lations." *Laboratory Investigation* 8:296–308.
Leduc, Elizabeth H.; Coons, Albert H.; and Connolly, Jeanne M.
    1955 "Studies on anti-body production. II. The primary and secondary responses
        in the popliteal lymph node of the rabbit." *Journal of Experimental Medi-
        cine* 102:61–71.
Lee, M. Raymond.
    1969 "A widely applicable technic for direct processing of bone marrow for
        chromosomes of vertebrates." *Stain Technology* 44:155–158.
Lehmann, F. E., and Mancuso, V.
    1957 "Improved fixative for astral rays and nuclear membrane of Tubifex em-
        bryos." *Experimental Cell Research* 13:161–164.
Lendrum, Alan C.
    1944 "On the cutting of tough and hard tissues embedded in paraffin." *Stain
        Technology* 19:143–144.
Lendrum, Alan C., and McFarlane, David.
    1940 "A controllable modification of Mallory's trichrome staining method." *Jour-
        nal of Pathology and Bacteriology* 50:38–40.
Lennox, Bernard.
    1956 "A ribonuclease-gallocyanin stain for sexing skin biopsies" *Skin Technology*
        31:167–172.
Leske, R. and Von Mayersbach, H.
    1969 "The role of histochemical and biochemical preparation methods for the
        detection of glycogen." *Journal of Histochemistry and Cytochemistry* 17:
        527–538.
Lev, Maurice, and Thomas, John T.
    1955 "An improved method for fixing and staining frozen sections." *American
        Journal of Clinical Pathology* 25:465.
Lev, Robert, and Stoward, Peter J.
    1969 "On the use of eosin as a fluorescent dye to demonstrate mucous cells and
        other structures in tissue sections." *Histochemie* 20:363–377.
Levine, N. D., and Morrill, C. C.
    1941 "Chlorazol black E, a simple connective tissue stain." *Stain Technology*
        16:121–122.
Lewis, Louis W.
    1945 "Method for affixing celloidin sections." *Stain Technology* 20:138.
Lewis, P. R., and Shute, C. C. D.
    1963 "Alginate gel; an embedding medium for facilitating the cutting and han-
        dling of frozen sections." *Stain Technology* 38:307–310.
Lhotka, J. F., and Davenport, H. A.
    1947 "Differential staining of tissue in the block with picric acid and the Feulgen
        reaction." *Stain Technology* 22:139–144.
    1949 "Deterioration of Schiff's reagent." *Stain Technology* 24:237–239.
    1951 "Aldehyde reactions in tissues in relation to the Feulgen technic." *Stain
        Technology* 26:35–41.
Lhotka, J. F., and Ferreira, Alberto Vaz.
    1950 "A comparison of deformalinizing technics." *Stain Technology* 25:27–32.

Lhotka, J. F., and Myhre, Byron A.
   1953 "Periodic acid-Foot stain for connective tissue." *Stain Technology* 28:129–133.
Lieb, Ethel.
   1947 "Permanent stain for amyloid." *American Journal of Clinical Pathology* 17:413–414.
   1948 "A modified phosphotungstic acid-hematoxylin." *Archives of Pathology* 45:559–560.
   1959 "The plastic (mylar) sack as an aid in the teaching of pathology." *American Journal of Clinical Pathology* 32:385–392.
Liebman, Emil.
   1951 "Permanent preparations with the Thomas arginine histochemical test." *Stain Technology* 26:261–263.
Lillie, R. D.
   1928 "Gram stain. A quick method for staining Gram positive bacteria in tissues." *Archives of Pathology* 5:828.
   1929 "A brief method for the demonstration of mucin." *Journal of Technical Methods and Bulletin of International Association of Medical Museums* 12:120–121.
   1940 "Further experiments with the Masson trichrome modifications of the Mallory connective tissue stain." *Stain Technology* 15:17–22.
   1944 "Acetic methylene blue counterstain in staining tissues for acid fast bacilli." *Stain Technology* 18:45.
   1945 "Studies on selective staining of collagen with acid anilin dyes." *Journal of Technical Methods and Bulletin of International Association of Medical Museums* 25:1.
   1946 "A simplified method of preparation of di-ammine-silver hydroxide for reticulum impregnation; comments on the nature of the so-called sensitization before impregnation." *Stain Technology* 21:69–72.
   1949 "Studies on the histochemistry of normal and pathologic mucin in man and laboratory animals." *Bulletin, International Association of Medical Museums* 29:1–53.
   1951a "Histochemical comparison of the Casella, Bauer, and periodic acid oxidation Schiff leucofuchsin technics." *Stain Technology* 26:123–136.
   1951b "Simplification of the manufacture of Schiff reagent for use in histochemical procedures." *Stain Technology* 26:163–165.
   1951c "The allochrome procedure. A differential method segregating the connective tissues, collagen, reticulum, and basement membranes into two groups." *American Journal of Clinical Pathology* 21:484–488.
   1952 "Staining of connective tissue." *Archives of Pathology* 54:220–233.
   1953 "Factors influencing periodic acid-Schiff reaction of collagen fibers." *Journal of Histochemistry and Cytochemistry* 1:353–361.
   1954a "Argentaffin and Schiff reactions after periodic acid oxidation and aldehyde blocking reactions." *Journal of Histochemistry and Cytochemistry* 2:127–136.
   1954b *Histopathologic Technic and Practical Histochemistry.* 2d ed. New York: Blakiston.
   1955 "The basophilia of melanins." *Journal of Histochemistry and Cytochemistry* 3:453–454.
   1956a "Nile blue staining technic of the differentiation of melanin and lipofuscins." *Stain Technology* 31:151–153.
   1956b "The mechanism of nile blue staining of lipofuscins." *Journal of Histochemistry and Cytochemistry.* 4:377–381.
   1956c "The p-dimethylaminobenzaldehyde reaction for pyrroles in histochemistry: melanins, enterochromaffin, zymogen granules, lens." *Journal of Histochemistry and Cytochemistry* 4:118–129.
   1957a "Ferrous ion uptake." *Archives of Pathology* 64:100–103.

1957b "The xanthydrol reaction for pyrroles and indoles in histochemistry: zymogen granules, lens, enterochromaffin, and melanins." *Journal of Histochemistry and Cytochemistry* 5:188–195.

1957c "Metal reduction reactions in melanins." *Journal of Histochemistry and Cytochemistry* 5:325–333.

1957d "Adaptation of the Morel Sisley protein diazotization procedure to the histochemical demonstration of protein bound tyrosine." *Journal of Histochemistry and Cytochemistry* 5:528–532.

1958 "Acetylation and nitrosation of tissue amines in histochemistry." *Journal of Histochemistry and Cytochemistry* 6:352–362.

1960 "Metal chelate reaction of enterochromaffin." *Journal of Histochemistry and Cytochemistry* 9:44–48.

1961 "Investigation on the structure of the enterochromaffin substance." *Journal of Histochemistry and Cytochemistry* 9:184–189.

1962 "Glycogen in decalcified tissue." *Journal of Histochemistry and Cytochemistry* 10:763–765.

1964a "Histochemical acylation of hydroxyl and aminogroups. Effect on the periodic acid-Schiff reaction, anionic and cationic dye and van Gieson collagen stains." *Journal of Histochemistry and Cytochemistry* 12:821–841.

1964b "Studies on histochemical acylation procedures. I. Phenols." *Journal Histochemistry and Cytochemistry* 12:522–529.

1965 *Histopathologic Technic and Practical Histochemistry.* 3d ed. New York: McGraw-Hill.

Lillie, R. D., and Burtner, H. J.

1953a "The ferric ferricyanide reduction test in histochemistry." *Journal of Histochemistry and Cytochemistry* 1:87–92.

1953b "Stable sudanophilia of human neutrophil leucocytes in relation to peroxidase and oxidase." *Journal of Histochemistry and Cytochemistry* 1:8–26.

Lillie, R. D.; Burtner, H. J.; and Henson, J. P.

1953 "Diazosafranin for staining enterochromaffin." *Journal of Histochemistry and Cytochemistry* 1:154–159.

Lillie, R. D., and Earle, W. R.

1939a "The use of Janssen's iron hematoxylin in place of the Weigert acid iron chloride hematoxylin." *Stain Technology* 14:53–54.

1939b "Iron hematoxylins containing ferric and ferrous iron." *American Journal of Pathology* 15:765–770.

Lillie, R. D., Gilmer, P. R., Jr.; and Welsh, R. A.

1961 "Black periodic and black Bauer methods for tissue polysaccharides." *Stain Technology* 36:361–363.

Lillie, R. D., and Glenner, G. G.

1957 "Histochemical aldehyde blockade by aniline in glacial acetic acid." *Journal of Histochemistry and Cytochemistry* 5:167–169.

Lillie, R. D.; Gutiérrez, Anselmo; Madden, Dolores; and Henderson, Raljean.

1968 "Acid orcein-iron and acid orcein-copper stains for elastin." *Stain Technology* 43:203–206.

Lillie, R. D., and Henderson, R.

1968 "A short chromic acid-hematoxylin stain for frozen sections of formolfixed brain and spinal cord." *Stain Technology* 43:121–122.

Lillie, R. D., Henson, Jacquelin P.; Greco, J.; and Burtner, Helen C. J.

1957 "Metal reduction reactions of melanins: silver and ferric ferricyanide reduction by various reagents in vitro." *Journal of Histochemistry and Cytochemistry* 5:311–324.

Lillie, R. D.; Henson, Jacquelin P., Greco, J., and Carson, Joseph C.

1960 "Azo-coupling rate of enterochromaffin with various diazonium salts." *Journal of Histochemistry and Cytochemistry* 9:11–21.

Lillie, R. D.; Lasky, A.; Greco, J.; Burtner, H.; Jacquier; and Jones, P.
  1951 "Decalcification of bone in relation to staining and phosphatase technics."
    American Journal of Clinical Pathology 21:711–722.
Lillie, R. D., and Pizzolato, P.
  1968 "Histochemical studies of oxidation and reduction reactions of the bile
    pigments in obstructive Icterus, with some notes on hematoidin." *Journal
    of Histochemistry and Cytochemistry* 16:17–28.
Lillie, R. D.; Windle, W. F., and Zirkle, Conway.
  1950 "Interim report of use committee on histologic mounting media: Resinous
    media." *Stain Technology* 25:1–9.
Lillie, R. D.; Zirkle, Conway; Dempsey, Edward; and Greco, Jacqueline F.
  1953 "Final report of the committee on histological mounting media." *Stain
    Technology* 28:57–80.
Lin, C. W., and Corlett, Michael.
  1969 "The use of a xylene spray for restoring compressed paraffin serial sec-
    tions." *Stain Technology* 44:159–160.
Lison, Lucien.
  1954 "Alcian blue 8G with chlorantine fast red 5B. A technic for selective
    staining of mucopolysaccharides." *Stain Technology* 29:131–138.
Liu, Winifred.
  1960 *An Introduction to Gynecological Exfoliative Cytology. A Working Manual
    for Cytotechnicians.* Springfield, Ill.: Charles C Thomas.
Lockard, Isabel, and Reers, Bernard L.
  1962 "Staining tissue of the central nervous system with luxol fast blue and
    neutral red." *Stain Technology* 37:13–16.
Lodin, Z.; Faltin J.; and Sharma, K.
  1967 "Attempts at standardization of a highly sensitive Schiff reagent." *Acta
    Histochemica* 26:244–254.
Lodin, Z.; Marés, V.; Karásek, J.; and Skřivanová, P.
  1967 "Studies on nervous tissue. II. Changes of sizes of nuclei of nervous cells
    after fixation and after further histological treatment of the nervous tissue."
    Acta Histochemica 28:297–312.
London Conference on the Normal Karotype.
  1964 [Report] *Annals of Human Genetics* 27:295–296.
Long, Margaret E.
  1948 "Differentiation of myofibrillae, reticular and collagenous fibrillae in verte-
    brates." *Stain Technology* 23:69–75.
Love, Robert.
  1957 "Distribution of ribonucleic acid in tumor cells during mitosis." *Nature*
    180:1338–1339.
  1962 "Improved staining of nucleoproteins of the nucleolus." *Journal of Histo-
    chemistry and Cytochemistry* 10:227.
Love, Robert, and Liles, Roscoe H.
  1959 "Differentiation of nucleoproteins by inactivation of protein-bound amino
    groups and staining with toluidine blue and ammonium molybdate." *Journal
    of Histochemistry and Cytochemistry* 7:164–181.
Love, Robert, and Rabotti, Giancarlo.
  1963 "Studies of the cytochemistry of nucleo-proteins. III. Demonstration of
    deoxyribonucleic-ribonucleic acid complexes in mammalian cells." *Journal
    of Histochemistry and Cytochemistry* 11:603–612.
Low, Frank N.
  1955 "A fixation method for the electron microscopy of endoplasmic reticulum."
    Anatomical Record 121:332–333.
Low, Frank N., and Freeman, James A.
  1956 "Some experiments with chromium compounds as fixers for electron micros-
    copy." *Journal of Biophysical and Biochemical Cytology* 2:629–631.

Lowry, Robert J.
  1963 "Aceto-iron-hematoxylin for mushroom chromosomes." *Stain Technology* 38:149–155.
Lucas, Alfred M., and Jamroz, Casimir.
  1961 *Atlas of Avian Hematology.* Agriculture Monograph 25. Washington, D.C.: U.S. Department of Agriculture.
Lucas, R. B.
  1952 "Observations on the electrolysis method of decalcification." *Journal of Pathology and Bacteriology* 64:654–657.
Luft, John H.
  1956 "Permanganate—a new fixative for electron microscopy." *Journal of Biophysical and Biochemical Cytology* 2:799–801.
  1959 "The use of acrolein as a fixative for light and electron microscopy." *Anatomical Record* 133:305.
  1961 "Improvements in epoxy resin embedding method." *Journal of Biophysical and Biochemical Cytology* 9:409–414.
Luna, Lee G.
  1968 *Manual of Histologic Staining Methods of the Armed Forces Institute of Pathology,* 3d ed. New York: McGraw-Hill.
Luna, Lee G., and Ballou, E. F.
  1959 "Better paraffin sections with the aid of vacuum." *American Journal of Medical Technology* 25:411–413.
Luthringer, David G., and Glenner, George G.
  1961 "The demonstration of endamoebae in tissue sections, by means of a modified methenamine-silver nitrate technic." *American Journal of Clinical Pathology* 36:376–378.
Lycette, R. M.; Danforth, W. F.; Koppel, J. L.; and Olwin, J. H.
  1970 "The binding of Luxol fast blue ARN by various biological lipids." *Stain Technology* 45:155–160.
Lynch, Mathew J., and Inwood, Martin J. H.
  1963 "Gold as a permanent stain for amyloid." *Stain Technology* 38:259–260.
Lynn, Joseph A.; Martin, James H.; and Race, George J.
  1966 "Recent improvements of histologic technics for the combined light and electron microscopic examination of surgical specimens." *American Journal of Clinical Pathology* 45:704–713.

McClintock, Barbara.
  1929 "A method for making aceto-carmine smears permanent." *Stain Technology* 4:53.
McClung, C. E.
  1939 *Handbook of Microscopical Technique.* New York: Hoeber.
McCormick, James B.
  1959a "Technic for restaining faded histopathologic slides." *Technical Bulletin of the Registry of Medical Technologists* 29:13–14.
  1959b "One hour paraffin processing technic for biopsy of bone marrow and other tissues, and specimens of fluid sediment." *American Journal of Clinical Pathology* 31:278–279.
  1961 "Color preservation of gross museum specimens." *Archives of Pathology* 72:82–85.
McCormick, William F., and Coleman, Sidney A.
  1962 "A membrane filter technic for cytology of spinal fluid." *Technical Bulletin of the Registry of Medical Technologists* 32:113–119.
McCurdy, L. E., and Burstone, M. S.
  1966 "Water-soluble plastic mounting media." *Journal of Histochemistry and Cytochemistry* 14:427–428.

McDonald, Donald M.
  1964 "Silver impregnation of Golgi apparatus with subsequent nitrocellulose embedding." *Stain Technology* 39:345–349.
McGee-Russell, S. M.
  1958 "Histochemical methods for calcium." *Journal of Histochemistry and Cytochemistry* 6:22–42.
MacLean, Kenneth S.
  1960 "Automation in staining of cytologic specimens." *Stain Technology* 35: 31–34.
McManus, J. F. A.
  1946 "The histological demonstration of mucin after periodic acid." *Nature* 158:202.
  1948 "Histological and histochemical uses of periodic acid." *Stain Technology* 23:99–108.
  1961 "Periodate oxidation techniques." *General Cytochemical Methods..* Ed. by J. F. Danielli. New York: Academic Press.
McManus, J. F. A., and Cason, J. E.
  1950 "Carbohydrate histochemistry studied by acetylation techniques. I. Periodic acid methods." *Journal of Experimental Medicine* 91:651.
McManus, J. F. A., and Mowry, Robert W.
  1958 "Effects of fixation on carbohydrate histochemistry." *Journal of Histochemistry and Cytochemistry* 6:309–316.
  1960 *Staining Methods: Histologic and Histochemical.* New York: Hoeber.
McNamara, W. L.; Murphy, Berta; and Gore, W. A.
  1940 "Method for simultaneous fixation and decalcification of bone." *Journal of Laboratory and Clinical Medicine* 25:874–875.
Maggi, V., and Riddle, P. N.
  1965 "Histochemistry of tissue culture cells: a study of the effects of freezing and of some fixatives." *Journal of Histochemistry and Cytochemistry* 13: 310–317.
Mahoney, Roy.
  1968 *Laboratory Techniques in Zoology.* London: Butterworth.
Makowski, E. L.; Prem, K. A.; and Kaiser, I. H.
  1956 "Detection of sex of fetuses by the incidence of sex chromatin body in nuclei of cells in amniotic fluid." *Science* 123:542–543.
Malhotra, S. K.
  1961 "Coloration of the Golgi-Nissl network in a vertebrate neuron by Sudan black." *Quarterly Journal of Microscopical Science* 102:387–389.
Malinin, Theodora I.
  1961 "Feulgen-oxidized tannin-azo (FOTA) technique for concomitant staining of DNA and protein." *Stain Technology* 36:198–200.
Mallory, Frank B.
  1944 *Pathological Technique.* Philadelphia: W. B. Saunders.
Manheimer, Leon H., and Seligman, Arnold.
  1949 "Improvement in the method for histochemical demonstration of alkaline phosphatase and its use in a study of normal and neoplastic tissues." *Journal of the National Cancer Institute* 9:181–199.
Manikas, S. G., and Umiker, W. O.
  1959 "An improved rapid method for mounting cytologic smears." *American Journal of Clinical Pathology* 31:414.
Manns, E.
  1960 "Combined myelin-Nissl stain." *Stain Technology* 35:349–351.
Manufacturing Chemists' Association.
  1957 *Technical Data on Plastics.* Washington, D.C.
Manwell, Reginald D.
  1945 "The JSB stain for blood parasites." *Journal of Laboratory and Clinical Medicine* 30:1078–1082.

Marberger, Eve; Boccabella, Rita A.; and Nelson, Warren O.
  1955 "Oral smear as a method of chromosomal sex detection." *Proceedings of the Society for Experimental Biology and Medicine* 89:488–489.
Marengo, P.
  1967 "Thin sectioning microtome knife edges from a modified automatic sharpener." *Stain Technology* 42:216–217.
Margolis, G., and Pickett, J. P.
  1956 "New applications of the luxol fast blue myelin stain." *Laboratory Investigation* 5:459–474.
Markert, Clement L., and Hunter, Robert L.
  1959 "The distribution of esterases in mouse tissue." *Journal of Histochemistry and Cytochemistry* 7:42–49.
Marshall, J. D.; Eveland, W. C.; and Smith, C. W.
  1958 "Superiority of fluorescein isothiocyanate (Riggs) for fluorescent-antibody technic with a modification of its application." *Proceedings of the Society for Experimental Biology and Medicine* 98:898.
Marshall, J. D.; Iverson, L.; Eveland, W. C.; and Kase, Alice.
  1961 "Application and limitations of the fluorescent antibody stain in the specific diagnosis of cryptococcosis." *Laboratory Investigation* 10:719–728.
Marti, Walter J., and Johnson, Berkley H.
  1951 "Acid fast staining technic for histological sections." *American Journal of Clinical Pathology* 21:793.
Martin, John H.
  1966 "A different organism for the demonstration of giant salivary chromosomes." *Turtox News* 44:178–180.
Martin, John H.; Lynn, J. A.; and Nickey, W. M.
  1966 "A rapid polychrome stain for epoxy-embedded tissue." *American Journal of Clinical Pathology* 46:250–251.
Marwah, A. S., and Weinmann, J. P.
  1955 "A sex difference in epithelial cells of human gingiva." *Journal of Periodontology* 26:11–13.
Masek, Bertha, and Birns, Monroe.
  1961 "Advantages of the polyvinyl alcohol-glycerol embedding method for enzyme histochemistry." *Journal of Histochemistry and Cytochemistry* 9:634–635.
Massignani, Adriana, and Malferrari, Rosella.
  1961 "Phosphotungstic acid-eosin combined with hematoxylin as a stain for Negri bodies in paraffin sections." *Stain Technology* 36:5–8.
Massignani, Adriana, and Refinetti, Eloisa Misasi.
  1958 "The Papanicolaou stain for Negri bodies in paraffin sections." *Stain Technology* 33:197–199.
Masson, P.
  1928 "Carcinoids (argentaffin cell tumors) and nerve hyperplasia of the appendicular mucosa." *American Journal of Pathology* 4:181–211.
Mayer, D. M.; Hampton, J. C.; and Rosario, B.
  1961 "A simple method for removing the resin from epoxy-embedded tissue." *Journal of Biophysical and Biochemical Cytology* 9:909–910.
Mayner, Doris A., and Ackerman, G. Adolph.
  1962 "Histochemical demonstration of tissue ribonuclease activity." *Journal of Histochemistry and Cytochemistry* 10:687.
  1963 "Tissue localization of ribonuclease activity by the substrate film technique." *Journal of Histochemistry and Cytochemistry* 11:573–577.
Mazia, Daniel; Brewer, Philip A.; and Alfert, Max.
  1953 "The cytochemical staining and measurement of protein with mercuric bromphenol blue." *Biological Bulletin* 104:57–67.

602   *References*

Melander, Yngve, and Wingstrand, Karl Georg.
  1953 "Gomori's hematoxylin as a chromosome stain." *Stain Technology* 28:217–223.
Mellors, Robert C.
  1959 *Analytical Cytology*. New York: McGraw-Hill.
Mellors, Robert C.; Siegel, Malcolm; and Pressman, David.
  1955 "Analytic pathology. Histochemical demonstration of antibody localization in tissues with special reference to the antigenic components of kidney and lung." *Laboratory Investigation* 4:69–89.
Melvin, Dorothy M., and Brooke, M. M.
  1955 "Triton X-100 in Giemsa staining of blood parasites." *Stain Technology* 30:269–275.
Mendelow, Harvey, and Hamilton, James B.
  1950 "A new technique for rapid freezing and dehydration of tisssues for histology and histochemistry." *Anatomical Record* 107:443–451.
Menzies, D. W.
  1959 "Picro-Gomori method." *Stain Technology* 34:294–295.
  1962a "A methylene blue-basic fuchsin stain for mast cells in paraffin sections." *Stain Technology* 37:43–44.
  1962b "Paraffin-beeswax-stearic acid: an embedding mass for thin sections." *Stain Technology* 37:235–238.
  1963a "Red-blue staining of hydrolysed nucleic acids in paraffin sections." *Stain Technology* 38:157–160.
  1963 "Romanowsky-type staining: enzymatic analysis of nuclear metachromasia in formalin-fixed paraffin sections." *Stain Technology* 38:161–163.
Mercer, E. H., and Birbeck, C.
  1961 *Electron Microscopy. A Handbook for Biologists*. Springfield, Ill.: Charles C Thomas.
Meriwether, L. S.
  1934 "Proceedings of Staff Meetings of Mayo Clinic." *Mayo Clinic Proceedings* 9:95–96.
Merton, H.
  1932 "Gestalterhaltende Fixierungsversuche an besonders kontractilen Infusorien." *Archiv für Protistenkunde* 77:449–521.
Meryman, H. T.
  1959 "Sublimation freeze-drying without vacuum. *Science* 130:628–629.
  1960 "Freezing and drying of biological materials. Part III. Principles of freeze-drying." *Annals of the New York Academy of Sciences* 85:501–734.
Metcalf, R. L., and Paton, R. L.
  1944 "Fluorescence microscopy applied to Entomology and allied fields." *Stain Technology* 19:11–27.
Mettler, Fred A.
  1932 "The Marchi method for demonstrating degenerated fiber connections within the central nervous system." *Stain Technology* 7:95–106.
Mettler, Fred A., and Hanada, Ruth E.
  1942 "The Marchi method." *Stain Technology* 17:111–116.
Mettler Fred A.; Mettler, Cecelia C.; and Strong F. C.
  1936 "The cellosolve-nitrocellulose technic." *Stain Technology* 11:165–166.
Mettler, Sidney, and Bartha, Alexander S.
  1948 "Brilliant cresyl blue as a stain for chromosome smear preparations." *Stain Technology* 23:27–28.
Meyer, James R.
  1945 "Prefixing with paradichlorobenzene to facilitate chromosome study." *Stain Technology* 20:121–125.
Meyer-Arendt, Jurgen R.
  1959 "Mylar film as carrier material for microradiography." *Journal of Histochemistry and Cytochemistry* 7:351–352.

Miller, Bernard Joseph.
1937 "The use of dioxane in the preparation of histological sections." *The Mendel Bulletin.* December:5–9.

Miller, O. L.; Stone, G. E.; and Prescott, D. M.
1964 "Autoradiography of water-soluble materials." *Methods in Cell Physiology,* I. Ed. by D. M. Prescott. New York: Academic Press.

Milligan, Mildred.
1946 "Trichrome stain for formalin-fixed tissue." *American Journal of Clinical Pathology, Technical Section.* 10:184–185.

Millonig, G.
1961 "Advantages of a phosphate buffer for OsO₄ solutions in fixation." *Journal of Applied Physiology* 32:1637.

Minick, O. T.
1963 "Low temperature storage of epoxy embedding resins." *Stain Technology* 38:131–133.

Mittwer, Tod; Bartholomew, T. W.; and Kallman, Burton J.
1950 "The mechanism of the Gram reaction. 2. The function of iodine in the Gram stain." *Stain Tcehnology* 25:169–179.

Moffat, D. B.
1958 "Demonstration of alkaline phosphate and periodic acid-Schiff positive material in the same section." *Stain Technology* 33:225–228.

Mohr, John L.
1950 "On the natural coloring matter, brazilin, and its use in microscopical technique." *Micro-notes* 5:4–16.

Mohr, John L., and Wehrle, William.
1942 "Notes on mounting media." *Stain Technology* 17:157–160.

Moldovanu, G.
1961 "L'Identification de la chromatine du sexe chez les chimères hématologiques canines." *Revue Francaise d'Etudes Cliniques et Biologiques* 6:165–167.

Moline, S. W., and Glenner, G. G.
1964 "Ultra-rapid tissue freezing in liquid nitrogen." *Journal of Histochemistry and Cytochemistry* 12:777–783.

Moliner, Enrique Ramon.
1957 "A chlorate-formaldehyde modification of the Golgi method." *Stain Technology* 32:105–116.

Mollenhauer, Hilton H.
1964 "Plastic embedding mixtures for use in electron microscopy." *Stain Technology* 39:111–114.

Moloney, William C.; McPherson, Kenneth; and Fliegelman, Lila.
1960 "Esterase activity in leukocytes demonstrated by use of napthol AS-D chloracetate substrate." *Journal of Histochemistry and Cytochemistry* 8:200–207.

Monk, Cecil R.
1938 "An aqueous media for mounting small objects." *Science* 88:174.

Monroe, C. W., and Spector, B.
1963 "Tannic acid, iron hematoxylin, alcian blue and basic fuchsin for staining islets and reticular fibers of the pancreas." *Stain Technology* 38:187–192.

Moody, M. D.; Ellis, E. C.; and Updyke, L.
1958 "Staining bacterial smears with fluorescent antibody. IV. Grouping streptococci with antibody." *Journal of Bacteriology* 75:553–560.

Moore, A. R.
1962 "Collecting and preserving the developmental stages of the Pacific coast Bat Starfish, *Patiria miniata.*" *Ward's Bulletin, 1.*

Moore, Keith L., and Barr, Murry L.
1955 "Smears from the oral mucosa in the detection of chromosomal sex." *Lancet* 269, vol. 2:57–58.

Moore, Keith L., Graham, Margaret A.; and Barr, Murray L.
    1953 "The detection of chromosomal sex in hermaphrodites from a skin biopsy."
        Surgery, Gynecology and Obstetrics 96:641–648.
Moorhead, P. S.; Nowell, P. C.; Mellman, W. J.; Batipps, D. M.; and Hungerford, D. A.
    1960 "Chromosome preparations of leukocytes from human peripheral blood."
        Experimental Cell Research 20:613–616.
Moorman, Albert E.
    1971 "A modification of Heidenhain's iron-haematoxylin technique, as used to
        stain smears of Drosophila larval polytene chromosomes and neuroblast
        cells." *Canadian Journal of Zoology* 49:132.
Moran, Thomas J.; Radcliffe, Merle L.; and Tevault, Isabelle H.
    1947 "Rapid method of staining tubercle bacilli with Tergitol." *American Journal
        of Clinical Pathology* 17:75–77.
Moree, R.
    1944 "Control of the ferric ion concentration in iron-acetocarmine staining." *Stain
        Technology* 19:103–108.
Morris, Russell E., and Benton, Robert S.
    1956 "Studies on demineralization of bone. IV. Evaluation of morphology and
        staining characteristics of tissues after demineralization." *American Journal
        of Clinical Pathology* 26:882–898.
Morrison, John H., and Kronheim, Suzanne.
    1962 "The cytochemical demonstration of succinic dehydrogenase in mouse
        leukocytes." *Journal of Histochemistry and Cytochemistry* 10:402–411.
Morrison, Maurice, and Samwick, A. A.
    1940 "Restoration of overstained Wright films and a new method of staining
        blood smears." *American Journal of Clinical Pathology, Technical Supple-
        ment* 4:92–93.
Mortreuil-Langlois, Micheline.
    1962 "Staining sections coated with radiographic emulsion; a nuclear fast red,
        indigo-carmine sequence." *Stain Technology* 37:175–177.
Mossman, H. W.
    1937 "The Dioxan technic." *Stain Technology* 12:147–156.
Movat, Henry Z.
    1955 "Demonstration of all connective tissue elements in a single section."
        Archives of Pathology 60:289–295.
Mowry, Robert W.
    1956 "Alcian blue technique for histochemical study of acidic carbohydrates."
        Journal of Histochemistry and Cytochemistry 4:407.
    1958 "Improved procedure for the staining of acidic polysaccharides by Müller's
        colloidal (hydrous) ferric oxide and its combination with the Feulgen and
        P.A.S. reactions." *Laboratory Investigation* 7:566–576.
    1959 "Effect of periodic acid used prior to chromic acid on the staining of poly-
        saccharides by Gomori's methenamine silver." *Journal of Histochemistry and
        Cytochemistry* 7:288.
    1960 "Revised method producing improved coloration of acidic mucopolysac-
        charides with alcian blue 8GX supplied currently. *Journal of Histochemistry
        and Cytochemistry* 8:323.
    1963 "The special value of methods that color both acidic and vicinal hydroxyl
        groups in the histochemical study of mucins, with revised directions for the
        colloidal iron stain, the use of alcian blue 8GX and their combinations with
        the periodic acid-Schiff reaction. *Annals of the New York Academy of
        Sciences* 106:402–423.
Mowry, Robert W., and Emmel, Victor M.
    1966 "The coloration of carbohydrate polyanions by National fast blue com-
        pared with that obtained with alcian blue 8GX." *Journal of Histochemistry
        and Cytochemistry* 14:799–800.

Mowry, Robert W., and Winkler, C. H.
    1956 "The coloration acidic carbohydrates of bacteria and fungi in tissue sections with special reference to capsules of *Crytococcus neoformans, Pneumococcus* and *Staphylococcus.*" *American Journal of Pathology* 32:628–629.
Mullen, John P., and McCarter, John C.
    1941 "A mordant preparing formaldehyde-fixed neuraxis tissue for phosphotungstic acid hematoxylin staining." *American Journal of Pathology* 17:289–291.
Mundkur, B., and Brauer, B.
    1966 "Selective localization of nucleolar protein with Amidoblack 10B." *Journal of Histochemistry and Cytochemistry* 14:94–103.
Munoz, Frank, and Charipper, Harry A.
    1943 *The Microscope and Its Use.* New York: Chemical Publishing Co.
Murgatroyd, L. B.
    1971 "Chemical and spectrometric evaluation of glycogen after routine histological fixatives." *Stain Technology* 46:111–119.
Murgatroyd, L. B., and Horobin, R. W.
    1969 "Specific staining of glycogen with hematoxylin and certain anthraquinone dyes." *Stain Technology* 44:59–62.
Murín, A.
    1960 "Substitution of cellophane for glass covers to facilitate preparation of permanent squashes and smears." *Stain Technology* 35:351–353.

Nachlas, Marvin; Tsou, Kwan-Chung; Desouza, Eustace; Cheng, Chao-Shing; and Seligman, Arnold M.
    1957 "Cytochemical demonstration of succinic dehydrogenase by the use of a new *p*-nitrophenyl substituted ditetrazole." *Journal of Histochemistry and Cytochemistry* 5:420–436.
Nairn, R. C.
    1962 *Florescent Protein Tracing.* Edinburgh and London: E. and S. Livingstone.
Nakane, Paul K., and Pierce, G. Barry.
    1967 "Enzyme labeled antibodies: Preparation and application for the localization of antigens." *Journal of Histochemistry and Cytochemistry* 14:929–931.
Nash, David, and Plaut, W.
    1964 "On the denaturation of chromosomal DNA *in situ.*" *Proceedings of the National Academy of Sciences* 51:731–735.
Nassar, Tamir K., and Shanklin, William M.
    1951 "Staining neuroglia with silver diamminohydroxide after sensitizing with sodium sulfite and embedding in paraffin." *Stain Technology* 26:13–18.
    1961 "Simplified procedure for staining reticulum." *Archives of Pathology* 71: 611/21–614/24.
Nassonov, D. N.
    1923 "Das Golgische Binnennetz and seine Beziehungen zu der Sekretion. Untersuchungen über einige Amphibiendrüsen." *Archiv für Mikroskopische Anatomie und Entwicklungsmechanik* 97:136–186.
    1924 "Das Golgische Binnennetz und seine Beziehungen zu der Sekretion. Morphologische und experimentelle Untersuchungen an einigen Säugetierdrüsen." *Archiv für Mikroskopische Anatomie und Entwicklungsmechanik* 100:433–472.
Nauman, R. V.; West, P. W.; Trou, F.; and Geake, G. C.
    1960 "A spectroscopic study of the Schiff reaction as applied to the quantitative determination of sulfur dioxide." *Annals of Chemistry* 32:1307–1311.
Nauta, W. J. H., and Gygax, P. A.
    1951 "Silver impregnation of degenerating axon terminals in the central nervous system. I. Technic. Chemical notes." *Stain Technology* 26:5–11.

1954 "Silver impregnation of degenerating axons in the central nervous system: a modified technic." *Stain Technology* 29:91–93.
Nauta, W. J. H., and Ryan, Lloyd F.
1952 "Selective silver impregnation of degenerating axons in the central nervous system." *Stain Technology* 27:175–179.
Navasquez, S. D.
1935 "Experimental tubular necrosis of the kidneys accompanied by liver changes due to dioxane poisoning." *Journal of Hygiene* 35:540–548.
Nedelkoff, B., and Christopherson, William M.
1962 "A millipore filter technic for cytologic examination of body fluids." *American Journal of Clinical Pathology* 37:97–103.
Needham, George Herbert.
1958 *The Practical Use of the Microscope.* Springfield, Ill.: Charles C Thomas.
Neurath, Peter W.
1968 "Computer-aided chromosome analysis system." *Laboratory Management,* March: 14–16.
1970 "Computer-aided chromosome analysis system [II]." *Laboratory Management,* September: 20–21.
Newcomer, Earl A.
1940 "An osmic impregnation method for mitochondria in plant cells." *Stain Technology* 15:89–90.
1959 "Feulgen staining of tissues prior to embedding and sectioning." *Stain Technology* 34:349–350.
Newcomer, Earl A., and Donnelly, Grace M.
1963 "Leukocyte culture for chromosome studies in the domestic fowl." *Stain Technology* 38:54–56.
Nickerson, Mark.
1944 "A dry ice freezing unit for rotary microtomes." *Science* 100:177–178.
Nito, Yuichi.
1961 "A carbonate-colloidal iron staining reagent." *Stain Technology* 36:323–325.
Nito, Yuichi, and Stokes, J. R.
1960 "An improved colloidal iron staining reagent." *Stain Technology* 35:103.
Noble, Glenn A.
1944 "A five-minute method for staining fecal smears." *Science* 100:37–38.
Nolte, D. J.
1948 "A modified technic for salivary gland chromosomes." *Stain Technology* 23:21–25.
Nonidez, José F.
1939 "Studies on the innervation of the heart. I. Distribution of the cardiac nerves, with special reference to the identification of the sympathetic and parasympathetic postganglionics." *American Journal of Anatomy* 65:361–413.
Norris, William P., and Jenkins, Phyllis.
1960 "Epoxy resin embedding in contrast radioautography of bones and teeth." *Stain Technology* 35:253–260.
Norton, William T.; Korey, Saul R.; and Brotz, Miriam.
1962 "Histochemical demonstration of unsaturated lipids by a bromine-silver method." *Journal of Histochemistry and Cytochemistry* 10:83–88.
Notenboom, C. D.; van de Veerdonk, F.; and de Kramer, J. C.
1967 "A fluorescent modification of the Sakaguchi reaction on arginine." *Histochemie* 9:117–121.
Novelli, Amato.
1962 "A short method for chondriome." *Journal of Histochemistry and Cytochemistry* 10:102–103.
Novikoff, Alex B.
1960 "Biochemical and staining reactions of cytoplasmic constituents." *Develop-*

ing Cell Systems and their Control. Ed. by Dorothea Rudnick. New York: Ronald Press.

1961a "*Mitochondria* (Chondriosomes)." *The Cell,* II. Ed by Jean Brachet and Alfred E. Mirsky. New York: Academic Press.

1961b "Lysosomes and Related Particles." The Cell, II. Ed. by Jean Brachet and Alfred E. Mirsky. New York: Academic Press.

Novikoff, Alex B.; Goldfischer, Sidney; and Essner, Edward.

1961 "The importance of fixation in a cytochemical method for the Golgi apparatus." *Journal of Histochemistry and Cytochemistry* 9:459–460.

Novikoff, Alex B., and Podber, Estelle.

1957 "The contribution of differential centrifugation to the intracellular localization of enzymes." *Journal of Histochemistry and Cytochemistry* 5:552–564.

Novikoff, Alex B.; Shin, Woo-Young; and Drucker, Joan.

1960 "Cold acetone fixation for enzyme localization in frozen sections." *Journal of Histochemistry and Cytochemistry* 8:37–40.

Nowell, Peter C.

1960 "Phytohemagglutin: an initiator of mitosis in cultures of normal human leucocytes." *Cancer Research* 20:462–466.

Noyes, Wilbur F.

1955 "Visualization of Egypt 101 virus in the mouse's brain and in cultured human carcinoma cells by means of fluorescent antibody." *Journal of Experimental Medicine* 102:243–247.

Noyes, Wilbur F., and Watson, Barbara K.

1955 "Studies on the increase of vaccine virus in cultured human cells by means of fluorescent antibody technique." *Journal of Experimental Medicine* 102:237–242.

Nurnberger, John J.

1955 "Ultraviolet microscopy and microspectroscopy." *Analytical Cytology.* Ed. by Robert C. Mellors. New York: Blakiston.

Ogawa, K.; Mizuno, N.; and Okamoto, M.

1961 "Cytochemistry of cultured neural tissue. III. Heterogeneity of lysosomes." *Journal of Histochemistry and Cytochemistry* 9:635.

Ohno, Susumo.

1965 "Direct handling of germ cells." *Human Chromosome Methodology.* Ed. by Jorge J. Yunis. New York: Academic Press.

Ojeda, J. L.; Barbosa, E.; and Bosque, P. Gomez.

1970 "Selective skeletal staining in whole chicken embryos; a rapid alcian blue technique." *Stain Technology* 45:137–138.

Oster, Gerald.

1955 "X-ray diffraction techniques and their application to the study of biomolecular structures." *Analytical Cytology.* Ed. by Robert C. Mellors. New York: Blakiston.

Owen, G.

1955 "Use of propylene phenoxetol as a relaxing agent." *Nature* 175:434.

Owen, G., and Steedman, H. F.

1956 "Preservation of animal tissues, with a note on staining solutions." *Quarterly Journal of Microscopical Science* 97:319–321.

1958 "Preservation of molluscs." *Proceedings of the Malacological Society of London* 33:101–103.

Padawar, Jacques.

1959 "A stain for mast cells and its application in various vertebrates and in a mastocytoma." *Journal of Histochemistry and Cytochemistry* 7:352–353.

Paddy, J. F.
   1970 "Metachromasy of dyes in solution." NATO Advanced Study Institute. St Margherita, [*Report*]. Ed. by E. A. Balasz. London: Academic Press.

Paget, G. E., and Eccleston, Enid.
   1959 "Aldehyde-thionin: a stain having similar properties to aldehyde-fuchsin." *Stain Technology* 34:223–226.
   1960 "Simultaneous specific demonstration of thyrotroph, gonadotroph and acidophil cells in the anterior hypophysis." *Stain Technology* 35:119–122.

Palade, G. E.
   1952 "A study of fixation for electron microscopy." *Journal of Experimental Medicine* 95:285–297.

Palkowsky, Walter.
   1960 "A new method of deep freezing pathologic specimens in a pliable state." *Technical Bulletin of the Registry of Medical Technologists* 30:187.

Palmer, M. Winnogene, and McDonald, L. W.
   1963 "A device for the continuous chilling of microtome knife and paraffin blocks." *American Journal of Clinical Pathology* 40:633.

Palmgren, Axel.
   1954 "Tape for microsectioning of very large, hard, or brittle specimens." *Nature* 174:46.

Pantin, C. F. A.
   1946 *Notes on Microscopical Technique for Zoologists.* Cambridge University Press.

Papamiltiades, M.
   1953 "Sur la composition de deux hematoxylines pour les colorations cytologiques." *Acta Anatomica* 19:25.

Papanicolaou, George N.
   1942 "A new procedure for staining vaginal smears." *Science* 95:438–439.
   1947 "The cytology of the gastric fluid of carcinoma of the stomach." *Journal of the National Cancer Institute* 7:357–360.
   1954 *Atlas of Exfoliative Cytology.* Cambridge, Mass.: Harvard University Press.
   1957 "The cancer diagnostic potential of uterine exfoliative cytology." *CA, Bulletin of Cancer Progress* 7:125–135.

Parsons, D. F.
   1970 *Some Biological Techniques in Electron Microscopy.* New York: Academic Press.

Past, Wallace L.
   1961 "The histologic demonstration of iron in osseous tissue." *American Journal of Pathology* 39:443–449.

Patau, K.
   1960 "The identification of individual chromosomes, especially in man." *American Journal of Human Genetics* 12:250–276.
   1961 "Chromosome identification and the Denver Report." *Lancet* 1:933–934.
   1965 "Identification of chromosomes." *Human Chromosome Methodology.* Ed. by Jorge J. Yunis. New York: Academic Press.

Patau, K.; Therman, Eva; Smith, David W.; Inhorn, Stanley L.; and Picken, Bruce F.
   1961 "Partial-trisomy syndromes. I. Sturge-Weber's disease." *American Journal of Human Genetics* 13:287–298.

Patten, Bradley M.
   1952 *Early Embryology of the Chick*, 4th ed. New York: Blakiston.

Patten, Stanley F., and Brown, Kenneth A.
   1958 "Freeze-solvent substitution technic. A review with application to fluorescence microscopy." *Laboratory Investigation* 7:209–223.

Pauly, H.
   1964 "Über die Konstitution des Histidins." *Hoppe-Seylers Zeitschrift für Physiologische Chemie* 42:508–518.

Pearse, A. G. Everson.
  1949 "The cytochemical demonstration of gonadotropic hormone in the human anterior hypophysis." *Journal of Pathology and Bacteriology* 61:195–202.
  1950 "Differential stain for human and animal anterior hypophysis." *Stain Technology* 25:95–102.
  1955 "Copper phthalocyanins as phospholipid stains." *Journal of Pathology and Bacteriology* 70:554–557.
  1969 *Histochemistry, Theoretical and Applied*, Vol. 1. 3d ed. Baltimore, Williams and Wilkins.
Pearson, Bjarne.
  1958 "Improvement in the histochemical localization of succinic dehydrogenase by the use of nitroneotetrazolium chloride." *Journals of Histochemistry and Cytochemistry* 6:112–121.
Pearson, Bjarne; Wolf, Paul; and Andrews, Marian.
  1963 "The histochemical demonstration of leucine aminopeptidase by means of a new indoyl compound." *Laboratory Investigation* 12:712–720.
Peary, Joseph Igor.
  1955 "Freeze dehydration for permanent mounts after aceto-orcein stain." *Stain Technology* 30:213–230.
Pease, Daniel C.
  1964 *Histological Techniques for Electron Microscopy*. New York: Academic Press.
Pease, Daniel C., and Baker, R. F.
  1948 "Sectioning techniques for electron microscopy using conventional microtome." *Proceedings of the Society for Experimental Biology and Medicine* 67:470–474.
  1950 "Electron microscopy of the kidney." *American Journal of Anatomy* 87:349–390.
Pelc, S. R.
  1956 "A stripping film technique for autoradiography." *International Journal of Applied Radiation and Isotopes* 1:172–177.
  1958 "Autoradiography as a cytochemical method, with special reference to C^{14} and S^{35}." *General Cytochemical Methods*, I. Ed. by J. F. Danielli. New York: Academic Press.
Peltier, Leonard F.
  1954 "The demonstration of fat emboli in tissue sections using phosphin 3R, a water-soluble fluorochrome." *Journal of Laboratory and Clinical Medicine* 43:321–323.
Pequeno, Rubens Alves.
  1960 "A simple pen method for preparing linear blood films for accurate differential leukocyte counts." *Technical Bulletin of the Registry of Medical Technologists* 30:193–196.
Perry, Robert P.
  1964 "Quantitative Autoradiography." *Methods in Cell Physiology*, I. Ed. by David M. Prescott. New York: Academic Press.
Persidsky, Maxim D.
  1954 "Restoration of deteriorated temporary aceto-carmine preparations." *Stain Technology* 29:278.
Peters, H.
  1961a "A glass sieve for carrying loose frozen and celloidin sections through all stages of processing." *Stain Technology* 36:201.
  1961b "An albumen air drying sequence for attaching celloidin sections to slides." *Stain Technology* 36:248–250.
Pickett, John Phillip; Bishop, Carl M.; Chick, Ernest W.; and Baker, Roger D.
  1960 "A simple fluorescent stain for fungi." *American Journal of Clinical Pathology* 34:197–202.

Pickett, John Phillip, and Klavins, J. V.
  1961 "Demonstration of elastic tissue and iron or elastic tissue and calcium simultaneously." *Stain Technology* 36:371–374.
Pickett, John Phillip, and Sommer, Joachim R.
  1960 "Thirty-five mm film as mounting base, and plastic spray as cover glass, for histologic sections." *Archives of Pathology* 69:13–21, 239–247.
Pickworth, J. W.; Cotton, K.; and Skyring, A. P.
  1963 "Double emulsion autoradiography for identifying tritium-labelled cells in sections." *Stain Technology* 38:237–244.
Pienaar, Uys de Villiers.
  1962 *Hematology of Some South African Reptiles.* Johannesburg: Witwatersrand University Press.
Pitelka, Dorothy Riggs.
  1945 "Morphology and taxonomy of flagellates of the genus *Peranema dujardin.*" *Journal of Morphology* 76:179–190.
Pizzolato, Philip, and McCrory, Phil.
  1962 "Light influence on von Kóssa's silver calcium reaction in the myocardium." *Journal of Histochemistry and Cytochemistry* 10:102.
Poirier, Louis J.; Ayotte, Robert A.; and Gauthier, Claude.
  1954 "Modification of the Marchi technic." *Stain Technology* 29:71–75.
Poley, R. W., and Fobes, C. D.
  1964 "Fuchsinophilia in early myocardial infarction." *Archives of Pathology* 77:325–329.
Pollak, O. J.
  1944 "A rapid trichrome stain." *Archives of Pathology* 37:294.
Pollister, Arthur W., and Pollister, Priscilla F.
  1957 "The structure of the Golgi apparatus." *International Review of Cytology,* VI. Ed. by G. H. Bourne and J. F. Danielli. New York: Academic Press.
Pool, Charlotte R.
  1969 "Hematoxylin-eosin staining of OsO_4-fixed epon-embedded tissue; prestaining oxidation by acidified H_2O_2." *Stain Technology* 44:75–79.
Popp, Raymond A.; Gude, William D.; and Popp, Diana M.
  1962 "Peroxidase staining combined with autoradiography for study of eosinophilic granules." *Stain Technology* 37:243–247.
Popper, Hans, and Szanto, Paul B.
  1950 "Fluorescence Microscopy." *Handbook of Microscopical Technique.* Ed. by Ruth McClung Jones. New York: Hoeber.
Porter, Keith R., and Blum, J.
  1953 "A study in microtomy for electron microscopy." *Anatomical Record* 117:685–710.
Pottz, Glenn; Rampey, James H.; and Furmandean, Benjamin.
  1964 "A method for rapid staining of acid-fast bacteria in smears and sections of tissue." *American Journal of Clinical Pathology* 42:552–554.
Powers, Margaret, and Clark, George
  1955 "An evaluation of cresyl echt violet as a Nissl stain." *Stain Technology* 30:83–92.
  1963 "A note on Darrow red." *Stain Technology* 38:289–290.
Powers, Margaret; Clark, George; Darrow, Mary; and Emmel, Victor M.
  1960 "Darrow red, a new basic dye." *Stain Technology* 35:19–22.
Preece, Ann.
  1959 *A Manual for Histologic Technicians.* Boston: Little Brown.
Prescott, D. M.
  1964 "Autoradiography with liquid emulsion." *Methods in Cell Physiology,* I. Ed. by D. M. Prescott. New York: Academic Press.
Prescott, D. M., and Bender, M. A.
  1964 "Preparation of mammalian metaphase chromosomes for autoradiography."

Methods in Cell Physiology, I. Ed. by D. M. Prescott. New York: Academic Press.

Prescott, D. M., and Carrier, R. F.
 1964 "Experimental procedures and cultural methods for *Euplotes eurystomus* and *Amoeba proteus." Methods in Cell Physiology.* I. Ed. by D. M. Prescott. New York: Academic Press.

Priman, Jacob.
 1954 "Blood serum as an adhesive for paraffin sections." *Stain Technology* 29: 105–107.

Procknow, John J.; Connelly, A. Philip, Jr.; and Ray, C. George.
 1962 "Fluorescent antibody technique in histoplasmosis." *Archives of Pathology* 73:313–324.

Proescher, Frederick.
 1933 "Pinacyanol as a histological stain." *Proceedings of the Society for Experimental Biology and Medicine* 31:79–81.
 1934 "Contribution to the staining of neuroglia." *Stain Technology* 9:33–38.

Proescher, Frederick, and Arkush, A. S.
 1928 "Metallic lakes of the oxazines (gallamin blue, gallocyanin, and coelestin blue) as nuclear stain substitutes for hematoxylin." *Stain Technology* 2: 28–38.

Puchtler, Holde.
 1958 "Significance of the iron hematoxylin method of Heidenhain." *Journal of Histochemistry and Cytochemistry* 6:401–402.

Puchtler, Holde; Chandler, A. B.; and Sweat, Faye.
 1961 "Demonstration of fibrin in tissue sections by the Rosindole method." *Journal of Histochemistry and Cytochemistry* 9:340.

Puchtler, Holde, and Leblond, C. P.
 1958 "Histochemical analysis of cell membranes and associated structures as seen in the intestinal epithelium." *American Journal of Anatomy* 102:1–31.

Puchtler, Holde; Meloan, Susan N.; and Terry, Mary S.
 1969 "On the history and mechanism of alizarin and alizarin red S stains for calcium." *Journal of Histochemistry and Cytochemistry* 17:110–124.

Puchtler, Holde; Rosenthal, Sanford I.; and Sweat, Faye.
 1964 "Revision of the amidoblack stain for hemoglobin." *Archives of Pathology* 78:76–78.

Puchtler, Holde, and Sweat, Faye.
 1960 "Commercial resorcin-fuchsin as a stain for elastic fibers." *Stain Technology* 35:347–348.
 1962 "Some comments on the ninhydrin-Schiff reaction." *Journal of Histochemistry and Cytochemistry* 10:365.
 1963a "A combined hemoglobin-hemosiderin stain." *Archives of Pathology* 75: 588–590.
 1963b "Influence of various pretreatments on the staining properties of connective tissue fibers." *Annales d'Histocheme* 8:189–198.
 1964a "On the mechanism of sequence iron-hematein stains." *Histochemie* 4:197–208.
 1964b "Histochemical specificity of stain-methods for connective tissue fibers: resorcin-fuchsin and van Gieson's picro-fuchsin." *Histochemie* 4:24–34.
 1965 "Congo red as a stain for fluorescence microscopy of amyloid." *Journal of Histochemistry and Cytochemistry* 13:693–694.
 1966 "A review of early concepts of amyloid in context with contemporary chemical literature from 1839 to 1859." *Journal of Histochemistry and Cytochemistry* 14:123–134.

Puchtler, Holde; Sweat, Faye; and Doss, Nancy O.
 1963 "A one-hour phosphotungstic acid-hematoxylin stain." *American Journal of Clinical Pathology* 40:334.

Puchtler, Holde; Sweat, Faye; and Levine M.
    1962 "On the binding of Congo red by amyloid." *Journal of Histochemistry and Cytochemistry* 10:355–364.
Puchtler, Holde; Waldrop, Faye Sweat; Meloan, Susan N.; Terry, Mary S.; and Conner, H. M.
    1970 "Methacarn (Methanol-Carnoy) fixation. Practical and theoretical considerations." *Histochemie* 21:97–116.
Puchtler, Holde; Waldrop, Faye Sweat; Terry, Mary S.; and Conner, H. M.
    1969 "A combined PAS-myofibril stain for demonstration of early lesions of striated muscle." *Journal of Microscopy* 89:329–338.
Puck, Theodore T.; Cieciura, Steven J.; and Robinson, Arthur.
    1958 "Genetics of somatic mammalian cells. III. Long term cultivation of euploid cells from human and animal subjects." *Journal of Experimental Medicine* 108:945–956.
Pugh, M. H., and Savchuck, W. B.
    1958 "Suggestions on the preparation of undecalcified bone for microradiography." *Stain Technology* 33:287–293.
Putt, F. A.
    1948 "Modified eosin counterstain for formaldehyde-fixed tissues." *Archives of Pathology* 45:72.
    1951a "A modified Ziehl-Neelsen method." *American Journal of Clinical Pathology* 21:92–95.
    1951b "A benzidine-thionine method for the demonstration of hemoglobin in formaldehyde-fixed, paraffin-embedded tissue." *Archives of Pathology* 52: 293.
    1956 "Flaming red as a dye for the demonstration of lipids." *Laboratory Investigation* 5:377–379.
Putt, F. A., and Huskill, Peter B.
    1962 "Alcian green, a routine stain for mucins." *Archives of Pathology* 74:169–170.

Quary, W. B.
    1957 "Experimental cyanine red, a new stain for nucleic acid and mucopolysaccharides." *Stain Technology* 32:175–182.

Radden, B. G.
    1961 "An investigation of the factors influencing the metachromatic staining of mast cells." *Journal of Histochemistry and Cytochemistry* 9:165–170.
Rae, C. A.
    1955 "Masson's trichrome stain after Petrunkewitsch or Susa fixation." *Stain Technology* 30:147–148.
Rafalko, Stanley.
    1946 "A modified Feulgen technic for small and diffuse chromatin elements." *Stain Technology* 21:91–94.
Ralph, P. H.
    1941 "The histochemical demonstration of hemoglobin in blood cells and tissue smears." *Stain Technology* 16:105–106.
Raman, K.
    1955 "A method of sectioning aspirated bone-marrow." *Journal of Clinical Pathology* 8:265–266.
Ramon y Cajal, Santiago.
    1903 "Un sencillo metodo de coloracion selectiva del reticulo proto-plasmico y sus efectos en los diversos organos nerviosos." *Trabajos del Laboratorio Investigaciones Biologica* 2:129–221.

1910 "Las formulas del proceder del nitrato de plata reducido." *Trabajos del Laboratorio Investigaciones Biologica* 8:1–26.

Ramon y Cajal, Santiago, and De Castro, F.
1933 *Elementos de tecnica micrografica del sistema nerviosa* Madrid: Tipografia Artistica.

Randolph, L. F.
1935 "A new fixing fluid and a revised schedule for the paraffin method in plant cytology." *Stain Technology* 10:95–96.

Rawlins, T. E., and Takahashi, William.
1947 "Elimination of distortion and poor staining in paraffin sections of plant tissues." *Stain Technology* 22:99–102.

Ray, H. N., and Hajra, B.
1962 "Hydrosulfite-Schiff; limitations in use as shown by periodic-Schiff reactions in protozoa." *Stain Technology* 37:75–77.

Reaven, Eve P., and Cox, Alvin J.
1963 "The histochemical localization of histidine in the human epidermis and its relationship to zinc binding." *Journal of Histochemistry and Cytochemistry* 11:782–790.

Reid, J. D., and Sarantakos, G.
1966 "Infiltrating and embedding with mixtures of polyethylene glycols and polyvinyl-acetate resins." *Stain Technology* 41:207–210.

Reid, J. D., and Taylor, D.
1964 "An improved method for embedding tissues, using polyethylene glycols, with incorporation of low-viscosity nitrocellulose." *American Journal of Clinical Pathology* 41:513–516.

Renaud, Serge
1959 "Superiority of alcoholic over aqueous fixation in the histochemical detection of calcium." Stain Technology 34:267–271.

Repak, Arthur J., and Levine, Alan B.
1967 "A new use for an old stain." *Turtox News* 45:227.

Richards, Oscar W.
1941 "An efficient method for the identification of M. tuberculosis with a simple fluorescence microscope." *American Journal of Clinical Pathology, Technical Section* 5:1–8.
1949 *The Effective Use and Proper Care of the Microtome.* American Optical Company.
1954 *The Effective Use and Proper Care of the Microscope.* American Optical Company.

Richards, Oscar W.; Kline, Edmund K.; and Leach, Raymond E.
1941 "Demonstration of tubercle bacilli by fluorescence microscopy." *American Review of Tuberculosis* 44:255–266.

Richardson, Howard L.
1960 "The advantages of a glycol fixative for the preparation of Papanicolaou smears." *Technical Bulletin of the Registry of Medical Technologists* 30:15.

Richardson, K. C.; Jarett, L.; and Finke, E. H.
1960 "Embedding in epoxy resins for ultrathin sectioning in electron microscopy." *Stain Technology* 35:313–323.

Richmond, Gordon W., and Bennett, Leslie.
1938 "Clearing and staining of embryos for demonstration of ossification." *Stain Technology* 13:77–79.

Riis, Povl.
1957 "The chromatin structure in lymphocyte cells from peripheral blood." *Acta Haematologica* 18:168–174.

Rinderknecht, H.
1960 "A new technique for the fluorescent labelling of proteins." *Experientia* 16:430–431.

Rinehart, J. F., and Abul-Haj, S. K.
    1951*a* "An improved method for histologic demonstration of acid mucopoly-
        saccharides in tissues." *Archives of Pathology* 52:189–194.
    1951*b* "Histological demonstration of lipids in tissue after dehydration and em-
        bedding in polyethylene glycol." *Archives of Pathology* 51:666–669.
Riopel, James L.
    1962 "Carbowax for embedding and serial sectioning of botanical material." *Stain
        Technology* 37:357–362.
Ritter, Carl; Di Stephano, H. S.; and Farah, A.
    1961 "A method for the cytophotometric estimation of ribonucleic acid." *Journal
        of Histochemistry and Cytochemistry* 9:97–102.
Ritter, H. R., and Oleson, J. J.
    1950 "Combined histochemical staining of acid mucopolysaccharides and 1,2
        glycol groupings in paraffin sections." *American Journal of Pathology* 26:
        639–645.
Rodriquez, Jose, and Deinhardt, Friedrich.
    1960 "Preparation of a semipermanent medium for fluorescent antibody studies."
        Virology 12:316–317.
Roels, F.; Schiller, B.; and Goldfischer, S.
    1970 "Peroxisomes and lysosomes in the toad, *Bufo marinus*." *Journal of Histo-
        chemistry and Cytochemistry* 18:681–682.
Roman, Nickolas; Perkins, Stephen F.; Perkins, Edwin M., Jr.; and Dolnick, Ethel H.
    1967 "Orcein-hematoxylin in iodized ferric chloride as a stain for elastic fibers,
        with methanil yellow counterstaining." *Stain Technology* 42:199–202.
Romeis, B.
    1948 *Mikroskopische Technik*. München: Leibniz Verlag.
Roozemond, R. C.
    1969 "The effect of calcium chloride and formaldehyde on the release and com-
        position of phospholipids from cryostat sections of rat hypothalamus."
        Journal of Histochemistry and Cytochemistry 17:273–279.
Roque, Augustine L.; Jafarey, Naeem A.; and Coulter, Priscilla.
    1965 "A stain for the histochemical demonstration of nucleic acids." *Experimental
        and Molecular Pathology* 4:266–274.
Rosa, Charles G.
    1953 "Preparation and use of aldehyde fuchsin stain in the dry form." *Stain
        Technology* 28:299–302.
Rosenthal, Sanford I.; Puchtler, Holde; and Sweat, Faye.
    1965 "Paper chromatography of dyes." *Archives of Pathology* 80:190–196.
Rosewater, Joseph.
    1963 "An effective anesthetic for giant clams and other molluscs." *Turtox News*
        41:300–302.
Rothenbacher, Hans J., and Hitchcock, Dorothy J.
    1962 "Heat fixation and Giemsa staining for flagella and other cellular structures
        of trichomonads." *Stain Technology* 37:111–113.
Rothfels, K. H., and Siminovitch, L.
    1958 "An air drying technique for flattening chromosomes in mammalian cells
        grown in vitro." *Stain Technology* 33:73–77.
Rothstein, N.
    1958 "Vital staining of blood parasites with acridine orange." *Journal of Para-
        sitology* 44:588–595.
Rothstein, N., and Brown, M. L.
    1960 "Vital staining and differentiation of microfilaria." *American Journal of
        Veterinary Research* 21:1090–1094.
Rouiller, C. H.
    1960 "Physiological-pathological changes in mitochondrial morphology." *Inter-
        national Review of Cytology, XI*, ed. G. H. Bourne and J. F. Danielli, New
        York: Academic Press.

Rubin, Robert.
1951 "A rapid method for making permanent mounts of nematodes." *Stain Technology* 26:257–260.
Ruch, Fritz.
1955 "Birefringence and dichroism of cells and tissues." *Physical Techniques in Biological Research, Vol. I.* Ed. by Gerald Oster and Arthur W. Pollister. New York: Academic Press.
Ruddell, Craig L.
1967a "Hydroxyethyl methacrylate combined with polyethylene glycol 400 and water; an embedding medium for routine 1-2 micron sectioning." *Stain Technology* 42:119–123.
1967b "Embedding media for 1-2 micron sectioning. 2. Hydroxyethyl methacrylate combined with 2-butoxyethanol." *Stain Technology* 42:253–255.
Rumpf, P.
1935 "Recherches physico-chimiques sur la reaction colorie des aldehydes, dite 'Reaction de Schiff'." *Annales de Chimie* 3:327.
Rutenberg, Alexander M.; Gofstein, Ralph; and Seligman, Arnold M.
1950 "Preparation of a new tetrazolium salt which yields a blue pigment on reduction and its use in the demonstration of enzymes in normal and neoplastic tissues." *Cancer Research* 10:113–121.
Rutenberg, Alexander M.; Wolman, Moshi; and Seligman, Arnold M.
1953 "Comparative distribution of succinic dehydrogenase in six mammals and modification in the histochemical technic." *Journal of Histochemistry and Cytochemistry* 1:66–81.
Ruth, E. B.
1946 "Demonstration of the ground substance of cartilage, bone, and teeth." *Stain Technology* 27–30.

Sabatini, D. D.; Bensch, K. G.; and Barrnett, R. J.
1963 "Cytochemistry and electron microscopy. The preservation of cellular ultrastructure and enzymatic activity by aldehyde fixation." *Journal of Cell Biology* 17:19–58.
Sabatini, D. D.; Miller, Fritz; and Barrnett, J.
1964 "Aldehyde fixation for morphological and enzyme histochemical studies with the electron microscope." *Journal of Histochemistry and Cytochemistry* 12:57–71.
Sachs, Leo.
1953 "Simple methods for mammalian chromosomes." *Stain Technology* 28:169–172.
Sacks, Jacob.
1965 "Tracer techniques: stable and radioactive isotopes." *Physical Techniques in Biological Research, II.* Ed. by Gerald Oster and Arthur W. Pollister. New York: Academic Press.
St. Amand, George Sims, and St. Amand, W.
1951 "Shortening maceration time for alizarine red S preparations." *Stain Technology* 26:271.
Sainte-Marie, Guy.
1962 "A paraffin embedding technique for studies employing immunofluorescence." *Journal of Histochemistry and Cytochemistry* 10:250–256.
Sakesela, E., and Moorhead, P. S.
1962 "Enhancement of secondary constrictions and the heterochromatic X in human cells." *Cytogenetics* 1:225–244.
Sallmen, Henry, and Sherman, Mary.
1961 "A method for the rapid processing of celloidin sections." *Technical Bulletin of the Registry of Medical Technologists* 31:189–191.

Salthouse, T. N.
    1958 "Tetrahydrofuran and its use in insect histology." *Canadian Entomologist* 90:555–557.
    1962 "Luxol-fast blue ARN: a new solvent dye with improved staining qualities for myelin and phospholipids." *Stain Technology* 37:313–316.
    1964 "Luxol-fast blue G as a myelin stain." *Stain Technology* 39:123.
Sams, Alice.
    1963 "A celloidin infiltration-frozen section sequence for enhanced preservation of phosphatases in bone." *Stain Technology* 38:1–8.
Sams, Alice, and Davies, Freda M. R.
    1967 "Commercial varieties of nuclear fast red; their behaviour in staining after autoradiography." *Stain Technology* 42:269–276.
Sanders, Phyllis C., and Humason, Gretchen L.
    1964 "Culture and slide preparation of leukocytes from blood of *Macaca. Stain Technology* 39:209–213.
Sandison, A. T.
    1955 "The histological examination of mummified material." *Stain Technology* 30:277–283.
Sandritter, W.
    1962 "Is gallocyanin chromalum a quantitative stain for nucleic acids?" *Journal of Histochemistry and Cytochemistry* 10:680–681.
Sani, Guelfo; Citti, Ugo; Caramazza, Giuliano; and Quinto, Pietro.
    1964 *Fluorescence Microscopy in the Cytodiagnosis of Cancer.* Springfield, Ill.: Charles C Thomas.
Santamarina, Enrique.
    1964 "A formalin-Wright staining technique for avian blood cells." *Stain Technology* 39:267–274.
Sasaki, Masao, Makeno, Sajiro.
    1963 "The demonstration of secondary constrictions in human chromosomes by means of a new technique." *American Journal of Human Genetics* 15:24–33.
Sasaki, Motomichi.
    1961 "Observations on the modification in size and shape of chromosomes due to technical procedures." *Chromosoma* 11:514–522.
Saunders, A. M.
    1962 "Acridine orange staining for the identification of acid mucopolysaccharides." *Journal of Histochemistry and Cytochemistry* 10:683.
    1964 "Histochemical identification of acid mucopolysaccharides with acridine orange." *Journal of Histochemistry and Cytochemistry* 12:164–170.
Sawicki, Wojciech, and Pawinska, Maria.
    1965 "Effect of drying on unexposed autoradiographic emulsion in relation to background." *Stain Technology* 40:67–68.
Saxena, P. N.
    1957 "Formalin-chloride fixation to improve silver impregnation of the Golgi apparatus." *Stain Technology* 32:203–207.
Schajowicz, F., and Cabrini, R. L.
    1955 "The effect of acids (decalcifying solutions) and enzymes on the histochemical behavior of bone and cartilage." *Journal of Histochemistry and Cytochemistry* 3:122–129.
    1956 "Chelating agents as histological and histochemical decalcifiers." *Stain Technology* 31:129–133.
    1959 "Histochemical demonstration of acid phosphatase in hard tissues." *Stain Technology* 34:59–63.
Schantz, A., and Schecter, A.
    1965 "Iron-hematoxylin and safranin O as a polychrome stain for epon sections." *Stain Technology* 40:479–482.

Scherz, Robert G., and Louro, Jose M.
1963 "A simple method for making chromosome slides." *Technical Bulletin of the Registry of Medical Technologists* 33:124–127.
Schiff, Robert; Quinn, Loyd Y.; and Bryan, J. H. D.
1967 "A safranin fast green stain for the differentiation of the nuclei of rumen protozoa." *Stain Technology* 42:75–80.
Schiffer, Lewis M., and Vaharu, Tiju.
1962 "Acridine orange as a useful chromosomal stain." *Technical Bulletin of the Registry of Medical Technologists* 32:91–92.
Schlegel, Jorgen Ulrik.
1949 "Demonstration of blood vessels and lymphatics with a fluorescent dye in ultraviolet light." *Anatomical Record* 105:433–443.
Schleicher, Emil Maro.
1951 "Floating solutions for mounting paraffin sections." *American Journal of Clinical Pathology* 21:900.
Schleifstein, J.
1937 "A rapid method for demonstrating Negri bodies in tissue sections." *American Journal of Public Health* 27:1283–1285.
Schmid, Werner.
1965 "Autoradiography of human chromosomes." *Human Chromosome Methodology.* Ed. by Jorge J. Yunis. New York: Academic Press.
Schmidt, Robert W.
1956 "Simultaneous fixation and decalcification of tissue." *Laboratory Investigation* 5:306–307.
Schneider, John D.
1963 "A simplified Gram stain for demonstrating fungi in tissues." *Technical Bulletin of the Registry of Medical Technologists* 33:195–197.
Schubert, Maxwell, and Hamerman, David.
1956 "Metachromasia: chemical theory and histochemical use." *Journal of Histochemistry and Cytochemistry* 4:159–189.
Schultz, J., and St. Lawrence, P.
1949 "A cytological basis for a map of the nucleolar chromosome in man." *Journal of Heredity* 40:31–38.
Schultze, Brigitte.
1969 "Autoradiography at the cellular level." *Physical Techniques in Biological Research,* Vol. 3. 2d ed. Ed. by Arthur W. Pollister. New York: Academic Press.
Schwind, Joseph L.
1950 "The supravital method in the study of the cytology of the blood and marrow cells." *Blood* 5:597–622.
Scott, Gordon H.
1928 "A method for making permanent preparations of supravitally stained blood cells." *Anatomical Record* 38:233–237.
Scott, H. R., and Clayton, B. P.
1953 "A comparison of the staining affinities of aldehyde-fuchsin and the Schiff reagent." *Journal of Histochemistry and Cytochemistry* 1:336–352.
Scott, Jesse F.
1955 "Ultraviolet absorption spectrophotometry." *Physical Techniques in Biological Research,* Vol. I. Ed. by Gerald Oster and Arthur W. Pollister. New York: Academic Press.
Scott, John E., and Mowry, Robert W.
1970 "Alcian blue—a consumer's guide." *Journal of Histochemistry and Cytochemistry* 18:842.
Seal, S. H.
1956 "A method for concentrating cancer cells suspended in large quantities of fluid." *Cancer* 9:866–868.

Seligman, Arnold M.; Wasserkrug, Hannah L., Deb. Chandicharan, and Hanker, Jacob S.
  1968 "Osmium-containing compounds with multiple basic or acidic groups as stains for ultrastructure." *Journal of Histochemistry and Cytochemistry* 16:87–101.
Seman, G.
  1961 "Technique de coloration des lobules hétérochromatiques des leucocytes (spécialement des mononucléés) valeur de cette technique pour la détermination du sexe." *Revue Française d'Études Cliniques et Biologiques* 6:161–165.
Serra, J. A.
  1946 "Histochemical tests for proteins and amino acids; the characterization of basic proteins." *Stain Technology* 21:5–18.
  1958 "Cytochemical demonstration of masked lipids." *Science* 128:28–29.
Shanklin, D. R., and Laite, Melville B.
  1963 "Pickett-Sommer film strip technique." *Archives of Pathology* 75:91–93.
Shanklin, W. M., and Nassar, T. K.
  1959 "Luxol fast blue combined with periodic acid Schiff procedure for cytological staining of kidney." *Stain Technology* 34:257–260.
Shanklin, W. M.; Nassar, T. K., and Issidorides, M.
  1959 "Luxol fast blue as a selective stain for alpha cells in the human pituitary." *Stain Technology* 34:55–58.
Shaver, Evelyn L.
  1962 "The chromosomes of the opossum, *Didelphis virginiana*." *Canadian Journal of Genetics and Cytology* 4:62–68.
Sheehan, Dezna.
  1960 "A comparative study of the histologic techniques for demonstrating chromaffin cells." *American Journal of Medical Technology* 26:237–240.
Shelley, Walter B.
  1963 "Rapid technic for obtaining leukocytes from small samples of blood." *American Journal of Clinical Pathology* 39:433–435.
  1969 "Fluorescent staining of elastic tissue with rhodamine B and related xanthene dyes." *Histochemie* 20:244–249.
Shillaber, Charles Patten.
  1944 *Photomicrography, in Theory and Practice.* New York: John Wiley.
Shires, T. K.; Johnson, M.; and Richter, K. M.
  1969 "Hematoxylin staining of tissues embedded in epoxy resins." *Stain Technology* 44:21–25.
Sidman, R. L.; Mottla, P. A.; and Feder, N.
  1961 "Improved polyester wax embedding for histology." *Stain Technology* 36:279–284.
Sieracki, Joseph; Michael, James E.; and Clark, Daisy A.
  1960 "The demonstration of beta cells in pancreatic islets and their tumors." *Stain Technology* 35:67–69.
Sikov, Melvin R.
  1965 "A commercial encapsulated desiccant for use in slide containers during autoradiographic exposure." *Stain Technology* 40:239.
Sills, Bernard, and Marsh, Walton H.
  1959 "A simple technic for staining fat with oil-soluble azo dyes." *Laboratory Investigation* 8:1006–1009.
Silva, E. E. M.
  1961 "Periodic acid-Schiff staining combined with iron hematoxylin and light green for amoebae in sections." *Stain Technology* 36:196–198.
Silver, Maurice L.
  1942 "Colloidal factors controlling silver staining." *Anatomical Record* 82:507–529.

Silverstein, Arthur M.
  1957 "Contrasting fluorescent labels for two antibodies." *Journal of Histochemistry and Cytochemistry* 5:94–95.
Simmel, Eva B.
  1957 "The use of a fast, coarse grain stripping film for radioautography." *Stain Technology* 32:299–300.
Simmel, Eva B.; Fitzgerald, P. J.; and Godwin, J. T.
  1951 "Staining of radioautography with metanil yellow and iron hematoxylin." *Stain Technology* 26:25–28.
Simonds, Herbert R.; Weith, Archie J.; and Bigelow, M. H.
  1949 *Handbook of Plastics.* New York: Van Nostrand.
Simpson, Charles F.; Carlisle, J. W.; and Mallard, L.
  1970 "Rhodanile blue: a rapid and selective stain for Heinz bodies." *Stain Technology* 45:221–223.
Singer, Marcus.
  1952 "Factors which control the staining of tissues with acid and basic dyes." *International Review of Cytology*, Vol. 1. Ed. by G. H. Bourne and J. F. Danielli. New York: Academic Press.
Sinha, Ranendra Nath.
  1953 "Sectioning insects with sclerotized cutile." *Stain Technology* 28:249–253.
Sisken, Jesse E.
  1964 "Methods for measuring the length of the mitotic cycle and the timing of DNA synthesis for mammalian cells in culture." *Methods in Cell Physiology*, I. Ed. by D. M. Prescott. New York: Academic Press.
Slidders, W.; Fraser, D. S.; Smith, R.; and Lendrum, A. C.
  1958 "On staining the nucleus red." *Journal of Pathology and Bacteriology* 75:466–468.
Sloper, J. C., and Adams, C. W. M.
  1956 "The hypothalamic elaboration of posterior pituitary principles in man. Evidence derived from hypophysectomy." *Journal of Pathology and Bacteriology* 72:587–602.
Smith, Arthur.
  1962 "Tissue shrinkage caused by attachment of paraffin sections to slides: its effects on staining." *Stain Technology* 37:339–345.
Smith, C. W.; Marshall, J. D.; and Eveland, W. C.
  1959 "Use of contrasting fluorescent dyes as counterstain in fixed tissue preparations." *Proceedings of the Society for Experimental Biology and Medicine* 102:179–181.
Smith, E. J.; Puchtler, Holde; and Sweat, Faye.
  1966 "Investigation of the chemical mechanism of trichrome stain." *Laboratory Investigation* 15:1141–1142.
Smith, G. Stewart.
  1943 "A danger attending the use of ammoniacal solutions of silver." *Journal of Pathology and Bacteriology* 55:227–228.
Smith, Luther.
  1947 "The acetocarmine technique." *Stain Technology* 22:17–31.
Smith, Robert E.
  1970 "Comparative evaluation of two instruments and procedures to cut non-frozen sections." *Journal of Histochemistry and Cytochemistry* 18:590–591.
Smyth, J. D.
  1944 "A technic for mounting free-living protozoa." *Science* 100:62.
Snodgrass, Ann B., and Dorsey, Charles H.
  1963 "Egg albumin embedding: a procedure compatible with neurological staining techniques." *Stain Technology* 38:149–155.
Snodgrass, Ann B., and Lacey, Leonard B.
  1961 "Luxol fast blue staining of degenerating myelinated fibers." *Anatomical Record* 140:83–89.

Snow, Richard.
    1963 "Alcoholic hydrochloric acid-carmine as a stain for chromosomes in squash preparations." *Stain Technology* 38:9-13.
Solcia, E.; Capella, C.; and Vassallo, G.
    1969 "Lead hematoxylin as a stain for endocrine cells." *Histochemie* 20:116-126.
Solcia, E.; Vassallo, G.; and Capella, C.
    1968 "Selective staining of endocrine cells by basic dyes after acid hydrolysis." *Stain Technology* 43:257-263.
Sommer, Joachim R., and Pickett, John Phillip.
    1961 "Thirty-five mm picture film for cytologic smears." *Archives of Pathology* 71:79-81.669-671.
Spatz, Maria.
    1960 "Bismarck brown as a selective stain for mast cells." *American Journal of Clinical Pathology* 34:285-287.
Spicer, S. S.
    1960 "A correlative study of the histochemical properties of rodent acid mucopolysaccharides." *Journal of Histochemistry and Cytochemistry* 8:18-33.
    1961 "Differentiation of nucleic acids by staining at controlled *p*H and by a Schiff-methylene blue sequence." *Stain Technology* 36:337-340.
    1962 "Histochemically selective acidophilia of basic nucleoproteins in chromatin and nucleoli at alkaline *p*H." *Journal of Histochemistry and Cytochemistry* 10:691-703.
Spicer, S. S., and Lillie, R. D.
    1961 "Histochemical identification of basic proteins with Biebrich scarlet at alkaline *p*H. *Stain Technology* 36:365-370.
Spicer, S. S., and Meyer, D. B.
    1960 "Histochemical differentiation of acid mucopolysaccharides by means of combined aldehyde fuchsin-alcian blue staining." *American Journal of Clinical Pathology, Technical Section,* 33:453-456.
Spurlock, B. O.; Skinner, M. S.; and Kattine, A. A.
    1966 "A simple rapid method for staining epoxy-embedded specimens for light microscopy with the polychromatic stain, Paragon-1301." *American Journal of Clinical Pathology* 46:252-258.
Srivastava, Probodh K., and Lasley, J. F.
    1968 "Leucocyte culture and chromosome preparations from pig blood." *Stain Technology* 43:187-190.
Stafford, E. S.
    1934 "An eosin-methylene blue technique for rapid diagnosis." *Johns Hopkins Hospital Bulletin* 55:229.
Stafford, Marvin W.
    1960 "An improved method for labelling frosted-end microslides used in exfoliative cytology." *Technical Bulletin of the Registry of Medical Technologists* 30:26-27.
Stagg, Franklin B., and Tappen, N. C.
    1963 "A device for reducing condensation within the rotary microtome adapted for cutting frozen sections and for reducing tissue distortion." *Stain Technology* 38:352-354.
Stapp, Paul, and Cumley, Russell W.
    1936 "A technic for clearing large insects." *Stain Technology* 11:105-106.
Stearn, Allen E., and Stearn, Esther.
    1929 "The mechanism of staining explained on a chemical basis. I. The reaction between dyes, proteins, and nucleic acid." *Stain Technology* 4:111-119.
    1930 "The mechanism of staining explained on a chemical basis. II. General considerations." *Stain Technology* 5:17-24.
Steedman, H. F.
    1947 "Ester wax: a new embedding medium." *Quarterly Journal of Microscopical Science* 88:123-133.

1950 "Alcian blue 8G: a new stain for mucin." *Quarterly Journal of Microscopical Science* 91:477–479.

1960 *Section Cutting in Microscopy.* Springfield, Ill.: Charles C Thomas.

Steil, W.
1936 "Modified Wright's method for staining blood smears." *Stain Technology* 11:99–100.

Steinman, Irvin D.
1955 "A vinyl plastic coating for stained thin smears." *Stain Technology* 30: 49–50.

Stenram, Unne.
1962 "Loss of silver grains from radioautographs stained by gallocyanin-chrome alum." *Stain Technology* 37:231–234.

Sterling, Clarence, and Chichester, C. O.
1956 "Autoradiography of water-soluble materials in plant tissues." *Stain Technology* 31:227–230.

Sternberg, S. S.; Cronin, A. P.; and Philips, F. S.
1964 "Histochemical demonstration of zinc by the formation of a fluorescent chelate of 8-hydroxyquinoline(oxine): comparison with dithizone reactions." *Proceedings of the Federation of American Societies for Experimental Biology* 23:575.

Stone, G. E., and Cameron, I. L.
1964 "Methods for using *Tetrahymena* in studies of the normal cell cycle." *Methods in Cell Physiology, I.* Ed. by David M. Prescott. New York: Academic Press.

Stowell, Robert E.
1941 "Effect on tissue volume of various methods of fixation, dehydration, and embedding." *Stain Technology* 16:67–83.
1945 "Feulgen reaction for thymonucleic acid." *Stain Technology* 20:45–58.
1946 "The specificity of the Feulgen reaction for thymonucleic acid." *Stain Technology* 21:137–148.

Straus, Werner.
1964 "Factors affecting the cytochemical reaction of peroxidase with benzidine and the stability of the blue reaction product." *Journal of Histochemistry and Cytochemistry* 12:462–469.
1967 "Methods for the study of small phagosomes and their relationship to lysosomes with horseradish peroxidase as a 'marker' protein." *Journal of Histochemistry and Cytochemistry* 15:375–380.

Strike, Thomas A.
1962 "A device for adapting the rotary microtome to frozen sectioning." *Stain Technology* 37:187–189.

Structure and Biochemistry of Mitochondria Symposium 1953 [Proceedings] *Journal of Histochemistry and Cytochemistry* 1:179–276.

Stumpf, Walter, and Roth, Lloyd J.
1964 "Vacuum freeze drying of frozen sections for dry-mounting high-resolution autoradiography." *Stain Technology* 39:219–223.
1966 "Freeze-drying of small tissue samples and thin frozen sections below −60°C." *Journal of Histochemistry and Cytochemistry* 15:243–251.

Swank, R. L., and Davenport, H. A.
1934a "Marchi's staining method. I. Studies of some of underlying mechanisms involved." *Stain Technology* 9:11–19.
1934b "Marchi's staining method. II. Fixation." *Stain Technology* 9:129–135.
1935a "Marchi's staining method. III. Artifacts and effects of perfusion." *Stain Technology* 10:45–52.
1935b "Chlorate-osmic-formalin method for staining degenerating myelin." *Stain Technology* 10:87–90.

Swanson, David W., and McKee, Mary E.
    1964 "Direct cover-glass culture and staining for chromosome analysis." *Stain Technology* 39:117–119.
Swartz, Frank J., and Nagy, Ernest R.
    1963 "Feulgen stain stability in relation to three mounting media and exposure to light." *Stain Technology* 38:179–185.
Sweat, Faye; Meloan, Susan N.; and Puchtler, Holde.
    1968 "A modified one-step trichrome stain for demonstration of fine connective tissue fibers." *Stain Technology* 43:227–231.
Sweat, Faye; Puchtler, Holde, and Rosenthal, Sanford I.
    1964 "Sirius red F3BA as a stain for connective tissue." *Archives of Pathology* 78:69–72.
Sweat, Faye; Puchtler, Holde; and Sesta, J. J.
    1968 "PAS-phosphomolybdic acid-sirius supra blue FGL-CF." *Archives of Pathology* 86:33–39.
Sweat, Faye; Puchtler, Holde; and Terry, Mary S.
    1968 "Staining isolated proteins and polysaccharides on slides; phosphotungstic acid-hematoxylin versus the Rosindole reaction." *Stain Technology* 43:283–286.
Sweat, Faye; Puchtler, Holde; and Woo, Pauline.
    1964 "A light-fast modification of Lillie's allochrome stain." *Archives of Pathology*, 78:73–75.
Swigart, Richard H.; Wagner, Charles E.; and Atkinson, William B.
    1960 "The preservation of glycogen in fixed tissues and tissue sections." *Journal of Histochemistry and Cytochemistry* 8:74–75.
Symeonidis, A.
    1935 "Neue Anwendungsmöglichkeiten der Gefrierschneidemethode mit Messertiefkühlung bei fixierten Gewebe." *Centralblatt für allgemeine Pathologie und Pathologische Anatomie* 63:245–246.

Taft, E. B.
    1951 "The problem of standardized technique for the methyl-green-pyronin stain." *Stain Technology* 26:205–212.
Taft, P. D., and Arizaga-Cruz, J. M.
    1960 "A comparison of the cell block, Papanicolaou, and millipore filter technics for the cytological examination of serous fluids." *Technical Bulletin of the Registry of Medical Technologists* 30:189–192.
Taft, P. D., and Lojananond, Preecha.
    1962 "An evaluation of fluorescence microscopy in gynecologic exfoliative cytology." *American Journal of Clinical Pathology* 37:334–336.
Takaya, Kinichi.
    1967 "Luxol fast blue MBS and phloxine; a stain for mitochondria." *Stain Technology* 42:207–211.
Takeuchi, Jun.
    1962 "Staining sulfated mucopolysaccharides in sections by means of acriflavine." *Stain Technology* 37:105–107.
Tanaka, Ryuso.
    1961 "Aceto-basic fuchsin as a stain for nucleoli and chromosomes of plants." *Stain Technology* 36:325–327.
Tandler, C. J.
    1955 "The reaction of nucleoli with ammoniacal silver in darkness; additional data." *Journal of Histochemistry and Cytochemistry* 3:196–202.
Tarkham, A. A.
    1931 "The effect of fixatives and other reagents on cell size and tissue bulk." *Journal of the Royal Microscopical Society* 51:387–399.

Tauxe, W. Newlon; Moser, Arthur H.; and Boyd, George A.
1954 "Etymology of autoradiography." *Science* 120:149–150.
Taylor, A. I.
1963 "Sex chromatin in the newborn. *Lancet* 1:912–914.
Taylor, John D.
1965 "Gelatin embedding on the tissue carrier for thin serial sections in the cryostat." *Stain Technology* 40:29–31.
Taylor, Richard Dee.
1966 "Modification of the Brown and Green Gram stain for the differential staining of Gram-positive and Gram-negative bacteria in tissue sections." *American Journal of Clinical Pathology* 46:472–474.
Taylor, William Ralph.
1967 "An enzyme method of clearing and staining small vertebrates." *Proceedings of U.S. National Museum* 122:1–17.
Tepper, H. B., and Gifford, E. M.
1962 "Detection of ribonucleic acid with pyronin." *Stain Technology* 37:52–53.
Terner, Jacob Y., and Clark, George.
1960a "Gallocyanin-chrome alum: I. Technique and specificity." *Stain Technology* 35:167–168.
1960b "Gallocyanin-chrome alum: II. Histochemistry and specificity." *Stain Technology* 35:305–311.
Thieme, George.
1965 "Small tissue dryers with high capacity for rapid freeze-drying." *Journal of Histochemistry and Cytochemistry* 13:386–389.
Thomas, John T.
1953 "Phloxine-methylene blue staining of formalin fixed tissue." *Stain Technology* 28:311–312.
Thomas, Lloyd E.
1950 "An improved arginine histochemical method." *Stain Technology* 25:143–148.
Thomison, John B.
1961 "Combination of millipore filtration and fluorescence microscopy in cytologic examination." *American Journal of Clinical Pathology* 35:407–410.
Thompson, E. C.
1957 "A tray for staining frozen sections in quantity." *Stain Technology* 32:255–257.
1961 "Simultaneous staining of reticulocytes and Heinz bodies with new methylene blue N in dogs given iproniazid." *Stain Technology* 36:38–39.
Thompson, Samuel Wesley.
1966 *Selected Histochemical and Histological Methods.* Springfield, Ill.: Charles C Thomas.
Thurston, Jean M., and Joftes, David L.
1963 "Stains compatible with dipping radioautography." *Stain Technology* 38:231–234.
Tijo, J. H., and Levan, A.
1956 "The chromosome number of man." *Hereditas-Genetiskt Arkiv* 42:1–6.
Tijo, J. H., and Puck, T. T.
1958a "The somatic chromosomes of man." *Proceedings of the National Academy of Sciences* 44:1229–1237.
1958b "Genetics of somatic mammalian cells. II. Chromosomal constitution of cells in tissue culture." *Journal of Experimental Medicine* 108:259–269.
Tijo, J. H., and Wang, J.
1962 "Chromosome preparations of bone marrow cells without prior in vitro culture or in vivo colchicine administration." *Stain Technology* 37:17–20.
Tilden, I. L., and Tanaka, Masso.
1945 "Fite's fuchsin-formaldehyde method for acid fast bacilli applied to frozen sections." *American Journal of Clinical Pathology* 9:95–97.

Tobie, John E.
1958 "Certain technical asects of fluorescence microscoy and the Coons fluorescent antibody technique." *Journal of Histochemistry and Cytochemistry* 6:271–277.

Tomlinson, W. J., and Grocott, R. G.
1944 "A simple method of staining malaria protozoa and other parasites in paraffin sections." *American Journal of Clinical Pathology* 14:316–326.

Tonna, Edgar A.
1958 "Factors which influence the latent image in autoradiography." *Stain Technology* 33:255–260.
1959 "Decalcification of bone by means of an automatic continuous flow apparatus." *American Journal of Clinical Pathology* 31:160–164.

Tonna, Edgar A., and Love, Robert A.
1961 "High pressure plastic lamination for flexible mounts of cover-slip preparations." *Stain Technology* 36:213–218.

Toren, David A.
1963 "A Giemsa-trichrome stain for mast cells in paraffin sections." *Stain Technology* 38:249–250.

Trott, J. R.
1961a "The presence of glycogen in the rat liver following in vitro processing in decalcifying agents." *Journal of Histochemistry and Cytochemistry* 9: 699–702.
1961b "An evaluation of methods commonly used for the fixation and staining of glycogen." *Journal of Histochemistry and Cytochemistry* 9:703–710.

Trott, J. R.; Gorenstein, S. L.; and Peikoff, M. D.
1962 "A chemical and histochemical investigation of glycogen in rat liver and palate following treatment with various fixatives and ethylenediamine tetraacetic acid." *Journal of Histochemistry and Cytochemistry* 10:245–249.

Trump, Benjamin F.; Smuckler, Edward A.; and Benditt, Earl P.
1961 "A method for staining epoxy sections for light microscopy." *Journal of Ultrastructure Research* 5:343–348.

Tuan, Hsu-Chuan.
1930 "Picric acid as a destaining agent for iron alum hematoxylin." *Stain Technology* 5:135–138.

Turchini, Jean-Pascal, and Malet, Paul.
1965 "Long conservation of histoenzymatic activities of fresh tissues in glycerol." *Journal of Histochemistry and Cytochemistry* 13:405–406.

Uber, Fred M.
1936 "Microtome knife sharpeners operating on the abrasive ground glass principle." *Stain Technology* 11:93–98.

Ueckert, Edwin.
1960 "Polyester embedding for sectioning hard and soft tissues." *Stain Technology* 35:261–265.

Umiker, William, and Pickle, Larry.
1960 "The cytologic diagnosis of lung cancer by fluorescence microscopy: acidine orange fluorescence technic in routine screening and diagnosis." *Laboratory Investigation* 9:613–624.

Ungewitter, Luise H.
1943 "A controllable silver stain for nerve fibers and nerve endings." *Stain Technology* 18:183–186.
1951 "A urea silver nitrate method for nerve fibers and nerve endings." *Stain Technology* 26:73–76.

Upadhya, M. D.
1963 "The use of a α-bromo-naphthalene, rapid hot fixation, and distributed

pressure squashing for chromosomes of Triticinae." *Stain Technology* 38: 293–295.

Uzman, L. Lahut.
1956 "Histochemical localization of copper with rubeanic acid." *Laboratory Investigation* 5:299–305.

Vacek, Z., and Plackova, A.
1959 "Silver impregnation of nerve fibers in teeth after decalcification with ethylenediaminetetraacetic acid." *Stain Technology* 34:1–3.

Van Cleave, Harley J., and Ross, Jean A.
1947 "A method of reclaiming dried zoological specimens." *Science* 105:318.

Van Duijn, P.
1961 "Acrolein-Schiff, a new staining method for proteins." *Journal of Histochemistry and Cytochemistry* 9:234–241.

Vasquez, J. J., and Dixon, F. J.
1956 "Immunohistochemical analysis of amyloid by the fluorescent technique." *Journal of Experimental Medicine* 104:727.

Vassar, Philip S., and Culling, Charles F. A.
1959 "Fluorescent stains with special reference to amyloid and connective tissue." *Archives of Pathology* 68:487–498.
1962 "Fluorescent amyloid staining of casts in myeloma nephrosis." *Archives of Pathology* 73:59–63.

Vernino, David M., and Laskin, David M.
1960 "Sex chromatin in mammalian bone." *Science* 132:675–676.

Vickers, A. E. J.
1956 *Modern Methods of Microscopy*. New York: Interscience Publishers.

Vidal, O. R.; Aya, T.; and Sandberg, A. A.
1971 "Glutaraldehyde-ammoniacal silver carbonate-formaldehyde staining of histones of mitotic chromosomes." *Stain Technology* 46:89–92.

Villyanueva, A. R.; Hattner, R. S.; and Frost, H. M.
1964 "A tetrachrome stain for fresh, mineralized bone sections useful in the diagnosis of bone diseases." *Stain Technology* 39:87–94.

Vincent, W. S.
1955 "Structure and chemistry of nucleoli." *International Review of Cytology*, IV. Ed. by G. H. Bourne and J. F. Danielli. New York: Academic Press.

Vinegar, Ralph.
1961 "Metachromatic differential fluorochroming of living and dead ascites tumor cells with acridine orange." *Cancer Research* 16:900–906.

Vlachos, John.
1959 "Desaturation of staining solutions for lipids, a means of avoiding precipitation on stained sections. *Stain Technology* 34:292.

Von Bertalanffy, Ludwig, and Bickis, Ivar.
1956 "Identification of cytoplasmic basophilia (ribonucleic acid) by fluorescence microscopy." *Journal of Histochemistry and Cytochemistry* 4:481–493.

Von Bertalanffy, Ludwig; Masin, Marianna; and Masin, Francis.
1958 "A new and rapid method for diagnosis of vaginal and cervical cancer by fluorescence microscopy." *Cancer* 11:873–887.

Von Bertalanffy, Ludwig, and Von Bertalanffy, Felix D.
1960 "A new method for cytological diagnosis of pulmonary cancer." *Annals of the New York Academy of Sciences* 84: 225–238.

Wachstein, Max, and Meisel, Elizabeth.
1964 "Demonstration of peroxidase activity in tissue sections." *Journal of Histochemistry and Cytochemistry* 12:538–544.

Wade, W. H.
  1952 "Notes on the carbowax method of making tissue sections." *Stain Technology* 27:71–79.
  1957 "A modification of the Fite formaldehyde (Fite I) method for staining acid fast bacilli in paraffin sections." *Stain Technology* 32:287–292.
Wagner, Bernard M., and Shapiro, Sylvia H.
  1957 "Application of alcian blue as a histochemical method. *Laboratory Investigation* 6:472–477.
Walker, P. M. B.
  1958 "Ultraviolet microspectrophotometry." *General Cytochemical Methods*. Ed. by J. F. Danielli. New York: Academic Press.
Wall, Patrick.
  1950 "Staining and recognition of fine degenerating nerve fibers." *Stain Technology* 25:125–126.
Walls, G. L.
  1932 "The hot celloidin technic for animal tissues." *Stain Technology* 7:135–145.
  1936 "A rapid celloidin method for the rotary microtome." *Stain Technology* 11:89–92.
  1938 "The microtechnique of the eye with suggestions as to material." *Stain Technology* 13:69–72.
Walsh, Richard J., and Love, Robert.
  1963 "Studies of the cytochemistry of nucleoproteins. II. Improved staining methods with toluidine blue and ammonium molybdate." *Journal of Histochemistry and Cytochemistry* 11:188–196.
Warmke, Harry E.
  1935 "A permanent root tip smear method." *Stain Technology* 10:101–103.
  1946 "Precooling combined with chromosmo-acetic fixation in studies of somatic chromosomes in plants." *Stain Technology* 21:87–90.
Warren, O.
  1944 "Some facts about nerve ending preparations." *Turtox News* 22.
Wartman, W. B.
  1943 "Notes on the Field's method of staining parasites in thick blood films." *Army Medical Bulletin* 68:173–177.
Watson, Barbara K.
  1952 "Distribution of mumps virus in tissue cultures as determined by fluorescein-labeled antiserum." *Proceedings of the Society for Experimental Biology and Medicine* 79:222–224.
Watson, M. L.
  1956 "Carbon films and specimen stability." *Journal of Biophysical and Biochemical Cytology, Supplement* 2:31–37.
Weaver, Harry Lloyd.
  1955 "An improved gelatine adhesive for paraffin sections." *Stain Technology* 30:63–64.
Weiner, S.
  1957 "A simplified vacuum embedder." *Stain Technology* 32:195–196.
Weiss, Jules.
  1954 "The nature of the reaction between orcein and elastin." *Journal of Histochemistry and Cytochemistry* 2:21–28.
Weissmann, Gerald.
  1964 "Lysosomes." *Blood* 24:594–606.
  1968 "The many-faceted lysosome." *Hospital Practice* 31–39.
Welshons, W. J.; Gibson, B. H., and Scandyln, B. J.
  1962 Slide processing for the examination of male mammalian meiotic chromosomes." *Stain Technology* 37:1–5.
Werth, Von G.
  1953 "Fluoeszenz mikroskopische Beobachtungen an menschlichem Knochenmark." *Acta Haematologica* 10:209–222.

West, William T.
  1962 "The Pearse-Sless cryostat: improvement in the antiroll plate." *Stain Technology* 37:320.
Westfall, Jane A.
  1961 "Obtaining flat serial sections for electron microscopy." *Stain Technology* 36:36–38.
Wheatley, Walter B.
  1951 "A rapid procedure for intestinal amoebae and flagellates." *American Journal of Clinical Pathology* 21:990–991.
White, Lowell E., Jr.
  1960 "Enhanced reliability in silver impregnation of terminal axonal degeneration—original Nauta method." *Stain Technology* 35:5–9.
White, Robert G.; Coons, Albert H.; and Connolly, Jeanne M.
  1955 "Studies on antibody production. III. The alum granuloma. IV. The role of a wax fraction of myobacterium tuberculosis in adjuvant emulsions on the production of antibody to egg albumin." *Journal of Experimental Medicine* 102:73–103.
Wieland, H., and Scheuing, G.
  1921 "Die Farbreaktion mit Aldehyden." *Berichte der Deutschen Chemischen Gesellschaft* 54:2527–2555.
Wigglesworth, V. B.
  1959 "A simple method for cutting sections in the 0.5 to 1μ range, and for sections of chitin." *Quarterly Journal of Microscopical Science* 100:315–320.
Wijffels, C. C. B. M.
  1971 "Freeze-drying by cryosorption; a simple device having a tissue capacity of 500 milligrams." *Stain Technology* 46:33–34.
Wilcox, Aimee.
  1943 *Manual for the Microscopical Diagnosis of Malaria in Man.* National Institute of Public Health, Bulletin 180. Washington, D.C.: U.S. Government Printing Office.
Wilder, H. C.
  1935 "An improved technique for silver impregnation of reticular fibers." *American Journal of Pathology* 11:817–819.
Wilhelm, Walter E., and Smoot, Al.
  1966 "Permanent demonstration of food vacuoles in *Paramecium* stained in relief." *Turtox News* 44:158.
Williams, Dorothy L.; Lafferty, Diana A.; and Webb, Sondra L.
  1970 "An air drying method for the preparation of dictyotene chromosomes from ovaries of Chinese hamsters." *Stain Technology* 45:133–135.
Williams, G., and Jackson, D. S.
  1956 "Two organic fixatives for acid mucopolysaccharides." *Stain Technology* 31:189–191.
Williams, R. C., and Backus, R. C.
  1949 "The electron micrographic structure of shadow-cast films." *Journal of Applied Physics* 20:98–106.
Williams, R. C., and Kallman, F.
  1954 "Examination of tissue-cultured Hela cells by electron microscopy of serial sections." *Journal of Applied Physics* 25:1455.
Williams, R. C., and Wyckoff, R. W. G.
  1944 "Thickness of electron microscopic objects." *Journal of Applied Physics* 15:712–715.
Williams, T. D.
  1957 "Mounting and preserving serial celloidin sections." *Stain Technology* 32:97.
Williams, Terry.
  1962 "The staining of nervous elements by the Bodian method. I. The influence of factors preceding impregnation." *Quarterly Journal of Microscopical Science* 103:155–161.

Willigan, Donald A.; Garric, Anita; and Trosko, Beverly K.
    1961 "New fuchsin-hematoxylin-eosin; a stain for acid-fast bacilli and surrounding tissue." *Stain Technology* 36:319–321.
Winteringham, F. P. W.; Harrison A.; and Hammond, J. H.
    1950 "Autoradiography of water-soluble tracers in histological sections." *Nature* 165:149–152.
Witten, Victor H., and Holmstrom, Vera.
    1953 "New histologic technics for autoradiography." *Laboratory Investigation* 2:368–375.
Wittman, W.
    1962 "Aceto-iron-hematoxylin for staining chromosomes in squashes of plant material." *Stain Technology* 37:27.
    1963 "Permanent-type mounting of aceto-iron-hematoxylin squashes in corn syrup." *Stain Technology* 40:161–164.
Wolberg, William H.
    1965 "Darrow red-light green as a stain for autoradiographs." *Stain Technology* 40:90.
Wolfe, Hubert J.
    1964 "Techniques for the histochemical localization of extremely water soluble acid mucopolysaccharides." *Journal of Histochemistry and Cytochemistry* 12:217–218.
Wolman, M.
    1955 "Problems of fixation in cytology, histology, and *histochemistry*." *International Review of Cytology,* IV. Ed. by G. H. Bourne and J. F. Danielli. New York: Academic Press.
    1956 "A histochemical method for the differential staining of acidic tissue components (particularly ground-substance polysaccharides)." *Bulletin of the Research Council of Israel, Section E, Experimental Medicine* 6E:27–35.
    1961 "Differential staining of acidic tissue components by the improved bi-col method." *Stain Technology* 36:21–25.
Woodruff, Lois A., and Norris, William P.
    1955 "Sectioning of undecalcified bone; with special reference to radioautographic applications." *Stain Technology* 30:179–188.
Woods, Philip S., and Pollister, A. W.
    1955 "An ice-solvent method of drying frozen tissue for plant cytology." *Stain Technology* 30:123–131.
Wright, J. H.
    1902 "A rapid method for the differential staining of blood films and malarial parasites." *Journal of Medical Research* 7:138–144.
Wróblewska, Joanna.
    1969 "Chromosome preparations from mouse embryos during early organogenesis: dissociation after fixation, followed by air drying." *Stain Technology* 44:147–150.

Yaeger, James A.
    1958 "Methacrylate embedding and sectioning of calcified bone." *Stain Technology* 33:220–239.
Yamada, K.
    1963 "Staining of sulfated polysaccharides by means of alcian blue." *Nature* 197:789.
Yamaguchi, Ben T., Jr., and Braunstein, Herbert.
    1965 "Fluorescent stain for tubercle bacilli in histologic sections. II. Diagnostic efficiency in granulomatous lesions of the liver." *Technical Bulletin of the Registry of Medical Technologists* 35:184–187.

Yasuma, Akiyasu, and Ichikawa, Toyoki.
    1953 "Ninhydrin-Schiff and Alloxan-Schiff staining. A new histochemical staining method for protein." *Journal of Laboratory and Clinical Medicine* 41:296–299.
Yenson, Jack.
    1968 "Removal of epoxy resin from histological sections following halogenation." *Stain Technology* 43:344–346.
Yerganian, G.
    1957 "Cytologic maps of some isolated human pachytene chromosomes." *American Journal of Human Genetics* 9:42–54.
    1963 "Cytogenetic analysis." *Methodology in Mammalian Genetics*. Ed. by Walter J. Burdette. San Francisco: Holden-Day.
Yetwin, I. J.
    1944 "A simple permanent mounting medium for *Necator americanus*." *Journal of Parasitology* 30:201.
Young, Irving.
    1962 "The use of a sponge base as an aid in preparing frozen sections of small objects." *Technical Bulletin of the Registry of Medical Technologists* 32:76.
Youngpeter, John M.
    1964 "Spalteholz preparations." *Ward's Bulletin* 3:7.
    1967 "Spalteholz preparations." *Ward's Bulletin* 6:7.

Zambernard, Joseph; Block, Matthew; Vatter, Albert; and Trenner, Lew.
    1969 "An adaptation of methacrylate embedding for routine histologic use." *Blood* 33:444–450.
Zbar, Marcus J., and Winter, William J.
    1959 "A method of concentrating particles of bone marrow for paraffin sections." *American Journal of Clinical Pathology* 32:41–44.
Zeiger, K.; Harders, H.; and Müller, W.
    1951 "Der Strugger-effekt an der Nervenzelle." *Protoplasma* 40:76–84.
Zinn, Donald J., and Morin, Lorraine P.
    1962 "The use of commercial citric juices in gold chloride staining of nerve endings." *Stain Technology* 37:380–381.
Zirkle, Conway.
    1937 "Aceto-carmine mounting media." *Science* 85:528.
    1940 "Combined fixing, staining, and mounting media." *Stain Technology* 15:139–153.
Zlotnik, I.
    1960 "The initial cooling of tissues in the freezing-drying technique." *Quarterly Journal of Microscopical Science* 101:251–254.
Zuck, Robert K.
    1947 "Simplified permanent aceto-carmine smears with a water-miscible mountant." *Stain Technology* 22:109–110.
Zugibe, Frederick T.
    1970 *Diagnostic Histochemistry*. St. Louis: C. V. Mosby.
Zugibe, Frederick T.; Brown, Kenneth D., and Last, Jules H.
    1959 "A new technique for the simultaneous demonstration of lipid and acid mucopolysaccharides on the same tissue section." *Journal of Histochemistry and Cytochemistry* 7:101–106.
Zugibe, Frederick T.; Fink, Marilyn L.; and Brown, Kenneth W.
    1958 "Carbowax flotation method." *Journal of Histochemistry and Cytochemistry* 6:381.
    1959 "Carbowax 400, a new solvent for oil red O and sudan IV for staining carbowax embedded and frozen sections." *Stain Technology* 34:33–37.

Zugibe, Frederick T.; Kopaczyk, Krystyna C.; Cape, William E., and Last, Jules H.
  1958 "A new carbowax method for routinely performing lipid, hematoxylin, and eosin and elastic staining techniques on adjacent freeze dried or formalin fixed tissues." *Journal of Histochemistry and Cytochemistry* 6:133–138.
Zwaan, J., and Van Dam, A. F.
  1961 "Rapid separation of fluorescent antisera and unconjugated dye." *Acta Histochemica* 11:306–308.

Index

Abopon, 145
Accelerator, 128
Accentuator, 128
Acetate buffer, 539
Acetate acetic acid buffer, 540
Acetic acid fixation, 5
Acetic hematoxylin, 243
Acetocarmine, 465
Aceto-orcein, 465, 476
Acetone, 6; embedding, 426; fixation, 6, 419
Acetylation, 301
Acid alcohol, 177, 531
Acid alizarine blue, 179, 353
Acid decalcification. See Decalcification
Acid-fast staining, 385
Acid fuchsin, 155, 180, 181, 182, 197, 266; for mitochondria, 372
Acid group blocking, 304
Acid hematein: for mitochondria, 373; for phospholipids, 312
Acid mucopolysaccharides, 317
Acid phosphatase, 435
Acidified water, 178, 181, 183, 184, 227, 392
Acridine orange, 226, 319, 345, 378, 400, 458, 460
Acriflavine, 334
Acylation. See Acetylation
Adamstone-Taylor cold knife technique, 425
Adhesive solutions, 536

Affixing sections. See Mounting sections
Affixing solutions. See Adhesive solutions
Agar embedding, 71, 466, 504, 532
Albumen fixative, 536
Alcian blue, 317, 320, 355
Alcian green, 319
Alcohol: acid, 177, 531; alkaline, 531; dehydration, 35; dilutions, 36; fixation, 7; with formalin, 21; for invertebrates, 485
Aldehydes, 7; blocking, 282, 304. See also Formaldehyde; Glutaraldehyde
Aldehyde fuchsin, 189, 357
Aldehyde reactions. See Feulgen reaction; Schiff reaction
Alizarine red S, 195, 281
Alkaline alcohol, 531
Alkaline methylene blue, 389, 403
Alkaline phosphatase, 429
Allochromasia, 130
Alpha particle tracks, 516
Altmann method: fixative, 18; mitochondria, 370
Alum cochineal, 123
Alum hematoxylin. See Hematoxylin
Amidoblack, 260
Amines, 301
Amino acids, 286
Aminopeptidase, 439
Ammoniacal silver, 201; for histones, 291